Lectins
Biology, Biochemistry, Clinical Biochemistry
Volume 3

Lectins

Biology, Biochemistry, Clinical Biochemistry
Volume 3

Proceedings of the
Fifth Lectin Meeting
Bern, May 31 – June 5, 1982

Editors T. C. Bøg-Hansen
G. A. Spengler

W
DE
G

Walter de Gruyter · Berlin · New York 1983

Editors
Thorkild Christian Bøg-Hansen, cand. scient., lic. techn.
The Protein Laboratory
University of Copenhagen
Sigurdsgade 34

DK-2200 Copenhagen N

Georg Alexander Spengler, PD Dr. med.
Institute for Clinical Protein Research
University of Bern
Tiefenauspital

CH-3004 Bern

Library of Congress Cataloging in Publication Data

Library of Congress Cataloging in Publication Data
Lectin Meeting (5th: 1982: Bern, Switzerland)
Proceedings of the Fifth Lectin Meeting, Bern, May 31 – June 5, 1982.
(Lectins-biology, biochemistry, clinical biochemistry; v. 3)
Bibliography: p.
Includes indexes.
1. Lectins-Congresses. I. Bøg-Hansen, T. C.
(Thorkild Christian), 1939- . II. Spengler, G. A.
(Georg Alexander), 1928- . III. Title.
IV. Title: Proceedings of the 5th Lectin Meeting.
V. Series.
QP552.L42L4 1982 599'.029 83-5134
ISBN 3-11-009504-1

CIP-Kurztitelaufnahme der Deutschen Bibliothek

Lectins, biology, biochemistry, clinical biochemistry:
proceedings of the . . . Lectin Meeting. – Berlin; New York: de Gruyter.
NE: Lectin Meeting
Vol. 3 = 5. Bern, May 31 – June 5, 1982. – 1983.
ISBN 3-11-009504-1

 Printing: Gerike, Berlin. – Binding: Dieter Mikolai, Berlin. – Printed in Germany.

Preface

At the end of the Fourth Lectin Meeting in Copenhagen in June 1981 we decided that future meetings should be organized at varying locations and on a collaborative basis. Therefore, from May 31 through June 5, 1982, the Fifth International Lectin Meeting 'INTERLEC 5' was held in Switzerland in cooperation of the Universities of Copenhagen and of Bern.

Over 100 participants attended the following sessions: Detection of Cancer Markers; Biological Function and Biological Effects of Lectins; and Lectins as Specific Tools. Eighty five communications (lectures and poster presentations) were given, the majority of which appears in this volume. Although the scope of topics in the present book is wide again - as it should be from a meeting covering the whole range of recent progress in the rapidly developing field of lectins - it is noteworthy that a great many contributions were centered on interactions of lectins with normal or malignant cells, opening new and exciting aspects for the application of these natural tools, e. g. in transplantation immunology and tumor diagnosis.

In addition to the scientific sessions, the Third Meeting of the International Working Party on Standardization of Lectins for Diagnosis was held, and a test program for con A from different sources was set up.

After four successive meetings in Copenhagen, it was indeed a challenge to organize 'INTERLEC 5' in Bern. It was for the large number and the scientific quality of the presentations, as well as for the high spirits that prevailed throughout the entire meeting that this event was particularly rewarding and unforgettable for the organizers. We wish to express our sincere appreciation to all participants of 'INTERLEC 5' and to thank Dr. B.Nordbring-Hertz, Lund, for her helpful assistance in planning the meeting.

'INTERLEC 5' was generously supported by the following institutions and companies (in alphabetical order): The Authorities of the State and of the City of Bern; Behringwerke AG, Marburg, Lahn; Boehringer Mannheim GmbH, Hamburg; Hoechst-Pharma AG, Zurich; Medac GmbH, Hamburg; Miles Italiana S.p.A., Cavenago Brianza, Milano; Riact, Copenhagen. Their help has been greatly appreciated.

'INTERLEC 5' will be held in 1984, and we look forward to meet you again at our next lectin meeting.

Copenhagen and Bern, January, 1983

T.C.Bøg-Hansen and G.A.Spengler

Obituary

During preparation of this volume, we recieved notice that one of the devoted participants of the INTERLEC meetings had unexpectedly deceased. Prof.Dr.med. Gunther Hermann, Docteur de l'Université de Paris, mention Sciences, Chief of the Immunological Division at the Surgical University Hospital in Cologne, was born September 20, 1924, in Stuttgart and died August 20, 1982, in Avignon.

We mourn for Gunther Hermann and we shall always remember him as a distinguished scientist and a devoted friend of the lectinologic community.

Copenhagen and Bern, January, 1983

T.C.Bøg-Hansen and G.A.Spengler

TABLE OF CONTENTS

INTRODUCTION

DEFINITION OF LECTIN

A NOTE ON THE RECENT DISCUSSION ON DEFINITION OF THE TERM "LECTIN"

Jan Kocourek, Václav Hořejší
Department of Biochemistry, Charles University, Albertov 2030, 128 40 Praha 2, Czechoslovakia

Recent years brought a rapid increase of our knowledge of lectins. In spite of their widespread occurrence in Nature, their extensive study as model ligand-binding proteins and their frequent use as specific tools in biological research, at present there exists some ambiguity as to the accurate delineation of the term "lectin". The original Boyd's definition (1) is rather broad and at present it seems necessary to re-define the term more specifically. Efforts in this direction were initiated by Goldstein et al. (2) who stressed sugar-binding and cell agglutinating properties as a basis for the definition. Their definition supplemented by several comments was also published by the Nomenclature Committee of IUB in its Newsletter 1981 (3). The comments specify in a more detail the scope of the definition; sugar-binding proteins with a single binding site are not considered as lectins, whereas sugar-specific enzymes, transport proteins and toxins may be classified as lectins, provided they have multiple sugar-binding sites.

The authors of the present paper have suggested a different definition (4), in which emphasis is put on the sugar-binding properties and the lack of enzymatic activity toward carbohydrate structures, whereas the number of sugar-binding sites or agglutination/precipitation activities are not considered. The Nomenclature Committee of IUB has given preference to the Goldstein's et al. definition (5) essentially because they considered the criterion of cell agglutination

Lectins, Vol. III

and/or glycoconjugate precipitation as operational and practical to apply; our definition was found too broad and the criterion of absence of enzymatic activity difficult to apply. However, we still believe that the definition should be based solely on the carbohydrate-binding and not on agglutination or precipitation properties and, also, that sugar-specific enzymes and transport proteins (as well as carbohydrate-specific immunoglobulins) should not be considered as lectins, mainly because:

(1) Cell agglutination is a complex phenomenon depending on the properties of the protein, the cells and of the medium used. Thus the selection of agglutinable cells and optimum conditions may be difficult. Moreover, cell agglutination does not represent - at least in a vast majority of cases - the true physiological activity of lectins. This renders the agglutination-criterion both impractical for general use and rather artifactitious.

(2) Exclusion of monovalent carbohydrate-binding proteins is not logical as they may be structurally and functionally closely related to other multivalent proteins or they may form multivalent aggregates.

(3) Carbohydrate-specific enzymes and transport proteins (regardless of the number of their binding sites) constitute distinct groups of proteins with well defined specialized functions; their classification as lectins might lead to confusion.

(4) The definition should reflect the assumed (and in many cases well documented) carbohydrate-recognition functions of lectins in vivo.

Thus, we suggest a modified version of our original definition (4) which is based on the above considerations: Lectins are proteins of non-immunoglobulin nature capable of specific recognition and reversible binding to carbohydrate moieties of complex carbohydrates without altering covalent structure of any of the recognized glycosyl ligands.

Comments:

1. A lectin molecule contains one or more sugar-binding sites; lectins may (but do not have to) agglutinate cells or precipitate glycoconjugates as a consequence of binding to their glycosyl moieties.

2. Lectins may possess various biological activities - they can be toxins, hormones etc.; they can be enzymes activity of which is not directed towards sugar molecules.

3. Lectins will usually bind sugars (i.e. low-molecular-weight carbohydrates) which specifically inhibit their interaction with soluble or cell-bound glycosylated macromolecules.

4. Lectins may occur in any type of organisms; they may be soluble or membrane-bound, may be glycoproteins or other posttranslationally modified protein derivatives.

5. Carbohydrate-specific immunoglobulins and any carbohydrate-specific enzymes are not to be classified as lectins. Also carbohydrate-binding proteins complexing only with free sugars but not with complex carbohydrates (some transport proteins, chemotaxis receptors, repressors of operons of carbohydrate metabolism etc.) are not lectins.

References

1. Boyd, W.C. in The Proteins, Vol. 2, Part B (Neurath, H. and Bailey, K., Eds.), pp. 756-844, Academic Press, New York 1954.

2. Goldstein, I. J., Hughes, R. C., Monsigny, M., Osawa, T., Sharon, N.: Nature (London) 285, 66 (1980).

3. Newsletter 1981 of the Nomenclature Committee of IUB (NC-IUB) and IUPAC-IUB Joint Commission on Biochemical Nomenclature (JCBN): Eur. J. Biochem. 114, 1-4 (1981; Arch. Biochem. Biophys. 206, 458-462 (1981); Hoppe-Seyler's Z. Physiol. Chem. 362, I-IV (1981); J. Biol. Chem. 256, 12-14 (1981).

4. Kocourek, J., Hořejší, V. : Nature (London) 290, 188 (1981).

5. Dixon, H. B. F. : Nature (London) 292, 192 (1981).

Part I

Biological Effects of Lectins

MECHANISMS OF INDUCTION OF TRANSPLANTATION TOLERANCE IN MICE BY LECTIN TREATMENT

Ivan Hilgert, Václav Horejsí, Hana Kristofová, Pavla Angelisová
Institute of Molecular Genetics, Czechoslovak Academy of Sciences, Vídenská 1083, Praha 4, Czechoslovakia

Administration of some lectins, e.g. Con A or PHA in vivo, results in prolongation of skin or organ graft survival time in such lectin treated recipients (1, 2). Lentil lectin (LCA) was found to be especially efficient in induction of transplantation tolerance in mice (3). In our initial experiments we were able to induce permanent survival of the B10.D2 mouse strain skin grafts in B10.D2(M504) recipients (abbreviated designation M504) after a twenty-dose treatment with LCA. This remarkable effect qualifies LCA as one of the most efficient immunosuppressive (or tolerogenic) substances known so far; a recent study indicates that the efficiency of LCA in rats can be still markedly increased by combination with antithymocyte serum (Harel and Nelken, these proceedings). Therefore, it was of interest to study in more detail the mechanisms underlying these in vivo effects. We tried to answer especially the following questions: (1) What is the immunosuppressive efficiency of LCA in various genetically well defined mouse strain combinations with histocompatibility disparity of different strength? What is the most efficient schedule of lectin treatment? (2) What is the effect of LCA on various functional characteristics of the recipient's lymphocytes? (3) What is the role of active suppressor cell mechanisms in this type of immunological tolerance? (4) What is the fate of LCA in the organism? Has LCA special affinity toward some tissues?

Materials and methods

LCA and Con A were prepared by affinity chromatography on Sephadex G-100 (4, 5). Two to three months old female mice of various inbred strains were obtained from the animal house of our Institute. Skin grafting was performed by the standard method (6). In adoptive transfer experiments 10^8 spleen cells from the tolerant animals were taken 10 days after the last LCA dose and injected into sublethally irradiated recipients which were subsequently given a skin graft. LCA doses administered i.p. or i.v. were dissolved in 0.5 ml saline, the s.c. doses in 0.1 ml saline. Cytotoxicity of LCA toward spleen cells was estimated in vitro by the ^{51}Cr-release test (7). Stimulation of spleen cell suspensions with LCA was measured by incorporation of ^{3}H-thymidine. Mixed lymphocyte cultures (MLC) consisted usually of 3 components: responder spleen cells, stimulator (either 20 Gy irradiated allogeneic cells or a lectin) and modulator cells. Total culture volume was 0.2 ml; cultures were harvested 48 hours after a 16 h pulse with 36 kBq of ^{3}H thymidine and mean cpm of cell incorporated ^{3}H were measured. Stimulation index was calculated as a ratio of cpm from experimental culture to control unstimulated (or syngeneic cell "stimulated") culture. As a rule, MLC consisted of 10^6 stimulator cells, 5×10^5 - 10^6 responder cells and 5×10^5 modulator cells. Modulator cells were prepared by cultivation of 1.5×10^7/ml spleen cells for 40 hours in the presence of lectins (0.5 - 25 µg/ml) or without lectins (controls). After culture the cells were washed with 0.3 M methyl-α-D-glucopyranoside and pretreated with mitomycin C (50 µg/ml, 30 min, 37°C). Mitomycin treatment was omitted in the case of modulator cells which were subsequently used in the cultures tested for cell mediated cytotoxicity (CMC) development. Cultures for generation of cytotoxic lymphocytes were performed essentially according to Ferguson et al. (8), ^{51}Cr-labelled EL-4 tumor cells were used as target cells. The target vs. effector cell ratio was 1:20 - 1:2. LCA and bovine serum albumin (BSA) were labelled by ^{131}I using the chloramine T method (9). The rate of elimination of these radiolabelled proteins after injection into mice was estimated by the whole-body radioactivity measurement at various intervals. Distribution of the labelled proteins within the body was followed after killing the animals (24 hours after the injection of the labelled protein), dissection and separate measurement of radioactivity in different organs and tissues.

TABLE 1

Mouse strain combinations used (donor - recipient)

1	2R - B10.A	15	A.TL - A.TH
2	B10.AKM - B10.BR	16	A.TH - A.TL
3	B10.BR - B10.AKM	17	B10.AQR - B10.AKM
4	R103 - M504	18	B10.AKM - B10.AQR
5	B10.D2 - M504	19	B10.AQR - 4R
6	R103 - B10.D2	20	B10 - B10.A
7	B10.D2 - R103	21	B10.HTT - B10.D2
8	B10.A - B10.BR	22	B10.D2 - B10.A
9	B10 - 5R	23	B10.HTT - B10.A
10	B10.A - 4R	24	3R - B10.A
11	4R - B10.A	25	5R - B10.A
12	4R - B10.AQR	26	4R - B10
13	B10.AQR - B10.DBA/1	27	B10.AQR - B10.A
14	B10.DBA/1 - B10.AQR	28	B10.A - B10.AQR

Results

Efficiency of LCA in different strain combinations. 28 different mouse strain combinations with various genetic disparities at the H-2 loci were tested (Table 1).

The efficiency of LCA was clearly dependent on the nature of antigenic disparity between the donor and recepient. In some strain combinations (No. 1-12, Table 1) LCA treatment resulted in permanent survival of the grafts (>100 days) in 80-100% recipients if LCA was administered daily (1 mg, i.p., the doses on the days 1, 3, 5, 7, 9 and 11 were given i.v., in animals bearing the grafts for more than 40 days the lectin was administered alternatively i.p. and s.c. until the rejection of the graft). In all these cases the difference between the donor and recipient is at the H-2DL locus only (No. 1-7) or H-2DL+ some of the H-2J, E, C, and S loci (No. 8-11); only in strain combination No. 12 the donor and recipient differ at the H-2K, J, C, S, DL loci. In other strain combinations involving differences at multiple H-2 loci including H-2K and/or H-2I (No. 13-26) or H-2K only (No. 27, 28) LCA treatment resulted always in marked prolongation of graft survival times as compared to controls (2-4 fold increase of the mean survival time depending on the particular strain combination but permanent survival of skin grafts was achieved only exceptionally. Thus, LCA treatment can be used to overcome the H-2DL antigenic barrier but is less efficient in the case of H-2K or H-2A antigenic disparities.

Cellular mechanisms involved in maintenance of LCA-induced transplantation tolerance. Involvement of active suppressor cell-mediated mechanisms was tested by adoptive transfer experiments. B10.D2 grafts survived permanently in 70% of M504 recipients which were sublethally irradiated and reconstituted with 10^8 spleen cells from M504 recipients bearing B10.D2 grafts as a result of LCA treatment. Only approximately 10% of R103 strain grafts survived permanently in such reconstituted recipients indicating specificity of the state of tolerance but also some degree of nonspecific inhibitory mechanisms involved. Transplantation of third-party test allografts (R103) onto recipients bearing a B10-D2 graft for >100 days resulted in a rapid rejection of these grafts without affecting the survival of the original B10.D2 grafts.

Effect of LCA on spleen cells in vitro. Mitogenic stimulation of murine spleen cells with LCA was typically dose-dependent (maximum stimulation index approximately 51 at 5 µg/ml in a 5% serum containing medium). The apparent decrease of mitogenic activity at higher concentrations was at least in part due to the cytotoxicity of LCA (approximately 65% of cells were killed after a 24 hours culture in the presence of 0.5 mg/ml LCA; virtually all cells were killed by Con A under similar conditions). Mitogenic response to LCA and Con A of the spleen cells obtained from the M504 mice which were given single 1 mg i.p. dose of LCA (48 hours before testing) was decreased by 93% as compared to untreated controls. There was essentially no mitogenic response in the case of spleen cells from the animals which received a 10-dose LCA treatment (each i.p. dose 1 mg, administered daily, treatment ceased 48 hours before testing). Spleen cells from the B10.A mice which received single 1 mg i.p. dose of LCA manifested a higher proliferative response (1.7 fold as compared to untreated controls) after stimulation with allogeneic B10 cells in MLC. The effects of modulator cells (i.e. spleen cells pre-cultured in the presence of the lectin) on several in vitro characteristics of murine spleen cells were tested. There was a marked difference between the effects of B10.A modulator cells generated in vitro either by LCA or Con A. Whereas addition of Con A-generated modulator cells suppressed the mitogenic response of syngeneic responder cells to LCA stimulation (by 55%), LCA-generated modulator cells enhanced this response (7.5 fold). Similarly, Con A-induced B10.A modulator cells suppressed, whereas LCA-generated modulator cells enhanced the blastogenic response of syngeneic responder cells in MLC toward allogeneic B10 stimulator cells (Hilgert et al., to be published). Both Con A- and LCA-induced modulator cells suppressed the development of CMC in MLC (by 59% and 65%, respectively, as compared to the cultures containing "modulator" cells pre-cultivated in the absence of any lectin). Similar effect (40-45% suppression of CMC) was observed also when modulator cells generated in vivo were used, i.e. the cells obtained from the spleens of animals pretreated with a single 1 mg LCA dose 2 days before testing or after a 10-dose LCA treatment (ceased 6 days before testing).

Distribution of LCA in vivo and the rate of its elimination. The rate of elimination of ^{131}I-labelled LCA was similar to that of ^{131}I-labelled BSA (approximately 80% of a 1 mg dose administered i.v. was eliminated after 48 hours; LCA administered i.p. or s.c. was eliminated approximately 2-3 times more slowly). Both LCA and BSA were distributed relatively homogeneously throughout the body but LCA possessed increased affinity for spleen, lymph nodes and lung (relative concentration per mg of the tissue was 3.1, 1.5, and 2.9 times higher in these organs, respectively, than the mean concentration in the body), whereas the level of BSA was elevated in blood only.

Discussion

Our results provide at least to some extent the answers to the questions raised in Introduction. First, it seems clear that the immunosuppressive efficiency of LCA depends on apparent strength of histocompatibility barrier between the donor and recipient. The H-2DL antigenic disparity can be relatively easily overcome by LCA treatment, whereas H-2K and H-2A antigens represent stronger histocompatibility barrier; the effects of non-H-2 histocompatibility antigens were not followed in this study. Interestingly, long-term tolerance could be induced only in the B10.D2 - M504 (mutant) strain combination by a limited number of LCA doses. Two in vitro observed effects may be directly related to the in vivo efficiency of LCA: first, cytotoxicity of LCA at higher concentrations might cause partial depletion of alloreactive cells. Second, the suppressive effect of LCA-generated modulator cells on CMC development might indicate that also in vivo development of alloreactive cytotoxic T cells may be inhibited. This conclusion is also in agreement with our experiments on adoptive transfer of LCA-induced tolerance. This suppressor cell mechanism may be sufficiently efficient for maintenance of B10.D2 graft survival on the M504 recipient after a 20-dose LCA treatment, but the necessity for continuous supply of LCA in other strain combinations tested (with apparently stronger histocompatibility barrier) seems to indicate, that suppressor cells induced in such recipients are not able to "manage" the rejection reaction and that LCA might also block directly some membrane bound or soluble complex carbohydrate molecule(s) taking part in the rejection reaction. The lectin might bind to the histocompatibility antigens of the graft and thus block their recognition. We cannot rule out this effect of LCA in some strain combinations (it might be in agreement with our results on the rate of LCA elimination, which also indicate the necessity for continuous application of LCA to maintain its high (blocking?) level in organism) but the experiments on retransplantation of long-term tolerated B10.D2 grafts onto untreated M504 recipients, transplantation of grafts presoaked in LCA solution or transplantation of grafts from LCA pretreated donors did not reveal apparently any modification of normal immunogenicity of such "LCA-pretreated" grafts (I. Hilgert, unpublished results). Increased affinity of LCA for spleen and lymph nodes (and lung) may support rather the notion that the recipient's immune system is the prime target of LCA action.

In summary, induction of suppressor cells inhibiting graft rejection reaction and direct cytotoxicity seem to be involved in immunosuppressive activity of LCA observed in vivo. Significance of other possible effects such as direct blocking of some cell surface or soluble molecules by LCA and some kind of antigenic modulation of the graft remains to be evaluated as well as the roles of antigen-specific and non-specific suppressor mechanisms.

References

1. Markley, K., Thornton, S., Smallman, E., Markley, P.: Transplantations 8, 258-264 (1969).
2. Rosenau, W., Haber, J., Goldberg, M.: Transplantation 13, 624-626 (1972).
3. Hilgert, I., Horejsí, V., Angelisová, P., Kristofová, H.: Nature (London) 284, 273-275 (1980).
4. Tichá, M., Entlicher, G., Kostír, J., Kocourek, J.: Biochim. Biophys. Acta 221, 282-289 (1970).
5. Agrawal, B.B.L., Goldstein, I.J.: Biochim. Biophys. Acta 147, 262-271 (1967).
6. Billingham, R.E., Medawar, P.B.: J. Exp. Biol. 28, 385-392 (1951).
7. Wigzell, H.: Transplantation 3, 423-431 (1965).
8. Ferguson, R., Anderson, S.M., Schmidtke, J.R., Simmons, R.L.: J. Immunol. 117, 2150-2157 (1976).
9. Hunter, W.M., Greenwood, F.C.: Nature (London) 194, 495-496 (1962).

THE EFFECT OF LENTIL LECTIN TREATMENT ON SKIN ALLOGRAFT SURVIVAL IN MICE AND RATS

W. Harel and D. Nelken
Laboratory of Immunohematology and Department of Immunology, Hadassah Hebrew University Medical Center, Jerusalem, Israel

It is well known that concanavalin A (Con A) activated lymphoid cells may suppress in vitro primary and secondary plaque-forming cell responses of mouse spleen cells to heterologous red blood cells (1) and are capable of suppressing cell mediated immune responses in vitro (2). It was shown that Con A inhibited the generation of cytotoxic lymphocytes and that this inhibition was critically dependent upon the concentration and time of addition of Con A or Con A activated cells (2). In vivo experiments showed that intraperitoneal injections of Con A markedly prolonged skin allograft survival in mice (3) and heart allograft survival in rats (4). In a recent article Con A was shown to inhibit immunologically specific cytolytic activity, without any toxic effect on target or effector cells and without interfering with ^{51}Cr release once the lytic lesion has been induced. It was suggested that the inhibition of cytolysis was caused by immobilization in the lymphocyte membrane of various protein moieties by the lectin which directly interfered with events at the effector cell surface (5).

Others have found that the inhibitory effect of Con A on cytotoxic effector cells was dependent upon treatment of the target cells with the lectin. When the targets were preincubated with the lectin, or when the lectin was present in the medium, effector cells were able to kill the targets, with a dependence upon the presence of specific histocompatibility antigens on the target cells (6,7). Similar effects were also observed with phytohaemagglutinin (PHA). When injected intraperitoneally under appropriate conditions, it effectively inhibited the rejection of skin allografts, even across the H-2 barrier in mice (8,9). In a recent study, lentil lectin was shown to have a similar effect in preventing the rejection of skin allografts in mice (10). It was shown that the immuno-suppressive efficiency of lentil lectin depends on the strength of the histocompatibility barrier between donor and recipient in mice. Lentil lectin was able to overcome the H-2DL antigenic disparity much better than the H-2K and H-2A disparity (11).

The present report deals with skin allografting experiments in mice and rats treated with various combinations of lentil lectin and anti thymocyte serum. It was found that such a treatment may induce a state of allograft acceptance in closely related mice and rats. In adoptive transfer experiments, cells from the treated animals were able to prolong skin graft survival in transplanted secondary hosts.

Materials and Methods

Lentil lectin was prepared as described elsewhere (12). In short, 250g of lentils were homogenized in a Sorvall OMNI-mixer and centrifuged. The supernatant was fractionated by addition of solid ammonium sulfate to obtain the material which precipitated between one and two thirds saturation. The precipitate was dialized against Tris buffer (0.05M, pH 8.1), and passed through a column of DEAE cellulose (Whatman DE-52). The lectin appeared in the first protein peak (OD 280 nm) passing unbound through the column. Activity was detected by hemagglutination of rabbit erythrocytes (13). The product was compared with a commercial lentil lectin, obtained from Pharmacia Fine Chemicals, Uppsala, Sweden, and showed a similar potency in hemagglutination tests and an identical pattern in acrylamide gel electrophoresis.

Anti rat thymocyte serum (ATS) was prepared in rabbits by the injection of rat thymocytes emulsified in complete Freund's adjuvant at four week intervals. Two weeks after the third immunization the rabbits were bled, the serum separated, pooled and frozen at -20°C in 10ml aliquots until use. The cytotoxic titer against rat thymocytes with guinea-pig complement was adjusted to 1:64 in the lymphomicrocytotoxicity test employing the eosin staining method (14). Antisera with weaker titers were not used. A single serum pool was used in each series of experiments. Inbred male Lewis (RT-1^l), Fisher (RT-1^l) rats and inbred BALB/c (H-2^d) and C57Bl (H-2^b) mice were used throughout this study.

Results

In a recent report it was shown that lentil lectin was able to induce skin allograft tolerance in mice (10). In our laboratory, treatment with 8-10 daily doses of lentil lectin alone following transplantation, resulted in prolonged skin allograft survival but a state of tolerance was never achieved (15,16). Using two mice strains with strong H-2 differences, BALB/c and C57Bl, the longest mean skin graft survival time in lentil lectin treated recipients was 22.3±3.81 days (Table 1 group 3) compared to 10.6±1.78 (Table 1 group 8) in untreated control mice. In experiments with rats differing in minor histocompatibility antigens (Lewis to Fisher) treatment with lentil lectin increased the mean survival time to up to 25.1±5.33 days (Table 2 group 2) compared to 10.3±1.46 (Table 2 group 6) in the untreated controls. In another series of experiments the syngergistic effect of rabbit anti rat thymocyte serum and lentil lectin was examined. ATS was injected at the time of the operation followed 48 hours later by the injection of lentil lectin and 48 hours thereafter by another injection of ATS and so on. A total of 8 injections was given, 4 of which were ATS and 4 lentil lectin. This combined treatment consistently resulted in prolonged skin graft survival. In mice the skin allografts survived for about 24 days (Table 1 groups 6,7) and in rats for an observation period over 75 days (Table 2 groups 4,5). In passive transfer experiments we found that spleen cells from ATS and lentil lectin treated Fisher rats with an established Lewis skin graft (Table 2 group 5) were able to prolong graft survival time when transferred to untreated syngeneic rats transplanted with Lewis skin grafts. The transplants were performed 24 hours after transfers of $5x10^7$ spleen cells (Fig. 1). The mean survival time of these transplants was

Table 1

Skin allograft survival in mice (C57Bl to BALB/c) treated with anti thymocyte serum and lentil lectin

Group no.	ATS	LCA	No. of mice	Mean survival time ± S.D.
1	-	5mg/kg, 10 i.p. injections every 24 h starting at transplantation time	25	14.2 ± 2.76
2	-	15mg/kg as group 1	18	16.8 ± 2.94
3	-	25mg/kg as group 1	15	22.3 ± 3.81
4	-	50mg/kg as group 1	13	15.1 ± 4.14
5	0.25ml/mouse 8-10 i.p. injections every 24-48 hours	-	15	18.3 ± 3.77
6	0.25ml x 5 i.p. injections; on days 0, 2, 4, 6, 8 after transplantation	25mg/kg x 5 i.p. injections on days 1, 3, 5, 7, 9 after transplantation	20	24.7 ± 5.12
7	0.25ml x 5 i.p. injections as group 6	25mg/kg 5 i.v. injections as group 6	14	23.6 ± 5.74
8	Control: 0.5ml of 0.9% NaCl 10 i.p. injections every 24 h		30	10.6 ± 1.78

Skin grafts (1 cm in diameter) prepared from the back of the donor mice were transplanted onto the backs of the tested recipients. Grafts with at least 50% necrosis were considered to be rejected.
$p < 0.001$ between groups 1,2,3,4,5,6,7 and group 8. No statistically significant difference was found between group 6 and 7.
S.D. = Standard Deviation
ATS = Anti Thymocyte Serum
LCA = Lens Culinaris Agglutinin (lentil lectin)
i.p. = intra peritoneal
i.v. = intra venous

Table 2

Skin allograft survival in rats (Lewis to Fisher) treated with anti thymocyte serum and lentil lectin

Group no.	ATS	LCA	No. of rats	Mean survival time ± S.D.
1	-	5mg/kg 8 i.p. injections on consecutive days	30	22.3 ± 4.72
2	-	5mg/kg 8 i.v. injections on consecutive days	16	25.1 ± 5.33
3	1ml/rat 8 i.p. injections on consecutive days	-	8	18.6 ± 4.94
4	1ml/rat 4 i.p. injections on days 0, 2, 6, 10 after transplantation	5mg/kg 4 i.p. injections on days 1, 4, 8, 12 after transplantation	10 9	32.1 ± 4.17 > 75
5	1ml/rat 4 i.p. injections as group 4	5mg/kg 4 i.v. injections as group 4	22	all grafts > 75
6	Control: 1ml of 0.9% NaCl, 8 i.p. injections on consecutive days		23	10.3 ± 1.46

Skin grafts (1 cm in diameter) prepared from the chest of the donor rats were transplanted to the chest of the recipients. Grafts with at least 50% necrosis were considered to be rejected.
$p < 0.0001$ between groups 1,2,3,4,5 and group 6.
S.D. = Standard Deviation
ATS = Anti Thymocyte Serum
LCA = Lens Culinaris Agglutinin (lentil lectin)
i.p. = intra peritoneal
i.v. = intra venous

22.2±2.50, compared with 15.0±0.82 in the control group in which transferred cells were from untreated and untransplanted Fisher rats. In several other Fisher rats which showed a perfect take of the Lewis skin allograft (Table 2 group 5), a second Lewis skin graft was performed 8-10 weeks after the first graft. The second graft was rejected in all cases and caused the simultaneous rejection of the first allograft 10-15 days after the operation. When rats, which showed a perfect take of the first Lewis skin allograft were transplanted with a second skin graft from an unrelated Sabra rat, the second graft was always rejected within 7-10 days. In these rats, however, the first skin graft was not rejected (Fig. 2).

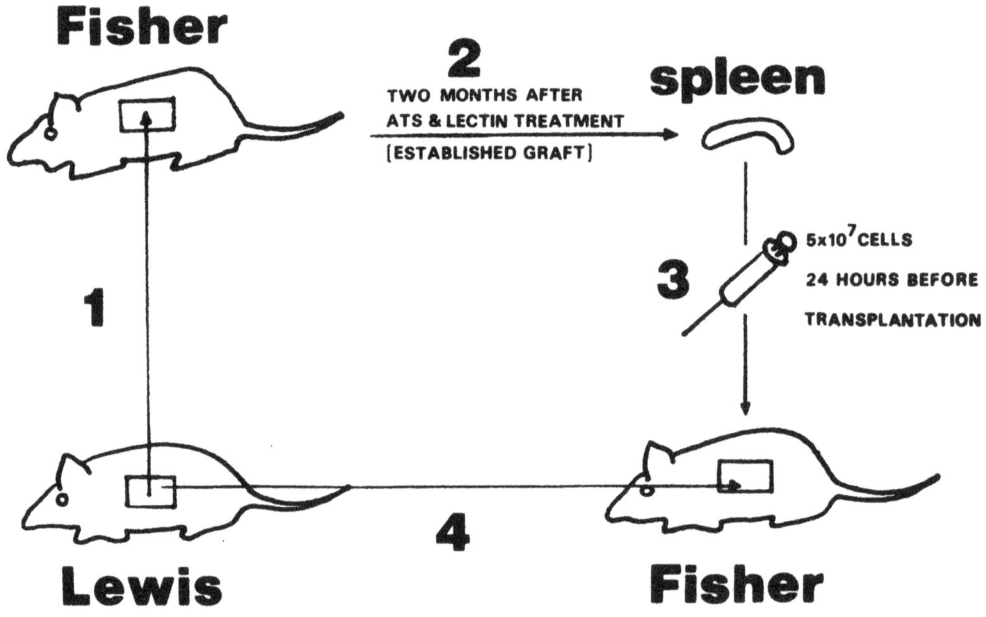

Fig. 1: Passive transfer of $5x10^7$ cells from treated rats with established grafts to newly transplanted rats, causes prolongation of the skin graft survival time.

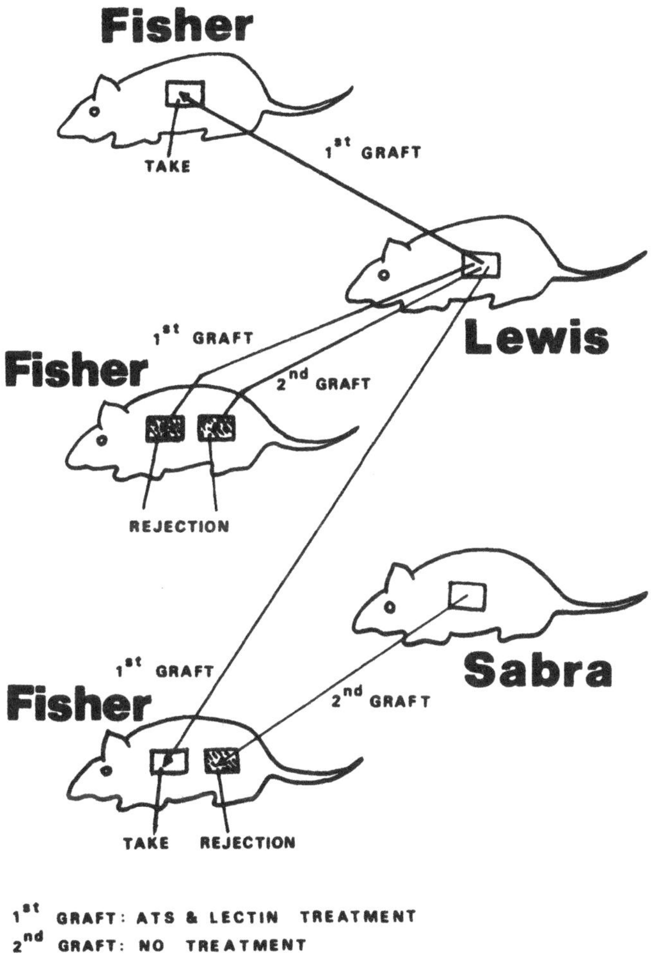

Fig. 2: Transplantation of second set skin grafts to treated rats with an established first graft. When the second graft is from an un unrelated donor (Sabra), it is rejected within 7-10 days, while the first graft survives. When the second graft is the same as the first, both grafts are rejected within 10-15 days.

Discussion

Our results indicate that the combination of anti thymocyte serum and lentil lectin causes a state of acceptance of skin allografts from Fisher to Lewis rats. Cells from the treated rats are able to prolong skin graft survival when transferred to syngeneic recipients transplanted with the same donor type grafts. No real tolerance is achieved, however, since a second allograft from the strain is rejected and seems to cause the rejection of the first otherwise surviving allograft.

In a combination of two mouse strains differing in all the H-2 loci (C57Bl/6 x BALB/c) we were unable to achieve permanent acceptance of the skin allografts by our lentil lectin treatment. These results are in agreement with the recent work of Hilgert, Horejsi et al. who have shown that only in certain genetic combinations lentil lectin treatment could overcome the histocompatibility barrier (11). The same authors suggest that a suppressor cell mechanism is at least partially responsible for the allograft acceptance in the lentil lectin treated mice. We are at present examining this point in rats and mice.

In our present report we were able to reduce the lentil lectin dose and at the same time significantly improve the results of the skin allografts by using a combination of lentil lectin and anti thymocyte serum. This synergistic effect of lentil lectin and anti thymocyte serum points to the eventual clinical feasibility of this treatment in an attempt to improve allograft transplantation.

References

1. Rich, R.R., Pierce, C.W.: J. Exp. Med. 137, 649-659 (1973).
2. Peary, D.L., Pierce, C.W.: J. Exp. Med. 140, 356-369 (1974).
3. Markowitz, H., Person, D.A., Gitnick, G.L., Ritts, R.E.: Science 163, 476 (1969).
4. Weil, R. III, Nozawa, M., Chernack, W., Weber, C., Reemtsma, K., McIntosh, R.: Transplantation 17, 600-604 (1974).
5. Tartof, D.: Cellular Immunology 50, 48-57 (1980).
6. Berke, G., Hu, V., McVey, E., Clark, W.R.: J. Immunol. 127, 776-781 (1981).

7. Berke, G., Hu, V., McVey, E., Clark, W.R.: J. Immunol. 127 782-787 (1981).

8. Stefani, S.S., Moore, C.D.: J. Immunol. 104, 780-784 (1970).

9. Rosenau, W., Habler, J., Goldberg, M.: Transplantation 13, 624-626 (1972).

10. Hilgert, I., Horejsi, V., Angelisova, P., Kristofova, H.: Nature 284, 273-275 (1980).

11. Hilgert, I., Horejsi, V., Kristofova, H., Angelisova, P.: Lectins, Biology, Biochemistry, Clinical Biochemistry, Vol. 3, 1982. T.C. Bog-Hansen and G.A. Spengler (Eds.), Walter de Gruyter, Berlin - New York.

12. Howard, I.K., Sage, H.J., Stein, M.D.: J. Biol. Chem. 246, 1590-1595 (1971).

13. Howard, I.K., Sage, H.J.: Biochemistry 8, 2436-2441 (1969).

14. Mittal, K.K., Mickey, M.R., Singal, D.P., Terasaki, P.I.: Transplantation 6, 913-927 (1968).

15. Harel, W., Nelken, D.: Transplantation 32, 69-71 (1981).

16. Nelken, D., Harel, W.: Transplantation 33, 560-561 (1982).

Supported by grants from the Lil and Ben Stein Transplantation Foundation, from Alan Liker and the Shaklee Fund.

POSSIBLE INVOLVEMENT OF LECTINS IN BACTERIA-INDUCED HISTAMINE RELEASE IN INTRINSIC ASTHMA

P. Stahl Skov, C. Jensen, S. Norn, T.C. Bøg-Hansen, H. Løwenstein, Chr. Koch, H. Permin, N. Høiby

Department of Pharmacology and Protein Laboratory, University of Copenhagen, Department of Paediatrics TG, Rigshospitalet and Statens Seruminstitut, Copenhagen, Denmark

A pathogenic role of microorganisms is suspected in the intrinsic asthma disease, since children with intrinsic asthma show a close temporal relationship between exacerbations of symptoms and acute respiratory tract infections (1-3). In our previous study of patients with intrinsic asthma we found that several strains of bacteria caused release of histamine from basophil leukocytes <u>in vitro</u> (3-4). The involvement of histamine release in the disease is supported by the finding of increased urinary excretion of histamine metabolites during acute attacks (5). Bacteria-induced histamine release might therefore contribute to the asthmatic symptoms. The mechanism of bacterial histamine release has been investigated (4, 6). Two different mechanisms seem to exist - an IgE-mediated reaction and a non-immunological mechanism consisting of a direct interaction with the basophil cell surface. Since many bacteria are known to contain surface lectins (7) (see also Wadström and Trust, Eshdat and Sharon, Gilboa-Garber in these proceedings), we studied whether the non-immunological mechanism of the release could be lectin-mediated.

Materials

Lectins. Con A, Pharmacia AB, Sweden, phytohemagglutinin (PHA), Difco Laboratories, Detroit, USA, poke weed mitogen (PWM) and lens culinaris agglutinin (LCA), IBF, Clichy, France, and wheat germ agglutinin (WGA), peanut lectin (PNA) and ricinus communis agglutinin (RCA) from E.Y. Laboratories, San Mateo, USA.

Carbohydrates. D-(+)-glucose, D-(+)-galactose, D-(+)-mannose and α-methyl-D-mannoside from Sigma, St. Louis, USA, D-(-)-ribose and L-(+)-rhamnose from Merck, Darmstadt, GFR, α-methyl-D-glucoside, Calbiochem AG, Lucerne, Switzerland, and N-acetyl-D-glucosamine, Fluka AG, Switzerland.

Bacteria. Staphylococcus aureus (protein A-free Wood 46) were cultured on solid media and suspended in PBS, pH 7.38 containing 3% formalin. Thereafter the bacteria were washed three times with PBS in the absence of formalin and resuspended in this medium in a final concentration of 120 mg wet weight/ml. The bacteria stock solution was kept frozen at -20°C.

Patients and Methods

Six children with intrinsic asthma (3) and six normal individuals were included in the study.

Preparation of the leukocyte suspension. By venipuncture 10 ml blood samples were collected in plastic tubes containing 0.5 ml 0.20 M EDTA (pH 7.0). 20 ml of 0.9% NaCl was then added to the samples, and the leukocytes were isolated according to the method of Day (8). 10 ml Ficoll-Hypaque solution (12.8 g Ficoll + 21 g Hypaque (sodium diatrizoate) dissolved in water to a final volume of 200 ml). This solution has a density of 1.080 g/ml at 20°C. It was carefully placed below the blood sample. The sample was centrifuged (400 g for 40 min) and the cotton wool-like interphase consisting of monocytes with approx. 2% basophils was mixed with 10 ml Tris-AMC

(Tris (hydroxymethyl) aminomethane 25 mM at pH 7.6, 0.12 M NaCl, 5 mM KCl, 0.6 mM $CaCl_2$, 1.1 mM $MgCl_2$, human serum albumin 0.3 mg/ml). The mixture was centrifuged (107 g for 10 min) and the cell sediment was resuspended in 10 ml Tris-AMC. The suspension was centrifuged (107 g for 10 min) and the supernatant containing thrombocytes was discarded. The cell sediment was finally resuspended in 2.5 ml Tris-AMC.

Histamine release

90 µl leukocyte suspensions containing 2000-8000 basophils were incubated in plastic tubes at 37°C for 40 min with 10 µl Staph. aureus (120 mg wet weight/ml) or 10 µl of the individual lectins in the following concentrations: Con A 10, 30, and 100 µg/ml; PHA 1, 10, 100, and 1000 µg/ml, and 1000 µg/ml of PWM, LCA, WGA, PNA or RCA. Bacteria and lectins were omitted for spontaneous histamine release and for determination of the total histamine content; in the latter case the samples were left at 4°C. Thereafter 3 ml ice-cold Tris-AMC was added to the samples and they were centrifuged at 2000 g for 10 min. Besides released histamine, the supernatant contains also protein and other substances, which interfere with the histamine analysis. It was therefore discarded and the residual histamine was determined in the cell sediment. To the cell sediment was added 400 µl of an OPT mixture (2 mg O-phthaldialdehyde, Fluka, dissolved in 1 ml methanol and 9 ml 0.05 N NaOH). After 4 min 400 µl 0.175 N H_3PO_4 was added and the sample was centrifuged at 3000 g for 10 min. The histamine content in the supernatant was determined spectrofluorometrically in an Aminco-Fluoro-Colorimeter provided with a micro-flow cell. Samples were read at 355/455 nm (excitation/emission). The release of histamine was expressed as a percentage of the total histamine content of the sample. The spontaneous histamine release amounted to 1-10%.

Inhibition studies for examining a lectin-mediated reaction were performed by addition of specific carbohydrates: α-methyl-D-glucoside, α-methyl-D-mannoside, N-acetyl-D-glucosamine, D-(+)-galactose, D-(+)-glucose, D-(+)-mannose, D-(-)-ribose, and L-(+)-rhamnose. The carbohydrates were given individually in concentrations of 0.2 to 20 mg/ml or as a mixture in a final concentration of 0.3 to 2.5 mg of each carbohydrate/ml. They were added to Staph. aureus or to the lectin 10 min before the addition of 80 µl leukocyte suspension.
The presence on the basophils of surface-bound IgE or IgG was shown by histamine release following incubation of the leukocyte suspension with anti-IgE or anti-IgG. Column-immunoabsorbed, no nonsense rabbit-antihuman IgE (9) or IgG (directed against all four subclasses) were used. Cell-bound IgE and possibly also IgG was removed by exposure of the leukocytes to pH 3.8 for 5 min at 4°C (10).

Results and Discussion

Lectin-mediated histamine release from basophil leukocytes was examined in vitro in both patients and normal individuals. Within both groups release of histamine was caused by Con A, PHA, PWM, LCA, WGA, PNA, and RCA. Fig. 1a shows the typical Con A response obtained in all patients and normal individuals. However, this typical Con A response was demonstrated in two different types of patients and normal individuals, i.e. by anti-IgE response cell-bound IgE was only found in one part of the subjects, but not in the other part (Fig. 1b). None of them showed surface IgG. A partial removal of the IgE (Fig. 1b) did not change the Con A response (Fig. 1a). These findings indicate that a lectin-induced histamine release can be independent of the presence of IgE on the basophil cell surface. Another example of a lectin response is given in Fig. 2. The PHA response was low, showing only a 14% release of histamine at the highest concentration of PHA (100 µg/ml). IgE, but not

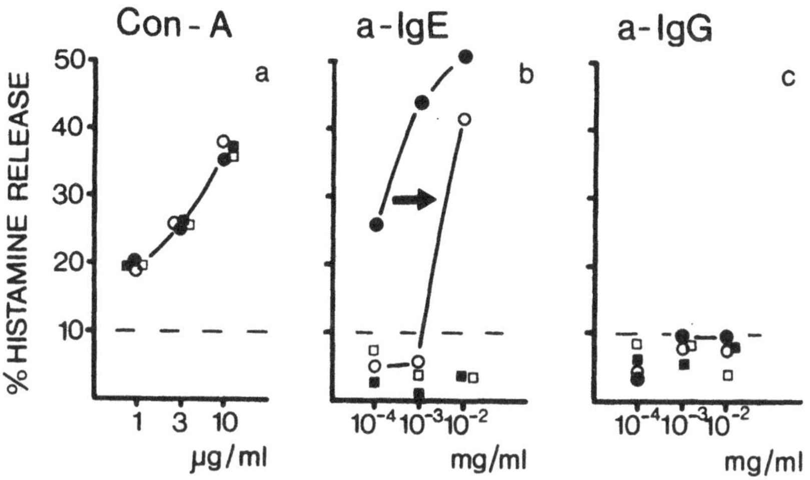

Fig. 1: Histamine release induced by Con A, anti-IgE or anti-IgG in leukocyte suspensions from intrinsic asthma patients or normal individuals. (●) anti-IgE responders, (■) non-anti-IgE responders, (o □) cells treated at pH 3.8 to remove Ig, and (● ■) intact cells.

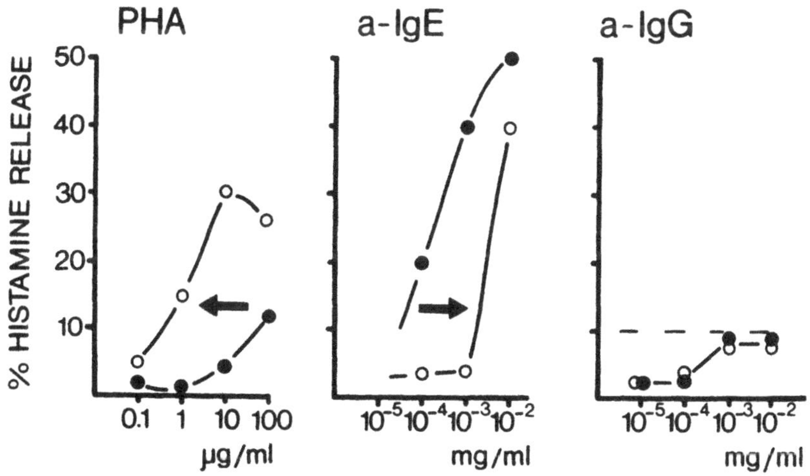

Fig. 2: Histamine release induced by PHA, anti-IgE or anti-IgG in leukocyte suspensions from intrinsic asthma patients or normal individuals. (o) cells treated at pH 3.8 to remove Ig and (●) intact cells.

IgG, was demonstrated on the cell surface. However, a partial removal of the IgE caused a significant increase in the PHA response. By the removal more receptors for PHA binding might be available on the cell surface, and this is possibly the reason for the enhanced response. This pattern was found in both patients and normal individuals. The experiments with Con A and PHA show a lectin-mediated histamine release where lectin bypasses the surface immunoglobulins and interacts directly with the basophil cell membrane leading to release of histamine.

In our previous study of intrinsic asthma we found that bacteria ultrasonicates were able to release histamine by two different mechanisms, an IgE-mediated reaction and a non-immunological mechanism consisting of a direct interaction with the cell membrane (4, 6). The non-immunological mechanism might depend on a lectin-mediated reaction, since many bacteria surfaces are known to contain lectins (see Wadström and Trust, these proceedings), and since lectins by an interaction with carbohydrate moieties on the cell membrane might lead to biological responses such as agglutination, mitogenic activity or release of mediators (see 7, 11, and other chapters in these proceedings). An indirect proof of a lectin-mediated response is the inhibition of the response by specific carbohydrates, which compete with the lectin binding sites on the cell membrane.

The inhibition of the bacteria-induced histamine release by specific carbohydrates was therefore examined *in vitro* in both patients and normal individuals. Formaldehyde-killed protein A-free *Staph. aureus* were used. Fig. 3 shows the results of the inhibition studies. A significant decrease in histamine release was obtained when a mixture of eight carbohydrates was included in the medium. In each of the 12 subjects the decrease was significant in the range of 0.3 to 1.2 mg/ml of each carbohydrate ($p < 0.01$ by Student's t-test), and a maximal inhibition of 70 to 90% was obtained. Different patterns of inhibition were obtained in different subjects, and in the day to day experiments, and the inhibition would sometimes be lacking in the same donor. These variations are possibly due to dyna-

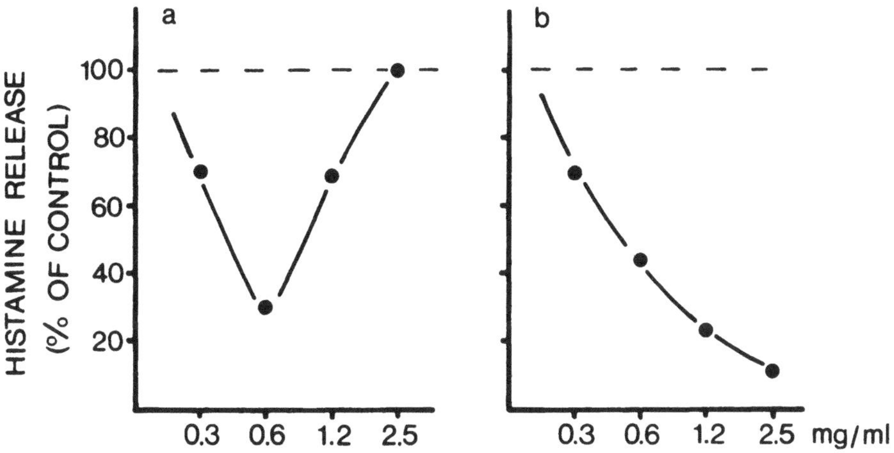

Fig. 3: Representative examples of the two main types of inhibition patterns (a and b) of Staph. aureus-induced histamine release in leukocyte suspensions from intrinsic asthma patients or normal individuals were obtained by the carbohydrate mixture. The mixture included α-methyl-D-glucoside, α-methyl-D-mannoside, N-acetyl-D-glucosamine, D-(+)-galactose, D-(+)-glucose, D-(+)-mannose, D-(-)-ribose, and L-(+)-rhamnose, each in final concentrations from 0.3 to 2.5 mg/ml. In the absence of carbohydrates (controls) the histamine release varied from 2o to 35%.

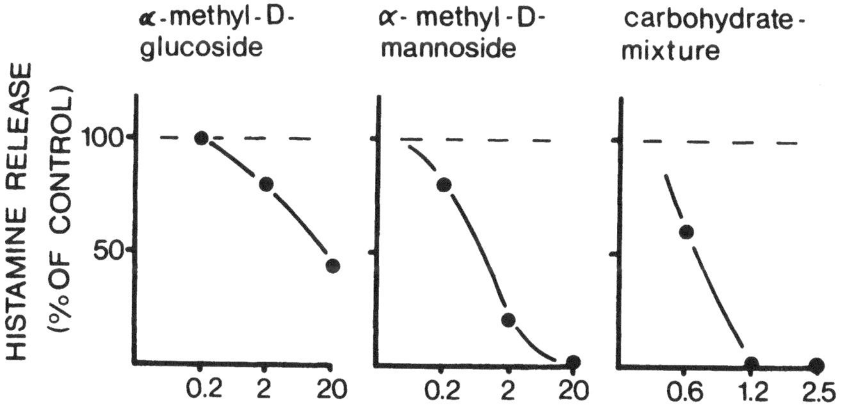

Fig. 4: Characteristic inhibition curves of Con A-induced histamine release in leukocyte suspensions from intrinsic asthma patients and normal individuals by the carbohydrate mixture and by single carbohydrates. The final concentration of the individual carbohydrates is given. In the absence of carbohydrates (controls) the histamine release varied between 2o and 45%.

mic changes in the carbohydrate content of the basophil cell membrane. Fig. 3a and b show the two main types of inhibition, which were found in both patients and normal individuals. In contrast to these experiments performed with the mixture of carbohydrates, no inhibition was obtained when the single carbohydrates were used.
The validity of these inhibition studies for a lectin-mediated reaction was investigated by similar experiments with Con A and PHA. Fig. 4 shows that Con A-induced histamine release was reduced by the carbohydrate mixture and that the inhibition pattern was similar to that obtained with the bacteria (cf. Fig. 3b). However, the release was also inhibited by the individual carbohydrates: α-methyl-D-glucoside and α-methyl-D-mannoside (Fig. 4). Similarly, also the PHA response was found to be reduced by the carbohydrate mixture (Fig. 5), but the effect of individual carbohydrates was not investigated.
In contrast to these lectin experiments the anti-IgE-induced histamine release was not influenced by the individual carbohydrates or by the mixture (data not shown). Nor did the carbo-

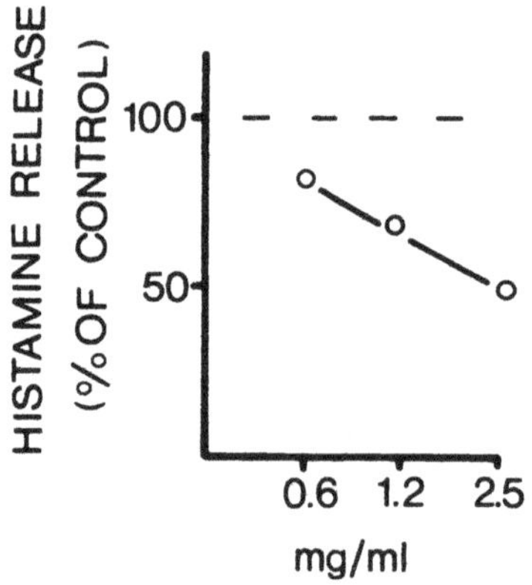

Fig. 5: Inhibition of PHA-induced histamine release in leukocyte suspensions from intrinsic asthma patients and normal individuals by the carbohydrate mixture. In the absence of carbohydrates (controls) the release of histamine varied between 2o and 30%.

hydrates themselves cause any release of histamine. The inhibition studies therefore seem to be an indirect proof of a lectin-mediated reaction.
As mentioned, the carbohydrate mixture inhibited both the bacteria- and the plant lectin-induced histamine release, whereas the single carbohydrate only decreased the plant lectin response. We therefore suggest that the non-immunological mechanism of bacterial histamine release might be caused by a multi-lectin reaction involving several types of lectins on the bacterial membranes reacting with different carbohydrate moieties on the basophil cell surface. On the other hand, the reverse reaction involving basophil surface lectins reacting with bacterial carbohydrates cannot be excluded as a co-operating factor.

Acknowledgements

This study was supported by the Danish Medical Research Council (Grant Nos. 12-2344 and 512-20724), Sygekassernes Helsefond (Grant Nos. H.11/31-79 and H.11/52-80), Fortsættelsessygekassen Danmarks Sundhedsfond, Nationalforeningen til Bekæmpelse af Tuberkulose og Sygdomme i Åndedrætsorganerne, Illum-Fondet, P. Carl Petersens Fond, Lundbeck-Fonden, and Løvens Kemiske Fabrik.

References

1. Flensborg, E.W.: Acta Allergol. 3, 66-74 (1950).
2. McIntosh, K., Elliot, F.E., Hofman, L.S., Lybass, T.G., Eller, J.J., Fulginiti, V.A.: J. Pediatr. 82, 578-590 (1973).
3. Kock, Chr., Andersen, P., Hertz, J.B., Høiby, N., Kappelgaard, E., Møller, N.E., Norn, S., Pedersen, M., Pedersen, P.S., Stahl Skov, P., Tønnesen, P.: Allergy 37, 191-201 (1982).
4. Norn, S., Stahl Skov, P., Kock, Chr., Andersen, P., Pedersen, M., Tønnesen, P., Pedersen, P.S., Møller, N.E., Hertz, J., Høiby, N.: Agents and Actions 12, 101-102 (1982).

5. Löwhagen, O., Granerus, G., Wetterqvist, H.: Allergy 34, 395-404 (1979).

6. Norn, S., Stahl Skov, P., Jensen, C., Kock, Chr., Permin, H., Bøg-Hansen, T.C., Løwenstein, H., Høiby, N.: Agents and Actions, in press (1982).

7. Wadström, T., Trust, T.J.: Medical Microbiology, in press (1982).

8. Day, R.P.: Clin. Allergy 2, 205-212 (1972).

9. Løwenstein, H., Weeke, B.: Scand. J. Immunol. 4, Suppl. 3, 459-466 (1975).

10. Stahl Skov, P., Permin, H., Malling, H.-J.: Scand. J. Immunol. 6, 1021-1028 (1977).

11. Sullivan, T.J., Greene, W.C., Parker, C.W.: J. Immunol. 115, 278-282 (1975).

T CELL STIMULATING LECTINS ARE MITOGEN INDUCERS BUT NOT MITOGENS

Beda M. Stadler, Jürg Liechti and Guy D. Bonnard

Institute of Clinical Immunology, University of Bern, Inselspital, CH-3010 Bern, Switzerland

The discovery that plant lectins, like phytohaemagglutinin or concanavalin A, initiated lymphocyte proliferation (1) has lead to the initial assumption that these lectins were the actual mitogens for lymphocytes. Subsequently, soluble mitogenic factors have been found in culture supernatants of cells stimulated with lectins (2,3). Thus, these endogeneous cell-derived factors, as well as the plant lectins, have been considered to be mitogens. With the observation that lectin-stimulated human peripheral blood lymphocytes secreted an activity that could maintain long-term T cell growth in cultures of human leukemic blood or bone marrow cells (4), it became apparent that the lymphocyte-derived T cell mitogenic activity was actually a growth factor. This mediator was renamed Interleukin 2 (IL-2) in view of its diverse biological activities described earlier, such as lymphocyte mitogenic factor, killer helper factor and helper activity for antibody production. The importance of IL-2 consisted of its ability to support long-term growth of functionally as well as phenotypically normal T cells, leading eventually to the development of clones (5,6,7). In contrast, lectins were found to be an obligatory first signal to activate T cells, but they were not able to cause proliferation by themselves. Rather, lectins were found to cause IL-2 production and

Lectins, Vol. III

also the appearance of receptors for IL-2 on T cells (7,8,9). In the present study, we present further evidence from our laboratories that IL-2 provides the second signal more directly responsible for T cell proliferation.

Material and Methods

The production of IL-2 by human mononuclear cells and by the human T cell line HSB-2 as well as the quantitative determination of IL-2 by mouse IL-2 dependent T cell lines have been described earlier (7,10). The monoclonal antibody against activated T cells (anti-Tac) was supplied by Dr. T. A. Waldman (NIH, Bethesda, MD, USA) and has been recently described (11). The monoclonal antibodies against human IL-2 were purified by ammonium sulfate precipitation and subsequent anion exchange chromatography as described (12). Lectins were purchased from Pharmacia Fine Chemicals (Uppsala, Sweden) and were immobilized with 2,5% glutaraldehyde on plastic tissue culture dishes (Costar, Cambridge, MA, USA).

Results

It has been widely recognized that lectins are mitogenic within a rather narrow dose-range, with higher doses resulting in cell death. Using acridine orange staining of RNA and flow-cytometry we have recently shown (10) two different effects of the lectins (Table 1). Optimal doses of lectins caused a G_0-G_1 shift in the cell cycle, production of IL-2, and subsequent optimal proliferation. However, higher than "optimal" concentrations of the lectins resulted in a block of the cells in G_1 and no proliferation. Nevertheless, production of IL-2 was the greatest at these high concentrations of lectins, suggesting that

TABLE 1

Con A µg/ml	TdR Uptake cpm[a]	IL-2 Units[b]	PHA µg/ml	TdR Uptake cpm	IL-2 Units
0	277	0	0	324	0
1	4,761	4	0.1	2,448	2
10	24,998	6	1.0	36,342	29
50	796	27	10.0	3,074	55
100	728	25	50.0	488	78

a Mean ^{3}H-TdR uptake by triplicate cultures containing 2×10^5 cells pulsed for 4 hr at 44 hr of culture.

b Supernant IL-2 activity of 2×10^6 cells cultured for 48 hr. IL-2 activity was determined on mouse CT6 cells as recently described. (10)

IL-2 was secreted by cells within the G_1 phase of the cell cycle. This correlate between high doses of lectins, G_1 block and increased IL-2 production was confirmed using the mouse EL-4 and the human HSB-2 leukemic T cell lines (10). Contrary to normal resting T cells, these leukemic cells were always in the exponential growth phase, and the lectins were not needed to shift the cells from G_0 to G_1. Therefore, the second effect of the lectins, when high doses were used, was clearly seen: block of the cells in G_1 sufficient to generate supernatant IL-2 activity not seen in the absence of lectins. The correlate between G_1 block and IL-2 production was reproduced in fresh lymphocytes and the cell lines, using the G_1 blocker phorbol myristate acetate. These data suggest that lectins not only are T cell activators, but they also arrest T cells in the G_1 phase of the cell cycle, which eventually leads to accumulation of supernatant IL-2 activity.

In addition to the production of IL-2, the first activation signal mediated by lectins also resulted in the appearance of cell-surface receptors for IL-2. Thus, only lectin-activated

T cells were able to proliferate to IL-2, and they were also able to absorb IL-2 from supernatants (8,9). Thus, the IL-2-receptors appear following the G_0-G_1 shift, and this is similarly dependent on the first activation signal by lectins (10). Binding studies of IL-2 to its IL-2-receptor have not been readily feasible, due to the lack of sufficient highly purified radiolabeled IL-2. However, rabbit antisera and mouse monoclonal antibodies, made by injections of cultured human T cells, could be used to further demonstrate the role of IL-2 and IL-2-receptors (9). Recently, the monoclonal anti-Tac antibody, raised by injections of human cultured T cells, was able to completely block cultured T cells' proliferation to IL-2 at very high dilution of the antibody. The antibody was also able to block the IL-2-receptors on lectin-activated T cells, thus rendering them incapable of absorbing IL-2 activity from IL-2 containing supernatants (Table 2).

TABLE 2

IL-2 Preparation Absorbed on:	TdR Uptake cpm	Comment
RM-1 cells (B cells)	42,700	Control, no absorption
PHA-blasts	25,000	Absorption of IL-2
PHA-blasts, pretreated	42,300	Absorption blocked

Anti-Tac monoclonal inhibits absorption of IL-2 by PHA-blasts. Absorption of IL-2 by PHA-blasts (9), or PHA-blasts pretreated for 2 hr with a 1:2,000 dilution of the anti-Tac. IL-2 assay of the resulting supernatants on secondary cultures of IL-2-dependent cells (TdR uptake).

We have recently described monoclonal anti-human IL-2 antibodies (12), which abrogated the proliferation of IL-2 dependent cultured T cells. Table 3 shows a dose-dependent inhibition of the proliferation. Lectins have no measurable effect on the proliferation of these IL-2 dependent cells. Thus, the only mitogenic signal for IL-2-dependent cells is IL-2 itself.

TABLE 3

Interleukin 2 units/ml	% INHIBITION OF PROLIFERATION[a] dilution of ascitic fluid antibody[b]			
	1/20	1/40	1/80	1/160
10	55	17	0	0
5	49	35	9	0
2.5	80	52	37	17
1.25	90	70	64	45

a IL-2 dependent mouse CT6 cells were cultured in the presence of different concentrations of human IL-2 (= 0% inhibition).

b Ascitic fluid anti-human IL-2 antibody were obtained by ammonium sulfate precipitation and were (before adding to the cultures) extensively dialyzed against culture medium (12).

These monoclonal anti IL-2 antibodies were further used to study the role of IL-2 in the lectin induced proliferation of normal resting cells. As shown in Table 4 the addition of monoclonal antibody inhibited also the concanavalin A-induced proliferation. PHA-induced proliferation peripheral blood mononuclear cells was also inhibited by these monoclonal antibodies (data not shown). The monoclonal anti-IL-2 antibodies were not cytotoxic as assayed by dye exclusion. Thus, these data suggest that the IL-2 activity generated by lectin stimulated cells was neutralized by anti-IL-2 antibodies.

Finally we studied whether the T cell stimulating lectins could have an affinity for IL-2. Therefore, we coupled the following lectins to plastic tissue culture plates: concanavalin A, helix pomatia lectin, lentil lectin, leucoagglutinin, phytohaemagglutinin, pokeweed mitogen, soybean lectin and wheat germ lectin. These lectin coated plates were then incubated with 100 units of IL-2. None of the coupled lectins absorbed IL-2 activity. Thus, lectins that stimulate T cells do not act as a molecular complex with IL-2 to initiate proliferation.

TABLE 4

Hybridoma designation	% inhibition of thymidine uptake[a]		
	1/20[b]	1/180	1/540
B-1	93	71	34
A-1	87	72	68
C-1	77	14	2
D-1	85	59	46
D-2	98	50	56
Control	5	0	0

a % inhibition of maximal cpm (^{3}H-TdR uptake) by 2 x 10^5 human mononuclear cells stimulated with 1 μg/ml concanavalin A.

b dilution of ammonium sulfate precipitated ascitic fluid

Discussion

Several lines of evidence speak in favour of our hypothesis that "mitogenic" lectins act indirectly by inducing hormone-like mitogenic factors or growth factors. It is obvious that lectins are needed to shift T cells from the resting phase (G_0) to the G_1 phase of the cell cycle. Lectins, in fact, act as the first signal that induces the production of the mitogen (= IL-2), as well as the expression of the mitogen receptor (= IL-2-receptor). This hypothesis is also confirmed by the finding that IL-2 dependent cells do most likely not have a resting phase and grow without a lectin stimulus. The growth factor IL-2 alone provides this necessary second signal for proliferation. Therefore, if this mitogenic signal is removed by the addition of anti-IL-2 antibody, or, if the receptor for IL-2 is blocked by a monoclonal antibody, IL-2 dependent growth will stop. Finally, lectins can block the proliferation of leukemic T cell lines in G_1, and thus lead some of them to produce IL-2.

References

1. Nowell, P.D.: Cancer Res. 20, 462 (1960)
2. Gordon, J., McLean, L.D.: Nature 208, 795 (1965)
3. Kasakura, S., Lowenstein, L.: Nature 208, 794 (1965)
4. Morgan, D.A., Ruscetti, F.W., Gallo, R.: Science 193, 1007 (1976)
5. Ruscetti, F.W., Morgan, D.A., Gallo, R.: J. Immunol. 119, 131 (1977)
6. Bonnard, G.D., Yasaka, K., Schendel et al.: In Kaplan, J.G. (ed): "The Molecular Basis of Immune Cell Function". Amsterdam: North-Holland, 1979, p. 239.
7. Stadler, B.M., Oppenheim, J.J.: Lymphokines 6, 117 (1982)
8. Bonnard, G.D., Yasaka, K., Jacobson, D. et al.: In Quastel M.R. (ed): "Cell Biology and Immunology of Leukocyte Function". New York: Academic Press, 1978, p. 569.
9. Bonnard, G.D., Yasaka, K., Jacobson, D.: J. Immunol. 123, 2704 (1979)
10. Stadler, B.M., Dougherty, S.F., Farrar, J.J., Oppenheim, J.J.: J. Immunol. 127, 1936 (1981)
11. Uchiyama, R., Broder, S., Waldmann, T.A.: J. Immunol. 126, 1393 (1981)
12. Stadler, B.M., Berenstein, E.H., Siraganian, R.P., Oppenheim, J.J.: J. Immunol. 128, 1620 (1982)

MEMBRANE ALTERATIONS AND INDUCTION OF RESPONSIVENESS TO INTERLEUCIN 2 IN LYMPHOCYTES BY LIMA BEAN LECTINS

Wolfgang G. Beßler, Hildegard Kraut, Dagmar Büsing
Lehrstuhl für Mikrobiologie II der Universität Tübingen
D-7400 Tübingen, Germany

Werner Müller-Hermes
Institut für Genetik der Universität Köln
D-5000 Köln, Germany

Hinrich Peters
Lehrstuhl für Immunologie der Universität Göttingen
D-3400 Göttingen, Germany

Lectins from lima beans have been isolated and characterized by several groups (cf. 1, 2, 3). They specifically agglutinate human blood group A erythrocytes, this agglutination can be inhibited by the addition of the lectin specific sugar N-acetyl-D-galactosamine (1, 2). Lima beans contain several isolectins, which are composed of subunits of MW 31,000. The isolectins of MW 124000 (LIMA 124) and 247000 (LIMA 247) were shown by equilibrium dialysis to possess 2 resp. 4 sugar binding sites (4). Reichert et al. showed in 1973, that a mixture of lima bean isolectins exhibits lymphocyte stimulatory activity (5). When investigating the mitogenicity of 2 purified isolectins towards normal and leukemic human lymphocytes, LIMA 247 could be shown to be a more potent mitogen than LIMA 124 (6). This valency-dependency of the mitogenic properties of the lima bean isolectins could be confirmed in a variety of species, e.g. in bovine, rabbit, rat and mouse lymphocytes (7). It also could be shown, that 2 further isolectins of higher molecular weight (LIMA poly I and LIMA poly II) were able to act as potent mitogens toward human peripheral blood lymphocytes (8). The molecular mechanism of the activation of lymphocytes by lectins is still widely unknown. For lima bean lectin, we could demonstrate binding proteins in the plasma membrane of a human T-cell line (8). The present study shows, that after the binding of lima bean lectin mitogens to lymphocytes, early alterations in the phospholipid metabolism of the cells are induced. Simultaneously to the membrane changes occuring, lectin-treated lymphocytes became responsive to the action of interleukin 2.

Experimental

Lectins. Lima bean isolectins were isolated from bean extracts by fractionated ammonium sulphate precipitation, affinity chromatography on Con A-Sepharose, and gel permeation chromatography as described earlier (3). Concanavalin A (Con A) was prepared from beans (*Canavalia ensiformis*) by the method of Agrawal and Goldstein (9).

Galactosamine derived Sepharose. A galactosamine derived Sepharose column was prepared as follows: 1 g of D-galactosamine was dissolved in 75 ml water at pH 7.5. To this solution, 15 g CH-Sepharose 4B (PHARMACIA FINE CHEMICALS, Uppsala, Sweden) suspended in water were added. Coupling was done by adding 1-ethyl-3(3-dimethylaminopropyl)-carbodiimide at 0.1 M. The pH of the suspension was adjusted to 5.3 and the mixture was incubated at 4°C for 12 h with gentle shaking.

Lymphocytes. Bovine lymph node lymphocytes were obtained from the local slaughter house and lymphocytes were prepared as described (7). Balb/c mice were from the Institut für Genetik, University of Köln. Mice were sacrificed by cervical dislocation , spleen cell suspensions were depleted of erythrocytes by lysing the red cells (10), and depleted of adherent cells by removing them after carbonyl iron (FLUKA, Buchs, Switzerland) treatment (2 h, 37°C) with the aid of a magnet.

Interleukin 2. (rat T-cell growth factor) was a gift from Dr. M. RETH, Köln, prepared according to Grönvic and Andersson (11) with slight modifications.

Incorporation of Phospholipid Precursors. 10^7 cells of a mixture of spleen and lymph node lymphocytes were incubated in 1 ml Hepes-buffered Dulbecco's modified Eagles medium supplemented with 2.5 mg/ml defatted albumin (12). Precursors of phospholipids were added at the following concentrations: 10 nmol ^{14}C-oleate (specific activity 2 Ci/mol) made by mixing ^{14}C-oleate (AMERSHAM-BUCHLER, Braunschweig, Germany) and oleate (SIGMA, Münic, Germany); and 25 nmol ^{14}C-acetate (specific activity 40 Ci/mol, AMERSHAM-BUCHLER, Braunschweig, Germany). After culturing the cells in the presence of mitogen the lipids were extracted and separated on thin-layer chromatography as described previously (12). Incorporation was measured in total phospholipids and in a fraction containing phosphatidycholine and somé contaminants of phosphatidyl-inositol and phosphatidylserine into which more than 80 % of the phospholipid precursors converted.

Results and Discussion

Lima bean isolectins were purified by fractionated ammonium sulphate precipitation, affinity chromatography, and gel permeation chromatography. Fig. 1 shows the elution diagram of a lima bean lectin purification experiment performed on galactosamine derived Sepharose 4B. As shown in the figure, the isolectins bound to the column could be eluted by a solution of 1 M galactose in PBS. This affinity column yielded very pure lectin preparations, since the purification step involves the interaction of the lectin with its specific sugar. Thus, this technique is superior to the affinity chromatographic purification on Con A-Sepharose which we published previously (3). Using the purification procedure described, an enrichment of the hemagglutinating activity up to 310-fold was achieved. The hemagglutination titers (2) of the preparations were 70 (LIMA 124), 1460 (LIMA 247), 3100 (LIMA poly II), and 2130 (LIMA poly I). In agreement with the literature, the tetravalent and the polymeric lectins were by far more active agglutinins than the divalent compound (1, 2, 8). The purification steps were controlled by polyacrylamide gel electrophoresis (Fig. 2): In the presence of sodium dodecyl sulphate, the 4 isolectins each exhibited their 31,000 MW subunits as prominent band. Furthermore the isolectins were checked for homogeneity by analytical ultracentrifugation. From the sedimentation patterns of the pure isolectins, the sedimentation coefficients of the substances were determined as 6.26 for LIMA 124, 9.88 for LIMA 247, 11.88 for LIMA poly II, and 13.61 for LIMA poly I. The

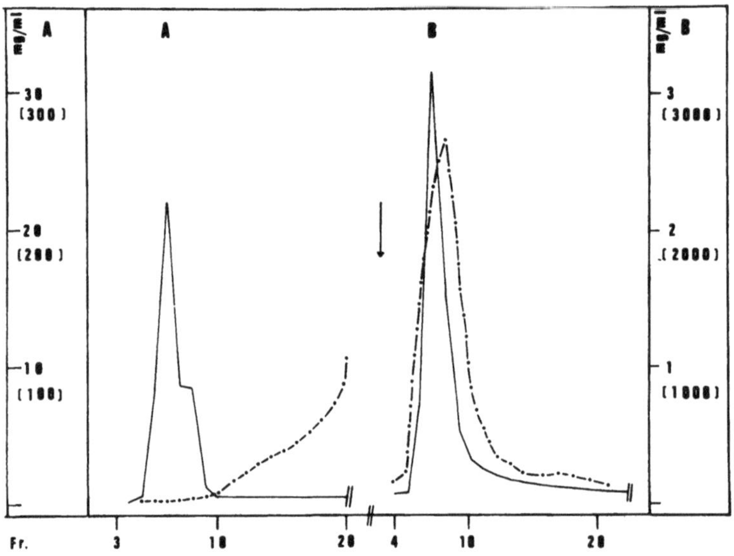

Fig. 1: Purification of lima bean lectins by affinity chromatography on galactosamine derived CH-Sepharose 4B. 800 mg of protein were applied to the column (l = 17.5 cm, d = 2 cm) in phosphate buffered saline (PBS), and the material not attached to the column was eluted (A). Lectins bound to the column were eluted with PBS containing 1 M galactose (B). Solid line: protein content of the eluted fractions (mg/ml). Broken line: hemagglutination titers. The preparation of the column was performed according to Allen and Neuberger (16) with modifications as described in Experimental.

results presented in figures 1 and 2 show that by the purification method employed, 50 mg amounts of the lima bean isolectins exhibiting a high purity can be isolated from 60 g of lima beans in a quick and reproducible manner. The combined method is also applicable for the purification of other galactosamine specific lectins (comp. 3, 16).

We had shown previously that the mitogenicity of the lima bean isolectins towards lymphocytes of a variety of species is dependent from their saccharide binding valency, and that the polymeric isolectins activate macromolecular syntheses to a degree superior to the isolectin LIMA 124 (6-8). These findings

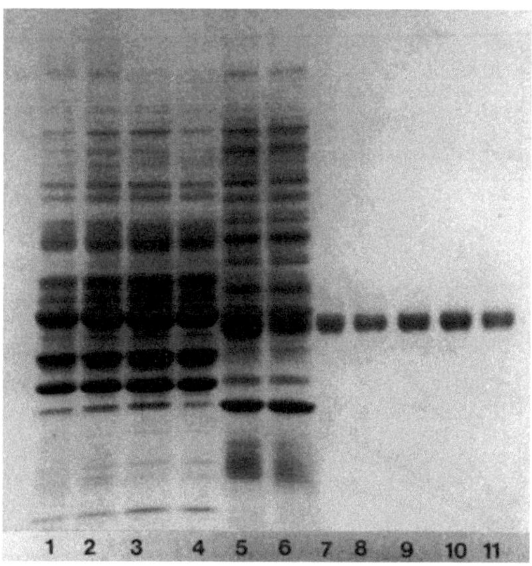

Fig. 2. Purification of the lima bean isolectins as controlled by polyacrylamide gel electrophoresis in the presence of sodium dodecyl sulphate (SDS). Gradient slab gels (10-15 % acrylamide) prepared according to Laemmli (17).
Crude lectin extract (lane 1), supernatants (lanes 2-4) and final sediment (lane 5) of the fractionated ammonium sulphate precipitation (cf. ref. 3). Fractions non binding (lane 6) and lectins binding (lane 7) to the galactosamine derived Sepharose 4B column (see Experimental). Isolectins after their separation by gel permeation chromatography on Sephadex G200 superfine (cf. ref. 3): LIMA poly I (lane 8), LIMA poly II (lane 9), LIMA 247 (lane 10), and LIMA 124 (lane 11).

were confirmed in our experiments measuring the early alterations in the phospholipid metabolism of lectin stimulated lymphocytes. Since relatively large amounts of cells were necessary for these assays, bovine lymph node lymphocytes were used. In the assays, the effects of divalent LIMA 124 and tetravalent LIMA 247 were compared; according to gel electrophoresis LIMA 247 contained some aggregated polymeric material (cf. 1, 3, 7). LIMA 124 and LIMA 247 both stimulated the incorporation of ^{14}C-oleate into phosphatidyl choline starting at concentrations around 10 µg/ml, the stimulation optimum was around 150 µg/ml for LIMA 124, and around 50-100 µg/ml for LIMA 247. For concanavalin A, the optimum response was found at concentrations

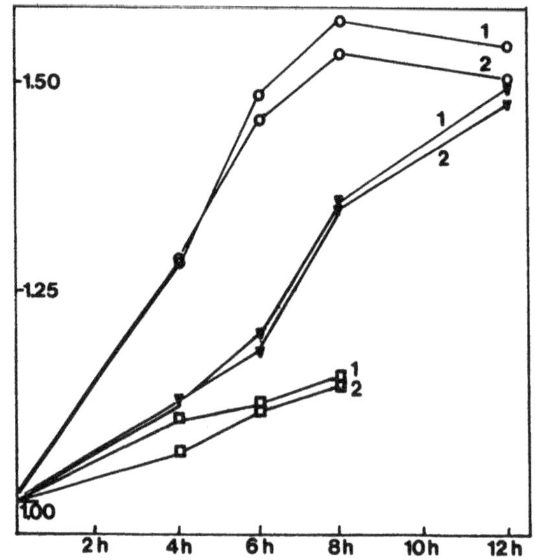

Fig. 3. Kinetics of incorporation of ^{14}C-oleate into membrane phosphatidyl choline (1) and into membrane phospholipids (2) in lymph node lymphocytes after various times of incubation of lymphocyte cultures with 1.6 µg/ml concanavalin A (o), 100 µg/ml LIMA 247 (▼), and 150 µg/ml LIMA 124 (□). Results are means of triplicate or quadruplicate assays. Controls without addtion of lectins: 4065 cpm for phosphatidyl choline, and 5037 cpm for membrane phospholipids. Ordinate: stimulation indices = cpm of mitogen stimulated cultures/cpm of control cultures. Abscissa: time of incubation.

around 1 µg/ml, which is in agreement with our results published for mouse lymphocytes (12). The summarizing fig. 3 shows the kinetics of oleate incorporation induced by the lectins. As seen from the figure, both lima bean isolectins stimulated the incorporation of ^{14}C-oleate into membrane phospholipids and membrane phosphatidylcholine already 4 hours after their addition to the cultures. LIMA 247 exhibited a much more pronounced effect than the divalent isolectin, and after 8-12 h, the effect of LIMA 247 became comparable to concanavalin A; the incorporation at this late time, however, is partly due to de novo net synthesis of phospholipids due to the mitogenic activation of the cells. Our results show that the membrane events induced by the lima bean isolectins are dependent from their valency,

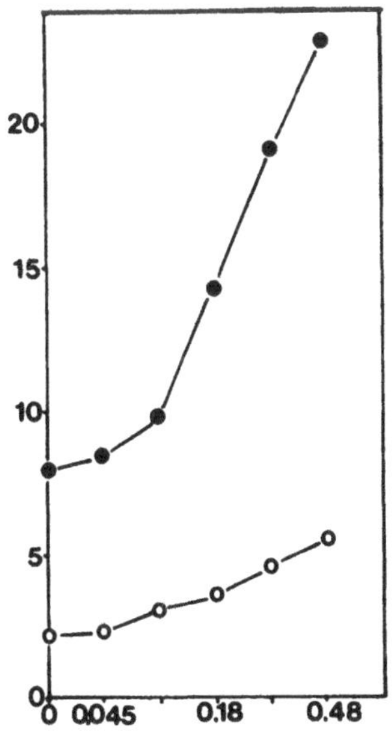

Fig. 4. Induction of responsiveness to interleukin 2 in lymphocytes by LIMA 124. Balb/c mouse lymphocytes were preincubated at a cell concentration of 1.2 x 10^7/ml with 0.4 U/ml neuraminidase (30 min, 37°C). After washing, LIMA 124 was added to the cell cultures at different concentrations. Cultures were performed in COSTAR microtiter plates at 3 x 10^6 cells/ml for 42 h; 18 h before harvesting, 0.2 µCi ^{3}H-thymidine were added to the wells. (o) Cultures performed without interleukin 2. (•) Cultures performed in the presence of 10 % interleukin 2 containing supernatant (see Experimental). Ordinate: ^{3}H-thymidine incorporation (cpm x 10^{-3}). Abscissa: lectin concentration (mg/ml).

as shown before for their mitogenic activity (6-8). A similar valency dependent stimulation of phospholipid metabolism by lima bean isolectins could also be shown, in preliminary assays, for human, rat and mouse lymphocytes (unpublished). These lectin induced changes, which we also could demonstrate for B-lymphocytes using the B-lymphocyte mitogens lipoprotein and lipopeptide (12), involve an enhanced uptake of unsaturated fatty acids into membrane phospholipids. The induced fluidization of the plasma membrane brings about a modulation of the activities of membrane associated enzymes, which could lead ultimately to lymphocyte proliferation and differentiation (13, 14, 15).

Simultaneously to the induction of membrane events, the divalent lima bean isolectin, which is only marginally mitogenic towards mouse lymphocytes, was able to induce responsiveness to interleukin 2 in Balb/c mouse lymphocytes. The effect could be further enhanced by pretreatment of the lymphocytes with neuraminidase. Fig. 4 shows, that in the presence of 0.2 - 0.4 mg/ml LIMA 124, ^{3}H-thymidine incorporation could be enhanced 4-5 fold by the addition of 10 % of rat interleukin 2 supernatant to the cultures. Similar assays employing LIMA 124 have been used by us to test for interleukin from different sources without the aid of interleukin-dependent cell lines or mitogenic concanavalin A (cf. ref. 19).

The results presented here might suggest that LIMA 124, which exhibits only a marginal mitogenicity towards mouse lymphocytes (7), induces membrane alterations in T-lymphocytes followed by the functional appearance of the receptor for interleukin 2 proposed by several groups (cf. ref. 19-21). On the other hand, the divalent LIMA 124 might be unable to induce, by itself, the production of the interleukins necessary for the mitogenic activation of the T-lymphocytes (21). A further investigation of these hypotheses is in progress in our laboratories.

Acknowledgement
This work was supported by the Deutsche Forschungsgemeinschaft.

References

1. Galbraith, W., Goldstein, I.J.: Biochemistry 11,3976-3984 (1972).

2. Gould, N.R., Scheinberg, S.L.: Arch. Biochem. Biophys. 137, 1-11 (1970).

3. Bessler, W.G., Goldstein, I.J.: FEBS Lett. 34, 58-62 (1973).

4. Bessler, W.G., Goldstein, I.J.: Arch. Biochem. Biophys. 165, 444-445 (1974).

5. Reichert, C.F., Pan, P.M., Mathews, K.P., Goldstein, I.J.: Nature New Biol. 242, 146-147 (1973).

6. Ruddon, R.W., Weisenthal, L.M., Lundeen, D.E., Bessler, W., Goldstein, I.J.: Proc. Nat. Acad. Sci. USA 71, 1848-1851 (1974).

7. Bessler, W.G., Resch, K., Ferber, E.: Biochem. Biophys. Res. Comm. 69, 578 (1976).

8. Kraut, H., Resch, K., Bessler, W.G. in: Protides of Biological Fluids (H. Peeters, ed.), Pergamon Press Oxford · New York, 619-622 (1979).

9. Agrawal, B.B.L., Goldstein, I.J.: Biochem. J. 96, 23C-25C (1965).

10. Boyle, W.: Transplantation 6, 761-770 (1968).

11. Grönvic, K., Andersson, J.: Imm. Rev. 51, 35-59 (1980).

12. Resch, K., Bessler, W.: Eur. J. Biochem. 115, 247-252 (1981).

13. Ferber, E., de Pasquale, C.G., Resch, K.: Biochim. Biophys. Acta 226, 494 (1975).

14. Rode, H., Szamel, M., Schneider, S., Resch, K. in: Mechanisms of lymphocyte activation (K. Resch and H. Kirchner, eds.), Elsevier/North Holland, 242-244 (1981).

15. Resch, K., Szamel, M. in: Mechanims of lymphocyte activation (K. Resch and H. Kirchner, eds.), Elsevier/North Holland, 206-208 (1981).

16. Allen, A.K., Neuberger, A.: FEBS Lett. 50, 362-364 (1975).

17. Laemmli, U.K.: Nature 227, 680-685 (1970).

18. Fischer, D.B., Mueller, G.C.: Biochim. Biophys. Acta 176, 316-329 (1969).

19. Coutinho, A., Larsson, E.L., Grönuik, K.D., and Andersson, J.: Eur. J. Immunol.9, 587 (1979).

20. Smith, K.A., Gillis, S., Baker, P.E., McKenzie, D., and Ruscetti, F.W.:Ann. N. Y. Acad. Sci. 332, 423 (1979).

21. Farrar, J.J., Benjamin, W.R., Hilfiker, M.L., Howard, M., Farrar, W.L., and Fuller-Farrar, J.: Immunological Rev. 63, 129 (1982)

T LYMPHOCYTE-MEDIATED CYTOLYSIS

IV. How do Mitogenic and Non-Mitogenic Lectins Mediate Lymphocyte - Target Interactions?

Gideon Berke

The Department of Cell Biology
The Weizmann Institute of Science
Rehovot, 76100 Israel

Cytotoxic T lymphocytes (CTL) can lyse target cells (TC) nonspecifically in the presence of certain plant lectins, notably Con A and PHA. Lectin-dependent cytotoxicity (LDCC) has been attributed to CTL-TC interactions mediated by intercellular 'bridging' of multivalent, carbohydrate-specific lectins, followed by 'activation' of the CTL by the lectin (1-3) (see scheme)

$$CTL + TC + L \rightarrow CTL^{*} - L - TC \rightarrow Lysis + TC$$

(L, lectin; CTL*, activated CTL)

Because the mechanism of LDCC is highly relevant to specific, CTL-mediated lysis, we investigated in detail how lectins mediate CTL-TC interactions. In previous communications of this series (4-6), we presented evidence that the lectin Concanavalin A does not act simply to 'bridge' CTL and TC, nor does it 'activate' the CTL. Instead, we proposed that lectin modify or cluster cell-surface determinants of the TC, rendering them nonspecifically recognizable through CTL surface receptors other than lectin-binding sites. The present paper delineates possible mechanisms whereby some mitogenic and non-mitogenic lectins mediate CTL-TC interaction in LDCC.

Lectins, Vol. III

Materials and Methods

These have been described in detail previously (4,5). As CTL we used either Con A-activated lymphoblasts or alloimmune peritoneal exudate CTL. As TC we used lymphoma EL4 of C57BL/6 or P815 of DBA/2 mice. CTL-TC binding and lysis were assessed by the conjugation and ^{51}Cr-release assays, respectively.

Results and Discussion

The observation that in the presence of certain plant lectins CTL can lyse virtually any TC, including syngeneic and even antologous cells (7-9), is in sharp contrast to experiments showing selective inactivation of only B anti-C CTL following their interaction with A anti-B CTL and, likewise, lack of simultaneous, mutual lysis when A anti-B CTL are confronted with B anti-A CTL (see refs. 10, 11 and Fig. 1).

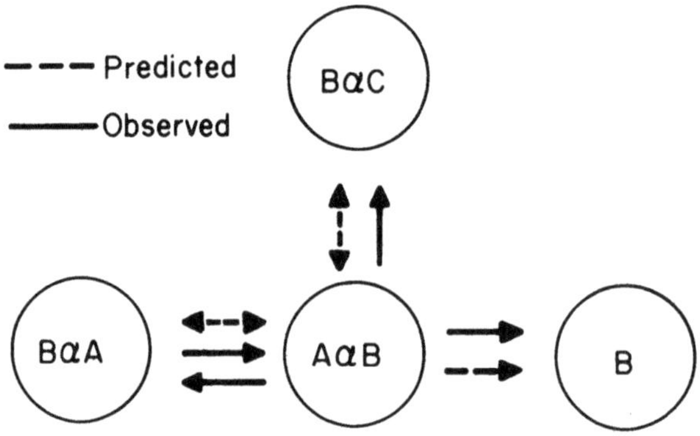

Figure 1.
Interaction of CTL and TC which are themselves CTL.

These killer- anti-killer experiments have indicated that TC binding through the CTL receptor is a prerequisite for subsequent lysis, suggesting that 'passive' binding of a CTL to its TC, although required, is not sufficient for affecting lysis of the latter. The previous interpretations of the mode of action of lectins in LDCC, (1-3) suggesting a 'dual' role of lectin stem primarily from the finding that only T cell mitogenic lectins such as Con A PHA and LCA can mediate LDCC while non-mitogenic lectins such as PNA and WGA do not (2). Although it has been assumed that non-mitogenic lectins can mediate 'effective' CTL-TC binding not leading to lysis, this assumption has never been substantiated by a CTL-TC conjugation assay, for example. In fact, direct examination of the CTL-TC conjugating ability of mitogenic (LDCC-supportive) and non-mitogenic (LDCC-nonsupportive) lectins revealed that while effective CTL-TC conjugation can be easily demonstrated with mitogenic lectins such as Con A, PHA and LCA, virtually no conjugates could be detected with non-mitogenic lectins (Table 1). These results provide a new basis for the lack of LDCC activity of non-mitogenic lectins such as PNA, WGA and SBA. Hence previous explanations of the dual role of Con A in LDCC

Target cells	Immune CTL	
	BALB/C anti EL4	C57BL/6 anti-P815
	(Number of conjugated/0.5 microliter	
EL4	60	2
P815	4	87
Con A·EL4	-	86
Con A·P815	60	-
PNA·EL4	-	5 (2)
PNA·P815	1 (3)	-
SBA·EL4	-	23 (2)
SBA·P815	3 (6)	-
WGA·EL4	-	2 (3)
WGA·P815	2 (3)	-

Table 1. CTL-TC conjugation by mitogenic and non-mitogenic lectins

BALB/C anti-EL4 or C57BL/6 anti-P815 alloimmune CTL were mixed with lectin-treated or untreated EL4 or P815 TC ($1x10^6$ cells each), centrifuged 170 xg, 10 min at room temperature to affect conjugate formation. Pellets were resuspended vigorously with a Pasteur pipette and the number of conjugates per 0.5 microliters was determined microscopically (see ref. 12). Numbers in parentheses show number of conjugates in the absence of lectins.

(1-3,13) must be revised as it rests on the assumption that both mitogenic and non-mitogenic lectins are equally effective in mediated CTL-TC conjugation.

Binding (even temporary) of CTL to TC is an essential, early step in a multistep process, ultimately leading to TC lysis (see ref. 12 for review). Although morphologically indistinguishable conjugates have been detected in

LDCC and in specific CTL-TC binding (Fig. 2) the mechanism of CTL-TC binding in LDCC has not been known.

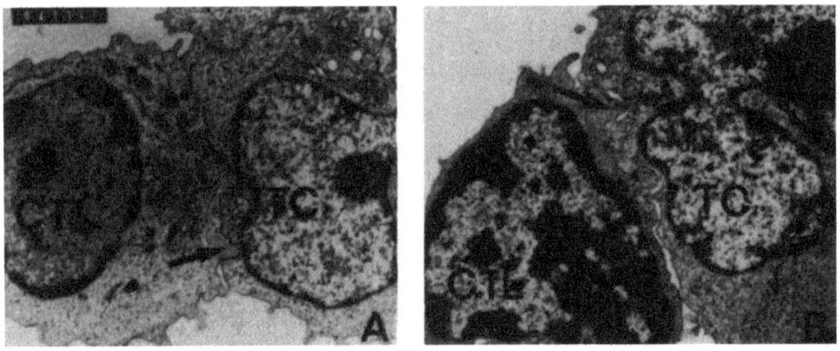

Figure 2.
Electron micrograph of (A) lectin mediated and (B) specific CTL-TC conjugates. Projections (arrows) can be seen in both types.

Contrary to previous suggestions that lectins mediate LDCC by first 'bridging' CTL and TC and then 'triggering' the TC-bound CTL (1-3, 13), it has been proposed (4, 5) that Con A mediates CTL-TC interactions by modifying the TC, to affect their nonspecific recognition by the CTL receptors, and that these CTL receptors are distinct from simple Con A binding sites. An argument against Con A functioning simply as an intercellular 'bridge' stems from experiments showing that Con A pretreatment of the TC but not of the CTL leads to optimal LDCC (4, see scheme).

	Lysis of TC	
	Expected	Observed
CTL + TC (non-specific)	-	-
L·CTL + TC	+	-
CTL + L·TC	+	+
L·CTL + L·TC	+	+
CTL + TC + L	+	+

(CTL, cytotoxic T lymphocyte; TC, target cell; L-lectin (Con A))

These studies have now been extended to additional lectins, mitogenic and non-mitogenic. Lectin-treated or untreated CTL or TC were allowed to react with or without additional lectin and the degree of ^{51}Cr release from labelled TC was monitored (Table 2). As reported (4, scheme), optimal lysis was obtained when Con A-treated TC (L·TC) were reacted with untreated CTL, but not when Con A-treated CTL (L·CTL) were reacted with untreated TC. Similar results were obtained when PHA was used instead of Con A, although PHA-CTL lysed untreated TC to a higher degree than did Con A-treated CTL. With LCA,

Lectin	CTL + TC	L·CTL + TC	CTL + L·TC	L·CTL + L·TC	CTL+ TC + L
Mitogenic					
Con A	4.6	7.6	95	51	49
PHA	5	23	99	46	80
LCA	6.8	32	58	35	85
nonmitogenic					
PNA	5	5.6	5.7	6	5
SBA	3.7	4.3	12	11	16
WGA	4.3	5.3	1.6	1.6	11

Table 2. LDCC by mitogenic and non-mitogenic lectins
The cytolytic activity of lectin-treated or untreated CTL (L·CTL, CTL) against lectin treated or untreated ^{51}Cr-labelled TC (L·TC, TC). Effector-to-target ratio 10:1, assay 2 hrs at 37°. Mitogenic lectin were used at 10 μg/ml, non-mitogenic lectins at 20 μg/ml. CTL were polyclonally activated lymphoblasts (4 5); similar results have been obtained with specifically immunized CTL.

however, unlike Con A and PHA, a clear-cut TC preference has not been shown. LCA, however, was less effective than Con A and PHA when either CTL or TC were pretreated with the lectin; optimal killing was obtained only if the lectin was present for the entire duration of the assay (for details see ref. 6).

The precise mechanism whereby lectins condition TC to affect their lysis by antigenically irrelevant CTL is not known at present. We have proposed (4, 5) that in LDCC, major histocompatibility complex (MHC) antigens of the TC were effective sites of lectin action, rendering the TC recognizable, non-specifically, by the CTL receptors. It was suggested that the TC-MHC determinants were affected by certain lectins in a manner causing their modification or redistribution (microclustering) on the cell surface. Such a modification or microclustering of MHC-Ag could stabilize interaction with non-specific (low affinity) receptors on the CTL surface, by increasing the avidity of interaction. Microclustering may be the result of cross-linking of MHC components alone, or together with other cell surface constituents, by the lectins Con A, PHA and LCA, known to bind effectively to MHC

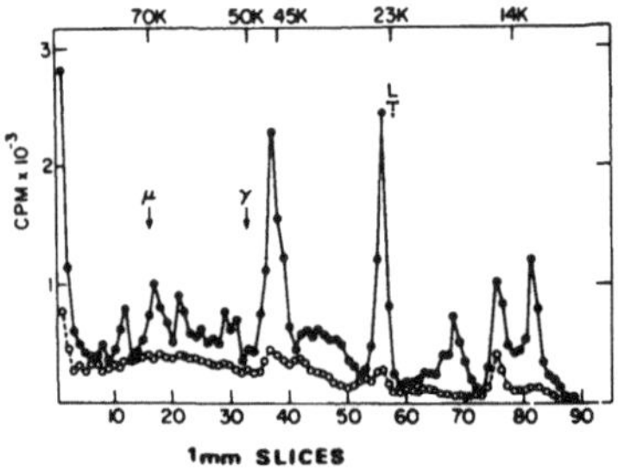

Figure 3.
Analysis of ^{125}I-labelled cell surface component which bind to Con A by the method of Choi and Jensen (16). α-methylmannoside induced elution from Con A sepharose (●—●), or control BSA-sepharose (O—O). L, γ and μ are labelled immunoglobulin light and heavy chain markers; main peaks are MHC class I and II components.

components (see binding of Con A, Fig. 3) and be effective mediators of LDCC (Table 2). It may be argued that the lectins SBA, WGA and PNA, that effectively bind to cells (6), but which do not support LDCC (Table 2) are either not reacting with MHC components or do so ineffectively. In fact, evidence has recently been presented suggesting reduced affinity for MHC-glycoproteins of lectins such as SBA, PNA and WGA (14), shown (table 2) to be ineffective mediators of LDCC.

Conditioning of the TC by lectins is time and temperature dependent (30-45 min at 37°), influenced by lectin concentration (optimum for Con A, 10 μg/ml) and by the metabolic activity of the cells (inhibitable by metabolic inhibitors). A general increase in the susceptibility of TC to lysis could explain some of the effects of lectin in LDCC. For example, the lectin can induce clustering of cell-surface glycoproteins and glycolipids, exposing the lipid bilayer hence increasing TC fragility. Careful analysis of the individual steps of non-specific LDCC, carried out by Gately and Martz (15), however, revealed only little deviation from the mechanism of specific, CTL-mediated binding and lysis of TC. Furthermore, in many (but not all) instances, the inclusion of lectin in a CTL-TC interaction had only a minimal effect on the effect and extent and rate of the lytic reaction and often none at all. Furthermore, we have tested the susceptibility of Con A-treated TC to osmotic and detergent-mediated lysis (Fig. 4) and found that lectin-treated TC were as susceptible to lysis as lectin-untreated cells. Thus a general non-selective increase in TC fragility following lectin treatment may be ruled out.

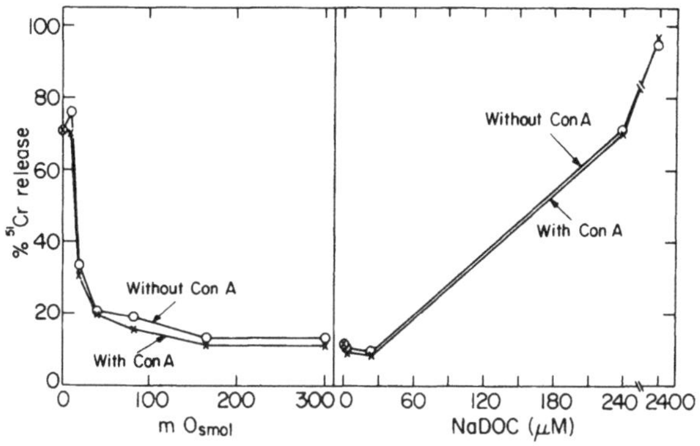

Figure 4.
Osmotic fragility and susceptibility to sodium deoxycholate induced lysis of Con A treated and untreated TC tagged with ^{51}Cr.

Summary

In the present study, we delineate the mechanism(s) of lymphocyte-mediated, lectin-dependent cytotoxicity (LDCC) by mitogenic and nonmitogenic lectins. We show that the nonmitogenic lectins SBA, PAN and WGA, unlike the mitogenic lectins Con A, PHA and LCA are ineffective mediators of LDCC, and suggest this to be due to their inability to mediate CTL-TC binding (conjugation) and probably not to their inability to 'activate' TC-bound CTL, as proposed before. We show that PHA, not unlike Con A, exert its primary effects in LDCC by affecting the TC rather than the CTL. LCA, on the other hand, unlike Con A and PHA does not show a clear TC preference. The three LDCC supportive lectins (PHA, Con A and LCA) show a similar TC preference when tested in the CTL-TC binding (conjugation) assay (6). We suggest that the potential of mitogenic lectins to mediate LDCC is related to their ability to interact with the TC determinants, to affect their recognition by the CTL receptor and not to a general increase in the susceptibility of TC to lysis.

Acknowledgements

I thank Dr. Dalia Rosen for the electrom micrography and for expert technical assistance.

References

1. Green, W., Ballas, Z. and Henney, C. J. Immunol. 121, 1566, (1978).
2. Parker, W.L. and Martz, E. J. Immunol. 124, 25, (1980).
3. Bradley, T.P. and Bonavida, B. J. Immunol. 127, 208, (1980).
4. Berke, G., Hu, V., McVey, E. and Clark, W.R. J. Immunol. 127, 776, (1981).
5. Berke, G., McVey, E., Hu, V. and Clark, W. J. Immunol. 127, 782, (1981).
6. Berke, G., Rosen, D. and Moscovitch, M. (1981) submitted.
7. Forman, J. and Möller, G. J. Exp. Med. 138, 672, (1973).
8. Bevan, M.J. and Cohn, M. J. Immunol. 114, 559, (1975).
9. Bonavida, B. and Bradley, T.P. Transplantation 21, 94, (1976).
10. Kuppers, R.C. and Henney, C.S. J. Immunol. 118, 71, (1977).
11. Fishelson, Z. and Berke, G. J. Immunol. 120, 1121, (1978).
12. Berke, G. Prog. Allergy 27, 69, (1980).
13. Green , W.R., Adv. Exptl. Biol. Med., Clark, W.R. and Golstein, P. eds. Plenum Press, New York, in press (1982).
14. Kimura, A. and Ersson, B. Eur. J. Immunol. 11, 475, (1981).
15. Gately, M.K., Marty, E. J. Immunol. 119, 1171, (1977).
16. Choi, Y.W., Jensen, J.C. J. Exp. Med. 140, 597, (1974).

LECTIN COATED BEADS STIMULATE SINGLE CELLS FROM MOUSE EMBRYOS TO SPREAD

Susan J. Kimber
MRC Laboratories, Woodmansterne Road, Carshalton, Surrey, UK.

M. Azim, H. Surani
ARC Research Station, 307 Huntingdon Road, Cambridge, UK.

Lectins have been used in studying the cell surface during the development of embryos of a variety of animals including those of mammals (1-8) and much of this work has been reviewed (1). More recent studies have revealed that fluorescent lectins bind in a polarized manner to embryonic cells from 8-cell stage morulae (9, 10). Inhibiting glycoprotein synthesis reduces the amount of PNA and Con A bound to morulae but does not abolish the polarity of binding (11). Con A labelled total cell proteins and cell surface components have also recently been isolated from preimplantation mouse embryos of different stages (12). Thus soluble lectin probes have provided much information about the development of the embryo. However, apart from a brief report (4), the potential of bound lectins has not been exploited in studies on mammalian embryos. We have used agarose bound lectins as a model system to probe the spreading behaviour characteristic of cells of the 8-cell stage mouse embryo. At this stage the previously rounded cells flatten on one another in a process termed compaction. In separate studies we found that the ability of embryonic cells to spread over an appropriately adhesive substrate arises first at the 8-cell stage (13). Here we show that agarose beads coated with certain lectins will support spreading by 1/8 - blastomeres (single cells from 8-cell embryos) although substrates coated with the monosaccharides that these lectins bind will not.

Materials and Methods

Embryos were flushed from oviducts of MF_1 mice at the 2-cell stage, 44-47 h post injection of human chorionic gonadotrophin, and cultured in Brinster's medium (14) to which 4 mg/ml bovine serum albumin (BSA, Sigma) was added. Lectin coated agarose beads were obtained from Vector, USA. The concentrations of lectin (mg/ml settled gel) were as follows: peanut agglutinin (PNA) 5, wheat germ agglutinin (WGA) 7, and Concanavalin A (Con A) 6. Monosaccharide coated agarose beads were obtained from P-L Biochemicals Inc., USA. The concentration of N-acetyl glucosamine (agarose p- aminophenyl β - D- glucosamine N-acetyl) fucose (agarose -p- aminophenyl - α - L-Fucopyranoside) and mannose (agarose - p-aminophenyl - α-D - mannopyranoside, was 1-5 μ moles saccharide/ml agarose, of galactosamine (agarose - aminocaproyl - β - D - galactosamine) was 1-6μ mole/ml agarose and of fucosylamine (agarose - aminocaproyl - α -L-fucosylamine) was at least 1μ mole/ml agarose. Beads were washed in phosphate buffered saline (PBS:136 mM NaCl, 2.7 mM KCl, 1 mM $CaCl_2$, 1 mM $MgCl_2$, 9.6 mM Na_2HPO_4, 1.5 mM KH_2PO_4) to remove saccharide or azide present, and then in Brinster's medium with albumin. To obtain single blastomeres from 8-cell precompacted embryos the zona pellucida was removed with acid tyrode (pH 2) (2), the cells were dissociated in medium without added calcium and magnesium (15) and plated out individually with the beads. When sugars were added (see below) the beads and cells were washed in several changes of medium with the appropriate

TABLE

Effect of saccharides on spreading of 1/8 blastomeres on PNA and WGA-agarose beads

Saccharide added	Average linear contact*			
	PNA		WGA	
————	100± 4.2	(55)	100± 2.8	(27)
D-galactose	47± 0.7	(42)	113±10.1	(12)
β-D-lactose	46± 1.0	(17)	100± 1.5	(11)
α-D melibiose	59± 9.5	(36)	90± 2.5	(18)
Yeast Mannan	————		78± 8.0	(9)
Galactan	57± 8.0	(37)	————	
L-fucose	103± 2.0	(32)	138± 2.5	(25)
D-mannose	87±10.0	(35)	88± 9.3	(11)
N-acetyl-D glucosamine	97± 5.0	(30)	62± 4.0	(17)
N-acetyl-D galactosamine	99± 3.0	(34)	73± 1.4	(24)
N, N^1-diacetyl chitobiose	————		27± 5.0	(18)

* Percentage relative to controls ± SE (no. cell-bead pairs analysed). See materials and methods for details.

addition before transfer to individual drops of the same composition. Bead and cell were gently pushed together using a glass needle and observed after 18-22 h. Mono- and di-saccharides were used at 25 mM, a concentration shown (for all the saccharides used) to allow reaggregation of disaggregated 8-cell morulae, compaction and formation of the blastocyst (15). All saccharides were from Sigma (USA) except galactan (P-L Biochemicals Inc., Wisconsin, USA) used at 5 mg/ml. Yeast mannan was used at 2 mg/ml. In order to achieve a relative measure of the area of contact between 1/8 blastomeres and beads, groups of cell-bead pairs were photographed under a Leitz Diavert microscope at right angles to an axis running through cell and bead. Photographs were all enlarged to the same magnification and the interface arc between cells and beads measured using a mapping wheel. The consistency of our results (see table) suggests that this is a good method of obtaining a relative measure of cell spreading.

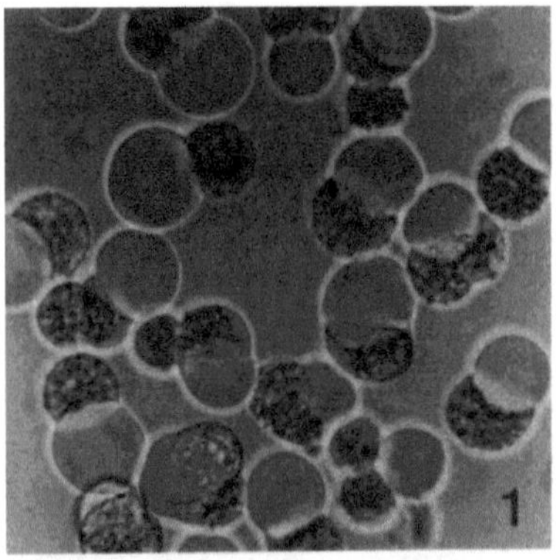

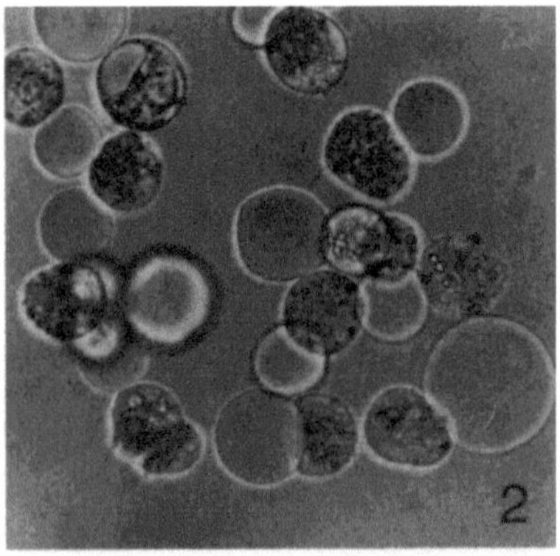

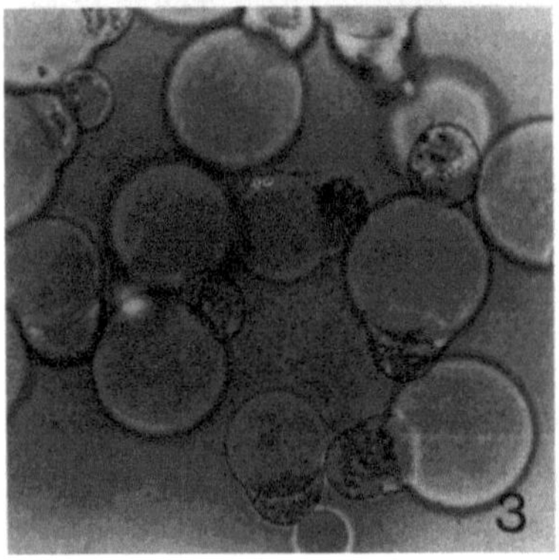

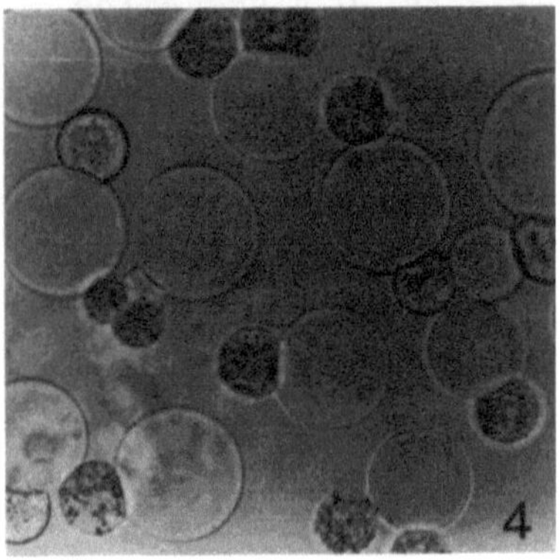

Fig. 1. Single cells from 8-cell mouse embryos spreading over PNA-agarose beads after 18 h culture.

Fig. 2. Single cells from 8-cell mouse embryos adhering to PNA-agarose beads after 18 h culture in the presence of β-lactose.

Fig. 3. Single cells from 8-cell mouse embryos spreading over WGA-agarose beads after 18 h culture.

Fig. 4. Single cells from 8-cell mouse embryos adhering to WGA-agarose beads after 18 h culture in the presence of N-acetyl-D-glucosamine.

Results

Single blastomeres from 8-cell embryos will spread on PNA and WGA coated agarose beads (Table and figures 1 and 3). Sugars which bind to PNA and WGA but not other sugars, inhibit spreading although the cells still adhere to the beads in the presence of the competing sugar (Table and figures 2 and 4). Blastomeres would not stick to protein A coated agarose beads used as an additional control for the specificity of the cell-bead interaction. Cells did adhere to and spread on Con A coated agarose beads. However, linear analysis revealed that the spreading interaction was non-specific since it could not be inhibited by saccharides for which the lectin has high affinity (glucose, mannose, yeast mannan and α-methyl mannoside). Further, spreading on Con A beads was not as extensive or reproducible as that on PNA and WGA beads. Spreading of cells on different beads could not strictly be compared because of the different amounts of lectin bound per ml settled gel (see Materials and Methods). However, spreading on WGA and PNA beads was within a closely overlapping range of arbitrary units (PNA: 0.85-1.05; WGA 0.92-1.21). Only in one case was spreading promoted by the addition of saccharide (table). Adding fucose to the culture medium increased spreading of 1/8-blastomeres on WGA beads by 38+2.5%. Fucose had no affect on spreading of blastomeres on PNA or Con A beads. We have also looked at the behaviour of cells towards agarose beads coated with Ricinus communis agglutinin (RCA-1; β -D-Gal) Ulex Europaeus agglutinin (UEA-1; α -L-fucose) Limulus polyphemus lectin (sialic acid) soybean agglutinin (SBA; α -D-Gal NAC-) and Lens culinarius lectin (α-D-Mannose). Blastomeres would not stick to UEA-1; Limulus or SBA coated beads. They stuck to but did not spread on RCA-1 and Lens culinarius coated beads. However, the concentration of these lectins per ml settled gel was approximately half that on WGA conjugated beads. In order to gain evidence about whether an endogenous lectin is involved in the spreading of one cell on another at compaction we looked at the interaction of 1/8-blastomeres with agarose beads coated with the monosaccharides, D-galactosamine, N-acetyl-D-glucosamine, L-fucose, L-fucosylamine and D-mannose. In no case did cells spread on the saccharide coated beads. Indeed, in most cases the cells would not adhere to the beads. When adhesion did occur (5 out of 26 on fucose coated beads, 1 out of 28 on fucosylamine coated beads, 6 out of 24 on mannose coated beads, 6 out of 33 on galactosamine coated beads, 5 out of 58 on N-acetyl glucosamine coated beads) it was tenuous (point contact) and could easily be disrupted by gently blowing medium over the cell with a mouth pipette. Because fucosylation of a cell surface glycolipid probably occurs at the 8-cell stage (see Discussion) we tried the effect of adding fucose (10-25 mM) on binding to galactosamine and N-acetyl glucosamine coated beads. No adhesion of cells to the beads was observed in the presence of fucose.

Discussion

We have found that PNA and WGA coated agarose beads are excellent model surfaces to stimulate spreading of cells from 8-cell mouse embryos. Spreading of cells on PNA and WGA beads appeared to rely on specific interaction of cell membrane components with these lectins since it could be inhibited by saccharides for which the lectins have high affinity. However, the cells remained attached to the beads in the presence of these sugars. Spreading of cells over Con A beads does not appear to involve specific binding because it is not inhibited by sugars for which the lectin has high affinity including yeast mannan and the non metabolisable sugar α- methyl mannoside. Non specific interactions between other types of cells and protein coated surfaces have been documented (16). One might expect cells to stick to lectin coated beads if appropriate receptors are present on their surfaces even if the beads are not sufficiently adhesive, or do not provide the appropriate signal, to trigger spreading. Thus the lack of adherence of cells to SBA, Limulus and UEA-1 coated agarose beads may reflect lack of suitable (exposed) receptors for these lectins. The one case of spreading being enhanced by a saccharide was seen with fucose which enhanced spreading on WGA beads. At the 8-cell stage in mouse embryogenesis the stage specific antigen SSEA-1 (17) arises by the fucosylation of Gal β 1 → 4 GLc NAc moieties of a glycolipid (18). The increased spreading of 1/8 blastomeres seen in the presence of fucose might suggest increased availability of N-acetyl glucosamine for interaction with WGA after fucosylation. A connection has been suggested between fucosylation of this glycolipid and the spreading of cells over one another during compaction (18). However, if there is a receptor for N-acetyl glucosamine on the surface of the 1/8 blastomeres it does not recognise N-acetyl glucosamine linked to agarose beads. This could be due to steric hinderence, incorrect molecular microenvironment or the type of linkage (β to p-aminophenyl) of the monosaccharide to the bead. However, neither N-acetyl glucosamine nor N, N^1- diacetyl chitobiose is affective up to 25 mM in inhibiting reaggregation of the cells of disaggregated 8-cell morulae, or their compaction (15). Indeed we have not found any saccharide (of various mono-,di- and oligosaccharides) used singly and in combination, or glycopeptide which will inhibit this process (21). The sensitivity of spreading on WGA and PNA beads is greater to cytochalasin D than to colcemid (15) as is also true for compaction (19, 20, 21). Removing calcium from the medium both inhibits early 8-cell embryos from compacting and causes decompaction of embryos which have already compacted (19). In contrast spreading of 1/8-blastomeres on lectin coated beads is insensitive to lack of calcium (15). These observations indicate that a single sugar-receptor interaction alone is not responsible for the cell-spreading seen in whole embryos at compaction. It is quite possible that a protein-protein interaction is involved, such as seems to be the case for the laminin mediated spreading of hepatocytes (22). Indeed there is now evidence that cells of various types will spread on laminin coated surfaces (23, 24) and this molecule is found between the cells of the mouse embryo by the 16-cell (compacted) stage (25). However, our data do not entirely rule out a saccharide-receptor interaction also being involved. Indeed compaction is dependent on protein glycosylation (26). It appears that adhesion of many types of cell involves a calcium dependent and calcium independent mechanisms (27, 28, 29). Thus the calcium dependent mechanism of compaction might involve a calcium controlled conformational change in a protein such as that proposed by Hyafil et al (30) while calcium independent interaction between cell surface saccharide containing moieties might also play a role. Our observation of calcium insensitive spreading of 1/8-blastomeres on PNA and WGA beads suggest that this must remain a possibility.

References

1. Johnson, L.V., Calarco, P.G.: Anat. Rec. 196, 201-219 (1980).

2. Nicolson, G.L., Yanagimachi, R., Yanagimachi, H.: J. Cell Biol. 66, 263-274 (1975).

3. Yanagimachi, R., Nicholson, G.L.: Exp. Cell Res. 100, 249-257 (1976).

4. Pienkowski, M.: Proc.Soc. exp. Biol. Med. 145, 464-469 (1974).

5. Magnuson, T., Stackpole, C.W.: Exp.Cell Res. 116, 466-469 (1978)

6. Rowinski, J., Solter, D., Koprowski, H.: Exp. Cell Res. 100, 404-408 (1976).

7. Sobel, J.S., Nebel, L.: J. Reprod. Fert. 52, 239-248 (1978).

8. Wu, J.T., Chang, M.C.: J. Exp. Zool. 205, 447-453 (1978).

9. Handyside, A.H.: J. Embryol. exp. Morph. 60, 99-116 (1980).

10. Handyside, A.H.: J. Reprod. Immunol. 2, 339-350 (1981).

11. Surani, M.A.H., Kimber, S.J., Handyside, A.H.: Exp. Cell Res. 133, 331-339 (1981)

12. Magnuson, T., Epstein, C.J.: Dev. Biol. 81, 193-199 (1980)

13. Kimber, S.J., Surani, M.A.H., Barton, S.C.: J. Embryol. exp. Morph. 70, in press (1982)

14. Brinster, R.L.: Advances in the Biosciences 4, 199-234 (1970)

15. Kimber, S.J.,Surani, M.A.H.: J. Cell Sci. in press (1982)

16. Rauvala, H., Carter, W.G., Hakomori, S.-I.: J. Cell Biol. 88, 127-137 (1981)

17. Solter, D., Knowles, B.B.: Proc. Natn. Acad. Sci. USA 75, 5565-5569 (1978)

18. Gooi, H.C., Feizi, T., Kapadia, A., Knowles, B.B., Solter, D., Evans, M.J.: Nature Lond. 292, 156-158 (1981)

19. Ducibella, T., Anderson, E.: Devl. Biol. 47, 45-58 (1975).

20. Surani, M.A.H., Barton, S.C., Burling, A.: Exp. Cell Res. 125, 275-286 (1980)

21. Ducibella, T.: Exptl. Cell Res. 138, 31-38 (1982)

22. Johansson, S., Kjellen, L, Höök, M., Timpl. R.: J. Cell Biol. 90, 201-219 (1981)

23. Carlsson, R., Engvall, E., Freeman, A., Ruoslahti, E.: Proc. Natn. Acad. Sci. USA 78, 2403-2406 (1981)

24. Kleinman, H.K., Klebe, R.J., Martin, G.R.: J. Cell Biol. 88, 473-485 (1981)

25. Leivo, I., Vaheri, A., Timpl. R., Wartiovaara, J.: Dev. Biol. 76, 100-114 (1980)

26. Surani, M.A.H.: Cell 18, 217-227 (1979)

27. Grunwald, G.B., Geller, R.L., Lilien, J.: J. Cell Biol. 85, 766-776 (1980)

28. Magnani, J.L., Thomas, W.A., Steinberg, M.S.: Dev. Biol. 81, 96-105 (1981)

29. Yoshida, C., Takeichi, M.: Cell 28, 217-224 (1982)

30. Hyafil, F., Babinet, C., Jacob, F.: Cell 26, 447-454 (1981)

LECTINRECEPTOR - MEDIATED ENDOCYTOSIS OF HELIX POMATIA LECTIN BY ZAJDELA HEPATOMA CELLS

Karl Walter Lempfrid, Hans-Peter Weil, Günther Hermann
Immunologische Abteilung der Chirurgischen Universitätsklinik, Robert-Koch-Str. 10, 5000 Köln 41, Germany

The uptake of extracellular material into the cell can be mediated by different mechanisms. Small molecules and ions pass the cell membrane by passive diffusion, wheras macromolecules and large particles make use of different endocytotic pathways: phagocytosis, macropinocytosis, fluid-phase micropinocytosis and adsorptive pinocytosis, also known as receptor-mediated endocytosis. This process is characterized by the selective uptake of macromolecules, bound to specific receptors of the cell membrane. After binding of the ligands to the receptors, the ligand-receptor complexes move to distinct areas of the cell surface. The internalization of the complex occurs via coated pits and coated vesicles, which fuse with lysosomes, where the ligand is degraded (for reviews see 9,10,11,20).
The uptake of lectins, such as Con A (16,22,25,33,34), ricin (12,13,14,22,27), WGA (8,19) and Wistaria floribunda agglutinin (32) by different cells has been investigated either morphologically with lectin-enzyme conjugates or quantitatively by ^{125}I-labeled lectins. Frequently the lectins were demonstrated in the Golgi apparatus - endoplasmatic reticulum - lysosome compartment. This could indicate, that the lectin internalization follows the same route as described for receptor-mediated endocytosis. Earlier we have shown, that

the uptake of HP-peroxidase conjugate by rat hepatoma cells is a specific, temperature dependent process, which does not occur at 4°C. It is inhibited in the presence of GalNAc, the blocking sugar of HP-lectin. The half time for the internalization of surface-bound HP is 120 min. Parallel determinations of internal and surface-bound lectin show an increase of internalized lectin immediately after the beginning of the incubation, whereas the amount of surface-bound lectin decreases after 60 min (47).
Whether the internalization of the lectin is indeed receptor-mediated or an unique process is analyzed in this paper.
We find, that the characteristics of HP-uptake are different compared to receptor-mediated endocytosis and should be called lectinreceptor-mediated endocytosis.

Material and methods

Isolation and neuraminidase treatment of Zajdela hepatoma cells, preparation and purification of the HP-peroxidase conjugate, enzyme assay, binding studies and Scatchard analysis were described in full detail previously (47).

Measurement of internalized HP-peroxidase

$5x10^5$ hepatoma cells were incubated at 37°C in 1 ml HP-peroxidase, dissolved in tissue culture medium (RPMI 1640, Seromed, Munich) containing BSA (Behringwerke, Marburg, 5mg/ml). The lectin concentration varied between 5 and 7 µg/ml. After indicated times the cells were washed with medium containing BSA (5mg/ml) and N-acetyl-D-galactosamine (GalNAc, Sigma, Munich, 0,1 M) and incubated for 10 min at 37°C. After this step the surface-bound lectin was removed by washing the cells twice with ice-cold medium. The cells were suspended in 1,2ml PBS (Seromed, Munich) containing BSA (5mg/ml), Triton X 100

(Merck, Darmstadt, 5 µl/ml) and GalNAc (0,1 M). The suspension was incubated for 15 min at 4°C and gently shaken. After centrifugation for 15 min at 4200x g 1 ml of the supernatant, containing the internalized lectin, was carefully removed and the peroxidase activity was measured.

Inhibition of lectin uptake

For studying the inhibition of lectin uptake, the cells were incubated with different drugs at 37°C for 30 min. The inhibitory substance was present during the whole experiment. The following inhibitors were used at concentrations, which had no cytotoxic effect on the cells as determined by trypan blue dye exclusion test: GalNAc (100 mM), procaine (10 mM), chloroquine (0,1 mM), cytochalasin B (0,1 mM), colcemid (0,1 mM), cycloheximide (1 mM), chlorpromazine (0,01 mM), (all purchased from Sigma, Munich), EDTA (10 mM, Seromed, Munich) and NH_4Cl (10 mM, Merck, Darmstadt).

Results

Time and concentration dependence of HP-peroxidase uptake

The uptake of HP-peroxidase as a function of time is a saturable process reaching its plateau after 480 min (Fig. 1). The increase of the extracellular lectin concentration leads to an unspecific lectin uptake at very high concentrations. The non-specificity of this process is demonstrated by the internalization of sugar-blocked lectin (Fig. 2). The difference between the amount of totally internalized lectin and the amount of internalized sugar-blocked lectin shows a saturable kinetic of the specific uptake.

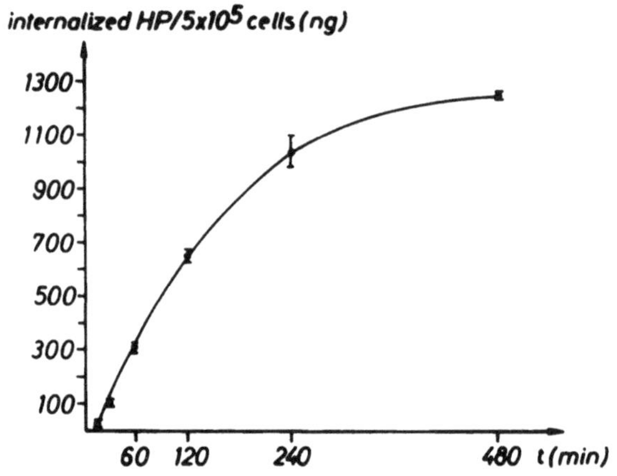

Fig. 1. Time dependence of HP-peroxidase uptake. Samples of $5x10^5$ neuraminidase treated hepatoma cells were incubated with 1 ml HP-peroxidase containing 4,3 µg HP at 37°C. After the indicated times HP-peroxidase uptake was measured.

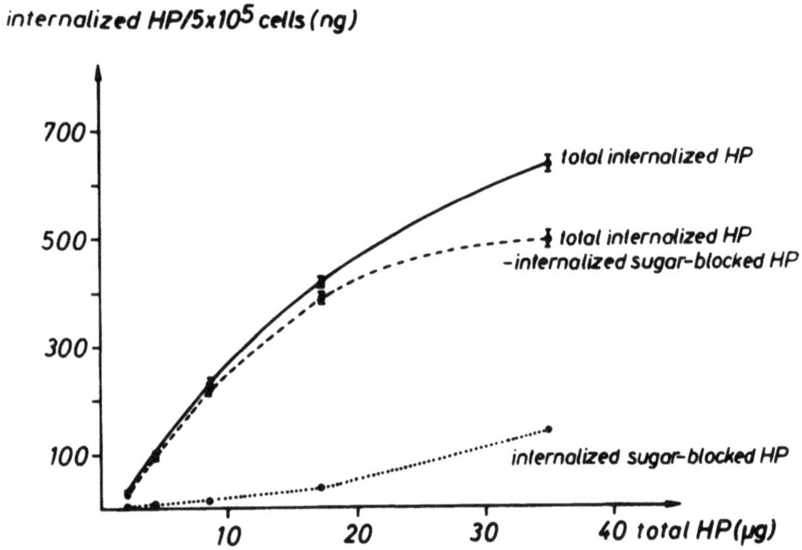

Fig. 2. Concentration dependence of HP-peroxidase uptake. $5x10^5$ neuraminidase treated hepatoma cells were incubated with 1 ml HP-peroxidase containing the indicated amounts of HP at 37°C for 15 min. For the measurement of sugar-blocked HP uptake the cells were incubated with 1 ml of the indicated lectin concentration containing GalNAc 0,1 M.

Table 1. Inhibition of HP-peroxidase uptake

Inhibition experiment	Inhibition of HP-peroxidase uptake
$4^{o}C$ incubation temperature	+++
GalNAc (100 mM)	+++
procaine (10 mM)	++
chloroquine (0,1 mM)	++
cytochalasin B (0,1 mM)	+
NH_4Cl (10 mM)	+
colcemid (0,1 mM)	-
cycloheximide (1 mM)	-
chlorpromazine (0,01 mM)	-
EDTA (10 mM)	-

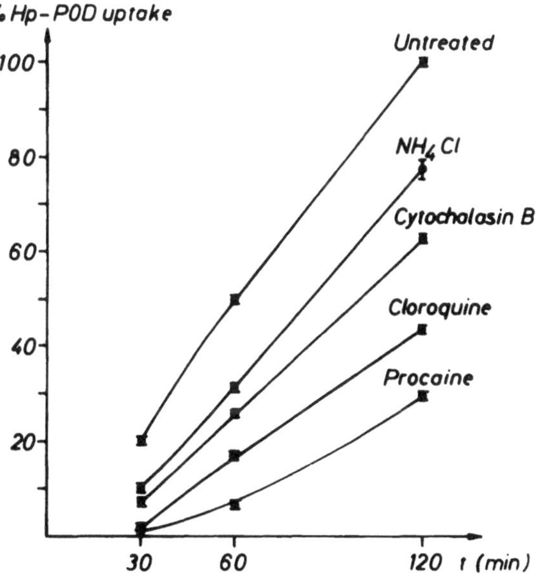

Fig. 3. Inhibitory capacity of different drugs. $5x10^5$ neuraminidase treated hepatoma cells were incubated with 0,5 ml medium containing the drugs at concentrations as indicated in table 1 at $37^{o}C$ for 60 min. 0,5 ml HP-peroxidase containing 5 µg HP was added and HP-peroxidase uptake was measured at the indicated times after HP-peroxidase addition.

Inhibition of HP-peroxidase uptake

A partial inhibition of lectin uptake is demonstrable in the presence of cytochalasin B and the lysosomotropic agents chloroquine, procaine and NH_4Cl. Inhibition is not achieved by drugs such as colcemid, cycloheximide, EDTA and chlorpromazine (table 1). The inhibitory capacity of the different drugs is shown in Fig. 3.

We analyzed further the influence of the inhibitory substances on the number of HP-cell surface receptors by Scatchard analysis performed in the presence of several drugs. Only for cytochalasin B a slight decrease of cell surface receptors can be demonstrated, whereas the lysosomotropic drugs chloroquine, procaine and NH_4Cl lead to a slight increase compared with untreated cells.

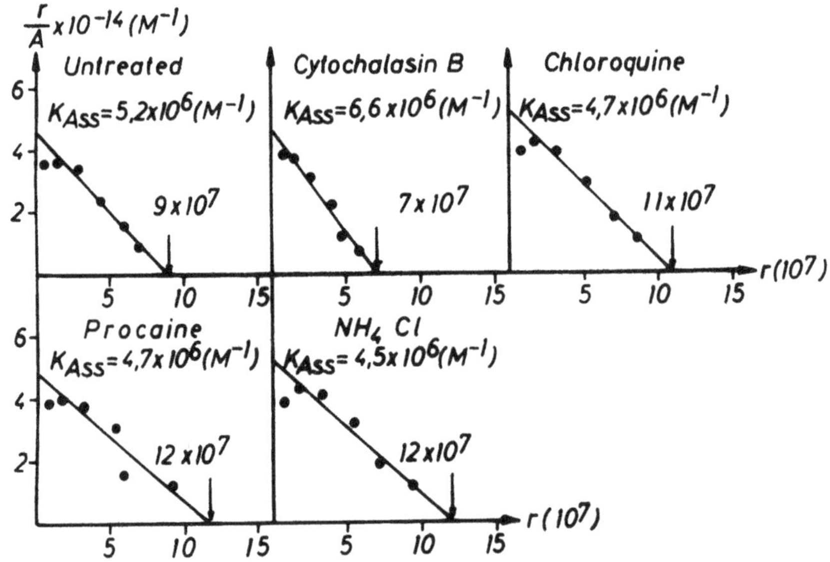

Fig. 4. Scatchard plot analysis of HP-peroxidase binding to hepatoma cells in presence of different drugs. 5×10^5 neuraminidase treated hepatoma cells were incubated with 1 ml HP-peroxidase containing 2,5 - 80 µg HP and the drugs at concentrations as indicated in table 1 at 4°C for 18 h. The Scatchard plots were calculated as described in (47). The correlation coefficients ranged from 0,929 to 0,989.

Table 2. Comparison between lectinreceptor-mediated endocytosis and receptor-mediated endocytosis

Experimental finding	lectinreceptor-mediated endocytosis	receptor-mediated endocytosis	(Ref.)
binding necessary for uptake	+	+	(11)
physiological effect of the ligand	not known	+	(11)
$T_{1/2}$ of uptake	120 min	10 min	(11)
coated pits involved in uptake	not known	+	(11)
lysosomal degradation of the ligand	indirect evidence	+	(11)
saturation as a function of time and extracellular ligand concentration	+	+	(2,3,15, 30,37, 38,49)
Ca^{2+}-dependence	-	+	(21,30,40, 42,45,49)
temperature-dependence	+	+	(1,5,17,21, 30,41,46)
Inhibitory capacity of cytochalasin B	+	+ -	(33,34) (13,18)
Inhibitory capacity of colcemid	-	+ -	(18) (13,34)
Inhibitory capacity of lysosomotropic agents	+	+ -	(26,28,36, 39,48) (49)
Receptor recycling	indirect evidence	+	(29,31,35, 43,44)

Discussion

A summarized comparison between lectinreceptor-mediated endocytosis and receptor-mediated endocytosis is shown in table 2. The receptor-mediated endocytosis in strict sense is defined by following criteria (11) :

1. Binding of the ligand is necessary for its uptake.
2. The ligand mediates a physiological effect.
3. The half-time for the uptake of surface-bound ligand is less than 10 min.
4. The internalization of the ligand-receptor complex occurs via coated pits and coated vesicles.
5. The degradation of the ligand is located in the lysosomes.

Ad 1. Internalization of the lectin occurs specifically only after binding to membrane receptors as shown by the complete inhibition of binding and uptake in the presence of GalNAc. An unspecific uptake of HP-lectin occurs only at very high extracellular lectin concentrations (Fig. 2), which were avoided for analysis of the specific lectinreceptor-mediated endocytosis.

Ad 2. We cannot conclude from our experiments, whether HP-lectin mediates a physiological effect on hepatoma cells. Although lectin receptors per se have not a physiological effect, binding to endocytosis mediating receptors may occur and lead to specific cellular responses. For example WGA competes with insulin for its receptors and provokes an insulin-like effect (4,6,7).

Ad 3. The surface-bound lectin disappears from the cell membrane with a half-time of 120 min (47), which is in contrast to the findings of true receptor-mediated endocytosis. Since there is indirect evidence for receptor-recycling, one has to consider that the receptor-ligand complex returns to the cell-surface, interfering with the internalization process. However the investigation of ricin-peroxidase uptake into neuroblastoma cells could not demonstrate any ligand recycling (12).

Ad 4. Up to now we have not yet studied the uptake of HP-lectin morphologically by electron microscopy, further studies are necessary to answer this question.

Ad 5. The inhibition of HP-peroxidase uptake by lysosomotropic agents such as chloroquine, procaine and NH_4Cl gives an indirect evidence that lysosomes are involved in lectin internalization.

Saturation of the endocytotic process depending on incubation time and extracellular ligand concentration are further characteristics of receptor-mediated endocytosis (2,3,15,30,37,38, 49). Both can be demonstrated by our experiments (Fig. 1,2). The unspecific HP-uptake at very high extracellular lectin concentrations has to be considered, evaluating the saturation curve for the specific HP-uptake (Fig. 2).

Binding of the ligand to the cell as a condition for its uptake depends on the presence of Ca^{2+} and can therefore be inhibited by EDTA (21,30,40,42,45,49). According to our experiments inhibition of lectin uptake could not be achieved by addition of EDTA or chlorpromazine, which competes with Ca^{2+} for membrane sites (24).

The participation of the cytoskeleton in uptake processes is controverse. The inhibition of cytochalasin B and the ineffectivness of colcemid argue for an involvment of microfilaments, whereas microtubuli seem not to be involved in lectinreceptor-mediated endocytosis. The same results are reported for the Con A uptake by fibroblasts (33,34). On the other hand lectin uptake into neuronal cells cannot be inhibited by either of the drugs (13). Contradictory findings, i.e. inhibition by colcemid and not by cytochalasin B is described for fluid phase endocytosis of PVP (23) and receptor-mediated endocytosis of asialoglycoproteins (18) by hepatocytes. The slight decrease of HP-receptors on the cell surface by cytochalasin B as demonstrated by Scatchard analysis (Fig. 4) could explain the inhibitory capacity (Fig. 3) of this drug.

The inhibition of endocytotic processes by lysosomotropic amines such as chloroquine, procaine and NH_4Cl is often found

(17,26,36,37,[illegible],48). However, for the uptake of transcobalamin II (49) by human fibroblasts no chloroquine effect is observed. Inhibition of receptor-recycling (36), reduction of cell surface receptors (26,39) or competition for membrane binding sites (48) are possible explanations for their effectivness. No dramatic change of cell surface receptors can be shown by means of Scatchard analysis performed in presence of these agents (Fig. 4).

The failure of cycloheximide to inhibit HP-uptake and the increase of intracellular lectin at the simultaneous constancy of surface-bound lectin during the first 60 min give indirect evidence for a receptor supply or a receptor recycling as described for several receptor-mediated endocytotic processes (29,31,35,43,44).

Concluding these studies may be of importance for modern approaches to chemotherapy, when drugs or toxins are focussed to tumor cells by specific carriers, such as monoclonal antibodies. Internalization of these conjugates will perhaps show a similar behaviour as our lectin-enzyme conjugate.

References

1. Anderson, Richard G.W., Brown, Michael S., Goldstein, Joseph L. : Cell 10, 351-364 (1977).
2. Anderson, Richard G.W., Brown, Michael S., Goldstein, Joseph L. : J. Cell Biol. 88, 441-452 (1981)
3. Brown, Michael S., Goldstein, Joseph L. : Proc. Nat. Acad. Sci. USA 76, 3330-3337 (1979)
4. Capeau, J., Picard, J., Caron, M. : Cancer Res. 38, 3930-3937 (1978)
5. Carpenter, G., Cohen, S.: J. Cell Biol. 71, 159-171 (1976)
6. Cherqui, G., Capeau, J., Caron, M., Picard, J. in: Bøg-Hansen, T.C.(Ed.): Lectins, Biology-Biochemistry-Clinical Biochemistry 2, Walter de Gruyter, Berlin-New York 1982

7. Cuatrecasas, P., Tell, G.P.E.: Proc. Nat. Acad. Sci. USA 70, 485-489 (1973).
8. Desbuquois, B., Huet de Froberville, A., Willeput, J.: Biol. Cellulaire 36, 175-184 (1979).
9. Gabrion, J.: Biochimie 63, 325-345 (1981).
10. Geisow, M.: Nature 288, 434-436 (1980).
11. Goldstein, J.L., Anderson, R.G.W., Brown, M.S. : Nature 279, 679-685 (1979).
12. Gonatas, J., Stieber, A., Olsnes, S., Gonatas, N.K. : J. Cell Biol. 87, 579-588 (1980).
13. Gonatas, N.K., Kim, S.U., Stieber, A., Avrameas, S.: J. Cell Biol. 73, 1-13 (1977).
14. Gonatas, N.K., Stieber, A., Kim, S.U., Graham, D.I., Avrameas, S.: Exp. Cell Res. 94, 426-431 (1975).
15. Goud, B., Antoine, J.C., Gonatas, N.K., Stieber, A., Avrameas, S.: Exp. Cell Res. 132, 375-386 (1981).
16. Hatae, T., Benedetti, E.L.: Cell Biol. Int. Rep. 2, 129-137 (1978).
17. Karin, M., Mintz, B.: J. Biol. Chem. 256, 3245-3252 (1981)
18. Kolset, S.O., Tolleshaug, H., Berg, T.: Exp. Cell Res. 122, 159-167 (1979).
19. Kramer, R.H., Canellakis, E.S.: Biochim. Biophys. Acta 551, 328-348 (1979).
20. Michl, J.: Amer. J. clin. Nutr. 33, 2462-2471 (1980).
21. Oka, J., Weigel, P.H.: Fed. Proc. 39, 2101 (1980).
22. Oliver, J.M., Ukena, T.E., Berlin, R.D.: Proc. Nat. Acad. Sci. USA 71, 394-398 (1974).
23. Ose, L., Ose, T., Reinertsen, R., Berg, T.: Exp. Cell Res. 126, 109-119 (1980).
24. Poste, G., Allison, A.C.: Biochim. Biophys. Acta 300, 421 (1973).
25. Roth, J., Binder, M.: Exp. Path. 16, 43-59 (1978).
26. Sando, G.N., Titus-Dillon, P., Hall, C.W., Neufeld, E.: Exp. Cell Res. 119, 359-364 (1979).
27. Sandvig, K., Olsnes, S., Pihl, A.: Eur. J. Biochem. 82, 13-23 (1978).
28. Schlesinger, P., Tietze, C., Stahl, P.: Fed. Proc. 40, 1653 (1981).
29. Schneider, Y., Tulkens, P., de Duve, C., Trouet, A.: J. Cell Biol. 82, 466-474 (1979).

30. Stahl, P., Schlesinger, P.H., Sigardson, E., Rodman, J.S., Lee, Y.C.: Cell 19, 207-215 (1980).

31. Steer, C.J., Ashwell, G.: J. Biol. Chem. 255, 3008-3013 (1980).

32. St John, A.C., Merion, M., Triemer, R.E., Kuchler, R.J., Bird, J.W.C., Carter, J.H., Poretz, R.D.: Exp. Cell Res. 128, 143-150 (1980).

33. Storrie, B., Edelson, P.J.: Cell 11, 707-717 (1977).

34. Storrie, B.: Exp. Cell Res. 118, 135-141 (1979).

35. Tanabe, T., Pricer, W.E., Jr., Ashwell, G.: J. Biol. Chem. 254, 1038-1043 (1979).

36. Tietze, C., Schlesinger, P., Stahl, P.: Biochem. Biophys. Res. Commun. 93, 1-8 (1980).

37. Tolleshaug, H., Berg, T.: Acta biol. med. germ. 36, 1753-1762 (1977).

38. Tolleshaug, H., Berg, T., Nilsson, M., Norum, K.R.: Biochim. Biophys. Acta 499, 73-84 (1977).

39. Tolleshaug, H., Berg, T.: Biochem. Pharmacol. 28, 2919-2922 (1979).

40. Tolleshaug, H., Berg, T., Frölich, W., Norum, K.R.: Biochim. Biophys. Acta 585, 71-84 (1979).

41. Tolleshaug, H., Berg, T., Holte, K.: Eur. J. Cell Biol. 23, 104-109 (1980).

42. Tolleshaug, H., Berg, T.: Hoppe-Seyler's Z. Physiol. Chem. 361, 1155-1164 (1980).

43. Tulkens, P., Schneider, Y.J., Trouet, A.: Acta biol. med. germ. 36, 1681-1690 (1977).

44. Tulkens, P., Schneider, Y.J., Trouet, A.: Biochem. Trans. Soc. 5, 1809 (1977).

45. Weigel, P.H.: J. Biol. Chem. 255, 6111-6120 (1980).

46. Weigel, P.H., Oka, J.: J. Biol. Chem. 256, 2615-2617 (1981).

47. Weil, H.-P., Lempfrid, K.W., Hermann, G. in : Bøg-Hansen, T.C. (Ed.): Lectins, Biology-Biochemistry-Clinical Biochemistry 2, Walter de Gruyter, Berlin - New York 1982

48. Wiesmann, U., Di Donato, S., Herschkowitz, N.N.: Biochem. Biophys. Res. Commun. 66, 1338-1343 (1975).

49. Youngdahl-Turner, P., Rosenberg, L.E., Allen, R.H.: J. clin. Invest. 61, 133-141 (1978).

Part II

Lectins as Specific Tools

LECTINS IN DIAGNOSTIC PATHOLOGY

Guido Coggi, Patrizia Dell'Orto, Emanuela Bonoldi, Patrizia Doi, Giuseppe Viale

IV Department of Pathology, University of Milan Medical School
Milan, Italy

In the recent years, the increasing need for more accurate and objective diagnostic criteria in surgical pathology has led to the development of new techniques of morphological investigation: among others, those based upon the use of labeled substances which specifically react with tissue or cell components, appear to be the most valuable ones.

Indeed, these techniques can identify specific cell markers which are either preserved or lost, or acquired in pathological conditions, both neoplastic and non-neoplastic, thus allowing the recognition of the histogenesis, of the degree of cell differentiation and even of the type of cellular deviation.

These properties, which are peculiar features of the labeled antibodies used in immunohistochemistry (25), account for the extraordinary development of immunohistochemical procedures for diagnostic purposes (6).

Recently, the results obtained by some authors with labeled lectins on tissue sections (2,3,7-10,13-16,18-24,26-38) have suggested their possible use as meaningful tools in diagnostic pathology.

Indeed, the high affinity of lectins for sugar residues and the possibility of labeling lectins by means of fluorochromes (10), ferritin (32,37), peroxidase (31) or biotin (1), represent important technical prerequisites for their practical use.

The purpose of this investigation has been to verify the diagnostic relevance of lectin histochemistry in human surgical pathology.

Material and methods

From the surgical pathology files of our department we have examined the following cases, the paraffin blocks of which had been stored for as long as two years:

a) gastrointestinal tract: chronic superficial gastritis (3 cases), chronic gastritis with intestinal metaplasia (24), gastric adenocarcinoma (20), gastric diffuse carcinoma (8), Crohn's disease (3), ulcerative colitis (2), hyperplastic colonic polyp (9), adenomatous colonic polyp (22) and colonic adenocarcinoma (8). Biopsy specimens of normal gastric and intestinal mucosa have been used as controls.

b) skin: Keratoacanthoma (4 cases), Bowen's disease (2), squamous cell carcinoma (3), basal cell carcinoma (54), basal-squamous cell carcinoma (2) malignant melanoma (3), cavernous hemangioma (2), Kaposi's sarcoma (2) and fibrous hystiocytoma (1). Normal skin biopsies have been used as controls.

c) breast: fibroadenoma (3 cases), cystic disease (3), carcinoma (4).

d) other diseases: malignant lymphomas other than Hodgkin's disease (11 cases), lymphoid follicular hyperplasia (4), carcinomas of lung (10), thyroid gland (4), kidney (2), uterus (7) and ovary (4), persistent and active chronic hepatitis (5).

From each paraffin block 10 serial sections (3 μ thick) were cut, dewaxed and brought to water.

To inhibit the endogenous peroxidase slides were treated with 1% H_2O_2 in absolute methanol for 30 min, washed in tap and distilled water and then transferred to 0.01M PBS pH 7.6, containing Ca^{++} 1mM and Mg^{++} 1mM, for 5 min.

The serial sections of each case were then incubated with one of the following biotin-labeled lectins (Vector Lab.) at the dilutions indicated below:

CON A and WGA	0.03 mg/ml
PNA	0.04 mg/ml
SBA, DBA, UEA, RCA	0.06 mg/ml

After being thoroughly washed in buffer, the sections were incubated in the avidin-biotinylated peroxidase complex (17,18) (Vector Lab.) for 30 min, and then washed. The peroxidase activity was developed with 50 mg Diamino-Benzidine in 100 ml TRIS-HCl 0.05M; pH 7.6 containing 0.01% H_2O_2 for 5 min according to Graham and Karnovsky (11).

Finally the slides were counterstained with hematoxilin, dehydrated and mounted in synthetic mounting medium(DPX,Gurr).

Some of the sections were treated, before incubation with PNA, with Vibrio Cholerae neuraminidase (Behringwerke) (0.01 U/ml in acetate buffer 0.05M containing 0.01% $CaCl_2$) for 1 hour at 37°C.

In control sections lectin binding was inhibited with the use of the corresponding sugar at 0.01M concentration.

Double blind evaluation of the results was done independently by two of us, and the results were expressed in a semiquantitative scale.

Results

The procedure used in the present investigation has yielded consistently specific cellular staining, as proved by the negativity of control reactions and by the low level of background. No adverse effect could be attributed to either formalin fixation or paraffin-embedding.

Gastrointestinal Tract

Table 1 shows the staining behaviour of different portions of human gastro intestinal tract in normal conditions.

TABLE 1

Staining patterns of human gastrointestinal mucosa in normal conditions. Intensity of staining is graded from absent (-) to weak (+/-) to strong (+).

	CON A	SBA	WGA	DBA	UEA	RCA	PNA	PNA+Neur*
Stomach								
Body mucosa								
Glandular epith.	+	+	+	+	+	+	-	+/-
Superficial epith.	+	+/-	+	-	+	-	-	-
Antral mucosa								
Glandular epith.	+	+	+	+	+	+	+	-
Superficial epith.	+	+	+	+	+	-	-	-
Ileum								
Absorptive epith.	+	+	+	+	-	-	-	-
Goblet cells	-	-	-	-	+/-	+/-	-	+
Paneth cells	-	-	-	+	-	+	-	-
Colon								
Glandular epith.	+	+	+	+	-	+	-	-
Mucus	+	-	-	-	-	+	-	-

* Neuraminidase digestion

Staining patterns consistently different from normal were observed in all cases of chronic gastritis with intestinal metaplasia. In fact, while in normal mucosa the affinity for PNA is weak after neuraminidase digestion, and is limited to the luminal borders of the glandular epithelium in the fundus, and to the supranuclear area of the cells in the antrum, in intestinal metaplasia the goblet cells show an intense cytoplasmic staining for PNA after neuraminidase digestion (Figs. 1-4). On the contrary, RCA, which strongly stains normal epithelium, fails to bind to metaplastic cells.

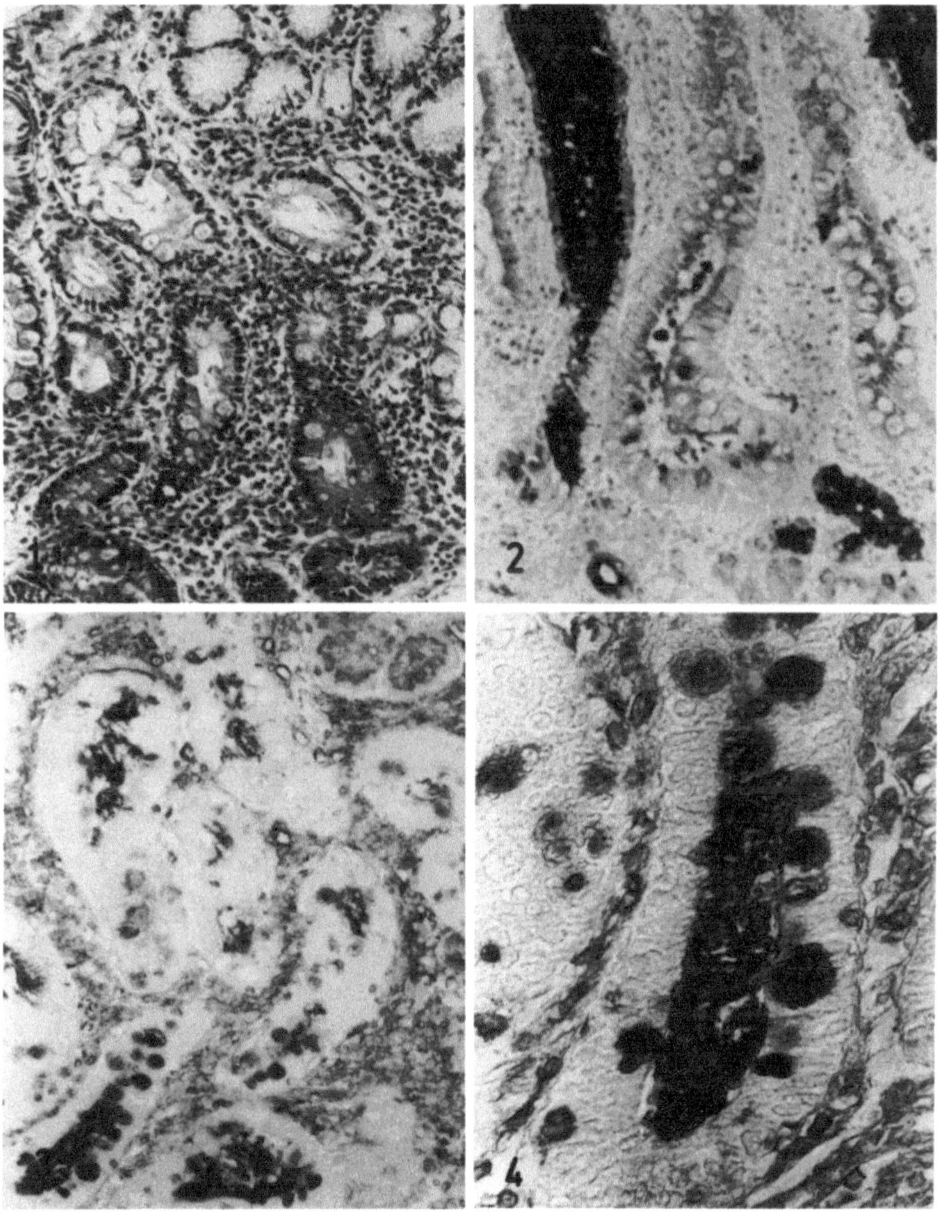

Figs.1-4: Chronic gastritis with intestinal metaplasia stained with H. & E. (Fig.1,x250), with PNA (Fig.2,x250), and with PNA after neuraminidase digestion (Fig.3,x250; Fig.4,x400).

Gastric adenocarcinoma differs from normal mucosa in that extracellular mucus and cell borders are strongly stained by PNA after neuraminidase digestion (Figs. 5-8) and in 50% of cases even without enzymatic treatment.
In Crohn's disease, SBA diffusely stains the entire cytoplasm of epithelial cells, whereas in normal ileum, the staining is limited to the central region of the cells.
Colonic hyperplastic polyps, when stained with PNA, show a peculiar pattern which is not affected by neuraminidase digestion: the Golgi region of the cytoplasm of the epithelial cells reveals the presence of several strongly positive granules. The intracellular mucus, which is located in the apical portion of the cells, is constantly negative, and so are the goblet cells (Figs. 9-10).
In active ulcerative colitis, the superficial glandular cells are strongly PNA positive after neuraminidase digestion: this pattern is in striking contrast with the constant negativity of normal colonic mucosa for PNA.
Adenomatous polyps are strongly positive for DBA, a lectin which is not bound by normal mucosal cells (Figs. 11-12): the dysplastic epithelium is easily identified even at low power microscopic examination. However, if the degree of dysplasia reaches the level of actual neoplasia, then the cells do not retain their affinity for DBA. Finally, in nearly 50% of colonic adenocarcinomas, PNA stains the neoplastic epithelial cells with and without enzymatic pre-treatment. The reaction is more intense on the cell borders and in the mucus, both intra- and extra-cellular (Figs. 13-14).

Skin

Table 2 shows the staining patterns of human skin: rather sharp differences in lectin affinity can be observed among the different layers.
In basal cell carcinoma the neoplastic cells are definitely negative for DBA, which binds normal basal cells in about 50% of the cases (Figs. 15-16).
In squamous cell carcinoma the behaviour of SBA and PNA varies according

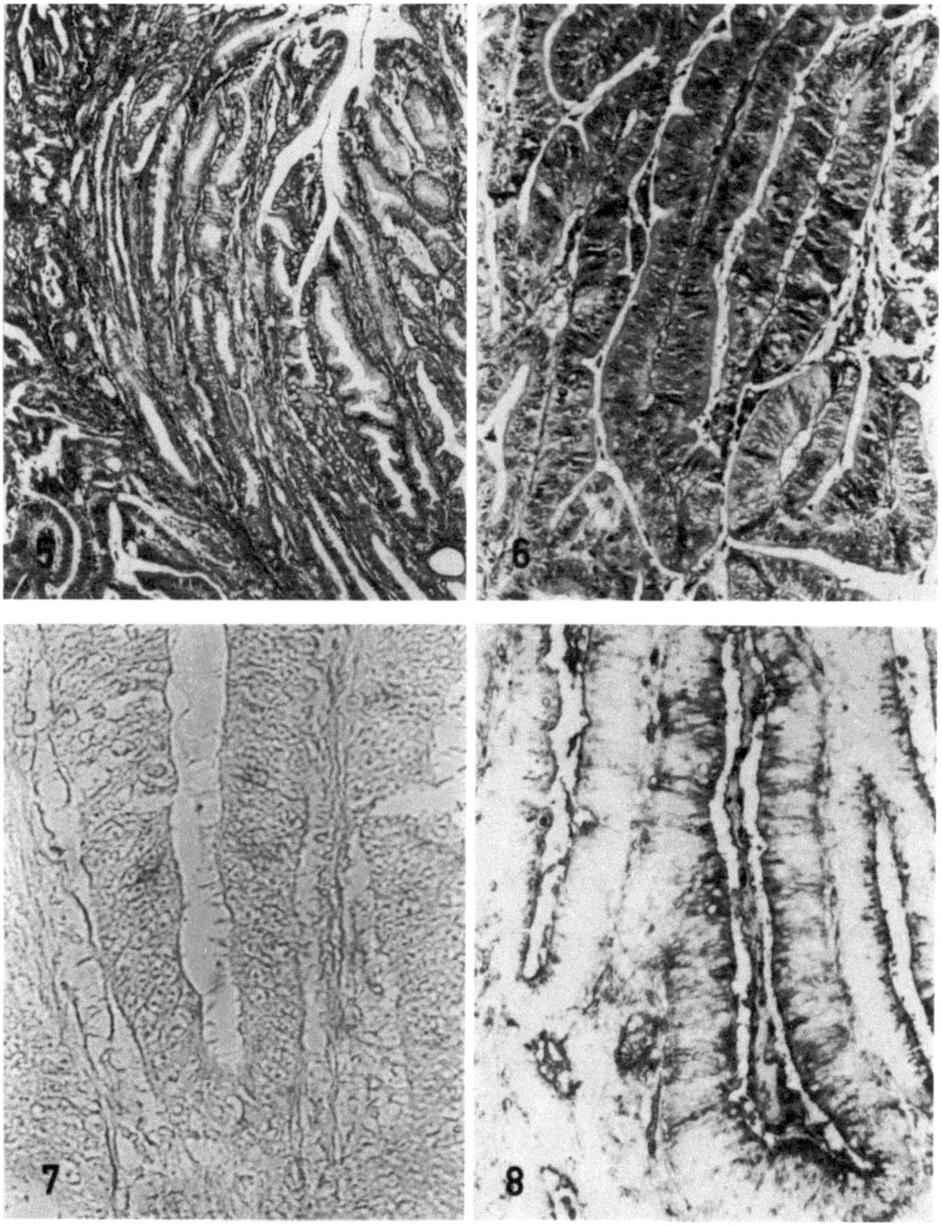

Figs.5-8: Gastric adenocarcinoma stained with H. & E. (Fig.5,x100; Fig.6,x400), with PNA (Fig.7,x400) and with Pna after neuraminidase digestion (Fig.8,x400).

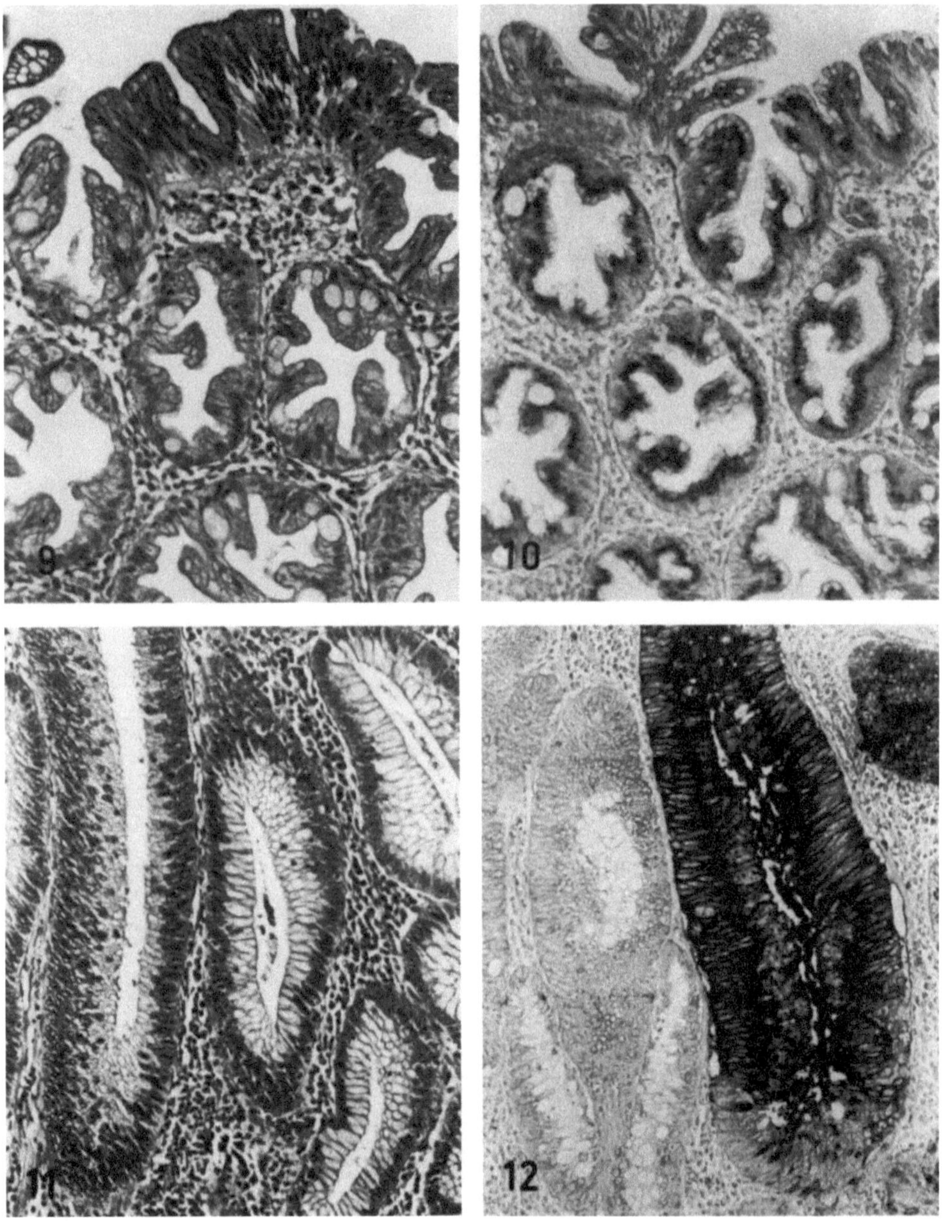

Figs.9–10: Colonic hyperplastic polyp (Fig.9: H. & E.,x250; Fig.10: PNA,x250).

Figs.11–12: Colonic adenomatous polyp (Fig.11: H. & E.,x250; Fig.12: DBA,x250).

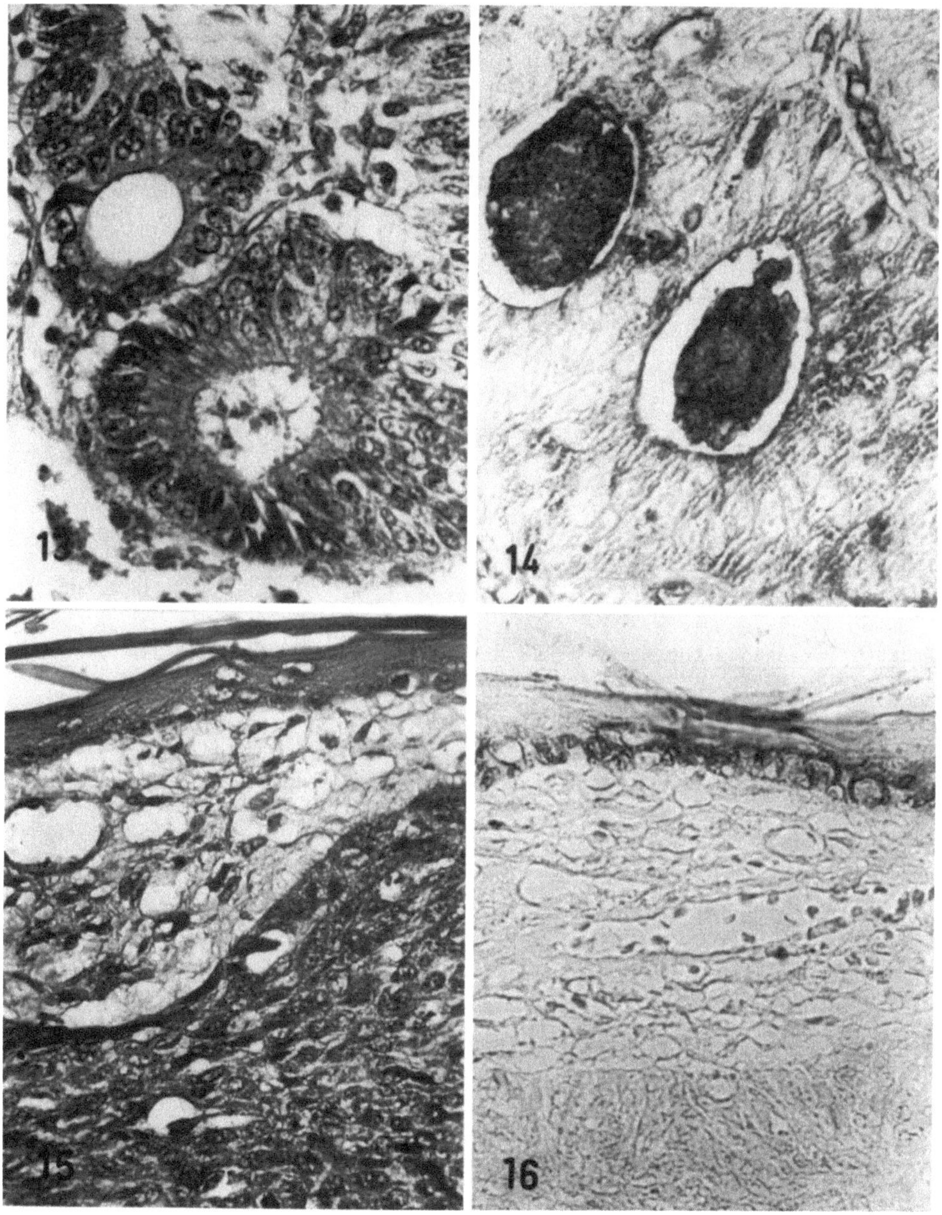

Figs.13-14: Colonic adenocarcinoma (Fig.13: H. & E.,x250; Fig.14: PNA,x250).

Figs.15-16: Basal cell carcinoma of the skin (Fig.15: H. & E.,x400; Fig.16: DBA,x400).

TABLE 2

Staining patterns of human normal skin. Intensity of staining is graded from absent (-), to weak (+/-) to strong (+).

	CON A	SBA	WGA	DBA	UEA	RCA	PNA	PNA+Neur*
Horny layer	+	+	+	+/-	-	+/-	+	+
Granular layer	+	+/-	+	-	-	+/-	+	+
Prickle cell layer	+	-	+	+/-	-	-	+	+
Basal layer	+	-	+	+	-	-	-	+

* Neuraminidase digestion

to the degree of keratinization of tumor cells (Figs. 17-18).

In basal-squamous cell carcinomas only the squamous component is decorated by PNA and SBA.

In keratoacanthoma the tumor cells are definitely negative for PNA, and become positive after neuraminidase digestion.

Finally in Bowen's disease the atypical cells do not show any affinity for DBA, SBA and PNA.

Lymphoid Tissue

The cells of the germinal center in lymph nodes with follicular hyperplasia are CON A, WGA, RCA positive and DBA, SBA, UEA and PNA negative.

In malignant lymphomas centroblasts and centrocytes characteristically bind PNA in absence of any enzymatic treatment (Figs. 19-20).

Other Tissues

No consistent staining pattern, both in normal and pathological conditions, was observed in the other organs investigated, including breast.

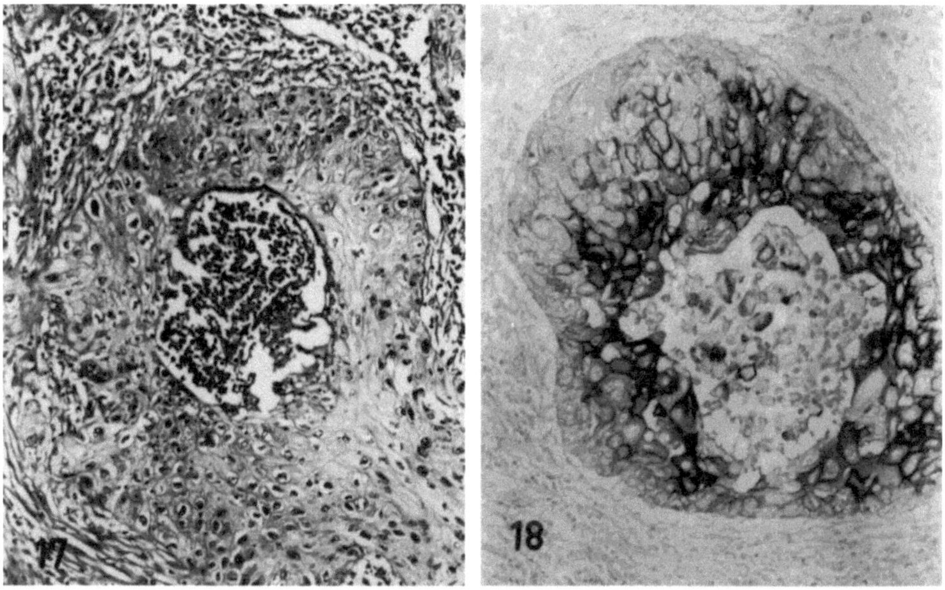

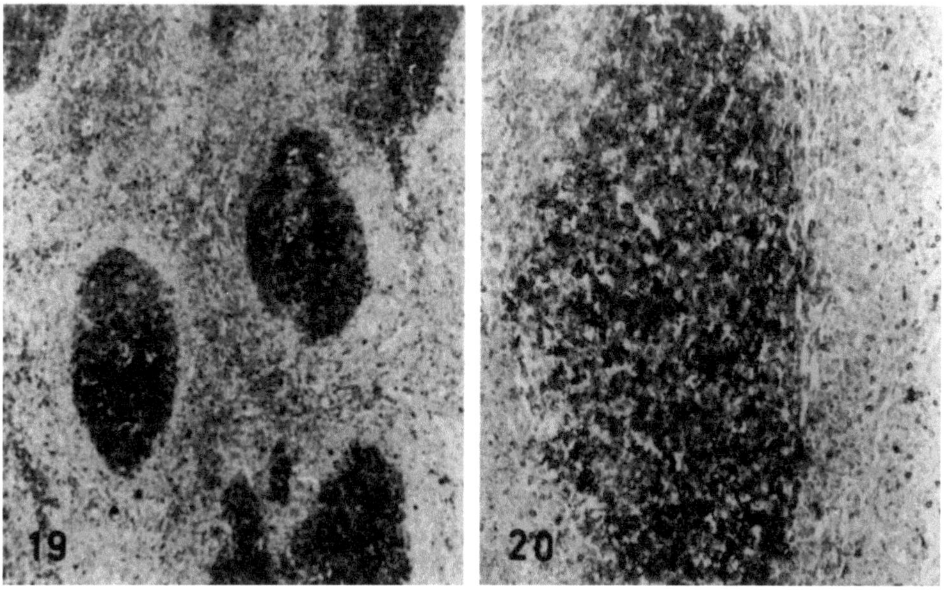

Figs.17-18: Squamous cell carcinoma of the skin (Fig.17: H. & E.,x250; Fig.18:PNA,x250).

Figs.19-20: Centroblastic-centrocytic follicular lymphoma (Fig.19: PNA, x100; Fig.20: PNA,x250).

Discussion

Although some reports (16,21,22,24,29) have described the use of lectins for the detection of specific cell markers for diagnostic purposes, no conclusion has been so far drawn about their actual suitability in routine surgical pathology. This holds true with regard both to the methodological aspects of lectin histochemistry and to its diagnostic value.

From the methodological point of view, labeled lectins are now easily available, even from commercial sources, both for light and fluorescence microscopy. For practical purposes, nevertheless, fluorochrome-labeled lectins are definitely less suitable than those labeled with peroxidase, since the latter can be easily detected with Diamino-Benzidine or other chromogens (31,36) so overcoming the need for the fluorescence microscope.

A considerable improvement is represented by the use of lectins labeled with biotin since, due to the low molecular weight of this molecule (1,12), the chemical reactivity of the lectin itself is not modified; moreover, multiple labeling can be performed. The subsequent use of an avidin-biotinylated peroxidase complex allows the decoration of the lectin already bound to its sugar in the tissue section. This avidin-biotin system, which is widely used in immunohistochemistry (17), has been applied to lectin histochemistry in the present investigation, resulting in a higher sensitivity than that shown by lectins directly labeled with peroxidase.

Finally, our investigation has confirmed that the affinity of cellular structures for lectins is not significantly affected by formalin fixation and paraffin embedding (31,33,36).

In conclusion, no technical or practical problem is encountered in the use of lectins on tissues routinely processed for surgical pathology.

With regard to the diagnostic value of lectin histochemistry, and to its actual effectiveness in increasing the accuracy of histopathological diagnosis, our results offer some interesting perspectives.

In spite of some lack of correlation in the staining patterns shown by two or more lectins, specific for the same glucidic residue, the reproducibility of the staining pattern strongly supports the "histological" specificity of the reaction itself.

Indeed, differences in the staining properties of lectins expected to react with the same sugar residue have been already observed by Holthoefer et al. (14), Vierbuchen et al. (34) and Freeman et al. (10).

Specific patterns are definitely observed in gastrointestinal pathology. Intestinal metaplasia is clearly identified by PNA after neuraminidase digestion: the staining pattern is comparable to that observed with Alcian blue (4,5) so pointing to the modifications of mucus production by the cells undergoing metaplastic changes. This holds also true for more than 50% of gastric adenocarcinomas of either type, the affinity of which for PNA is strikingly similar to that observed in metaplastic gastric mucosa.

In adenomatous polyps of the colon, the use of lectins allows a new insight into the cytochemistry of dysplastic cells. These are indeed clearly distinguishable on the ground of purely morphological parameters, but are not known to present specific histochemical staining patterns.

DBA, on the contrary, clearly identifies in the cytoplasm of dysplastic cells the presence of considerable amounts of galactosyl residues, which are absent from the normal epithelial cells of the colon.

Although no quantitative correlation between the staining intensity and the degree of dysplasia could be identified, it is noteworthy that, once the dysplastic cells turn into actual neoplastic cells, they lose their affinity for DBA. Thus, it would appear that the pre-neoplastic changes undergone by the dysplastic cells of colonic polyps are accompanied by a synthesis of galactosyl-containing sugars, which then declines in adenocarcinomas. In addition, in about 50% of colonic adenocarcinomas the neoplastic cells acquire a strong affinity for PNA, which is not a feature of normal or dysplastic colonic epithelium.

Formalin-fixed paraffin-embedded skin biopsies appear to be a suitable material for lectin histochemistry. Indeed, the use of several biotinylated lectins yields specific staining patterns for the different layers of epidermal cells.

In pathological conditions, DBA is consistently negative in basal cell carcinomas (54/54 cases), so pointing to a difference in sugar composition between neoplastic cells and their normal counterpart, which binds the lectin in as much as 50% of the cases.

On the other side, PNA and SBA do stain normal prickle cell layer and squamous cell carcinomas: in the latter, the intensity of the staining well correlates with the degree of cellular differentiation.

With regard to lymphoid tissue, previous studies on frozen sections (29) have shown the affinity of PNA for germinal center cells of human lymph nodes, both in normal conditions and in malignant lymphomas.

On formalin-fixed and paraffin-embedded lymph nodes, the behaviour of the same lectin changes strikingly: in fact, while the neoplastic centroblasts and centrocytes are intensely stained, the cell population of normal follicles is negative.

Thus the neoplastic follicles sharply contrast against a clear background so making it possible to exactly evaluate the extension of the neoplasia.

No definite explanation can be at the moment given for this phenomenon: one can only speculate that fixatives and embedding procedures affect the affinity of lymphoid cells for PNA: it is fortunate however, that such an effect can be of considerable help in the differential diagnosis of follicular lymphomas versus reactive follicular hyperplasia.

In conclusion, lectins do represent a valuable tool in histopathology, in that they allow not only to identify and localize the glucidic components of the cells, but also to follow their modifications in pathological conditions. The sensitivity and specificity of the techniques employing biotinylated lectins are extremely high and comparable to those shown by labeled antibodies.

The use of lectins in histopathology has yet to be investigated, but from our studies on gastrointestinal, cutaneous and lymphoid pathology, as well as from other investigations on breast cancer (20,21,23,26,35) and kidney diseases (15,34) it can be assumed that a considerable improvement in diagnostic accuracy will result from their use.

References

1. Bayer, E.A., Wilchek, M., Skutelsky, E.: FEBS Letters 68, 240-244 (1976).
2. Brabec, R.K., Peters, B.P., Bernstein, I.A., Gray, R.H., Goldstein, I.: Proc. Natl. Acad. Sci. 77, 477-479 (1980).
3. Bretton, R., Bariety, J.: J. Histochem. Cytochem. 24, 1092-1100 (1976).
4. Clamp, J.R.: Recent Advances in Gastrointestinal Pathology, R. Wright ed., W.B. Saunders Co., London 1980.
5. Culling, C.F.A., Reid, T.E., Worth, A.W.: Methods and Achievements in Experimental Pathology; Cell Markers, Jasmin, G.ed, S. Karger, Basel 1981.
6. DeLellis, R.A.: Diagnostic Immunohistochemistry, R.A. DeLellis, Masson Publishing USA, Inc. 1981.
7. Essner, E., Schreiber, J., Griewski, R.A.: J. Histochem. Cytochem. 26, 452-458 (1978).
8. Etzler, M.E., Branstrator, M.L.: J. Cell Biol. 62, 329-343 (1974).
9. Fischer,J., Klein, P.J., Vierbuchen, M., Fischer, R., Uhlenbruck, G.: These Proceedings.
10. Freeman, H.J., Lotan, R., Kim, Y.S.: Lab. Invest. 42, 405-412 (1980).
11. Graham, R.C., Karnovsky, M.J.: J. Histochem. Cytochem. 14, 291-296 (1966).
12. Green, N.M.: Biochem. J. 89, 585-591 (1963).
13. Hermann, G., Lembke, J., Weil, H.P., Gerecke, D.: These Proceedings.
14. Holthoefer, H., Virtanen, I., Pettersson, E., Toernroth, T.,Alfthan,O.,

Linder, E., Miettinen, A.: Lab. Invest. 45, 391-399 (1981).

15. Holthoefer, H., Miettinen, A., Virtanen, I.: These Proceedings.

16. Howard, D.R., Ferguson, P., Batsakis, J.G.: Cancer 47, 2872-2877 (1981).

17. Hsu, S., Raine, L., Fanger, H.: J. Histochem. Cytochem. 29, 577-580 (1981).

18. Hsu, S., Raine, L.: Am. J. Clin. Pathol. 77, 508-509 (1982).

19. Jacobson, W., Stoddart, R.W., Collins, R.D.: Histopathology 4, 491-506 (1980).

20. Klein, P.J., Vierbuchen, M., Schulz, K.D., Wuerz, H., Citoler, P., Uhlenbruck, G., Ortmann, M., Fischer, R.: TumorDiagnostik 2, 240-245 (1981).

21. Klein, P.J., Newmann, R.A., Mueller, P., Uhlenbruck, G., Citoler, P., Schaefer, H.E., Lennartz, K.J., Fischer, R.: J. Cancer Res. Clin. Oncol. 93, 205-214 (1979).

22. KLein, P.J., Osmers, R., Vierbuchen, M., Ortmann,M., Kania, J., Uhlenbruck, G.: Recent Results Cancer Res. 79, 1-9 (1981).

23. Klein, P.J., Vierbuchen, M., Farrar, G., Fischer, J., Uhlenbruck, G.: These Proceedings.

24. Louis, C.J., Wyllie, R.G., Chou, S.T., Stzynda, T.: Am. J. Clin. Pathol. 75, 642-647 (1981).

25. Nakane, P.K., Pierce, G.B.Jr.: J. Cell Biol. 33, 307- (1967).

26. Newman, R.A., Klein, P.J., Rudland, P.S.; JNCI 63, 1339-1346 (1979).

27. Pena, S.D.J., Gordon, B.B., Karpati, G., Carpenter, S.: J. Histochem. Cytochem. 29, 542-546 (1981).

28. Raedler, A., Raedler, E., Arndt, R., Thiele, H.G.: Cell Tissue Res. 218, 219-226 (1981).

29. Rose, M.L., Habeshaw, J.A., Kennedy, R., Sloane, J., Wiltshaw, E., Davies, A.J.S.: Br. J. Cancer 44, 68-74 (1981).

30. Spicer, S.S., Sannes, P.L., Katsuyama, T.: J. Histochem. Cytochem. 27, 1182-1184 (1979).

31. Stoward, P.J., Spicer, S.S., Miller, R.L.: J. Histochem. Cytochem. 28, 979-990 (1980).

32. Suzuki, S. Tsuyama, S., Suganuma, T. Yamamoto, N., Murata, F.: J. Histochem. Cytochem. 29, 946-952 (1981).

33. Vedtofte, P., Dabelsteen, E.: Acta Path. Microbiol. Scand. Sect. A:

89, 439-449 (1981).

34. Vierbuchen, M., Klein, P.J., Uhlenbruck, G., Klein, O.H., Schaefer, H.E., Fischer, R.: Recent Results Cancer Res. 75, 68-75 (1980).

35. Vierbuchen, M. Klein, P.J., Roesel, S., Fischer, J., Uhlenbruck, G.: These Proceedings.

36. Watanabe, M., Muramatsu, T., Shirane, H., Ugai, K.: J. Histochem. Cytochem. 29, 779-790 (1981).

37. Whyte, A.: Histochem. J. 12, 599-607 (1980).

38. Yamada, K., Shimizu, S.: Histochemistry 67, 31-41 (1980).

REACTIONS WITH LECTINS ON BENIGN AND MALIGNANT LESIONS OF THE HUMAN BREAST

Relationship with the Presence of Differentiation Antigens of the Mammary Gland.

Ph. Hageman, L. Bobrow, M.A. van der Valk, W. Misdorp and J. Hilkens
Division of Clinical Oncology and Division of Genetics, The Netherlands Cancer Institute, Amsterdam, The Netherlands.

Presence of receptors for peanut agglutinin (PNA) can be considered as a marker for the degree of differentiation of mammary epithelium, both in the rat and in man (1). Another tool to study differentiation markers can be found in monoclonal antibodies directed against antigens of the human mammary gland. Taylor-Papadimitriou et al. (2) prepared monoclonal antibodies against delipidated membranes of human milk fat globules (HMFG). They obtained two antibodies that react as well with the secretory mammary cell as with primary human mammary carcinoma cells, both in fresh and in formalin fixed tissues (3). Both antibodies reacted more strongly with highly differentiated carcinomas, and less so with poorly differentiated tumors. Both antibodies also reacted with the luminal cell surface and secretions in the lactating breast. Hilkens et al. (4) made monoclonal antibodies against native HMFG membranes. One of the antibodies: 67-D11 reacted also with thin sections of formalin fixed and paraffin embedded primary breast cancer, but not with the tumor cells of all mammary carcinomas.
Milk fat globule membranes are derived from the apical surface of the cells of the secreting mammary gland and contain anti-

gens of mammary gland cells in a highly differentiated state. The aim of this study was to look for a possible correlation between the presence of lectin receptors, especially those of PNA and the reaction with antibodies against HMFG membranes.

Materials and methods

Thin sections of formalin-fixed and paraffin or histowax embedded material from 29 mammary carcinomas were used: 12 infiltrating ductal carcinomas, 2 medullary carcinomas, 4 scirrhous carcinomas, 1 lobular carcinoma, 6 adenocarcinomas of low or medium differentiation, 3 carcinoma solidum and 1 mucinous carcinoma. Other mammary tumors (6 infiltrating ductal carcinomas, 3 adenocarcinomas, 1 carcinoma solidum and 1 medullary carcinoma) had been fixed in 2.5% buffered glutaraldehyde. They belonged to a series that was also studied electron microscopically (5).
One lactating mammary gland and 4 resting mammary glands and 30 benign mammary lesions were studied. Six of the patients with a benign lesion developed a mammary carcinoma within the next 6 years. 5 Benign lesions were obtained from women in the first part of the menstrual cycle and 5 from the second part of the menstrual cycle. In the other specimens the time of the cycle was not known, or the women were postmenopausal.
In preliminary experiments adjacent pieces of the same mammary tumors were frozen or fixed in formalin or in 2.5% glutaraldehyde. The lectins conjugated with horse radish peroxidase (E.Y. Laboratories Inc., San Mateo, Cal. U.S.A.) that were used were peanut agglutinin (PNA) in concentrations 1-10 µg/ml, Dolichos biflorus agglutinin (DBA) 10-20 µg/ml and soy bean agglutinin (SBA) 1-5 µg/ml.
Antibody 67-D11 was prepared against native HMFG membranes as described by Hilkens et al. (4). In this study we will refer to this antibody as anti-HMFG-membrane. Antibody 1.10.F3 and 3.14.A3 called anti-HMFG-1 and anti-HMFG-2 respectively were

kindly donated by J. Taylor-Papadimitriou of the Imperial Cancer Research Fund in London. Antibody CE6/2D3.1 directed against carcinoembryonic antigen was obtained from E.S. Lennox, Laboratory of Molecular Biology, Cambridge, U.K. This antibody is a singly cloned rat monoclonal antibody from a rat immunized with highly purified CEA and screened on a crude CEA (6). In preliminary immunohistochemical tests we found an identical distribution for the reaction with CE6/2D3.1 and a conventional rabbit anti-CEA from DAKO (Denmark). We will refer to antibody CE6/2D3.1 as anti-CEA.

In all tests Dulbecco's Buffered Saline (PBS) was used as a diluent.

The reactions were performed on the deparaffinized sections as described previously (7). As a control sections were incubated with lectin mixed with a blocking sugar in a final concentration of 2% sugar: PNA and SBA with galactose, and DBA with N-acetylgalactosamine. Thin sections of the benign lesions and of fifteen mammary tumors were incubated with 1 U/ml neuraminidase from Vibrio cholerae (Boehring, Germany) for 30 min at 37°C prior to the reaction with PNA. Sections that were to be tested with the monoclonal antibodies were incubated 10 min at 37° with undiluted goat serum; rapidly rinsed in PBS, covered with an appropriate dilution of the monoclonal antibody and incubated for 60 min at 37°C in a humid box. As a control sections were incubated with the dilution fluid for the immunological reactions (PBS + 10% normal goat serum). After thoroughly rinsing in PBS the endogenous peroxidase was blocked by immersing the sections for 15 min at room temperature in methanol + 1% H_2O_2 (8). After rinsing in PBS the sections were incubated for 30 min at room temperature with a 1:60 dilution of peroxidase conjugated goat anti-rat IgG (9). This second antibody reacted equally well with rat and with mouse monoclonal antibodies.

Staining was done by application of a solution of 5 mg diaminobenzidin-HCL (DAB) in 10 ml PBS + 0.003% H_2O_2 at room temperature for 10 min. The sections were counterstained in hematoxilin for 20-30 sec, dehydrated and mounted in DePex.

Results

Formalin fixation and embedding did not appreciably change the distribution pattern of the lectin receptors (7) or of the reaction with monoclonal antibodies. Glutaraldehyde fixation seems somewhat to decrease the amount of PNA receptors. As electron microscopy turned out to be a more sensitive technique than light microscopy for the study of lectin receptors this compensates for the possible loss of receptors due to the fixation. Receptors for other lectins and the mammary gland antigens were not apreciably affected by glutaraldehyde fixation. The control sections incubated with lectin and blocking sugar were negative. Anti-HMFG-1 and anti-HMFG-2 reacted with all mammary tumors studied in this series, but in a few benign lesions the reaction was absent or very weak. Anti-HMFG-2 reacted frequently with a smaller percentage of cells. 13 mammary tumors (45%) of the series of the formalin fixed material were positive with anti-HMFG-membrane but frequently only a small percentage of the cells of the tumor reacted. Pre-existing normal mammary ducts in carcinomas were frequently positive with anti-HMFG-membrane even if the tumor tissue was negative. 19 benign lesions (63%) were positive for anti-HMFG-membrane. A negative correlation with a small overlap existed in the reaction pattern of anti-HMFG-membrane and anti-HMFG-1, anti-HMFG-membrane frequently reacted with a part of the tumor that was negative for anti-HMFG-1.

The control sections incubated with the conjugate were completely negative.

In mammary tumors several reaction patterns of PNA were found: a) a rather uniformly distributed reaction in the cytoplasm; b) a strong reaction with the membrane of small vacuoles in the cytoplasm (Fig. 1); c) a reaction with the apical surface of the cells of tumors with a glandular or papillary differen-

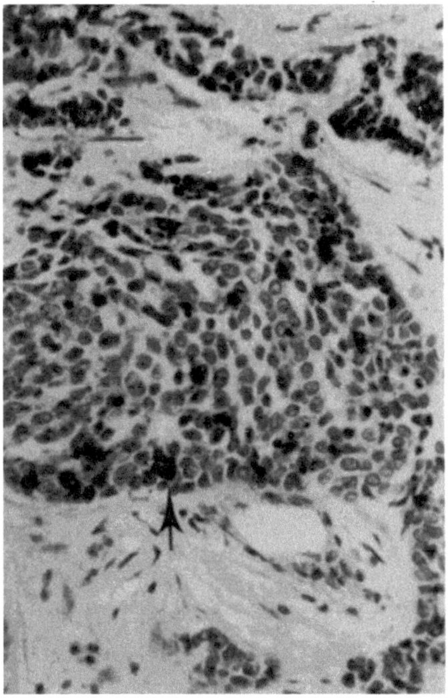

Fig. 1: Reaction of PNA on metastasis of infiltrating ductal carcinoma of the mamma is mainly located on intractyoplasmic vacuoles (arrow).

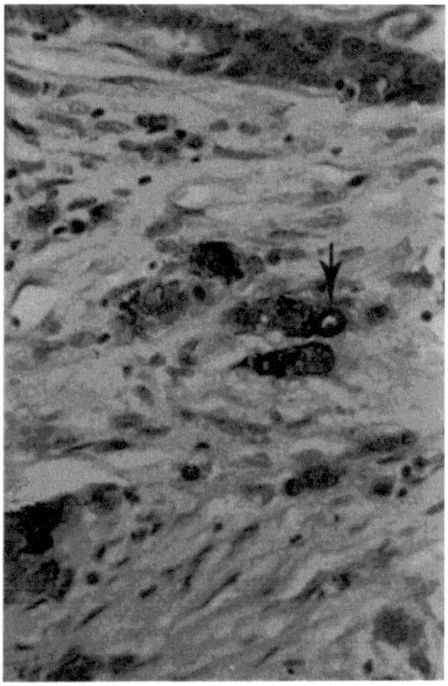

Fig. 2: Reaction of anti-HMFG-1 with intracytoplasmic vacuoles in infiltrating mammary carcinoma (arrow).

tiation (Fig. 3) and a reaction with the membrane of cells that are arranged in "Indian file" (Fig. 6). When no apical surface could be distinguished the tumor cells often were positive only on the surface that bordered the stroma.
66% of the mammary tumors were positive for PNA receptors, a percentage that may also be influenced by the size of the sample. In 11 tumors of which more than one, often rather large sample was studied no tumors were observed that were completely negative for PNA receptors.
In benign lesions the distribution of PNA receptors was often very heterogenous; ducts and acini in some lobuli reacted, others in the same section were completely negative. In some cases the binding of PNA was not restricted to the luminal

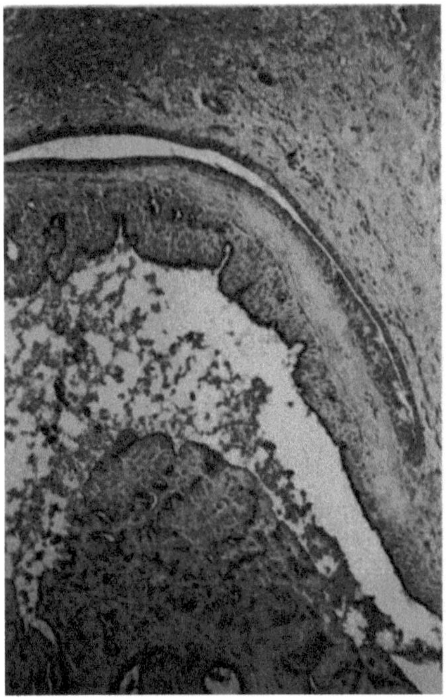

Fig. 3: Infiltrating papillary carcinoma of the mamma: PNA reacts strongly with the apical surface of the tumor cells, the duct is only weakly positive.

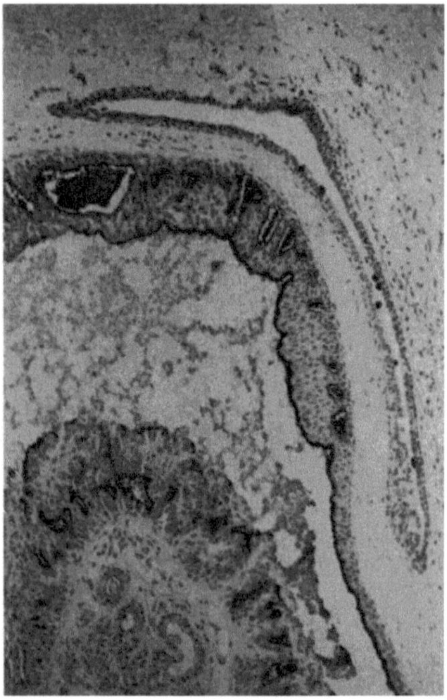

Fig. 4: Reaction of the mammary tumor of Fig. 3 with anti-HMFG-1, note the strong resemblance with the reaction of PNA.

surface of the cells of the acini and ducts and the secretion product but the entire cell membrane was positive. 3 benign lesions showed no or only a weak reaction with PNA.

Both in mammary tumors, benign lesions and mammary glands a strong binding of PNA to the stroma occurred, especially in cases where tumor cells or cells of normal acini were negative for PNA receptors. The DAB reaction product was often arranged in bundles but probably did not follow completely the collagen-fibre pattern that could be observed by their birefringent character. Probably the PNA receptors were not present on the fibres but on an amorphous matrix.

In benign lesions the reaction of PNA and DBA with intralobular stroma was considerably weaker than the reaction with the

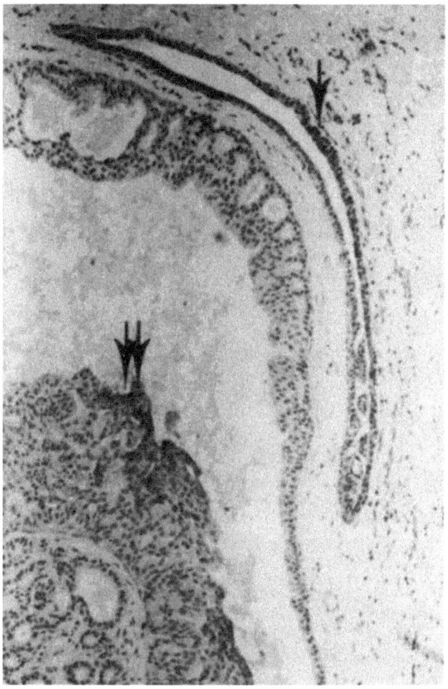

Fig. 5: Reaction of the mammary tumor of Fig. 3 with anti-HMFG-membrane. Reaction mainly with the cells of the duct (arrow) and only very locally with the tumor cells, mainly at the site of squamous metaplasia.

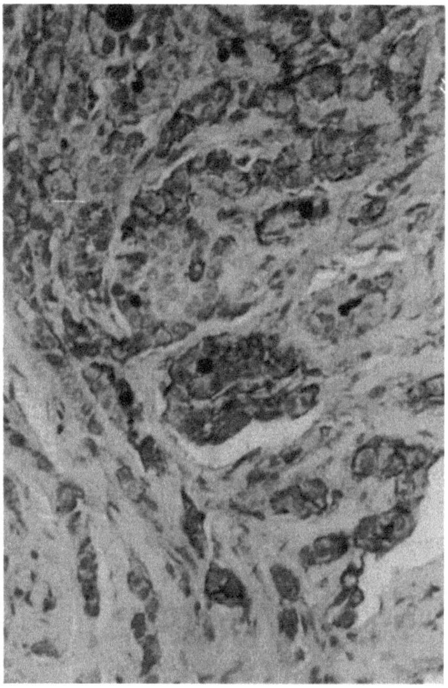

Fig. 6: Carcinoma solidum of the mamma: reaction with PNA of the membrane of tumor cells arranged in "Indian file" pattern.

interlobular stroma.

Though reaction with anti-HMFG-membrane and with PNA did not completely exclude each other, a negative correlation between these phenomena was found. In some tumors or lesions that were positive for both, different structures reacted with PNA and anti-HMFG-membrane (Fig. 3 and 5). A similarity between the reaction patterns of anti-HMFG-1 and the presence of PNA receptors was observed often in striking details (Fig. 1, 2, 3 and 4). In the tumors, membranes of the intracytoplasmic vacuoles reacted strongly with anti-HMFG-1. In benign lesions a special structure, showing snout-like protrusions and blebs was posi-

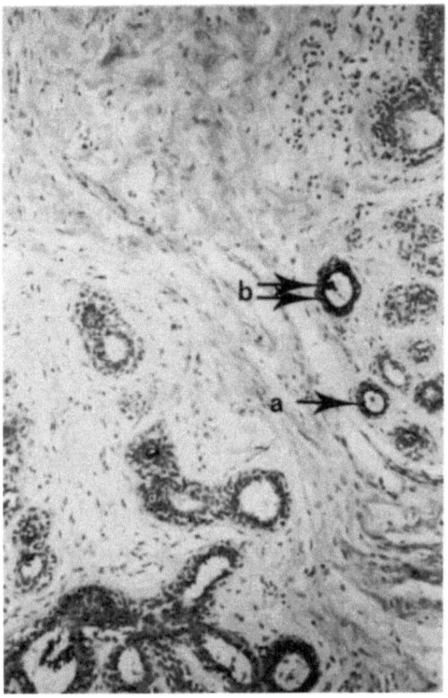

Fig. 7: Cells of some acini in a benign lesion of the human breast react with PNA only at their luminal surface (arrow a). In other ducts the entire surface of the cell is positive (arrow b), other ducts do not react with PNA.

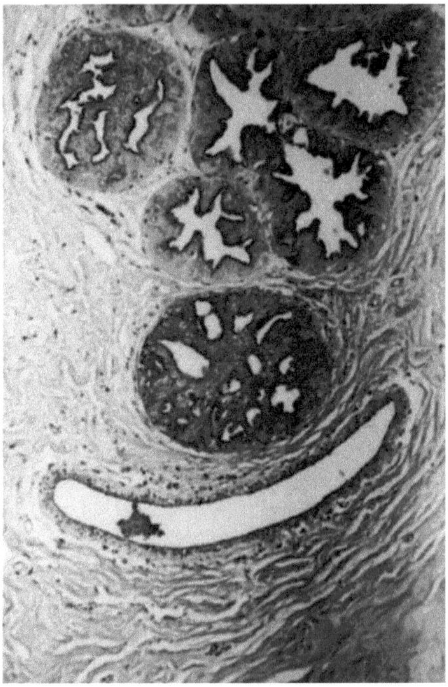

Fig. 8: Reaction with PNA of apocrine metaplasia in a breast lesion. Note some reaction in the connective tissue as well.

tive with anti-HMFG-1 and PNA. These structures were positive with SBA and negative with DBA. Apocrine metaplasia in mammary lesions, especially the apical (luminal) surface of the cells was invariably positive with anti-HMFG-1 and anti-HMFG-2 and negative with anti-HMFG-membrane. These sites reacted with PNA.

In lactating mammary glands PNA and anti-HMFG-1 reacted very strongly with the apical surface of secreting cells and only weakly with non-secreting alveoli. Though anti-HMFG-membrane had been raised against the milk fat globule membranes, the antibody probably is directed against a minor component of the

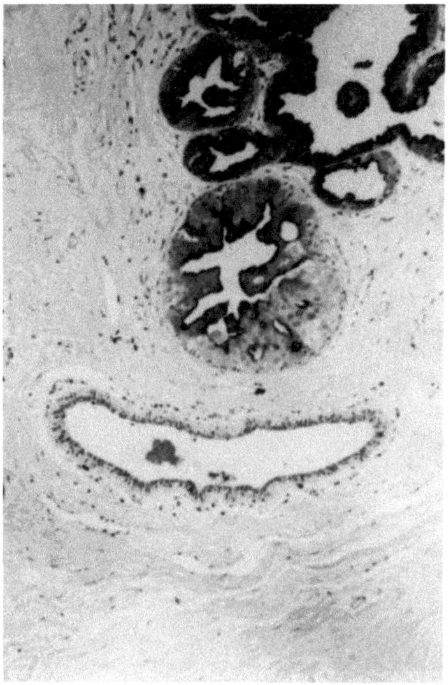

Fig. 9: Reaction with anti-HMFG-2 of the apocrine metaplasia of Fig. 8. Note the resemblance with the reaction pattern of PNA with the exception of the stromal reaction.

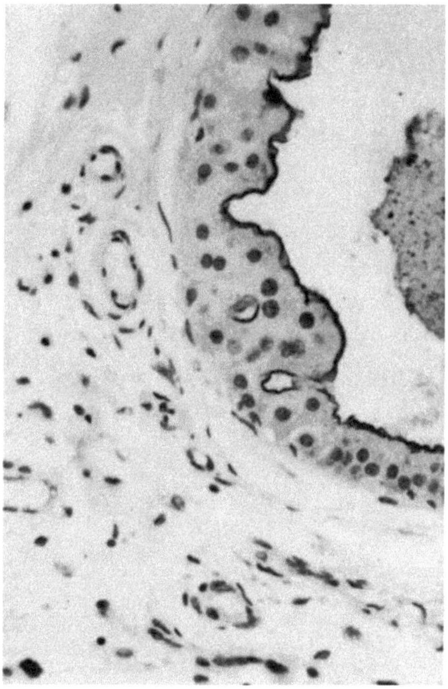

Fig. 10: Reaction of apocrine metaplasia with anti-HMFG-1 (detail).

milk, as the reaction with the secreting cells of the lactating mammary gland was weak, compared to the reaction on tumor cells or pre-existing normal ducts in tumors. Anti-HMFG-membrane preferentially reacted with parts of the tumors where a squamous differentiation of luminal situated epithelial cells occurred.

In 11 mammary carcinomas a negative correlation was observed between presence of PNA receptors and reaction with anti-CEA, though this series is too small to draw any definite conclusions from. No characteristic reaction pattern with any of the markers was observed in the benign lesions of patients who developed a mammary tumor later in life. Benign lesions of the breast in women in the second half of the menstrual cycle did

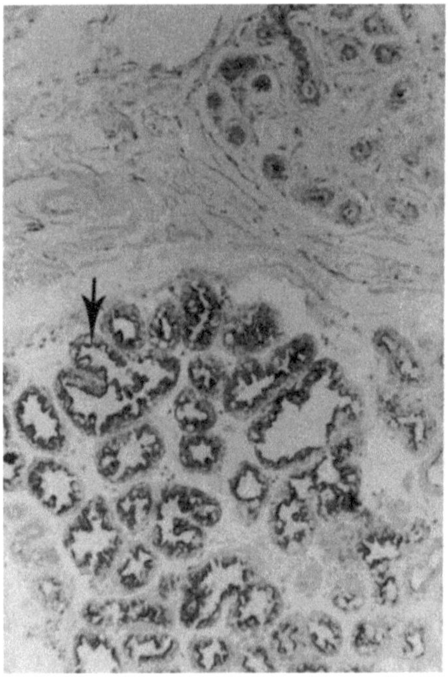

Fig. 11: PNA receptors on benign mammary lesions: presence on snout-like protrusions (arrow) on the cell surface.

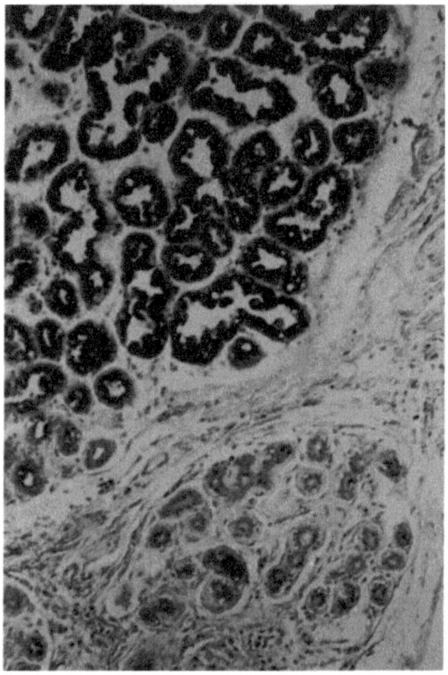

Fig. 12: Reaction of anti-HMFG-1 mainly on snout-like protrusions on the cell surface. Same lesions as in Fig. 11.

not show a remarkably stronger reaction with the antibodies or lectins that were used in this study, compared to the samples from the first half of the cycle.

Treatment of the thin sections with neuraminidase prior to the incubation with PNA revealed that in several specimen cryptic receptors for PNA were present beside the free receptors. After neuraminidase treatment the reaction with PNA frequently was stronger than without pretreatment; this was especially clear when a rather low concentration of PNA (1 µg/ml) was applied. We could not observe yet a large difference in the distribution of the cryptic and free receptors for PNA.

Discussion

The peroxidase-labeled lectins proved to be a useful tool to visualize the presence of lectin receptors in routine pathology specimens. It seems unlikely that the binding might be influenced by lectin-like substances in the thin sections that could bind directly to the peroxidase label (10). The control reactions with the peroxidase-labeled goat anti-rat serum, that in this respect are comparable with peroxidase-labeled lectin, were always completely negative.

Peanut lectin reacts with the Thomsen-Friedenreich (TF) antigen, that was supposed to be present on human mammary tumors in a free state and to be absent or masked by sialic acid residues on benign or normal breast tissue according to Springer (12, 13) and to Howard (14). In a later study Howard and coworkers (15) reported a reaction with PNA also on benign lesions without neuraminidase pretreatment, but they observed a completely different binding pattern in benign lesions. They found PNA receptors in benign lesions along the luminal cytoplasmic membranes and in mammary carcinomas diffusely in the cytoplasm.

We found the PNA receptors both in benign and malignant lesions. In benign lesions the reaction occurred in some cases on the entire surface of the cells and not only on their apical surface. Cells of malignant lesions had a diffuse reaction with PNA in their cytoplasm but also with small intracytoplasmic vacuoles, while especially in areas where papillary structures were present PNA binding was seen in the apical region of the cells, resembling the reaction pattern of benign lesions. This is much in agreement with the observations of Newman et al. (1), Klein et al. (18) and Stegner et al. (16). These last authors used two methods to detect the Thomsen-Friedenreich antigen in thin sections: reaction with antibody against TF-antigen visualized by a peroxidase-antiperoxidase procedure and treatment with PNA.

The results of these authors (1, 16, 17 18) and our own observations contradict the theory that free TF-antigen is a marker for malignancy of breast tissue.

PNA binding may be a marker for a certain degree of differentiation (16, 17). Newman (1) reported that the presence of masked PNA receptors was inversely related to the differentiation state of the rat mammary gland, more differentiated tissue had mainly free PNA receptors. The direction of the differentiation is towards secretion activity (16) and hormone dependency of the tumor (18, 19, 20). Moreover PNA binding is correlated with the presence of both estrogen and progesterone receptors as is shown in this volume by Klein and his colleagues (18). The partial similarity in reaction pattern of PNA and anti-HMFG-1 underlines the concept of PNA-receptors as marker of a differentiation stage of the tissue that has features in common with the lactating mammary gland. This is in agreement with the work of Klein (18) who has immunohistological evidence that the PNA receptor in human mammary tumors is located on a milk protein.

The distribution patterns of the reactions with PNA and anti-HMFG-1 in our study show enough differences to make it unlikely that both react with the same molecule. Anti-HMFG-1 reacts with another antigenic component that might indicate a differentiation in the direction of squamous cells.

The binding of PNA and the reaction with the antibodies to breast tissue illustrates the great variety that exists in stages of differentiation in structures in these tissues.

The fact that no significant increase in presence of any of the markers was found in benign lesions of women in the second half of the menstrual cycle seems surprising, because several authors (21, 22) report secretory activity in the breast tissue at that time. However, this phenomenon is not always present and its effect might be obscured by the individual variation of the specimens.

Why we and Calafat and Jansen (6) find in several specimens a binding of PNA to connective tissue, whereas Klein (18, 19, 20) does not so, remains a question that still has to be solved.

Acknowledgements

We thank the pathologists and pathological departments of the "Netherlands Cancer Institute", Amsterdam, The "Ziekenhuis Amsterdam Noord", Amsterdam and the "Stichting Samenwerkende Delftse Ziekenhuizen", Delft for providing us with the tissue material and for giving the diagnosis of the cases and Dr. J. Taylor-Papadimitriou and Dr. E.S. Lennox for sending the antibodies.
We thank Dr. J. Hilgers of the Netherlands Cancer Institute for his help in obtaining anti-HMFG-1, anti-HMFG-2 and anti-CEA and for his useful remarks.
We thank mr. D. Atsma and mr. D. Schol for technical assistence and mrs. N. van Nuland for typing the manuscript.

During the preparation of this manuscript Foster and his colleagues (23) published a study in which they compared the reactions of monoclonal antibodies to the human mammary gland with the binding of PNA to breast carcinomas.

References

1. Newman, R.A., Klein, P.J., Rudland, P.S.: J. Natl.Cancer Inst. 63, 1339-1345 (1979).
2. Taylor-Papadimitriou, J., Peterson, J.A., Arklie, J., Burchell, J., Ceriani, R.L., Bodmer, W.F.: Int. J. Cancer 28, 17-21 (1981).
3. Arklie, J., Taylor-Papadimitriou, J., Bodmer, W., Egan, M., Millis, R.: Int. J. Cancer 28, 23-29 (1981).
4. Hilkens, J., Buys, F., Hilgers, J., Hageman, Ph., Sonnenberg, A., Koldowsky, U., Karande, K., van Hoeven, R.P., Feltkamp, C., van de Rijn, J.M. In: Protides of the Biological Fluids 29, Proc. 29th Coll., Brussels 1981, p. 813-816.

5. Calafat, J., Janssen, H. This volume.
6. Lennox, E., Alderson, T., Newby, G.: Personal communication.
7. Van der Valk, M.A., Hageman, Ph.: Lectins, vol. 2. T.C. Bøg-Hansen Ed., de Gruyter, Berlin (1981).
8. Streefkerk, J.G.: J. Histochem. Cytochem. 20, 829-831 (1972).
9. Wilson, M.B., Nakane, P.K. In: Immunofluorescence and related staining techniques, W. Knapp ed., Elsevier (1978) p. 215-224.
10. Azzopardi, J.G.: Problems in breast Pathology. W.B. Saunders, London (1979).
11. Gross, D., Bruce, B. Challice, C.E., Schrevel, J.: J.Hist. Cytochem. 30, 193-200 (1982).
12. Springer, G.F., Desai, P.R., Banatwala, I.: J.Natl. Cancer Inst. 54, 335-339 (1975).
13. Springer, G.F., Desai, P.R., Scanlon, E.F.: Cancer 37, 169-176 (1976).
14. Howard, D.R., Taylor, C.R.: Cancer 43: 2279-2287 (1979).
15. Howard, D.R., Ferguson, P., Batsalis, J.G.: Cancer 47, 2872-2877 (1981).
16. Klein, P.H., Newman, R.A., Müller, P., Uhlenbruck, G., Citoler, P., Schaefer, H.E., Lennarts, K.J., Fischer, R.: J.Cancer Res. Clin. Oncol. 93: 205-214 (1979).
17. Stegner, H.E., Fischer, K., Poschman, A.: Tumor Diagnostik 3, 127-130 (1981).
18. Klein, P.J., Vierbuchen,, M., Farrar, G., Fischer, J., Uhlenbruck, G.: This volume.
19. Klein, P.J., Vierbuchen, M., Schulz, K.D., Würz, H., Citoler, P., Uhlenbruck, G., Ortmann, M., Fischer, R.: Tumor Diagnostik 2, 240-245 (1981).
20. Klein, P.H. Vierbuchen, M., Würz, H., Schulz, K.D., Newman, R.A.: Br. J. Cancer 44, 746-748 (1981).
21. Geschikter, Ch.F.: Diseases of the breast 2. Ed. J.B. Lippincot Comp. (1945).
22. Fanger, H., Jung Ree, H.: Cancer 34, 574-585 (1974).
23. Foster, C.S., Dinsdale, E.A., Edwards, P.A.W., Neville, A.M.: Virchows Archiv. A. Path. Anat. Hist. 394, 295 (1982).

ULTRASTRUCTURAL STUDY ON THE PRESENCE OF RECEPTORS FOR SEVERAL LECTINS IN HUMAN BREAST CANCER

Jero Calafat and Hans Janssen
The Netherlands Cancer Institute, Plesmanlaan 121,
1066 CX Amsterdam, The Netherlands

Introduction

The presence of lectin receptors in different types of human mammary tumors was studied with the electron microscope. Light microscopical studies concerning binding of lectins to human mammary carcinomas have been performed by several groups (1, 2, 3, 4, 5). In most cases the binding of peanut agglutinin was studied but the group of Klein (1, 4) pointed also at the significance of receptors for the lectins of Helix pomatia and Ricinus communis.

The ultrastructural study has the advantage that the receptors can be localized more precisely on the different cell types (6). The aim of this work was first an inventarisation of the localisation of lectin receptors in human mammary tumors and second the search of a marker that could help in the classification of these tumors.

Material and Methods

18 mammary carcinomas were studied of the following histological types according to our pathology department: 5 adenocarcinomas; 2 infiltrating lobular carcinomas; 7 infiltrating ductal carcinomas; 3 carcinomas solidum and 1 medullary carcinoma. These tumors were incubated with the lectin from Cana-

Lectins, Vol. III

TABLE 1

Distribution of lectins and hormone receptors in human breast cancer

patient blood group	or	pr	PNA T	PNA S	HPA T	HPA S	UEA T	UEA S	ConA T	ConA S	WGA T	WGA S	RCA T	RCA S
Adenocarcinoma:														
O	ND	ND	+,-	+++L	ND	ND	+,-	-	ND	ND	ND	ND	ND	ND
O	+	ND	-	++	+	+,-	ND	ND	ND	ND	ND	ND	+++	+++
B	+	+	+,-	++L	+,-	+,-	+,-	BV,L	ND	ND	++	++	ND	ND
O	+	-	-	+,-	-	-	++	+,-	ND	ND	++	++	ND	ND
O	ND	ND	+,-	+,-	+,-	+,-L	-	BV	ND	ND	ND	ND	ND	ND
Infiltrating lobular carcinoma:														
A	+	ND	+++	++	ND	ND	+,-	-	+++	+++	ND	ND	ND	ND
A	+	ND	+,-	++	ND	ND	-	-	ND	ND	ND	ND	ND	ND
Infiltrating ductal carcinoma:														
A	-	ND	+	+,-	ND	ND	+,-	-	ND	ND	ND	ND	ND	ND
A	ND	ND	+,-	++	-	-	-	-	+++	+++	++	++	ND	ND
A	+	ND	+,-	+L	-	BV	ND	ND	ND	ND	ND	ND	+++	+++
O	+	+	-	++	+	+	-	+	ND	ND	ND	ND	ND	ND
A	ND	ND	ND	ND	+	+	ND	ND	ND	ND	++	++	ND	ND
A	+	ND	-	++	-	+,-	+	BV,L	ND	ND	++	++	ND	ND
B	+	-	-	++PL	-	+,-	+,-	+,-	ND	ND	++	++	ND	ND
Carcinoma solidum:														
AB	+	ND	+,-	+++ L,PL	ND	ND	-	+,-	ND	ND	ND	ND	ND	ND
-	+	+	+,-	++L	+	+	+	BV	ND	ND	++	++	ND	ND
B	+	-	-	+++	+,-	BV	++	BV	ND	ND	++	++	ND	ND
Medullary carcinoma:														
O	+	-	+	++	+	+,-	-	BV	ND	ND	++	++	ND	ND

T, tumor cells; S, stroma; L, positive lymphocytes; PL, positive plasma cells; BV, positive blood vessels; ++ to +, indicates positive reaction all over these cells varying from strong to weak; +,-, reaction localized in some areas; -, no reaction; ND, not done; or, oestrogen receptor; pr, progesterone receptor.

valin ensiformis (ConA) and the following peroxidase-labeled lectins: peanut agglutinin (PNA), Helix pomatia agglutinin (HPA), wheat germ agglutinin (WGA), Ulex europaeus agglutinin (UEA) and Ricinus communis agglutinin 120 (RCA). Not all tumors were incubated with all lectins (see Table 1).

The procedure followed for the incubation of the tumors with the peroxidase-labeled lectins is described in the accompanying paper (6) and for the incubation with ConA in the text of figure 2. In all experiments one section was incubated with peroxidase-labeled lectin with addition of an inhibitory saccharide as control. All controls were always negative (Figs. 2b and 5b).

Results

The results are summarized in Table 1. The lectin-binding patterns did not depend on the type of blood group of the patient. In this table we have added also the presence of hormone receptors determined in our department of Clinical Chemistry with the charcoal method. Peroxidase-labeled WGA, ConA and RCA gave positive reactions on the cell surface of all cells in all carcinomas studied (Figs. 1 and 2a).

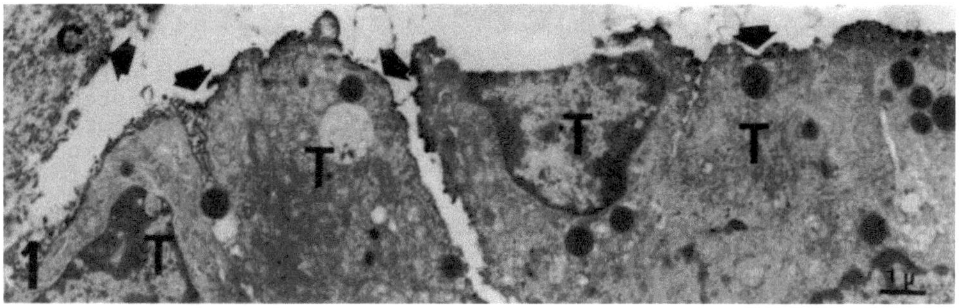

Figure 1. Adenocarcinoma incubated with peroxidase-labeled WGA. Labeling (arrows) on the surface of tumor cells (T) and on collagen bundles (C). Unstained thin section.

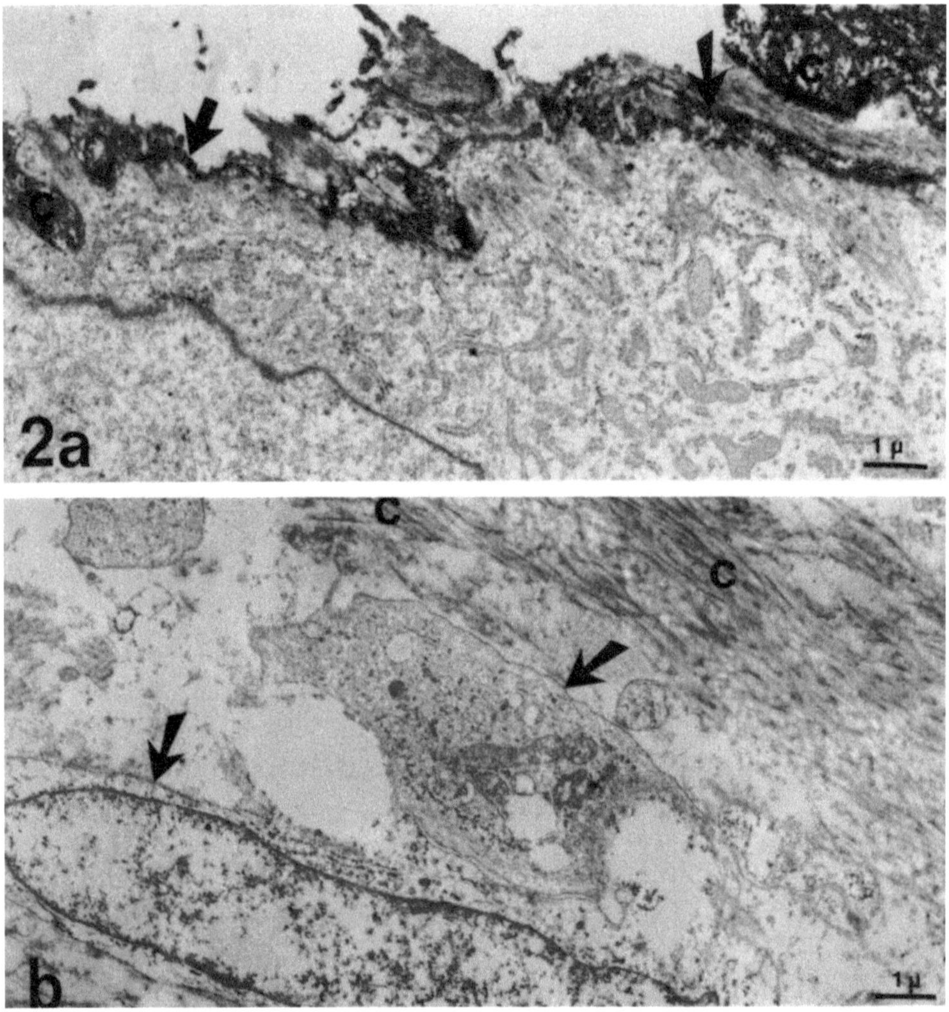

Figure 2. Infiltrating ductal carcinoma. a) incubated with ConA, thereafter with peroxidase, followed by the diaminobenzidine reaction. Strong labeling (arrows) on the surface of a fibroblast and along collagen bundles (C). b) the same as in a), but 2% α-methyl-D-glucoside was added to the solution of ConA. No labeling on the surface of the cells (arrows) and along collagen bundles (C). Unstained thin sections.

No correlation was found between the presence of PNA, HPA and UEA receptors with either hormone receptors or histological type of the tumors. Therefore the results will be presented without referring to the type of the tumor. The distribution of PNA receptors in these tumors has been extensively described in the accompanying paper (6).

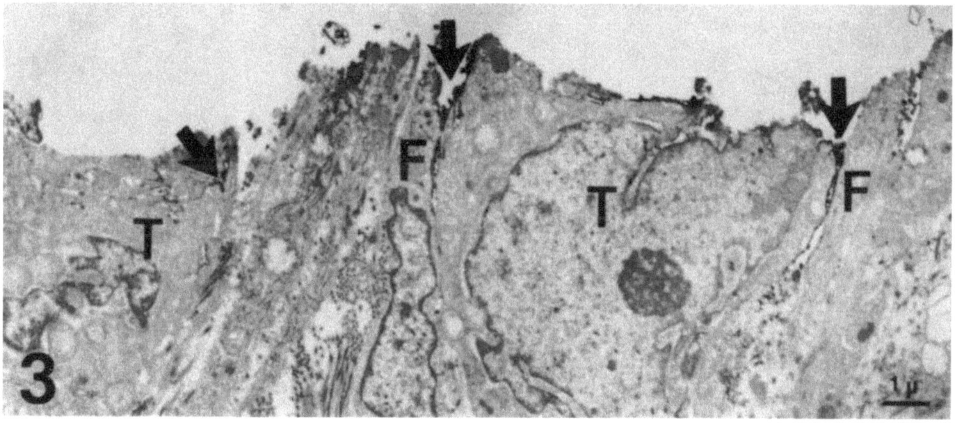

Figure 3. Infiltrating lobular carcinoma incubated with peroxidase-labeled HPA. Labeling (arrows) on the surface of tumor cells (T) and fibroblasts (F). Thin section stained with uranyl acetate and lead hydroxide.

HPA receptors: From the 13 carcinomas studied with this lectin 2 were negative; 7 showed a reaction on the tumor cells and in the stroma (Fig. 3). In some tumors all cells were positive, in others the reaction was restricted to some areas. Sometimes the reaction with tumor cells was confined to the surface of small lumina some of which were intracytoplasmic (Fig. 4). Of the 4 remaining positive tumors one carcinoma showed a reaction with some restricted surface areas of the tumor cells and at the same time on the vascular endothelia. Two tumors were only positive in the stroma and in the last one only blood vessels reacted.

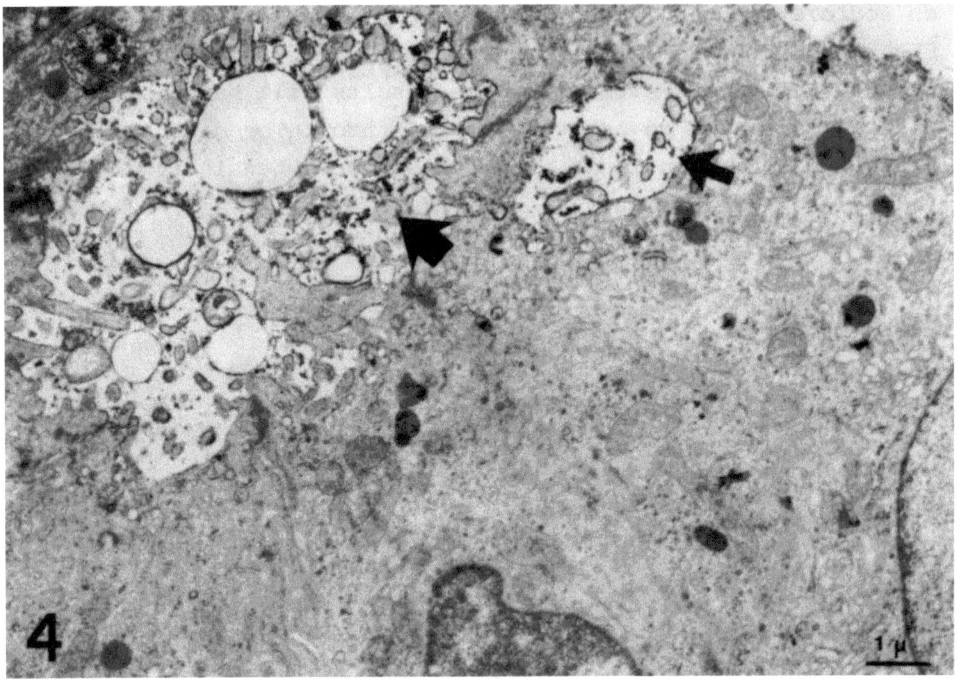

Figure 4. Adenocarcinoma incubated with peroxidase-labeled HPA. Labeling on the luminal surface of the cells and on the secretion product in the lumen (thick arrow). Intracytoplasmic lumen (thin arrow) is also positive. Thin section stained with uranyl acetate and lead hydroxide.

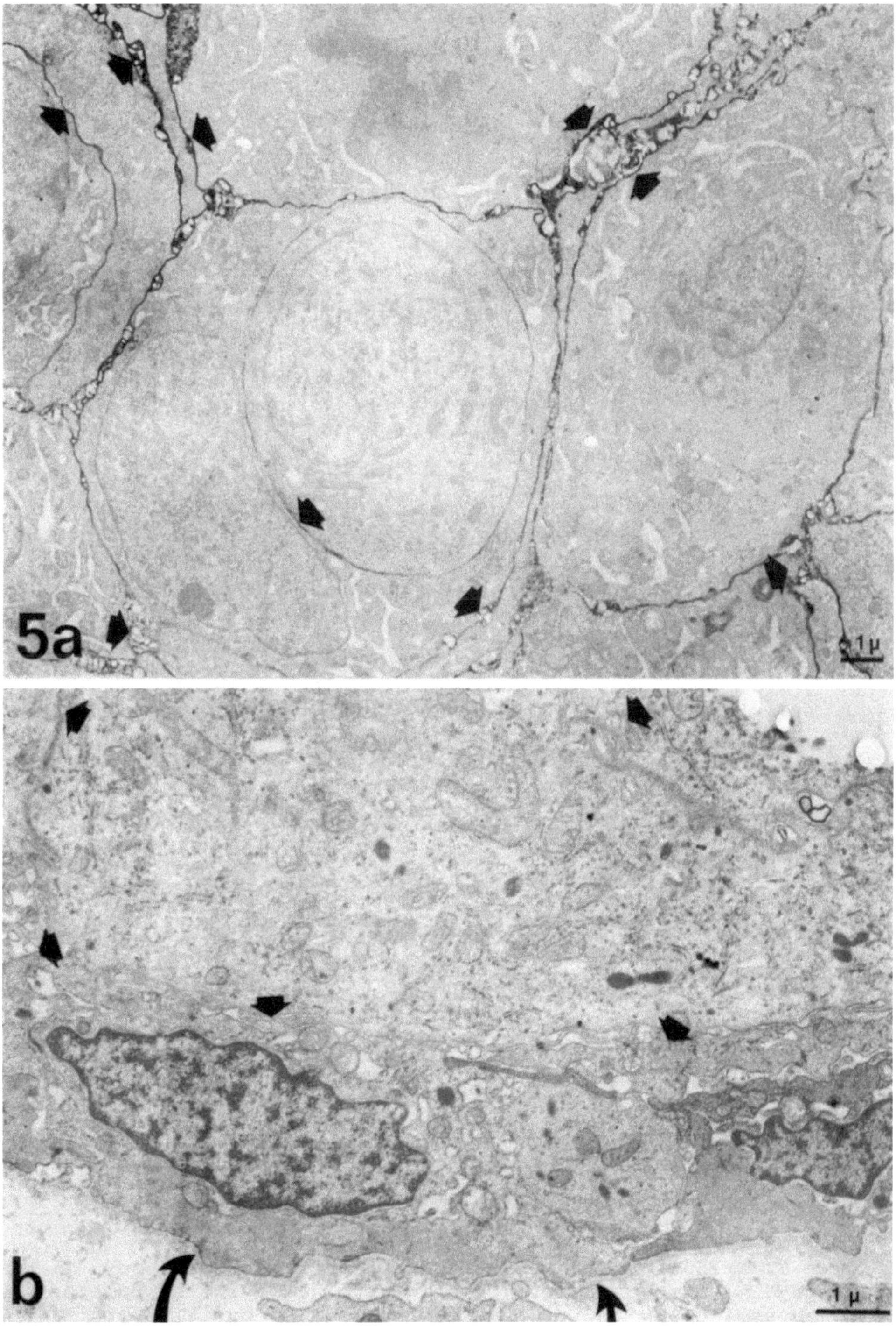

Figure 5. Carcinoma solidum. a) incubated with peroxidase-labeled UEA. The label fully surrounds the tumor cells (arrows). b) the same as in a), but 2% fucose was added to the solution of UEA. No labeling surrounding the tumor cells (thick arrows). Basal lamina (thin arrows) is also negative. Unstained thin sections.

UEA receptors: Fifteen carcinomas were studied for the presence of UEA receptors: 9 were positive on the tumor cells generally restricted to a few areas of the tumor where the label fully surrounded the cells (Fig. 5a); 2 tumors showed reaction only in the stroma and 4 tumors were negative. Vascular endothelia, when present in the section, always gave a strong reaction (Fig. 6).

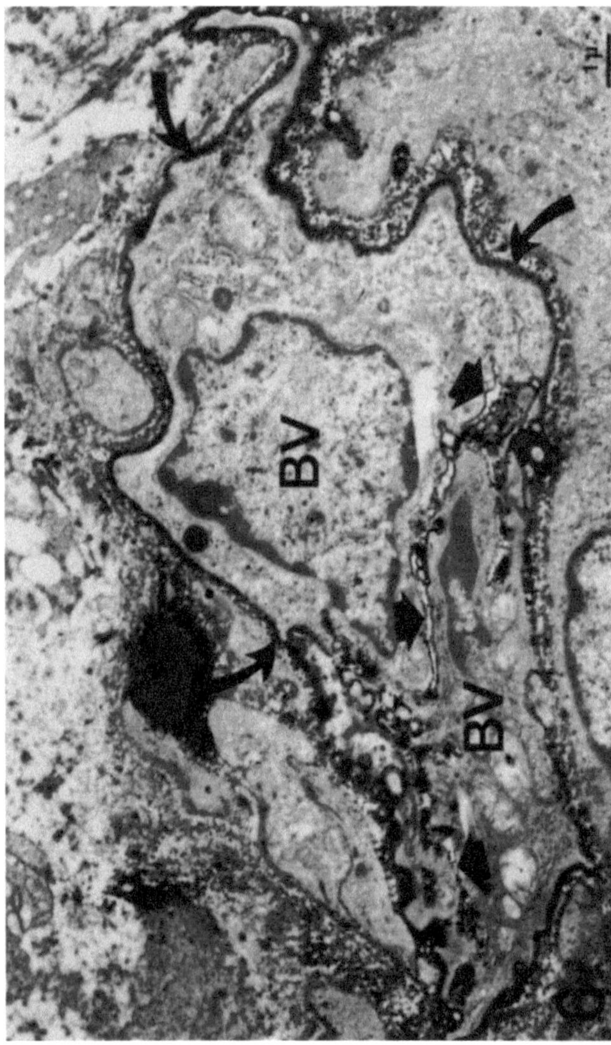

Figure 6. Medullary carcinoma incubated with peroxidase-labeled UEA. Strong reaction (thick arrows) on the surface of blood vessel cells (BV). Basal lamina is also positive (thin arrows). Thin section stained with uranyl acetate and lead hydroxide.

Discussion

In the hope to find markers helpful for diagnosis, recently histochemical work has been done using lectins on tissues in normal and pathological conditions (1-8). Some good results have been obtained in the field of gastrointestinal pathology and skin tumors (7, 8). More controversial are the results in the field of mammary gland pathology. Newman at al. (2) and Howard et al. (9) have found that PNA lectin is a good marker of mammary differentiation. In our study (6) the binding of PNA on mammary tumor cells did not always correlate with the stage of differentiation in these tumors.

In this paper we describe how human mammary carcinomas have been incubated with several lectins. WGA, RCA and ConA bound to most structures, and therefore are not interesting as markers, at least not qualitatively. This is in agreement with the findings of Holthöfer et al. (8, 10) that WGA, ConA and RCA showed uniform binding to most part of human kidney, Pena et al. (11) that ConA, RCA and WGA gave a general staining of human skeletal muscle, and Fischer et al. (12) that RCA stains all human gastrointestinal cells. In quantitative studies, i.e. determining the degree to which cells are agglutinated by these lectins, probably some of them can be used as prognostic marker, as has been found for the reactivity of ConA with breast cancer cells in tissue culture (13).

HPA and UEA reacted in a more selective way with human mammary tumors and each followed a typical pattern. HPA receptors are sometimes exclusively localized in small lumina, some of which were intracytoplasmic, whereas UEA binding to the positive areas of tumor cells gave a positive reaction all around the cell membrane. However, no correlation between the presence of each one of these two lectin receptors with either the histological type of the tumor or the presence of hormone receptors was found.

Klein et al. (5) have found a good correlation between the presence of PNA and hormone receptors. It is difficult to compare their results with our results because we have studied less tumors and only in seven out of the eighteen tumors both oestrogen and progesterone receptors were determined. Klein et al. (1) have also found a correlation between PNA and HPA binding to human mammary tumors, only the intensity of the reactions was different. All HPA positive tumors were also PNA positive, while some PNA positive tumors were HPA negative. However, we could not find any correlation between the presence of the receptors of PNA and HPA (Table 1).

UEA was shown to bind to all vascular endothelia in the breast tumors. Holthöfer et al. (8) and Pena et al. (11) have also shown the selectively staining of the blood vessels by UEA in human kidney tissue and in human skeletal muscle. It seems that UEA is a good marker for human vascular endothelia.

In conclusion, no correlation was found between the presence of lectin receptors and the histological type of the tumor. Moreover, no specificity of lectin binding to a determined morphologically distinct domain of the mammary tumor was found, with the exception of UEA binding with vascular endothelial tissue. However, until now we do not know the significance of the presence or absence of the different lectin receptors on these carcinomas because the biological functions of carbohydrates in glycoproteins are still not elucidated.

Acknowledgment

We thank Drs. C.A. Feltkamp and Ph.C. Hageman for critical reading of this manuscript; Mr. N. Ong for excellent technical assistance and Miss M.A. van Halem for typing the manuscript. We thank also the department óf Pathology for providing the material, specially Mr. J. Verrijk.

References

1. Klein, P.J., Vierbuchen, M., Schulz, K.-D., Würz, H., Citoler, P., Uhlenbruck, G., Ortmann, M., Fischer, R.: Symposium Carcinoembryonales Antigen (CEA) und Andere Tumormarker; Eds. G. Uhlenbruck and G. Wintzer. Tumor Diagnostik Verlag, Leonberg, 1981, in press.
2. Newman, R.A., Klein, P.A., Rudland, Ph.S.: J. Natl. Cancer Inst. 63, 1339-1346 (1979).
3. Stegner, H.E., Fischer, K., Poschmann, A.: Tumor Diagnostik 3, 127-130 (1981).
4. Klein, P.J., Citoler, P., Newman, R.A., Uhlenbruck, G., Schäfer, H.E., Fischer, R.: Verh. Dtsch. Ges. Path. 63, 591-596 (1979).
5. Klein, P.J., Vierbuchen, M., Farrar, G., Fischer, J., Uhlenbruck, G.: Lectins. Biology, Biochemistry, Clinical Biochemistry, Volume 3, T.C. Bøg-Hansen and G.A. Spengler, eds. Walter de Gruyter, Berlin, New York, 1982, in press.
6. Calafat, J., Hageman, Ph.C., Janssen, H.: Lectins. Biology, Biochemistry, Clinical Biochemistry. T.C. Bøg-Hansen and G.A. Spengler, eds. Walter de Gruyter, Berlin, New York, 1982, Volume 3, in press.
7. Coggi, G., Dell'Orto, P., Viale, G.: Lectins. Biology, Biochemistry, Clinical Biochemistry, Volume 3, T.C. Bøg-Hansen and G.A. Spengler, eds. Walter de Gruyter, Berlin, New York, 1982, in press.
8. Holthöfer, H., Virtanen, I., Pettersson, E., Törnroth, T., Alfthan, O., Linder, E., Miettinen, A.: Lab. Invest. 45, 391-398 (1981).
9. Howard, D.R., Ferguson, P., Batsakis, J.G.: Cancer 47, 2872-2877 (1981).
10. Holthöfer, H., Miettinen, A., Virtanen, I.: Lectins. Biology, Biochemistry, Clinical Biochemistry, Volume 3, T.C. Bøg-Hansen and G.A. Spengler, eds. Walter de Gruyter, Berlin, New York, 1982, in press.
11. Pena, S.D.J., Gordon, B.B., Karpati, G., Carpenter, S.: J. Histochem. Cytochem. 29, 512-516 (1981).
12. Fischer, J., Klein, P.J., Vierbuchen, M., Uhlenbruck, G., Fischer, R.: Lectins. Biology, Biochemistry, Clinical Biochemistry, Volume 3, T.C. Bøg-Hansen and G.A. Spengler, eds. Walter de Gruyter, Berlin, New York, 1982, in press.
13. Furmanski, Ph., Kirkland, W.L., Gargala, T., Rich, M.A., and the Breast Cancer Prognostic Study Clinical Associates: Cancer Res. 41, 4087-4092.

ULTRASTRUCTURAL LOCALIZATION OF PEANUT AGGLUTININ RECEPTORS ON CELLS OF HUMAN MAMMARY TUMORS

Jero Calafat, Philomena C. Hageman and Hans Janssen
The Netherlands Cancer Institute, Plesmanlaan 121, Amsterdam

In an attempt to use peroxidase labeled peanut lectin (PNA) as a marker for evaluation of mammary lesions with light microscopy we have found (1), in contrast to other groups (2, 3) that the reaction was not confined to the tumor cells but that also parts of the stroma reacted. Therefore we started an ultrastructural study to localize the PNA receptors on the different types of cells.

With light microscopy much work has already been done on PNA receptors in human breast carcinoma and a correlation has been found between the amount of PNA receptors and the degree of differentiation, judged by histologic criteria (2, 3). In this ultrastructural study we have not paid attention to the histological degree of differentiation of the tumors studied, because it is difficult to assess the degree of differentiation based only on the small vibratome sections used here. We are aware of the limitations of the ultrastructural studies but the advantage is that PNA receptors can be localized more precisely on the different cell types of human mammary tumors.

Lectins, Vol. III

Materials and Methods

Tissues.

Seventeen human mammary carcinomas were studied of the following histological types according to our pathology department: 5 adenocarcinomas; 2 infiltrating lobular carcinomas; 5 infiltrating ductal carcinomas; 3 carcinomas solidum, and 2 medullary carcinomas. Fresh tissue was fixed immediately in 4% paraformaldehyde and 0.5% glutaraldehyde in 0.1 M cacodylate buffer, pH 7.2 for at least 18 hrs. In a preliminary study using this fixative and periodate-lysine-paraformaldehyde fixative we have found that preservation of the PNA receptors was the same with both fixatives, while the preservation of the cellular structure was better with the first one. Fixed pieces about 3 mm thick were mounted on corks with Loctile brand cyanocrylate ester and sectioned with an Oxford vibratome. Sections about 20 µm in thickness were cut and soaked for 1 hr in 0.1 M lysine in PBS to block non-specific binding of subsequent reagents to cross-linked aldehydes. For the incubations the vibratome sections were floated on a drop of the reagents and kept in a moist chamber. The specimens were washed at least 3 times with PBS between incubations. The labeling was performed at room temperature with the following procedure: 1 hr incubation with peroxidase-labeled PNA (E.Y) diluted 20 µg/ml PBS; secondary fixation, 15 min in 2.5% glutaraldehyde in 0.1 M cacodylate buffer pH 7.2; neutralization 10 min in 300 mM glycine pH 10 to prevent non-specific oxidation of the diaminobenzidine used to detect peroxidase (4); incubation 30 min with diaminobenzidine 0.5 mg/ml and hydrogen peroxide 0.003%; postfixation 30 min in 1% osmium tetroxide in 0.1 M cacodylate buffer pH 7.2. The sections were dehydrated and embedded in a mixture of Epon and Araldite. Thin sections were not always counterstained with uranyl acetate and lead hydroxide. This allows easier detection of the diaminobenzidine reaction product. The preparations were examined with a Philips electron microscope EM 300 or EM 301.

For light microscopy 3-5 mm thick pieces of tissue were fixed embedded in Histowax (Histo Lab, Sweden) and cut in 5-6 µm sections. Histowax was removed from the sections by washing in xylene and in ethanol. Endogenous peroxidase was blocked by immersing the sections in PBS + 1% H_2O_2 for 10 min. The sections were then incubated with peroxidase-labeled PNA 5 µg/ml and 2 µg/ml, followed by incubation for 10 min with a solution containing 0.5 mg diaminobenzidine/ml and 0.003% H_2O_2.

Controls.
In all experiments one section was incubated with peroxidase-labeled PNA mixed with 2% galactose as control. The following control experiment was also done: incubation of the sections first with unlabeled PNA 100 µg/ml followed by incubation with peroxidase labeled PNA 20 µg/ml. All controls were always negative.

Results

No correlation was found between the presence of PNA receptors and the histological type of the tumor. Therefore the results will be presented without referring to the type of the tumor. In order to know whether the results, based on the incubation of only one or two vibratome sections from a small piece of each tumor, were representative, we have incubated vibratome sections from at least three different parts of three carcinomas and did not observe variations between the different sections of each carcinoma. Though the results obtained by electron microscopy and by light microscopy correlated well, electron microscopy seemed to be the more sensitive method, partly due to the fact that PNA incubation for electron microscopy was done prior to the embedding. On the other hand, we have found that for light microscopy, the paraformaldehyde-glutaraldehyde fixation used for electron microscopy seems to reduce slightly the reactivity with PNA receptors compared to the fixation in buffered formalin as used in routine histology.

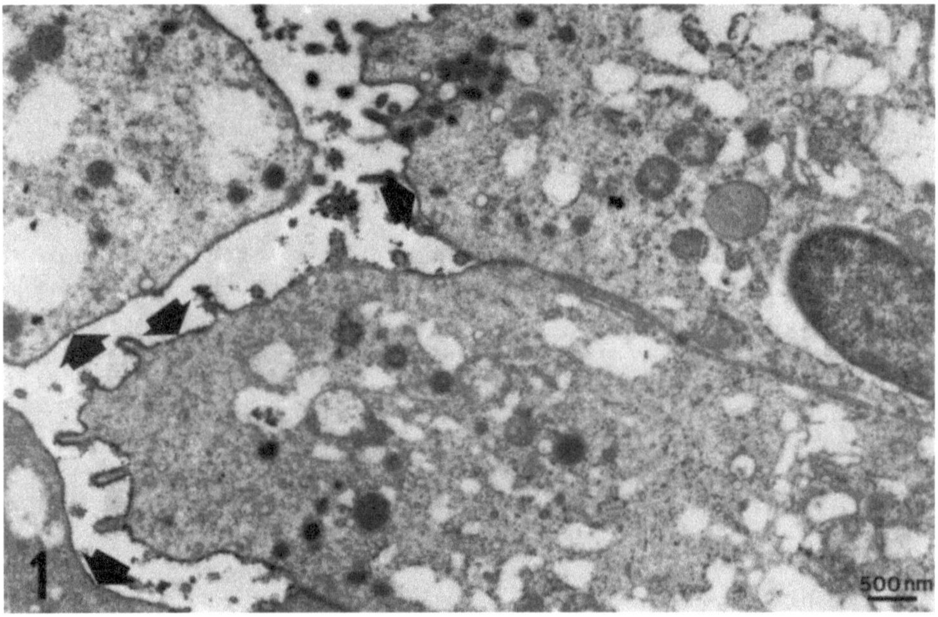

Figure 1. Infiltrating ductal carcinoma. Peroxidase reaction product on the luminal surface (arrows) of the tumor cells. Thin section stained with uranyl acetate and lead hydroxide.

Tumor cells.

Five carcinomas showed no evidence of PNA receptors on the tumor cells; the others showed reaction in a few areas of the tumor. In these positive areas the reaction was located mainly on the luminal surface of the cells (Fig. 1). However, in two positive carcinomas solidum the binding of PNA was more intense on the plasma membranes of the outermost tumor cells of the solid areas (Fig. 2). In one of these tumors the basal lamina was also stained (Fig. 2). This was, however, not always the case: in two other tumors of which the tumor cells were negative, the basal lamina were also negative.

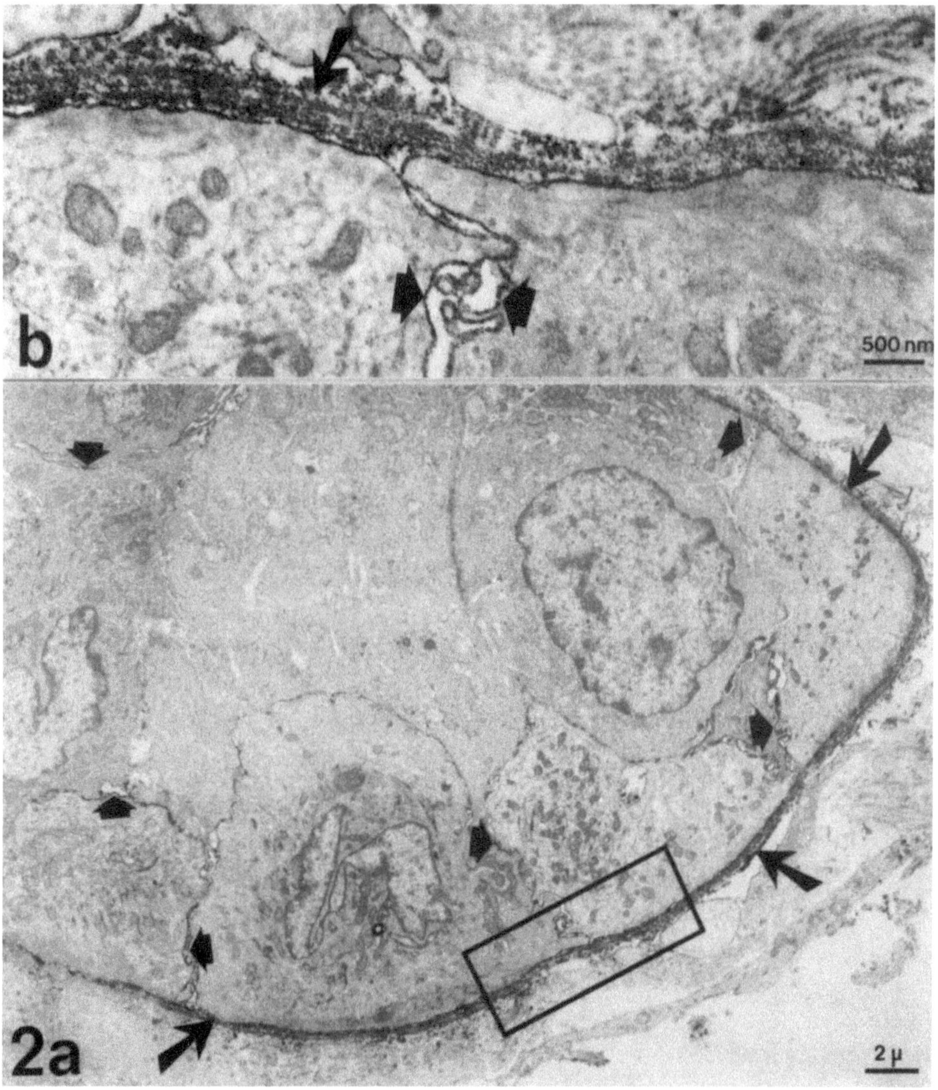

Figure 2. Carcinoma solidum. Peroxidase reaction product on the surface of the peripheral tumor cells (thick arrows). Basal lamina is also positive (thin arrows). b) higher magnification of the marked area in a. Unstained thin section.

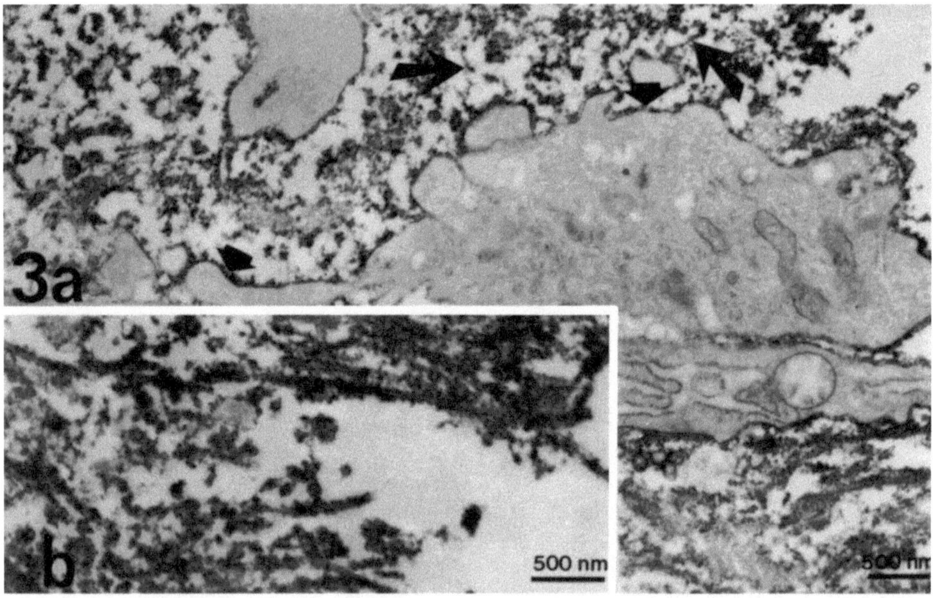

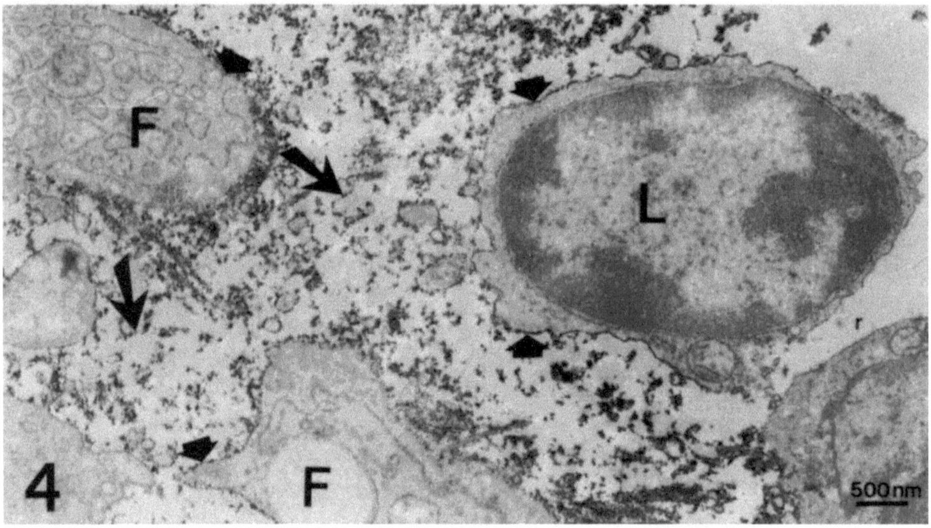

Figure 3. Carcinoma solidum. a) peroxidase reaction product on the surface of fibroblasts (thick arrows). Fibrils are also decorated (thin arrows). b) higher magnification of an area containing labeled collagen bundles. Unstained thin section.

Figure 4. Infiltrating lobular carcinoma. Thin section through the stroma showing labeling (thick arrows) on the surface of a lymphocyte (L) and on the surface of fibroblasts (F). Fibrils (thin arrows) are also positive. Unstained section.

Stroma.

In nearly all carcinomas studied, a positive reaction was found in the stroma. In general this reaction was stronger in those tumors where the tumor cells were negative; this was also observed with light microscopy. Positive reaction was found on the surface of fibroblasts, along collagen bundles and also decorating the interstitial fibrils (Figs. 3 & 4). In the stroma, as it was the case with the tumor cells, the different tumors showed different patterns of reaction both in intensity and in extension, i.e., in some tumors almost all fibroblasts and connective tissue was strongly positive, whereas in the stroma of other tumors only some areas were stained.

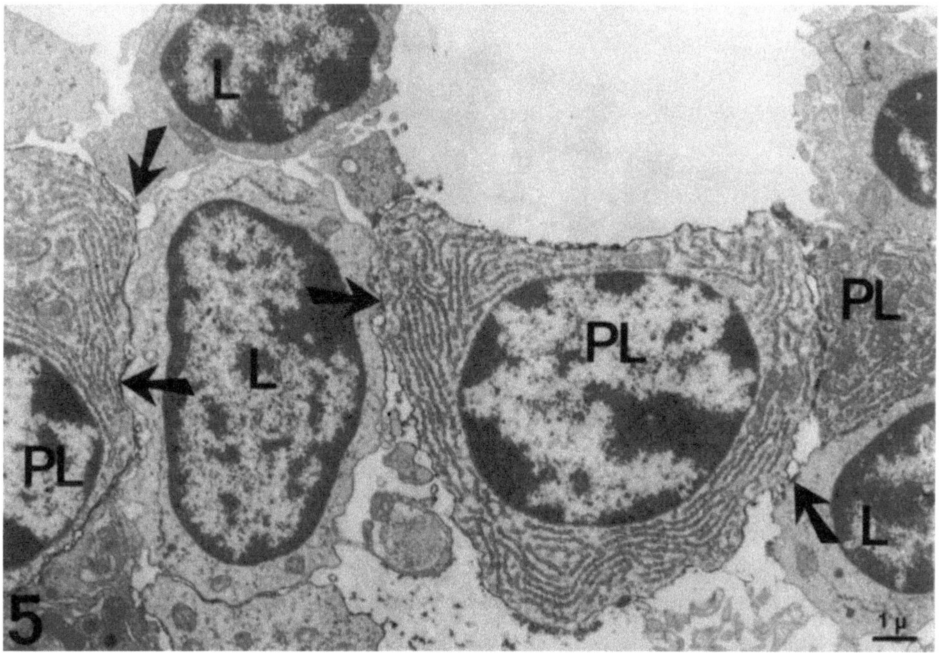

Figure 5. Lymphoid infiltrate in the stroma of a medullary carcinoma. Peroxidase reaction product all around the surface of plasma cells (PL). Lymphocytes (L) are positive on that part of their surface that is in contact with a plasma cell (arrows). Section stained with uranyl acetate and lead hydroxide.

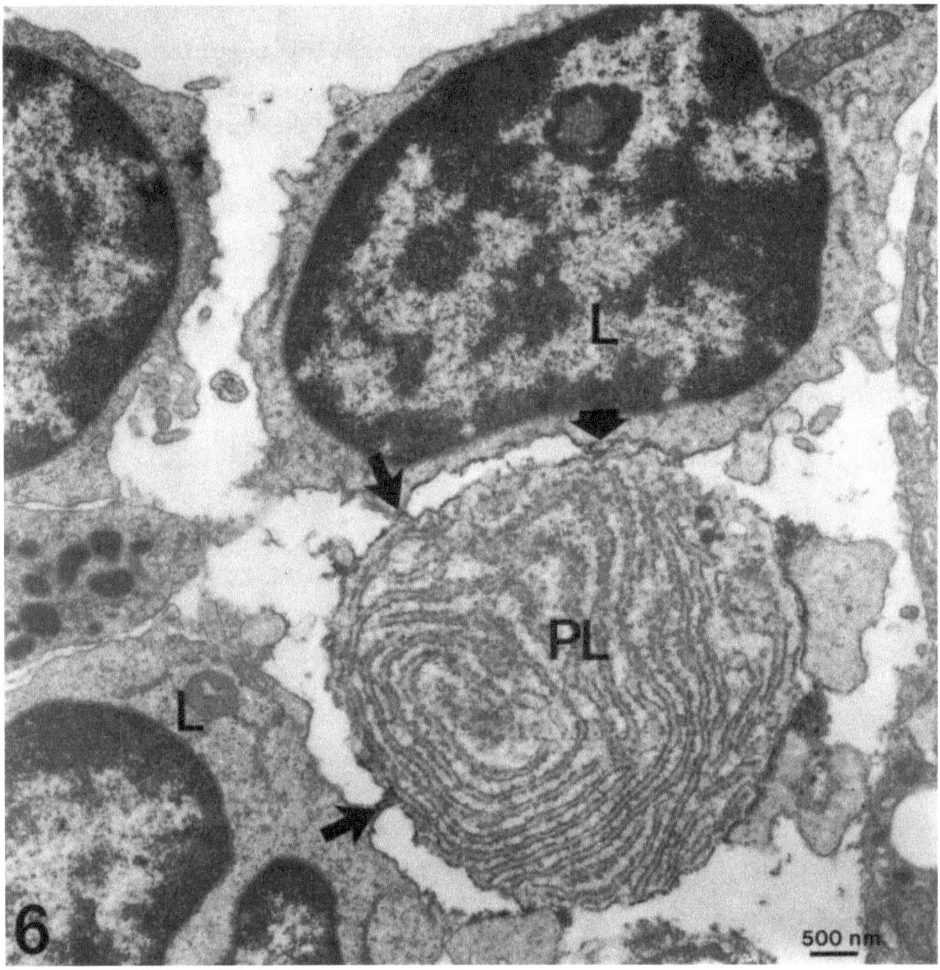

Figure 6. Thin section through the same specimen as in Fig. 5. Higher magnification showing the surface contact (arrows) between a plasma cell (PL) and lymphocytes (L).

In several carcinomas, positive lymphocytes were found in areas with a strong reaction in the stroma (Fig. 4). In one medullary carcinoma with prominent lymphoid infiltration, many plasma cells were present showing abundant PNA receptors on the surface (Figs. 5 & 6). Some of the otherwise negative lymphocytes in this infiltrate demonstrated a positive reaction on that part of their cell membrane that was in close contact with a plasma cells (Figs. 5 & 6).

In an infiltrating ductal carcinoma with a peripheral lymphoid infiltrate, abundant plasma cells with PNA receptors were shown.

Discussion

Newman et al. (2) and Howard et al. (3) have found in their histological studies that in differentiated mammary carcinomas PNA receptors were located mainly on the luminal cytoplasmic membrane and that undifferentiated tumors generally had no receptors. In the few positive undifferentiated carcinomas the reaction was often localized in the cytoplasm of the tumor cells. As mentioned above we have not classified the tumors according to the histological grading given by Bloom and Richardson (5). However, in our study the binding of PNA on tumor cells did not correlate with presence of lumina. In some adenocarcinomas no reaction was detected on the luminal surface of the cells, whereas two carcinomas solidum did show a positive reaction all over the plasma membrane of the outermost tumor cells of the solid areas, and light microscopy showed PNA receptors on the membrane of tumor cells lying in "Indian file" in the stroma.

In the studies of Newman et al. (2) and Howard et al. (3) with human mammary tumors, binding of PNA was only described on tumor cells. However, in nearly all carcinomas studied here the stroma was positive. As it was the case with the tumor cells, also in the stroma no correlation was found between PNA binding and the histological type of the tumor. This positive reaction in the stroma is not surprising, because glycoproteins are important constituents both of cell membranes and of the intercellular matrix. PNA binds to terminal non-reducing galactose residues often present in glycoproteins (6). Collagen type I is present in the stroma of mammary tumors and synthesized by the fibroblasts, and collagen type IV, a component of the basal lamina and synthesized by the myoepithelial and epithelial cells, are glycoproteins. Moreover, extracellular glycoproteins form a link between cells and collagen, e.g. fibronectin is involved in the attachment of fibroblasts to type I collagen and laminin is an attachment factor between epithelial cells and type IV collagen (7). Therefore PNA binding can be expected when these glycoproteins contain terminal non-reducing galactose.

The pictures shown in this study demonstrate that the peroxidase reaction product can be localized on the surface of the tumor cells and of the fibroblasts, probably detecting glycoproteins that are produced by these cells. We are not sure if the positive reaction on the basal lamina and along the collagen fibers in the stroma represents reaction of the collagen and/or of the proteoglycans in which the collagen is embedded.

Until now we do not know the biological significance of the presence or absence of the PNA receptors in these structures. Howard et al. (3) suggested that the T antigens that can be detected by PNA binding, have a role in the modulation of the host immune response to breast carcinoma. In their study they have found that in benign breast tissues and in well differentiated areas of neoplastic tissue, T antigen is located on the luminal membrane and they considered it "outside" the body or in an immunologically privileged site. They interpreted that the absence of the antigen in the poorly differentiated and in the more malignant carcinomas is associated with a failure in recognition of the tumor as foreign and thus the inability to mount an effective immune reaction. Our results show the presence of PNA receptors in the stroma. In this case the receptors can influence the host immune response by blocking of the antibodies directed against the T antigens from the tumor cells.

Lymphocyte infiltration is quite common in human breast cancer. Using immunohistochemical techniques Schoorl et al. (8) have shown that in nearly all human mammary carcinomas studied, T cells were found to predominate in and around nests of tumor cells. Moreover, Rose (9) has suggested that the lectin-binding properties of lymphocytes may reflect their tendency to recirculate or to be sessile. Therefore the PNA positive lymphocytes found in the stroma and in lymphoid infiltrate are probably sessile T cells. In two carcinomas plasma cells were found with PNA receptors. This confirms the observation of Rose (9).

Striking was the observation that the presence of PNA receptors on the lymphocytes was either restricted to that part of the surface that was in close contact with positive plasma cells or present all over their surface when they were situated in a strongly positive stroma. A direct contact with their surroundings might be a cause for the positive reactions on these lymphocytes.

PNA binding is supposed to be a marker for immature T cells (10). Though Rose et al. (11) did not find PNA receptors on in vitro stimulated mouse lymphocytes, she observed PNA binding in lymphocytes in germinal centers, both of mice and men (12, 13). Expression on activated lymphocytes of markers that are typical for immature lymphocytes, is also found in the T antigen system. The human thymic cell-surface antigen T_{10} is present on immature T cells, is not expressed on cells in the peripheral blood but reappears on stimulated human lymphocytes (14). Whether the presence of PNA receptor on the surface of lymphocytes in contact with plasma cells reflects a state of activation needs further confirmation.

Conclusion: 1) No correlation was found between the presence of PNA receptors and the histological type of the tumor. 2) The stroma was nearly always positive. 3) Plasma cells when present were strongly positive, and 4) in the lymphocyte infiltrates some positive lymphocytes were found situated in a strongly positive stroma or in close contact with positive plasma cells. These results show that the indicating value of the presence of PNA receptors in human mammary tumors as marker for evaluation of mammary lesions or to measure the production of T antigens is more complex than thought until now.

Acknowledgment

We thank Dr. C.A. Feltkamp for critical reading of this manuscript; Mr. N. Ong for excellent technical assistance and Miss M.A. van Halem for typing the manuscript. We thank also the department of Pathology for providing the material, specially Mr. J. Verrijk.

References

1. Hageman, Ph.C., Bobrow, L., Van der Valk, M.A., Misdorp, W.: Lectins. Biology, Biochemistry, Clinical Biochemistry. Volume 3. T.C. Bøg-Hansen and G.A. Spengler, eds. Walter de Gruyter, Berlin, New York, 1982, in press.
2. Newman, R.A., Klein, P.J., Rudland, P.S.: J. natl. Cancer Inst. 63, 1339-1345 (1979).
3. Howard, D.R., Ferguson, P., Batsakis, J.G.: Cancer 47, 2872-2877 (1981).
4. Willingham, M.C., Yamada, S.S.: J. Histochem. Cytochem. 27, 947-960 (1979).
5. Bloom, H.J.G., Richardson, W.W.: Br. J. Cancer 11, 359-377 (1957).
6. Sharon, N., Lis, H.: Chem. Engineering News 59, 21-44 (1981).

7. Kleinman, H.K., Klebe, R.J., Martin, G.R.: J. Cell Biol. 88, 473-485 (1981).

8. Schoorl, R., Brutel de la Rivière, A., Von dem Borne, A.E.G.Kr., Feltkamp-Vroom, T.M.: Am. J. Pathol. 84, 529-544 (1976).

9. Rose, M.L.: Immunology Today 3, 6-7 (1982).

10. Reisner, Y., Bibiaminov, M., Rosenthal, E., Sharon, N., Ramot, B.: Proc. natl. Acad. Sci. USA 76, 447-451 (1979).

11. Rose, M.L., Birbeck, M.S.C., Wallis, V.J., Forrester, J.A., Davies, A.J.S.: Nature 284, 364-366 (1980).

12. Rose, M.L., Malchiodi, F.: Immunology 42, 583-591 (1981).

13. Reisner, Y., Kapoor, N., Good, R.A., O'Reilly, R.J.: Lectins. Biology, Biochemistry, Clinical Biochemistry. Volume 3. T.C. Bøg-Hansen and G.A. Spengler, eds. Walter de Gruyter, Berlin, New York, 1982, in press.

14. Terhorst, C., Van Agthoven, A., LeClair, K., Snow, P., Reinherz, E., Schlossman, S.: Cell 23, 771-780 (1981).

ENDOCRINE THERAPEUTICAL STUDIES ON CARCINOGEN-INDUCED RAT MAMMARY CARCINOMAS CHARACTERIZED BY HORMONE-DEPENDENT LECTIN-BINDING SITES.

Mathias Vierbuchen, Peter Josef Klein, Siegfried Rösel

Institute of Pathology University of Cologne
D 5000 Cologne, Federal Republic of Germany

Janos Fischer

Humboldt-Fellow from Institute of Pathology, Med. School Pecs
Hungary

Gerhard Uhlenbruck

Department of Immunobiology University of Cologne
D 5000 Cologne, Federal Republic of Germany

Endocrine therapy represents an important concept for the management of metastazing breast cancer. About one third of disseminated breast cancers responds to alterations in hormonal balance. Until now the most useful way of identifying this group has been the measurement of the estrogen receptor (ER) content. In most studies about 70 % of the mammary carcinomas contain ER at a wide range of concentrations. However, only half of the estrogen-receptor-positive tumors respond to endocrine therapy (1), owing to defective hormone receptor mechanisms. Therefore, the demonstration of an estrogen induced posttranscriptional product may be useful for the selection of tumors for endocrine therapy. Secretory activities obviously represent a hormonally induced process in breast tissue which is accompanied by the formation of lipids, carbohydrates, proteins, and glycoproteins. Recent histochemical studies with labelled lectin from Arachis hypogaea (peanut agglutinin, PNA) on rat mammary gland have revealed that the expression of this lectin receptor which is represented by the disaccharide residue ß-D-Gal(1-3)NAcGal (2, 3) depends on the presence of estrogen and polactin (4). Furthermore, comparative studies on rat mammary gland have shown that this lectin receptor represents a marker of mammary differentiation (5). In addition, the diagnostical relevance of this lectin receptor is emphasized by its occurrence in normal and neoplastic human breast tissue (6, 7).

The growth of certain mammary tumors in rodents is hormone dependent and provides possible models for hormone dependent breast cancers in women. In this context many studies have been performed on 7,12-dimethylbenz(a)anthracene (8) induced rat mammary tumors. Recently, N-nitrosomethylurea (NMU) was used to induce mammary carcinomas in rats that would constitute a better model for human mammary carcinomas than other carcinogen induced tumors (9). Therefore, the present studies were performed on N-nitrosomethylurea induced rat mammary carcinomas to investigate the occurrence and distribution of PNA binding sites. Furthermore, the effects of estrogen administration, as well as hormonal ablation by ovariectomy or by treatment with the estrogen antagonist tamoxifen on the expression of PNA binding sites were analyzed. Finally the amount of secretory activity with positive PNA staining was quantified and correlated with tumor response to ovariectomy or tamoxifen treatment in order to evaluate the predictive significance of this lectin receptor for endocrine therapy.

Materials and Methods

Sources of the materials used were: Fluoresceine isothiocyanate labelled peanut agglutinin (FITC-PNA), unlabelled PNA, anti PNA antibodies (from rabbit) Medac (Hamburg, F.R.G.); peroxidase labelled anti-rabbit immunoglobuline antibodies (from swine) Dako (Denmark); bovine serum albumine, neuraminidase (vibrio cholerae, 1U/ml) Behringwerke (Marburg, F.R.G); N-nitrosomethylurea, ß-estradiol 3-benzoate Sigma (München, F.R.G.); tamoxifen ICI Pharma (Planckstadt, F.R.G.). Lectins and antibodies were dissolved in PBS (phosphate buffered saline, 0.01 M phosphate buffer pH 7.4 containing 0.15 M saline). Tamoxifen and ß-estradiol 3-benzoate were dissolved in sesame oil/ethanol 9/1. N-nitrosomethylurea was wetted with 3 % acetic acid and dissolved in an aqueous solution to a final concentration of 10mg/ml.

Rat mammary tumor induction with N-nitrosomethylurea was performed on 50 day-old female Sprague Dawley rats according to the procedure of Gullino et al. (9). Twice a week the animals were placed under light ether anesthesia and tumor size was measured with a caliper. Tumor volumes were calculated from the maximal tumor diameter and the perpendicualar diameter. Ovariectomy and portions of the tumors (before treatment) were surgically removed under ether anesthesia. The animals were divided into three groups which were subjected to ovariectomy (20 animals with 40 biopted tumors), tamoxifen treatment (20 animals with 51 biopted tumors), and estrogen treatment (7 animals with 15 biopted tumors). Tamoxifen (500 ug/animal) and ß-estradiol 3-benzoate (50ug/animal) were given by daily s.c. injections. After a three week treatment period the tumor responses were categorized as follow: Regression-decrease of initial tumor volume of more than 30 %; static growth ± 30 % of the initial tumor volume; progressive - increase of more than 30 % of the initial tumor volume. At the end of the treatment period which ranged in individual cases from (at least) 3 weeks to six month the animals were killed and another tumor sample taken, to analyze the therapeutic affects on tumor morphology and PNA binding sites. Tumor samples were fixed in 6 % neutral buffered formaldehyde. Sections were cut at 6 µ from paraffin blocks and stained with haematoxylin and eosin, for fat, as well as with the periodic acid Schiff's reaction (PAS-reaction).

Staining for free as well as sialic acid substituted PNA binding sites was in general performed by the use of FITC labelled PNA as described previously in detail (10). Twenty fields at high magnification (500x) were counted for positive stained secretion under the fluorescence microscope in order to quantify the amount of secretion associated PNA binding sites (secretory PNA index) for each tumor sample taken before therapy. Beside the direct fluorescence technique an indirect immunoperoxidase technique was used to demonstrate lectin binding sites as follow: 1. Treatment of deparaffinized rehydrated tissue sections with 0.5 % H_2O_2 to inhibit endogenous peroxidase followed by thorough washing with PBS. 2. Incubation with neuraminidase (10-20 mU/tissue section) at 37°C. Floating of tissue sections with 1 % bovine serum albumine (10 min) to reduce non-specific background staining. 4. Removal excess of bovine serum albumine. 5. Incubation with unlabelled PNA (1-2 µg/tissue section). 6. Incubation with rabbit anti-PNA antibody (dilution 1 in 50). 7. Incubation with peroxidase labelled swine anti-rabbit immunoglobuline antibodies (dilution 1 in 100). 8. Localization of the site of peroxidase by the 3-amino-9-ethyl-carbazole hydrogen peroxide reaction according to Graham et al. (11). 9. Weak counterstaining with haematoxylin and mounting in glycerol jelly. After incubation step 2 and between incubation steps 5 to 7 sections were washed in several changes of PBS. Unless otherwise stated incubations were performed at room temperature for 30 min. For the demonstration of free PNA binding sites incubation step 2 was omitted.

Results

Almost all of the tumors induced by NMU had the structures of adenopapillary carcinomas but differentiation varied from tumor to tumor and even from area to area in the same tumor. The epithelium of the acini and papillary structures varied inshape from columnar to flattened cuboidal and in thick ness from one to several layers (Fig. 1). Stroma between the tumor cell mas-

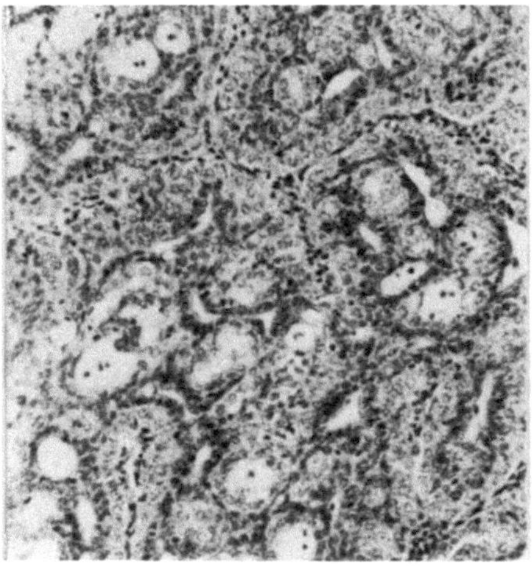

Fig. 1
N-nitrosomethylurea induced rat mammary adenocarcinoma; PAS reaction

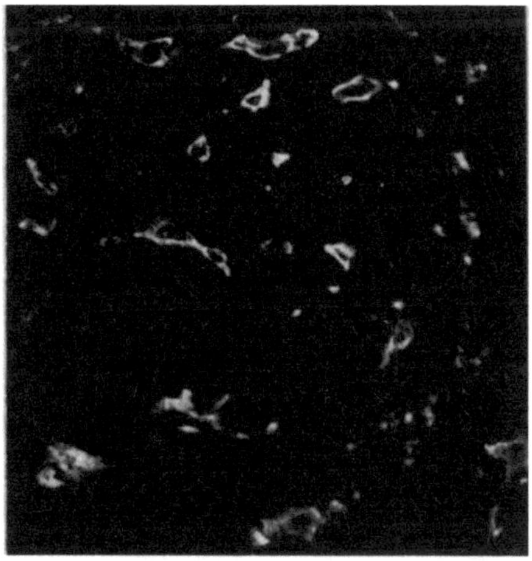

Fig. 2
Same tumor as in Fig. 1; demonstration of PNA binding sites at the apical surface of acinar and tubular structures, as well as in secretions (direct fluorescence technique)

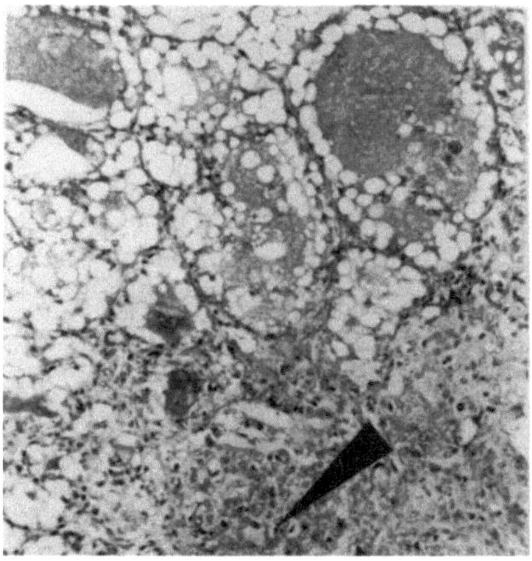

Fig. 3
Tumor of Fig. 1 after estrogen treatment producing extensive epithelial vacuolization and stop of proliferation; unaffected proliferating tumor cell complexes (PAS reaction)

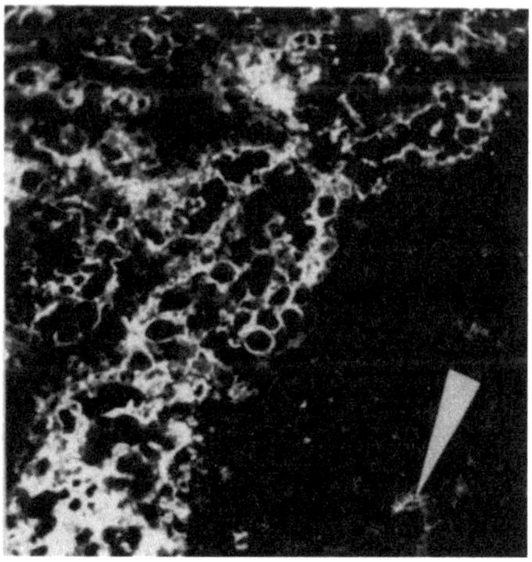

Fig. 4
Tumor of Fig. 3; estrogen induced secretory activities with positive PNA staining (direct fluorescence technique) unaffected tumor cell complexes possessing little PNA positive secretion

ses was scanty and contained various amounts of inflammatory cells. Binding sites for the PNA were detected in the luminal cytoplasm of tumor cells lining tubular and acinar structures (Fig. 2). In addition the secretion within some glandular structures gave positive PNA staining (Fig. 2). In both cases free as well as sialic acid substituted PNA binding sites. However, not all glandular structures and secretions were equally positive for PNA staining. There was a continuum from negative to very strong areas within the same tumor. In general neuraminidase treatment of tissue sections increased the number of PNA binding sites. In this respect PNA binding sites of the whole cell membrane were only observed after neuraminidase pretreatment of tissue sections.

Arrest of neoplastic growth after estrogen treatment resulted in secretion like changes with formation of intracytoplasmatic vacuoles (Fig. 3) which stain positively for fat. Beside the induction of lipid biosynthesis, estrogen treatment stimulated the formation and secretion of free as well as sialic acid substituted PNA binding sites (Fig. 4). However no entire tumor was affected uniformly by this hormone treatment and areas could be found where the epithelium was unaffected and continued to grow (Fig. 3). Beside high proliferative activity, these hormone independent tumor cell complexes contained little or no secretion associated PNA binding sites (Fig. 4).

In contrast to estrogen treatment, ovariectomy and tamoxifen administration induced regressive epithelial changes which resulted in the formation of cystic spaces (Fig. 5). With increasing time regressive changes were associated by the formation of abundant fibrous stroma. During tumor regression after ovariectomy or tamoxifen treatment the epithelial atrophy was associated with a loss of secretory activity with positive PNA staining (Fig. 5), indicating an arrest of PNA receptor biosynthesis. Chart 1 shows the growth records of two tumors which developed in the same animal but exhibited a different response to anti-estrogenic treatment with tamoxifen. The tumor which

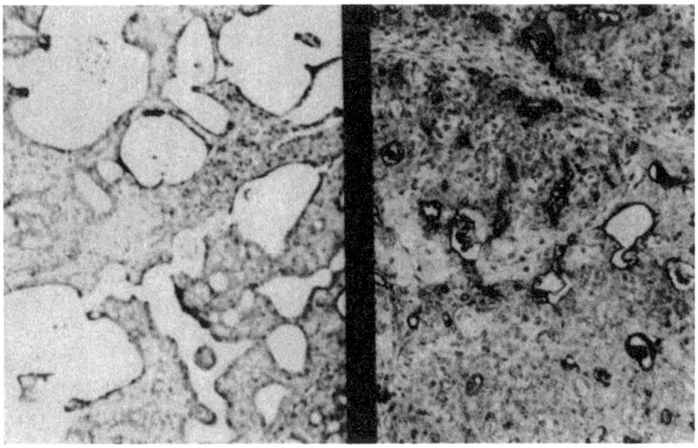

Fig. 5
Left side: biopsy sample from rat mammary adenocarcinoma before tamoxifen treatment with significant amounts of PNA binding sites at the apical surfaces of glandular structures and in secretions (indirect immunoperoxidase technique - weak counterstaining with haematoxylin); right side: biopsy from the same tumor after 3 weeks tamoxifen treatment causing regressive epithelial changes and loss of PNA staining (indirect immunoperoxidase technique - weak counterstaining with haematoxylin)

Table 1
Relation between the occurrence of secretion associated PNA binding sites (secretory PNA index) and response to ovariectomy or tamoxifen treatment

SECRETORY PNA INDEX	Tamoxifen group (N = 51) tumor response		
	progressive (N = 14)	static (N = 10)	regression (N = 27)
0 < 5	7(50.0%)	6(60.0%)	2(7.4%)
5 < 10	4(28.5%)	3(30.0%)	1(3.7%)
>10	3(21.5%)	1(10.0%)	24(88.9%)
SECRETORY PNA INDEX	Ovariectomy group (N = 40) tumor response		
	progressive (N = 8)	static (N = 4)	regression (N = 28)
0 < 5	5(62.5%)	1(25.0%)	1(3.5%)
5 < 10	2(25.0%)	1(25.0%)	1(3.5%)
>10	1(12.5%)	2(50.0%)	26(93.0%)

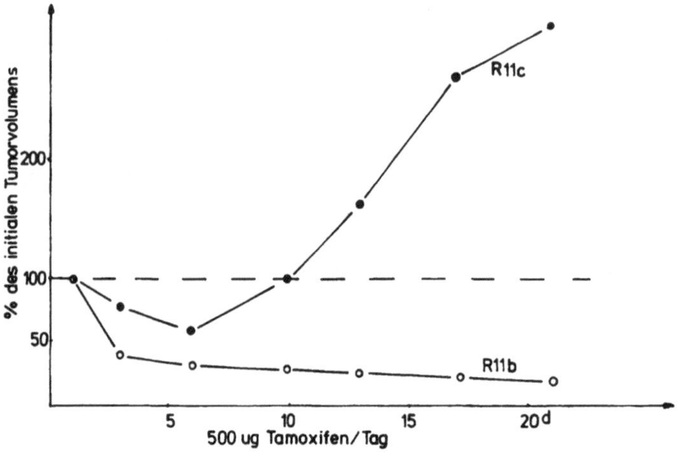

Chart 1
Growth records of two rat mammary carcinomas which developed in the same animal but exhibited different responses to antiestrogenic treatment with tamoxifen (R 11 b regressing tumor, R 11 c progressive tumor)

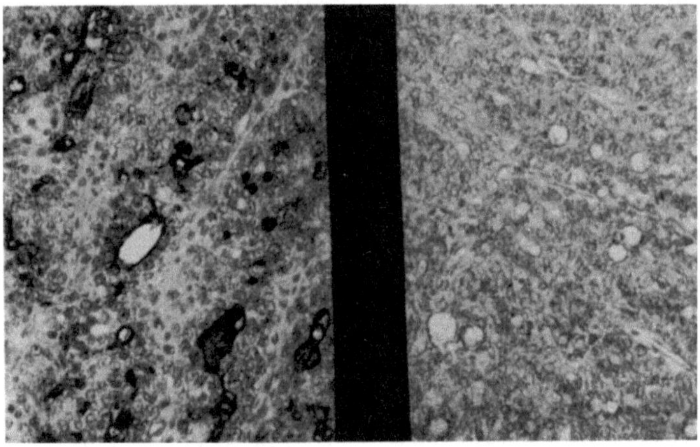

Fig. 6
Left side: biopsy sample (before treatment) from tumor R 11 b (s. chart 1) which possessed secretory activities with positive PNA staining; right side: biopsy sample (before treatment) from tumor R 11 c (s. chart 1) with no PNA positive secretion (indirect immunoperoxidase technique - weak counterstaining with haematoxylin)

regressed after tamoxifen administration (Fig. 6, R 11 b) contained significant amounts of PNA binding sites at the apical surfaces of glandular structures and within secretions. In contrast, the hormone independent tumor possessed no secretory activity with positive PNA staining (Fig. 6, R 11 c). Finally a greater number of tumors were analyzed for their content of secretion associated PNA binding sites. The values showed a continuum from very low to very high amounts. Tumors which responded to endocrine therapy possessed a greater amount of secretory activity with positive PNA staining than the hormone independent tumors (Table 1).

Discussion

N-nitrosomethylurea induced rat mammary carcinomas have been shown to possess binding sites for the lectin from *Arachis hypogaea* (peanut agglutinin, PNA). These lectin receptors occur at the apical cytoplasm of tumors cells lining glandular structures and within secretions as free and as sialic acid substituted ß-D-Gal(1-3)NAcGal residues of glycosubstances. In contrast the whole cell membrane in most tumors contains only sialic acid substituted PNA binding sites.

Inhibition of neoplastic growth after estrogen treatment was associated by marked secretory changes. Beside abundant formation of lipids, estrogens induced the biosynthesis and secretion of large amounts of PNA binding sites, which represent a marker of mammary differentiation (5). In this respect the formation of large amounts of PNA receptors indicates the transformation of the neoplastic tissue from a functionally undifferentiated state with high proliferative activity to a highly differentiated state with inhibition of mitotic activity which sometimes resembled lactating mammary gland.

On the other hand formation and secretion of PNA receptors was inhibited by ovariectomy which emphasizes the importance of

ovarian hormones for the expression of this lectin receptors in hormone dependent neoplastic mammary tissue. Furthermore, the inhibition of PNA receptor biosynthesis by tamoxifen which produces its effect by competitive binding to the estrogen receptor (12) indicates the involvement of the estrogen receptor mechanism in the induction of PNA binding sites in these mammary tumors. Based on these data one can conclude that the occurrence of PNA binding sites in this model of experimental breast cancer indicates an intact hormone receptor pathway as proposed for certain enzyme activities (13) or the progesterone receptor (14). However, some of the tumors possessed relatively high amounts of secretion associated PNA binding sites but did not respond to endocrine therapy. This may be explained by a dissociation of hormone dependent proliferative and secretory events. Growth of these tumors became hormone independent whereas secretory activities were still inducible by hormones. From the occurrence of tumor cell complexes in responding tumors which were unaffected by these endocrine manipulations it is reasonable to conclude that these tumors are mixed populations of hormone dependent and hormone independent cells. This concept is supported by the low secretory activities with positive PNA staining in unaffected areas.
In conclusion, these studies based on carcinogen induced rat mammary carcinomas have revealed that secretion associated PNA binding sites represent a useful marker distinguishing hormone dependent and hormone independent mammary tumors.

References

1. McGuire, E. L., Carbone, P. P., Sears, M. E., Escher, G. C.: In Estrogen Receptors in Human Breast Cancer (edited by McGuire, W. L., Carbone, P. P. and Vollmer, E. P.). p. 1 New York 1975

2. Uhlenbruck, G., Pardoe, G. I., Bird, G. W. G.: Z. Immunitaetsforsch. 138, 423 (1969)

3. Lotan, R., Skutelsky, E., Danon, D., Sharon, N.: J. Biol. Chem. 250, 8518 (1975)

4. Vierbuchen, M., Klein, P. J., Uhlenbruck, G., Fischer, R.: Tumordiagnostik 2, 235 (1981)

5. Newman, R. A., Klein, P. J., Rudland, R. S.: JNCI 63, 1339 (1979)

6. Klein, P. J., Newman, R. A., Müller, P., Uhlenbruck, G., Citoler, P., Schaefer, H. E., Lennartz, K. J., Fischer, R.: J. Cancer Res. Clin. Oncol. 93, 205 (1979)

7. Klein, P. J., Vierbuchen, M., Schulz, K. D., Würz, H., Citoler, P., Ortmann, M., Fischer, R.: Tumordiagnostik 2, 240 (1981)

8. Huggins, C., Grand, L. C., Brillantes, F. P.: Nature 189, 204 (1961)

9. Gullino, P. M., Pettigrew, H. M., Grantham, F. H.: JNCI 54, 401 (1975)

10. Klein, P. J., Newman, R. A., Müller, P., Uhlenbruck, G., Schaefer, H. E., Lennartz, K. J., Fischer, R.: Klin. Wschr. 56, 761 (1978)

11. Graham, R. C. Jr., Lundholm, U., Karnovsky, M. J.: J. Histochem. Cytochem. 13, 150 (1965)

12. Capony, F., Rochefort, H.: Mol. Cell. Endocrinol. 11, 181 (1978)

13. Salih, H., Flax, H., Hobbs, J. R.: Lancet 1, 1198 (1972)

14. Horwitz, K. B., McGuire, W. L.: J. Biol. Chem. 253, 2223 (1978)

THE CHARACTERIZATION OF MILK PROTEINS BY LECTINS AND ANTIBODIES IN HORMONE DEPENDENT BREAST CANCER

Peter Josef Klein, Mathias Vierbuchen

Institute of Pathology, University of Cologne, F.R.G.

Janos Fischer

Humboldt-Fellow from the Institute of Pathology, Med. School Pecs, Hungary

Graham Farrar, Gerhard Uhlenbruck

Department of Immunbiology of the Medical Clinic, University of Cologne, F.R.G.

Conventional histomorphology of breast cancer is based on cytological and structural criterias predominantly concerning nuclear changes and glandular differentiation. These parameters have been generally accepted to be of prognostic importance in mammary carcinoma, but they do not allow definite statements with respect to functional differentiation of tumour tissue. Therefore, for a better characterization of breast cancer tissue, additional parameters are needed which provide a more precise information on special functional processes with possible therapeutic relevance. In this context new histochemical methods may contribute to a better knowledge of the biology of breast cancer, especially with regard to tumour growth regulating mechanisms.

A possible approach to investigate the hormone dependence of breast cancer is the demonstration of peanut agglutinin (PNA) binding sites. We have found this lectin receptor to be mainly associated with secretion equivalents in normal as well as in pathologic conditions of human breast (3). Furthermore, experimental investigations have shown, that the expression of the PNA receptor is modulated by estrogen and prolactin (5, 8).

In the present paper the secretion associated PNA binding in breast cancer has been correlated with the response to endocrine therapy as well as with the steroid hormone receptor content of tumour tissue. Furthermore, the PNA receptor in mammary carcinomas will be proven as a constituent of a milk protein by immunohistochemistry. Finally, an antibody directed against human colostrum is additionally used to confirm the general potency of tumour cells for milk protein synthesis.

Material

Some tissue specimen of normal lactating breast and fibro-cystic disease have been exemplarily investigated. Furthermore, milk fat globules were isolated from human milk as is outlined in detail in the chapter methods and incubated with the FITC-labelled PNA. Additionally, in a retrospective study two groups of female patients with breast cancer were histochemically investigated. In the first group (n = 46) the histochemical findings were compared to the hormone receptor content in tumour tissue as has been estimated biochemically (9). In the second group (n = 47) the relationship between lectin-histochemistry and the response of the tumour tissue to endocrine treatment with anti-estrogen and -androgen as well as gestagen was analysed. In some cases cytotoxic therapy preceded endocrine treatment. From each patient one to five tissue samples were available that mainly included primary tumours besides loco-regionary recidives and axillary lymph node metastases. In some cases with fatal outcome of the disease metastases from different organs were investigated. The results of different tumour samples from one patient were summarized and graded from negative to strong positive as will be outlined in detail in the chapter methods. The response to endocrine therapy was classified according to the UICC rules.

Methods

For the histochemical investigations on formaldehyde fixed normal and benign as well as malignant breast tissue the FITC labelled lectin from peanut (Arachis hypogaea) which has a high affinity for D-galactosyl-(1-3)-N-acetyl-D-galactosamine (4, 7) was used. One tissue section from each block was prepared for demonstration of sialic acid-substituted PNA-binding sites and another slide for the visualization of free PNA-binding sites. Demonstration of free or sialic-acid substituted sites was performed as follows (2):

1. Tissue section was incubated with neuraminidase (Vibrio cholerae, Behringwerke, Marburg, F.R.G.) 10-20mU/section, in a moist chamber at 37°C for 30 min.
2. The desialylated tissue sections were then washed with PBS (phosphate buffered saline, pH 7,4).
3. Thereafter the slides were incubated with FITC-PNA (10-20 µg lectin/tissue section, obtained from Medac, Hamburg, F.R.G.) for 30 min. in the moist chamber.
4. The tissue sections were again washed with PBS and then examined under the fluorescence microscope (Zeiss, Epifluorescence, 435-490 nm spectrum filter, HBO 50 W mercury lamp).

For the demonstration of free PNA binding sites, the tissue sections were not treated with neuraminidase before incubation with FITC-PNA.

Beside FITC-labelled PNA, additionally, peroxidase conjugated PNA (Medac, Hamburg, F.R.G.) was used. Thereby, in the beginning of the incubation procedure the deparaffinised tissue sections were incubated with absolute methanol - 0,5 % H_2O_2 for 30 min at room temperature to block endogenous peroxidase activity, followed by thorough washing with PBS. The incubation steps 1 and 2 were identical to FITC-staining of tissue sections. Thereafter the slides were stained with 3-amino-9-ethyl carbazole (1), counterstained with haematoxylin, and mounted in glycerol jelly.

Tumour tissue exhibiting secretion-associated PNA binding sites in less than 10 % of tumour cells was graded as negative (0) with respect to hormone sensitivity. Breast cancer tissue with 10 to 30 % PNA positive tumour cells were judged as weak (+) positive. Tumours showing 30 to 60 % stained cells belonged to the moderate (++) graded neoplasms. Finally, mammary carcinomas demonstrating more than 60 % positive cells were graded as strong (+++) positive breast carcinomas.

For the isolation of milk fat globules fresh human milk was centrifugated with 3000 g for 15 min. The milk fat globules in the upper creamy layer were washed three times with physiological saline solution, and than incubated with neuraminidase (1 U/ML) at 30°C for 60 min. After washing (3x) thin smears were prepared for histochemical investigations.

In addition, to the histochemical studies with labelled PNA immunohistochemical investigations were performed with an antibody (from rabbit), which was directed against the protein portion of the glycoprotein exhibiting lectin binding sites. The isolation of this PNA receptor bearing glycoprotein and the induction of an antibody against the fraction will be published elsewhere (Farrar et al. in preparation).

For the immunohistochemical studies on breast carcinomas not only this antibody against a definite component of human milk has been employed, but also the tumours were incubated with an anticolostrum antibody (DAKO, Denmark). Thereby, the peroxidase-antiperoxidase (PAP) technique from Sternberger (1979) was used; working dilutions of antibodies were prepared with PBS: antimilk fat globule membrane glycoprotein from rabbit (1:100), anti-colostrum from rabbit (1:50), anti-rabbit -globulin from swine (1:20), PAP-complex from rabbit (1:100).

Results and discussion

Apical surfaces as well as secretion within glandular formations of lactating breast exhibited a high amount of PNA binding sites (Fig. 1) which were only slightly increased by pretreatment of tissue sections with neuraminidase. Thereby, PNA was predominantly bound to the milk fat globule membrane as was ascertained in smear preparations of those globular components from human milk.

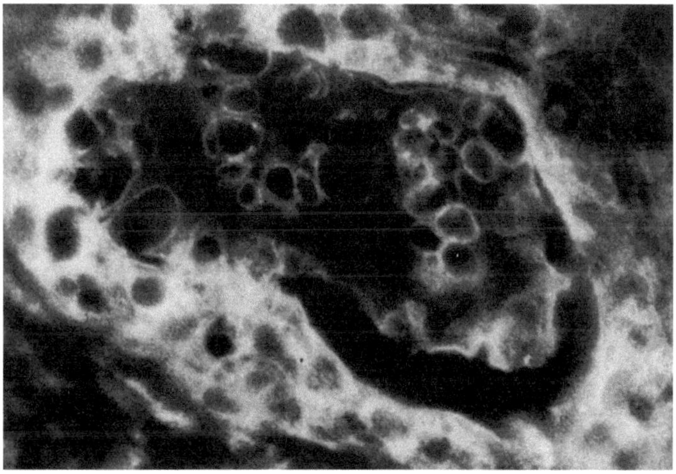

Fig. 1
Human lactating breast tissue. Demonstration of PNA binding sites on apical surface as well as in secretion within glandular structure by fluorescence microscopy.

The antibody directed against the milk protein that was isolated by PNA chromatography nearly showed the same distribution pattern as it was seen with the labelled lectin.

In contrast to the hormone stimulated lactating female breast the mammary gland of non pregnant women showed a reduced staining with labelled PNA, whereby the cells of the lobules and ducts (Fig. 2) expressed both free as well as sialic acid substituted PNA binding sites, which were restricted to the luminal surfaces.

As a consequence of hormonal dysregulation PNA binding in fibrocystic disease of the breast varied in wide ranges and was generally related to secretory conditions.

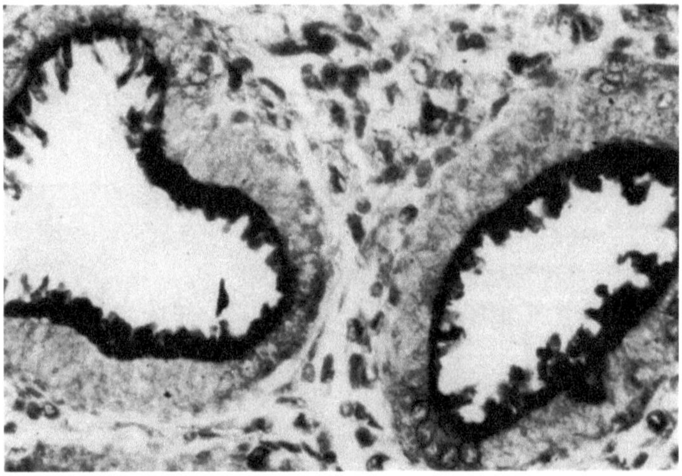

Fig. 2
Ductular structure in normal human female breast tissue. Strong staining of the luminal surfaces with peroxidase labelled PNA.

The staining pattern of breast carcinomas correlated to some degree with the morphological differentiation of the tumours. So differentiated mammary carcinomas showed in correlation to the secretory activity of the neoplastic tissue PNA receptors on apical surfaces and in secretions lying within glandular structures (Fig. 3 a). On the other hand the expression of lectin receptors to cytoplasmic vacuoles obviously represented secretory malfunction that was typically observed in carcinoma lobulare in situ (Fig. 3 b) and in infiltrating lobular carcinomas of the breast. Some times differences were observed between the lectin binding of primary tumours and metastases. Thereby, in some cases a minor component in the primary tumour occurred as predominant part in the metastases with respect to the lectin binding pattern. However, secretory activity was generally only slightly diminished in metastases when compared with the primary tumour. Additional studies with the antibody directed against the milk fat globule glycoprotein confirmed the PNA receptor as a carbohydrate component of a milk protein in breast carcinomas.

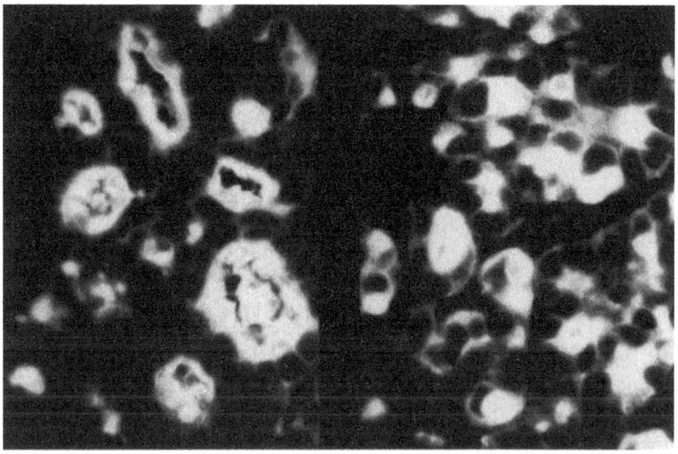

Fig. 3 a
Partly differentiated adenocarcinoma showing apical and secretion staining with FITC-PNA (left side). Partly vacuolar distribution of PNA binding sites in undifferentiated areas of the same tumour (right side).

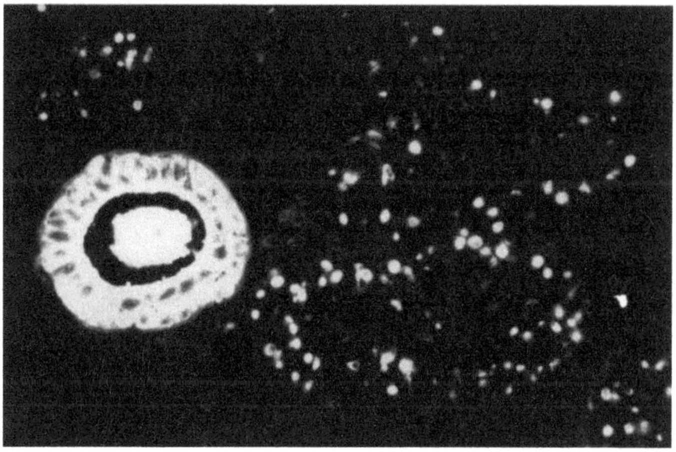

Fig. 3 b
Carcinoma lobulare in situ of the breast with typical vacuolar staining (FITC-PNA) of the cytoplasm. In addition ductular structure with staining of cytoplasm and cell membrane as well as of secretion.

Yet, in few cases negative staining of secretions with PNA contrasted with a positive antibody binding, which might be explained by disturbances in the glycosylation of the milk protein in these tumours. Furthermore, differences between the occurrence of PNA receptor and ligands for the antibody directed against colostrum also revealed some variations in the synthesis of milk proteins in mammary carcinomas in comparison to the great amount of glycoproteins synthesized in the lactating breast.

The expression of all free as well as neuraminic acid substituted PNA binding sites compared with the presence of hormone receptors determined biochemically showed a good correlation between the histochemical findings and the steroid hormone receptor content of mammary carcinomas (Table).

HISTOCHEMISTRY (Secretion)	N	ENDOCRINE THERAPY	
		RESPONSE*	NO RESPONSE
PNA pos.*	26	20(-77%)	6 (-23%)
PNA neg.**	21	3(14%)	18 (86%)

* > 10% positive tumour cells
** < 10% positive tumour cells
\+ = Complete and partial remission as well as stationary phase
PNA = peanut agglutinin

Table
Relation between the occurrence of secretory activity with positive PNA staining and tumour response to endocrine therapy.

These investigations revealed that tumours exhibiting both estrogen and progesterone receptors possessed secretion associated lectin binding sites in over 70 %. On the other hand, in breast cancers that failed to show steroid hormone receptors PNA binding sites were only seen in a small number of tumours. These findings were supported by clinicopathological results. These investigations revealed that breast cancers with more than 10 % PNA-positive tumour cells responded at about 77 % to endocrine treatment. In contrast mammary carcinomas with less than 10 % PNA-positive tumour cells showed only at about 14 % a positive therapeutic effect.

In conclusion, the lectin- and immunohistochemical investigations allow to distinguish between hormone dependent and hormone independent tumours. The advantage of the lectin-method over hormone receptor analysis is that it can be carried out on fixed tissue and on very small amounts of tumour tissue specimen. Thereby, not only a hormone induced milk protein is stained by the labelled lectins, but also disturbances in the biosynthesis and secretion mechanism of this glycoprotein can be demonstrated. Thus, the histochemical method provides a more detailed information about the functional state of mammary carcinomas as it can be predicted by hormone receptor analysis alone.

References

1. Graham, R. C., Lindholm, U., Karnovsky, M. I.: J. Histochem. Cytochem. 15, 150-152 (1965)

2. Klein, P. J., Newman, R. A., Müller, P., Uhlenbruck, G., Schaefer, H. E., Lennartz, K. J., Fischer, R.: Klin. Wschr. 56, 761-765 (1978)

3. Klein, P. J., Newman, R. A., Müller, P., Uhlenbruck, G., Citoler, P., Schaefer, H. E., Lennartz, K. J., Fischer, R.: J. Cancer Res. Clin. Oncol. 93, 205-214 (1979)

4. Lotan, R., Skutelsky, E., Danon, D., Sharon, N.: J. Biol. Chem. 250, 8518-8523 (1975)

5. Newman, R. A., Klein, P. J., Rudland, P. S.: JNCI 63, 1339

5. -1346 (1979)

6. Sternberger, L. A.: Immunocytochemistry (2nd ed.) John Wiley, New York, 1979

7. Uhlenbruck, G., Pardoe, G. I., Bird, G. W. G.: Z. Immunforsch. 138, 423-433 (1969)

8. Vierbuchen, M., Klein, P. J., Uhlenbruck, G., Fischer, R.: Tumordiagnostik 2, 235-239 (1981)

9. Wagner, R. K.: Hoppe-Seyler's Z. Physiol. Chem. 353, 1235-1245 (1972)

HISTOCHEMICAL AND BIOCHEMICAL CHARACTERIZATION OF GLYCOPROTEIN COMPONENTS IN NORMAL GASTRIC MUCOSA, INTESTINAL METAPLASIA AND GASTRIC CANCERS WITH LECTINS

Janos Fischer

Humboldt-Fellow from Institute of Pathology, Med. School Pecs Hungary

Peter Josef Klein, Mathias Vierbuchen, Robert Fischer

Institute of Pathology, University of Cologne, Germany

Gerhard Uhlenbruck

Department of Immunbiology, University of Cologne, Germany

Gastric epithelial cells are rich in secretory glycoproteins. Since most of the pathological alterations of the mucosa are accompanied by changes of this mucinous substances there is an increasing interest for the histochemical demonstration of glycosubstances in normal and neoplastic tissues. In this context conventional histochemical methods (PAS-reaction, Hale's reaction, alcianblue staining) revealed that the mature cell types of the gastric mucosa contain generally neutral mucin. On the other hand sialic acid substituted or sulfated glycoproteins were shown to be the main glycocomponents of the small and large intestine. These acid mucins normally absent in gastric mucosa were however often found in intestinal metaplasia and gastric cancers. (For a review see: 8). Biochemical studies demonstrated that the major component of the secreted mucus is a high mol. weight glycoprotein, the basic structure of which is common in all epithelial mucins. It has a well definied amino acid composition with large amounts of serine and threonine, to which the carbohydrate side chains are bound. The oligosaccharide chains (about 80 % of the molecule) are built up of fucose, galactose, galactosamine and glucosamine. They may be substituted by sialic acid or sulfate-groups and show some structural similarities to that of the blood-group determinant glycolipids (3, 13, 14). During the last years labelled lectins which exhibit a high affinity for individual carbohydrate structures have introduced in histochemistry of gastrointestinal tract (5, 6, 9, 12, 16, 17, 18) to demonstrate some distinct differences which are not demonstrable by the conventional methods. From this starting point, the aim of the present study was to compare the localization of lectin binding sites in normal, hyperplastic and malignant gastric tissue in order to get some informations regarding the histogenesis of intestinal metaplasia and gastric cancers. In addition the histochemical studies were supplemented by lectin-immunochemical analysis.

Material and Methods

The tissue samples used for the studies were derived from 70 surgical specimens of gastric resections or partial gastrectomies with the following diagnoses: ulcus (n = 27), chronic atrophic gastritis (n = 2), gastric polyp (n = 1) and gastric carcinoma (n = 40). Beside the macroscopically altered regions of the resection materials tissue blocks were always prepared from the surrounding cardial, antral, fundic and duodenal regions. 20 cases of them showing no micorscopical changes when investigated by the routine stainings were selected and used as controls. For comparative purposes 3 ileal, 3 coecal and 3 colonic biopsies and even specimens from 4 and 5 months old embryos were also analysed.

Lectin-histochemical investigations

Labelling of gastrointestinal glycoproteins was performed on formalin fixed and paraffin embedded tissue sections using either fluoresceinisothiocyanate (FITC) or peroxidase conjugated lectins (Medac, Hamburg). Cryptic, i.e. neuraminic acid substituted lectin binding sites, were demonstrated by preincubation of the tissue sections with neuraminidase (Vibrio cholerae), (Behring, Marburg) as described earlier (11). The lectins used and their carbohydrate binding specificities (19) are listed below:

Canavalia ensiformis (Con A)	D Man al-
Ulex europaeus I. (UEA I)	L Fuc al-2D Gal ß1-4D GlcNAc ß1
Lotus tetragonolobus (LTA)	L Fuc al-2D Gal ß1-4(Fuc al-3)-
Ricinus communis 120 (RCA I)	D Gal ß1-
Arachis hypogaea (PNA)	D Gal ß1-3 D GalNAc-
Helix pomatia (HP)	D GalNAc al-
Tritium vulgaris (WGA)	D GlcNAc ß1-4D GlcNAc ß1-4DGlcNAc

The terminal hexoses (underlined) were used in 0,2 M concentrations for the inhibition of lectin binding.

Biochemical procedures

For glycoprotein extraction tissue samples from fresh human antrum, corpus and gastric cancer were washed in saline to remove any loosely adherent material. The mucosal scrapings were dispersed in 10 vol. of buffer (10mM Tris, 1mM $CaCl_2$, 0,9 % NaCl, pH 7.3, containing 1 % Triton X-100, 2mM phenylmethanesulfohylfluoride (PMSF) and 1 % ethanol), homogenized in Potter glasteflon homogenizer. After centrifugation (20.000 g for 30 min.) the soluble glycoproteins were further purified by alcohol precipitation at 75 % ethanol concentration. The precipitate for formed in 1 h at 4°C was sedimented at 500 g for 15 min. and dissolved in extraction buffer. Protein content was determined using the modified method of Lowry (4).

Polyacrylamide gel electrophoresis in a linear gradient gel (10 to 20 %) was performed under reducing conditions (1 % 2-mercaptoethanol and 1 % SDS at 99°C for 5-10 min.) according to Anderson and Anderson (1). Gels were stained with Serva blue (Serva, Heidelberg) or alternatively by the periodic acid-Schiff reaction (7).

Ouchterlony double diffusion was made in 1 % agarose in the

same buffer as used for glycoprotein extraction.
For lectin-affinity chromatography PNA- and HP-agarose columns (Medac, Hamburg) were equilibrated and eluted with the extraction buffer. For Con A-agarose columns (Medac, Hamburg) 0,1M sodium acetate buffer, pH 6.0, containing 1M NaCl, 1mM $MnCl_2$, $MgCl_2$ and $CaCl_2$ were chosen. Therfore glycoprotein extracts were dialyzed against this buffer before application. The flow rate was maintained in every cases at 20 ml/h, and the specifically bound components were eluted with buffers containing 0,2 M D-Gal., D-GalNAc, or a-methyl-mannoside (Sigma, Munich).

Results

Lectin binding sites on normal human gastrointestinal cells

The characteristic lectin binding pattern of the different cell types of normal mucosa in the gastrointestinal tract are summarized in the table:

	CELL TYPES OF GASTRIC MUCOSA						INTESTINAL MUCOSA			
LECTINS	SURFACE MUCUS CELLS	ISTHMUS CELLS	MUCUS NECK CELLS	PARIETAL CELLS	CHIEF CELLS	ANTRAL CARDIAC GLANDS	BRUNNER GLANDS	COLUMNAR EPITHEL CELLS	GOBLET CELLS OF SMALL INTESTINE	GOBLET CELLS OF LARGE BOWEL
CANAVALIA ENSIFORMIS CON A	-	-	+ + + +	DIFFUSE	DIFFUSE	+ +	+ +	+ +	-	+ + +
ULEX EUROPEUS UEA I	+ + +	+ +	+ + +	5 + +	DIFFUSE	+ +	+ +	-	+	+ +
LOTUS TETRAGONOLUBUS LTA	-	+	+	-	-	+	+	-	-	-
RICINUS COMMUNIS RIC	13 + +	+ + + +	+	DIFFUSE	15 + +	MOSAIC LIKE	MOSAIC LIKE	GOLGI + + +	+ + +	+ +
ARACHIS HYPOGAEA PNA	-	GOLGI + + + +	+ + + +	11 + + +	10 + +	MOSAIC LIKE	MOSAIC LIKE	GOLGI + + +	+ +	-
HELIX POMATIA HP	14 + +	+ +	+ +	17 + + +	-	+ +	+ +	GOLGI + +	+ + +	-
TRITICUM VULGARIS WGA	11 + +	GOLGI + +	+ + +	-	-	MOSAIC LIKE	MOSAIC LIKE	-	+ +	+

In surface mucus cells the absence of Con A and PNA receptors was a very characteristic finding. PNA receptors could be de-

monstrated in the isthmus cells near to the golgi apparatus. The deep foveolar cells reacted predominantly with RCA I, but the amount of lectin binding gradually decreased towards the surface. Mucus secretions in the surface epithelial cells possessed high amounts of UEA I binding sites combined with variable staining intensities of HPA, WGA and RCA I lectins.
The cell types of the fundic glands differed from the surface epithelial cells by the appearance of Con A receptors within their cytoplasm. Besides the strong Con A labelling, mucus neck cells also bound high amounts of other lectins investigated (s. Table). The parietal cells were selectively marked on their invaginated membrane systems either with HPA and or with PNA and UEA I. The HPA and PNA binding was most pronounced in some cases after pretreatment of tissue sections with neuraminidase. Besides the diffuse intracytoplasmic Con A and UEA I staining chief cells showed a very intensive RCA I or/and PNA binding on the luminal membrane-surfaces of the cells.
The cells of cardiac, antral and Brunner glands showed quite similar lectin binding patterns in comparison to the mucus neck cells of the fundic region. It is remarkable that while Con A, HPA, UEA I and LTA receptors revealed a homogenous distribution in the whole gland, PNA, WGA and RCA I binding sites were localized to the golgi region in some tubules but also to the cytoplasmic granules in others. This resulted in a mosaik-like staining, probably representing different functional or secretory stages of the gland cells.
Columnar epithelial cells in the small intestine possessed distinct binding sites for PNA, HPA, RCA I and Con A nearly always localized at the golgi region of the cells. Thereby an increasing staining tendency was seen from the bottom to the top of the villi.
A different lectin binding pattern was observed between the goblet cells of the small intestine and of the large bowel. The former were characterized by HPA and PNA lectin binding which were partly substituted with sialic acid and by the absence of Con A receptors (Fig. 1). An inverse lectin distribu-

tion could be demonstrated in the goblet cells of the large intestine (s. Table).

Hyperplastic alterations of the gastric mucosa

Inflammatory gastric mucosa and the transitional zones near to ulcus or cancer are often characterized by hyperplasia of surface mucus cells and isthmus cells. The elongation of these parts of the glands was seen to be always accompained by an increasing PNA binding in the perinuclear or golgi region of cells in the isthmus. Similar PNA labelling could be demonstrated in developing embryonal stomach. With an increasing renewal of the surface mucus cells, however, also their mucus content showed a positive reaction with PNA which was accompained by the decrease of their UEA I labelling. Extreme cases of this type of alteration were observed in two patient with chronic callous gastric ulcer.

Intestinal metaplasia

The lectin receptors of intestinalized gastric mucosa resembled neither those of the small intestine nor those of the colon. They showed mainly an intermediate form in which the strong Con A binding of the goblet cells (typical for the colonic ones) may be associated with partly sialic acid substituted HPA and PNA binding sites (as has been seen in the intestinal ones), (Fig. 2).

Gastric cancer cells

The lectin binding pattern of the 40 gastric carcinomas investigated correlated with the degree of differentiation of these tumors. Detailed observations regarding these findings will be published elsewhere. In general the glycoprotein secreting tumor cells showed binding properties with various labelled lectins similar to those of the isthmus cells in normal human stomach. The lectin which showed a high grade labelling in the widest variety of tumor cells was PNA. It may be of special interest, that while the PNA labelling of intestinal type car-

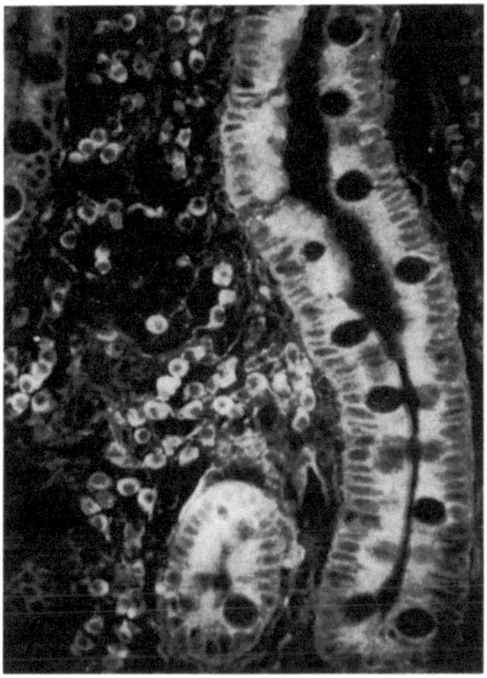

Fig. 1
Small intestine-Con A-FITC. Columnar epithelial cells and plasma cells show a diffuse cytoplasmic fluorescence, but the glycoprotein secretion in the goblet cells is negative

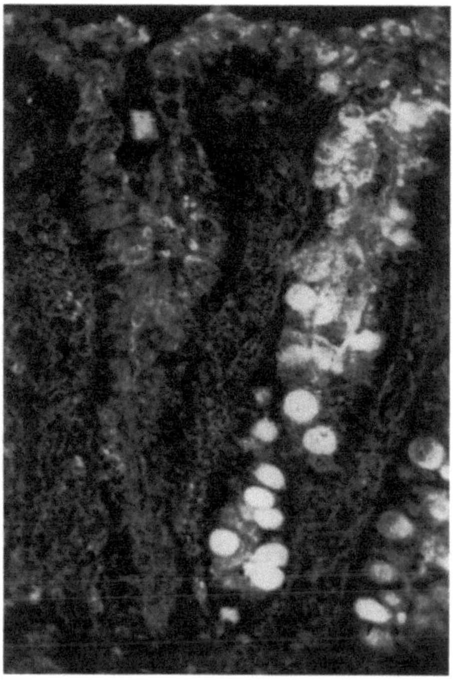

Fig. 2
Intestinal metaplasia-Con A FITC
Beside the negative surface epithelial cells (left) the goblet cells in intestinal metaplasia show an intensive fluorescence (right)

cinomas proved to be mainly unsensitive to neuraminidase pretreatment the PNA binding of the typical "signet ring cells" was increased. These types of cancer cells showed a different labelling with Con A, too. While adenocarcinomas possessed lectin binding sites on their plasmamembranes only, signet ring cells exhibited intensive staining within the cytoplasm (Fig. 3).

Lectin binding of extracted glycoproteins

The soluble glycoproteins in normal human corpus, antrum and small intestinal extracts gave marked precipitates against HPA and WGA and sometimes against RCA I lectins in double diffusion

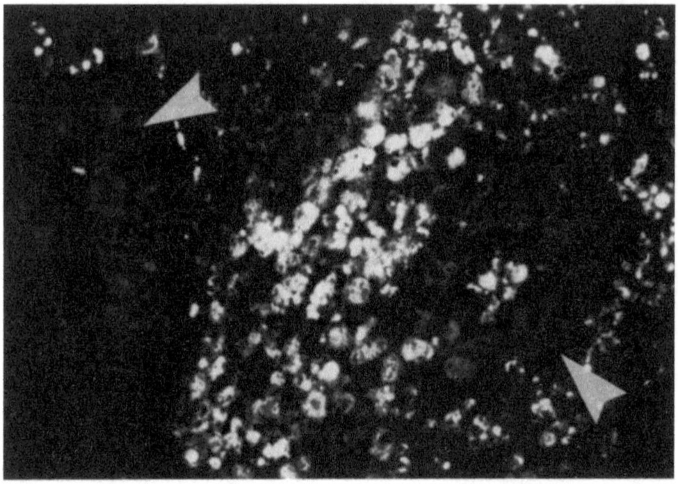

Fig. 3
Gastric carcinoma with partly adenomatous, partly diffuse localization of the tumor cells - Con A - FITC: Beside the almost negative adenoid areas (arrows) the undifferentiated tumor cells in the middle of the picture intensively labelled by the lectin

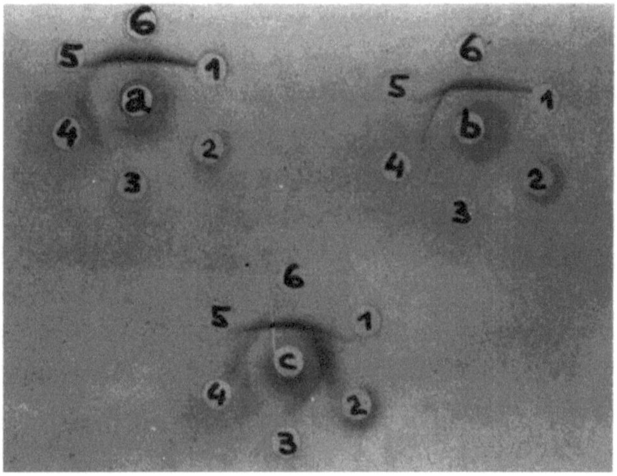

Fig. 4
Ouchterlony double-diffusion experiment. Lectins are directed against the soluble extracts of normal antrum (a), normal corpus (b) and adenocarcinoma (c) of the same patient.
Lectins used in clockwise order: 1. PNA, 2. Con A, 3. UEA I, 4. RCA I, 5. WGA, 6. HPA

experiments (Fig. 4a,b). In addition to that a very marked precipitate was formed against PNA in adenocarcinoma extract of the same patient (Fig. 4c). Con A and UEA I binding of the extract could only be demonstrated when the extracts were applied in a more concentrated form, but even in this case the precipitates formed were broad and diffuse, indicating the heterogenity of the reacting molecules.
Con A affinity chromatography of the normal human corpus and duodenal extracts resulted in a retention of some low mol. weight (25-80 K dalton) fractions, of which the 25 K dalton component was the most prominent. From the PNA-agarose column a marked protein peak with 30-35 K dalton and a smaller one with an apparent mol. weight of about 60 K dalton could be specifically eluted if adenocarcinoma extract was applied. Using HPA-agarose column only this latter peak could be isolated both from adenocarcinoma and normal corpus extracts.

Discussion

Our investigations showed that by using lectins some structural differences in glycoprotein composition of the various normal and pathologically altered cell types of human gastric mucosa and of intestinal epithelia can be detected. The different binding of Con A to the glycoprotein components of surface mucus cells and of cardiac, fundic and antral gland cells which was also demonstrated by Spicer et al. (17) indicates that there exsist definite differences in mucus composition of cells laying above and below the proliferation zone. Similar differences were demonstrated between the goblet cells of the small and large intestine, and it was found that goblet cells in intestinal metaplasia possessed Con A binding sites similar to the colonic ones. The close relation between these two cell types was also postulated by the high amount of their sulfated glycoprotein content in earlier histochemical investigations (10, 15). The binding of the mannose specific lectin to mucin secreting gastrointestinal cells is of special interest since the main, high-molecular-weight mucin does not contain mannose at all. Recently, however, Clamp et al. (3) isolated low-mol. weight glycoproteins from normal and pathologically altered colon with high mannose content. Moreover they suggested that such low-mol.-weight glycoproteins may be involved as cross linking agents in the maintainance of the high viscosity of native mucus. Changes in the viscosity of the mucus on the other hand are believed to be the main ethiologic factors in the most gastrointestinal damages (15). It was shown that using Con A affinity chromatography some low-mol.-weight glycoproteins can be isolated from gastric mucosa which may allow further investigation of this question. PNA binding of gastric epithelial cells seems to be strongly associated with the early phase of glycoprotein synthesis. This is most apparent in the surface epithelial cells, where PNA affinity was exclusively located at the golgi region of normal isthmus cells. Similary, there was a very intensive PNA binding to embryonic, hyperplastic and carcinomatous gastric mucosa. This concept is also supported by the earlier finding of Klein et al. (12) that PNA receptors were only demonstrable in transitional mucosa and carcinomas but were absent in the normal mucosa of the colon. Since it was shown that PNA receptor bearing low-mol.-weight glycoprotein can be isolated by PNA-affinity chromatography in high purity, a more accurate chemical analysis of this glycoprotein fraction can be performed. An interesting finding is the selective HPA and PNA labelling of the invaginated membrane systems of parietal cells,too.

The work was supported partly by the Alexander von Humboldt-Foundation.

References

1. Anderson, L., Anderson, N. G.: Proc. Natl. Acad. Sci. USA 74, 5421-5425 (1977)

2. Clamp, J. R., Allen, A., Gibbons, R. A., Roberts, G. P.: Br. Med. Bull. 34, 25-41 (1978)

3. Clamp, J. R., Fraser, G., Read, A. E.: Clin. Sci. 61, 229-234 (1981)

4. Dulley, J. R., Grieve, P. A.: Anal. Biochem. 64, 136-141 (1975)

5. Essner, E., Schreiber, J., Griewski, R. A.: J. Histochem. Cytochem. 26, 452-458 (1978)

6. Etzler, M. E., Branstrator, M. L.: J. Cell Biol. 62, 329-343 (1978)

7. Fairbanks, G., Steck, T. L., Wallach, D. F. H.: Biochemistry 10, 2606-2617 (1971)

8. Filipe, M. J.: Invest. Cell Pathol. 2, 195-216 (1979)

9. Hsu, S. M., Raine, L.: J. Histochem. Cytochem. 30, 157-161 (1982)

10. Jass, J. R., Filipe, M. J.: Histopathology 4, 271-279 (1980)

11. Klein, P. J., Newman, R. A., Müller, P., Uhlenbruck, G., Schaefer, H. E., Lennartz, K. J., Fischer, R.: Klin. Wochenschr. 56, 761-763 (1978)

12. Klein, P. J., Osmers, R., Vierbuchen, M., Ortmann, M., Kania, J., Uhlenbruck, G.: Recent Res. Cancer Res. 79, 1-9 (1981)

13. Picard, J., Edward, D. W., Feizi, T.: J. Clin. Lab. Immunol. 1, 119-128 (1978)

14. Schrager, J., Oastes, M. D. G.: Br. Med. Bull. 34, 79-82 (1978)

15. Sipponen, P., Seppälä, K., Varis, L., Hjelt, L., Ihamäki, T., Kekki, M., Siurala, M.: Acta path. microbiol. scand. Sect. A. 88, 217-224 (1980)

16. Spicer, S. S., Sannes, Ph. L., Katsuyama, T.: J. Histochem. Cytochem. 27, 1182-1184 (1979)

17. Spicer, S. S., Baron, D. A., Sato, A., Schulte, B. A.: J. Histochem. Cytochem. 29, 994-1002 (1981)

18. Stoward, P. J., Spicer, S. S., Miller, R. L.: J. Histochem. Cytochem. 28, 979-990 (1980)

19. Uhlenbruck, G.: Naturwissenschaften 68, 606-610 (1981)

Affinities of Ricin 120 and Mistletoe Lectin 1 for membrane components of liver and nervous tissue.

GROEGER, B.K.,[1] WILLIAMS, L.G.[2] PIGOTT, J.,[3] ZISKA., P.,[4] O'DELL,.D.S.,[1] WILLIAMS, D.J.,[2] FRANZ, H.[4] and DEBBAGE, P.L.,[3]
Departments of Anatomy,[3] Biochemistry[2] and Zoology,[1] University College London, U.K. and Staatliches Institut für Immunpräparate und Nährmedien,[4] Berlin, D.D.R.

Ricin 120 receptors are expressed selectively upon cell surface membranes. In cell cultures of nervous tissue their distribution may reflect an association with actively motile cells and membrane zones (2,5,6,8) and on hamster fibroblast surfaces may play a role in fibronectin binding, and be involved in cell-substrate adhesion (1). In liver, Ricin 120 receptors coat small cells of the non-parenchymal components of the tissue, but are absent from large parenchymal cells (9). We present here the distributions of Ricin 120 receptors visualised by lectin histochemistry in cultures of cerebellar tissue and in various types of preparation derived from liver tissue, and compare these Ricin 120 binding patterns with those observed using Mistletoe Lectin 1. We also present an initial characterisation of the Ricin 120 receptors in hepatic plasma membranes as visualised using a lectin blotting technique (4).

Ricin 120, the lectin of molecular weight 120,000 extracted from <u>Ricinus communis</u>, exhibits a carbohydrate selectivity involving terminal galactose moieties and β-linked galactose when tested in glycopeptide precipitation studies (3). Similar affinities are shared by other lectins, but amongst these are some of similar subunit structure to Ricin 120, and although these are derived from unrelated plant species they may share closely related affinities with Ricin 120. One such lectin is

Mistletoe lectin 1 (ML1), derived from Viscum album (7). Our comparison of Ricin 120 and ML1 used the two lectins conjugated to different fluorescent dyes in histochemical double-labelling procedures. We show that in different tissues and in different types of preparation the two lectins label identical structures.

Materials and Methods

Cultures: chick embryo cerebella (15-17 days in ovo) were dissociated mechanically by passage through nylon cloth of 120µm mesh size, and the resulting suspension plated at $2x10^6$ cells/ml in Eagle's medium supplememted with 10% horse serum, 20mM potassium, and antibiotics. Each culture resulted from plating approximately $3x10^5$ cells onto polylysinised glass coverslips. Cultures were fixed in 5% glutaraldehyde in PBS (phosphate-buffered saline) for 30 minutes at room temperature and rinsed in PBS, prior to staining (5).

Hepatic Cell Suspensions.

Livers of adult male Sprague-Dawley rats were collagenase perfused, disaggregated and pelleted, washed in collagenase-free perfusing medium, and repelleted twice (9).

Liver Sections.

Adult female Sprague-Dawley rat liver was snap-frozen in iso-pentane cooled in liquid nitrogen, and cryostat sections of 10 µm thickness were cut, mounted on albuminised glass slides and air-dried.

Hepatic Plasma Membranes.

Plasma membrane fragments were obtained from hepatic cell

suspensions, by homogenisation in 1mM bicarbonate buffer, followed by centrifugation in buffer over 45% sucrose. The interface material was washed twice, prior to staining (9).

Lectin Histochemistry.

Cerebellar cultures and liver sections were incubated with ML1-fluorescein conjugate, $4x10^{-7}M$ in PBS, for 30 minutes, rinsed in PBS, then incubated with Ricin 120-rhodamine conjugate $4x10^{-7}M$ in PBS for 30 minutes, rinsed, and viewed mounted in PBS. Hepatic cell suspensions and plasma membranes were stained for 30 minutes, at $4^{0}C$, in 1 ml ML1-fluorescein conjugate $4x10^{-7}M$ in PBS, then 9ml PBS added and the cells or membranes pelleted. After a similar incubation with Ricin 120-rhodamine conjugate $4x10^{-7}M$ in PBS, preparations were viewed in PBS. Fluorescein conjugates were visualised using 490nm excitation and 530 nm emission wavelengths. For rhodamine conjugates the corresponding wavelengths were 540nm and 625nm.

Lectin Blotting.

Rat serum was incubated with 10 units/ml neuraminidase from Vibrio cholerae at room temperature for 24 hours. Proteins were precipitated using 15% TCA (trichloracetic acid), pelleted, and albumin preferentially extracted from the pellet by addition of 1% TCA in absolute ethanol. Rat liver endoplasmic membranes were prepared from postmitochondrial supernatants. Rat liver plasma membranes were prepared as described above.

Serum and membrane samples were separated by vertical slab polyacrylamide gel electrophoresis (60V DC for 12 hours).Some gel strips were stained in Kenacid Blue, others were blotted onto nitrocellulose filters (0.5μm pore size) by electrophoresis at 0.4 amps DC for at least 3 hours. Blots were stained by a variety of techniques, including Kenacid Blue and Amido Black. Some blots were incubated with Ricin 120 conjugated to fluorescein, and viewed under long-wavelength ultraviolet (490nm) illumination. Others were incubated with Ricin 120-peroxidase conjugate (Miles), $1x10^{-8}M$ in PBS containing 1% BSA, for 30 minutes at room temperature. After rinsing in PBS, blots were incubated for 20 minutes in 20 ml substrate solution containing 10mg diaminobenzidine and 0.01% H_2O_2. Control experiments omitted the Ricin-peroxidase conjugate.

Results

Cell cultures of chick embryo cerebellum of various ages *in vitro* were double-labelled with ML1-fluorescein and rhodamine-conjugated Ricin 120, some incubations being performed with both lectin comjugates simultaneously, and others sequentially (Ricin 120 followed by ML1 and *vice versa*). Fig. 1 shows that both lectins labelled the same structures, microglial cells being most intensely stained by both lectins,other cell types remaining almost totally unstained. Neither lectin bound to neuronal cell bodies or to the flat membrane surfaces of flat cells. However, both lectins stained faintly a few pre-terminal neurite lengths and occasional folded regions at the edges of some flat cells. The patterns of staining produced by the two lectins were identical both in the distribution of stain, and also in the intensity with which different structures were labelled.

In cultures incubated live with the ML1-fluorescein conjugate, at $4^{0}C$, the fluorescence patterns observed were identical with those seen in glutaraldehyde-fixed cultures. Within 5 minutes

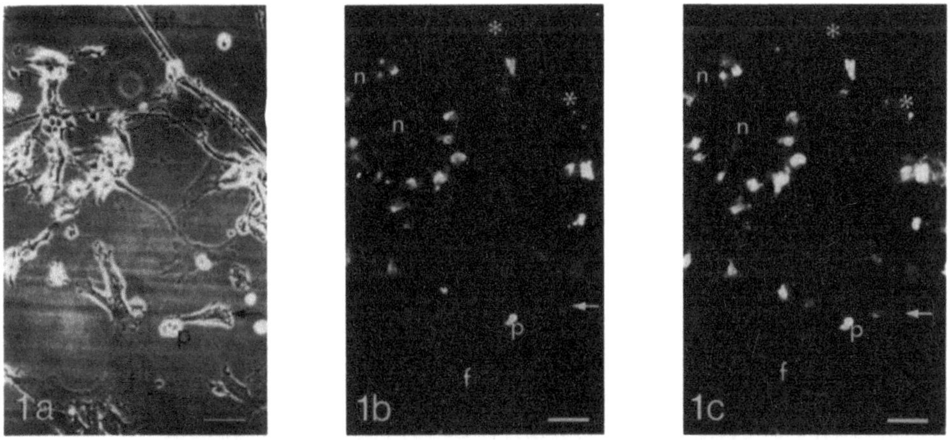

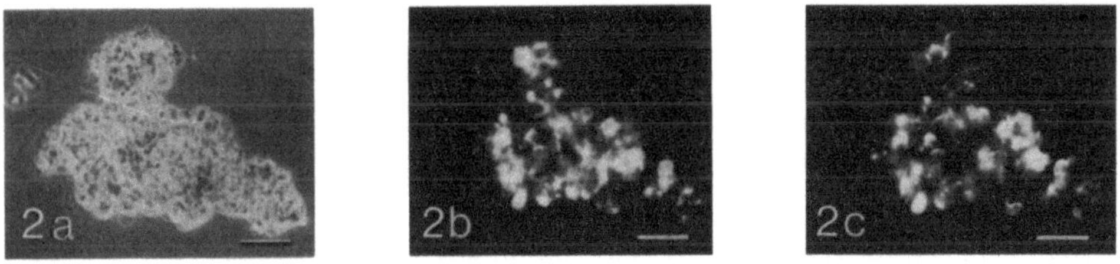

Figure 1 Cell culture of chick embryo cerebellum, 4 days *in vitro*, double-labelled with ML1-fluorescein and Ricin 120-rhodamine conjugates. Calibration bar, 30μm.

a) Phase contrast image. Neurons (n) occur both in isolation and in clusters. Bundles of neuronal fibres (bf) link clusters of neurons. Flat cells (f) include glial fibroblast cells. Phagocytic cells (p) appear haloed in phase contrast images.

b) This image reveals the distribution of ML1-fluorescein conjugate on the cells in a).

c) This image reveals the distribution of Ricin 120-rhodamine conjugate on the cells in a). It is identical with that for the ML1-fluorescein conjugate in b), though the rhodamine label is brighter.

Figure 2 Cells in a hepatic cell suspension double-labelled with ML1-fluorescein and Ricin 120-rhodamine conjugates. Calibration bar, 50μm.

a) Phase contrast image. Cell aggregates form in the suspension during the staining procedure, and the one shown here contains both large and small cells.

b) This image visualises the distribution of ML1-fluorescein binding sites in the aggregate in a).

c) This image visualises the distribution of Ricin 120-rhodamine binding in the aggregate in a). The distribution is identical to that shown for ML1 in b), though there is a slight difference in focal plane between b) and c).

of warming to 37^{0}C the microglia in these cultures internalised all the ML1 conjugate, leaving the cell surface devoid of stain, a result identical to that observed after similar treatment with Ricin 120 conjugate (5).

Figs. 2 and 4 compare the affinities of Ricin 120 and ML1 for liver cell membranes, and show that both lectins bind to identical liver components. Fig. 2 illustrates the results in double-labelled liver cell suspensions. Both lectins labelled intensely small non-parenchymatous cells and did not bind to the larger parenchymatous cells. In tissue sections labelled with both lectins (Fig. 3) only profiles of interlobular elements, corresponding to the walls of blood vessels and collecting ducts, were positive. Neither lectin labelled liver lobular cells. In preparations of hepatic plasma membranes (Fig. 4) selective staining was observed, both lectins binding to the same membrane fragments.

Fig. 5 illlustrates an initial biochemical analysis of hepatic plasma membrane proteins, performed using the blotting technique. The results obtained in blots stained with fluorescein and peroxidase conjugates of Ricin 120 were the same. Fig. 5b shows that Ricin 120 bound strongly to one band of low molecular weight (approximately 30,000 daltons), and weakly to other bands. Most bands remained unstained, compare Fig. 5a. The intensely positive band comprised a significant fraction of the total protein in the preparation, but comparison of the Ricin 120 labelled blot with the Kenacid Blue labelled gel shows that the prominence of this band after Ricin 120 labelling was not due solely to the quantity of protein in it. Hepatic smooth endoplasmic reticulum membranes however, did not contain any protein binding significant quantities of Ricin 120 (Fig. 5d), although a protein of similar electrophoretic mobility (Fig. 5c) to the Ricin-binding protein of plasma membranes was present.

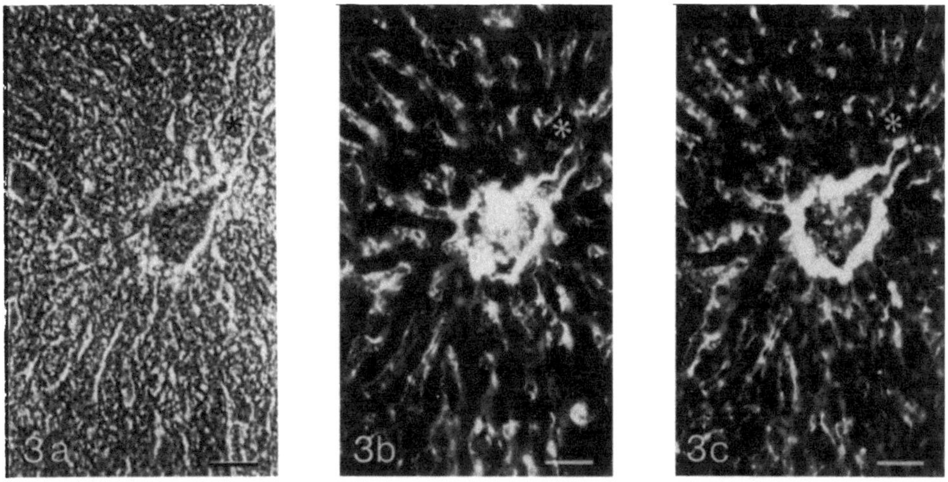

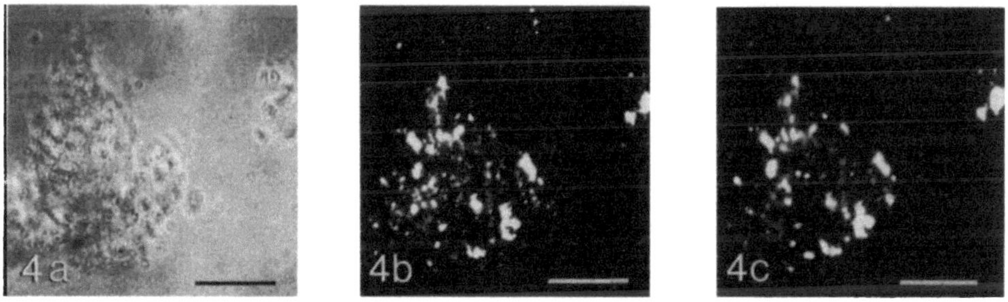

<u>Figure 3</u> Section of adult rat liver, double-labelled with ML1-fluorescein and Ricin 120-rhodamine conjugates. Calibration bar, 50μm.

a) Phase contrast image. A large profile may represent a vascular element, possibly a vein. Parenchyma (asterisk) lies between structures radiating from the large profile.

b) This visualises the distribution of ML1-fluorescein binding sites in the liver tissue in a).

c) This visualises the distribution of Ricin 120-rhodamine binding sites in the liver tissue in a). It is the same as that shown for the ML1-fluorescein conjugate in b). The faint background staining of the tissue is absent when similar sections are stained with a Ricin 120-fluorescein conjugate.

<u>Figure 4</u> Hepatic plasma membranes, double-labelled with ML1-fluorescein and Ricin 120-rhodamine conjugates. Calibration bar, 50μm.

a) Phase contrast image of aggregated plasma membranes.

b) ML1-fluorescein binding sites are visualised as bright areas in this aggregate.

c) The distribution of Ricin 120-rhodamine binding sites shown in this fluorescence image is identical with that for ML1-fluorescein, shown in b).

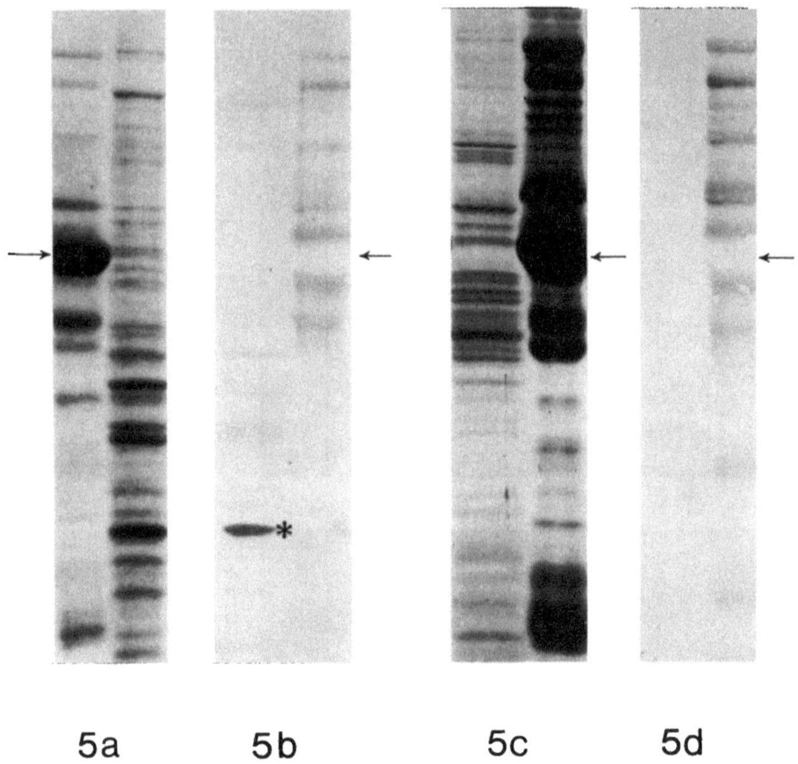

Figure 5 Affinities of hepatic plasma membrane components for Ricin 120.
a) Polyacrylamide gel stained with Kenacid Blue.
At right, rat liver plasma membrane preparation.
At left, neuraminidase-treated rat serum. The arrow marks the position of albumin, MW=60,000.
b) Blot taken from neighbouring channels of the gel shown in a), incubated with Ricin 120 conjugated to peroxidase, and developed in diaminobenzidine and H_2O_2.
At right, neuraminidase-treated rat serum. Arrow, albumin.
At left, rat liver plasma membrane preparation. One protein band (asterisk) is conspicuous due to its affinity for Ricin 120.
c) Polyacrylamide gel stained with Kenacid Blue.
At right, neuraminidase-treated rat serum. Arrow, albumin.
At left, rat liver smooth endoplasmic reticulum preparation.
d) Blot taken from neighbouring channels of the gel shown in c), incubated with Ricin 120 conjugated to peroxidase, and developed in diaminobenzidine and H_2O_2.
At left, rat liver smooth endoplasmic reticulum is without significant Ricin-binding bands.
At right, neuraminidase-treated rat serum (positive control) displays numerous Ricin-binding bands. Albumin (arrowed) excludes binding of Ricin-peroxidase conjugate.

CONCLUSIONS

1) Lectin histochemistry shows that Ricin 120 and ML1 have closely similar selectivities for tissue components in liver and nervous tissue. The hepatic lectin-binding cells may be endothelial, consistent with the histological patterns. Such cells possess actively motile membranes. The lectin-blotting technique used here will permit comparison of the affinities of these two lectins for hepatic plasma membrane proteins.
2) Our preliminary results using this method have indicated that Ricin 120 has a major affinity for only one protein in non-parenchymatous hepatic cell membranes. Possibly the lectin histological patterns found in nervous tissues and liver both map the distribution of this one protein. In either tissue type the distribution of Ricin 120 receptors is consistent with involvement of these receptors in endocytosis or in membrane motility of a more general nature. The analysis presented here suggests that Ricin 120 may bind selectively to a protein involved functionally in plasma membrane motility.

REFERENCES

1. Aplin, J.D., & Hughes, R.C., 1981. J.Cell Sci. 50, 89-103
2. Badley, R.A., Lloyd, C.W., Woods, A., Carruthers, L., Allcock, C., & Rees, D.A., 1978. Exp.Cell Res.117,231-244.
3. Baenziger, J.U., & Fiete, D., 1979. J. Biol. Chem. 254, 9795-9799.
4. Burnette, W.N., 1981. Analytical Biochem., 112, 195-203.
5. Debbage, P.L., O'Dell, D.S., & James, D.W., 1981. Cell & Tissue Res., 220, 313-323.
6. Debbage, P.L., O'Dell, D.S., & James, D.W., 1981. In: "Lectins, Vol. 2., Proc., 4th Lectin Conference", Ed. T.C. Bog-Hansen, W. de Gruyter, Berlin, 273-283.
7. Franz, H., Ziska, P., & Kindt, A., 1981. Biochem. J. 195 481-484.
8. Pfenninger, K.H., & Maylié-Pfenninger, M.F., 1981. J. Cell Biol. 89, 547-559.
9. Williams, L.G., Williams, D.J., O'Dell, D.S., & Debbage, P.L., 1982. Biochem. Soc. Trans., 10, 38.

STABILITY AND STERIC HINDRANCE OF LECTIN-LABELLED GOLD MARKERS IN TRANSMISSION AND SCANNING ELECTRON MICROSCOPY

Marc Horisberger and Monique Tacchini-Vonlanthen
Nestlé Products Technical Assistance Co. Ltd., Research Dept., CH-1814 La Tour-de-Peilz, Switzerland

It has become apparent that the accessibility of probes such as antibodies and enzymes to membrane components may depend upon the size of the probe. These differences in reactivity have been attributed, among other factors, to the spacing of glycoprotein bushes on membranes. Recently we have also suggested that the binding of lectin-labelled gold particles to membrane components was primarily determined by the granules size (1).

Gold particle probes are very interesting as they are rigid, can be labelled with a variety of macromolecules, presumably through a non-covalent process, and are easily detected by transmission or scanning electron microscopy (1). Gold particles can be made in a variety of sizes from 5 to 150 nm. In order to avoid spurious interpretation in transmission and scanning electron microscopy, the stability and steric hindrance of lectin-labelled gold granules was studied as a function of particle size.

Methods

Gold colloids (Au) were prepared as described previously by reducing $HAuCl_4$ with white phophorus (Au_5), sodium ascorbate (Au_{12}) or sodium citrate (Au_{17}-Au_{64}) depending upon the size required (1,2). The code following each Au preparation refers to the average diameter of the gold particle in nm when measured from electron micrographs (100 particles counted). Colloids were labelled with lectins and protein A by standard procedures (2) and suspended in 0.15 M NaCl-0,01 M Tris, pH 7.2 containing 0.5 mg/ml Carbowax 20-M (Union Carbide) and sodium azide. For Concanavalin A preparation (ConA-Au) the buffer was made 0.001M in $CaCl_2$ and $MnCl_2$. These buffers were used throughout.

Human red blood cells (RBC) from O Rh^+ donor were fixed in 2.5% glutaraldehyde as described previously (3). RBC were also treated with neuraminidase (Vibrio cholerae) (4) and fixed in glutaraldehyde.

RBC were marked by the gold markers as described elsewhere (2,3). For transmission electron microscopy (TEM) cells were embedded according to Spurr (5). Thin sections were post-stained with 0.5% uranyl acetate and lead citrate and examined in a Philips EM 300 electron microscope. For scanning electron microscopy (SEM), the specimens were dehydrated on the aluminium stub by increasing the concentration of ethanol and then subjected to the critical point method with CO_2. All specimens were examined without metal coating in a Cambridge S 4-10 Stereoscan at an accelerating voltage of 30 kV. All SEM figures are secondary electron micrographs.

Results and Discussion

Au_{40} was labelled with ^{125}I-ConA and centrifuged 3 times to remove the residual radioactivity of the supernatant and stored at 4°C. Following an initial release of 1.7% of the total radioactivity in the supernatant during the first 2 days, no further release was noticed during 45 days. It is concluded that a protein such as ConA is irreversibly adsorbed onto the gold particles.

Candida utilis CBS 567 cells were incubated with increasing amount of the marker (2). The number of particles bound onto the yeast cell surface was determined by spectrometry (2) and radioactivity measurements. By both methods the binding constant of the particles to the cell surface mannan was found to be 2.1×10^{9} M^{-1} (Scatchard plot analysis). A control experiment in the presence of 0.2 M methyl-α-D-mannopyranoside indicated that the binding was specific. The binding constant of ConA-Au_{40} is therefore several order of magnitude higher than that of ConA to the monosaccharide 4-methylumbelliferyl-α-D-mannopyranoside (4.26×10^{4} M^{-1}) (6) and to the polysaccharide mannan (1.0×10^{7} M^{-1}) (7) and results from multivalent interactions.

Markers were prepared with gold particles of increasing sizes (5-64 nm) labelled with ConA, soya bean agglutinin (SBA), Ricinus communis, agglutinin (M_r 120'000, RCA_I) and wheat germ agglutinin (WGA). RBC were incubated with an excess of the markers and the decrease in absorbance of the supernatant was determined at the maximum of absorbance. The number of bound particles was calculated by difference (2). In control experiments marking was found to be specific using as inhibitors penta-N-acetyl-chitopentaose (2 mg/ml) for WGA, N-acetyl-D-galactosamine (2 mg/ml) for SBA, D-galactose (5 mg/ml) for RCA_I and methyl-α-D-mannopyranoside (50 mg/ml) for ConA markers. The results are shown in Table I.

Table I : Binding of lectin and lectin-labelled gold markers to red blood cells

Lectin and lectin markers	Molecules or particles bound/RBC	Lectin and lectin markers	Molecules or particles bound/RBC
ConA (11)[a]	1.2-1.7 x 10^5	SBA (13)	1.0 x 10^6
ConA-Au_5 (3)	20'000	SBA-Au_{17}	12'300
ConA-$Au_{12,32}$	NB[b]	SBA-Au_{32}	10'000
		SBA-$Au_{50,64}$	NB
RCA_I (12)	1.2 x 10^6	WGA (12)	7.9 x 10^6
RCA_I-Au_{32}	16'000	WGA-Au_{26}	31'000
RCA_I-Au_{45}	5'600	WGA-Au_{50}	5'600
RCA_I-Au_{50}	1'200	WGA-Au_{64}	2'000
RCA_I-Au_{64}	NB		

(a) References for data are given in parenthesis;
(b) NB = no binding.

No direct relationship was found between the number of bound particles and the number of lectin binding sites since binding occurred only when the markers had a size below 12, 50 and 64 nm for ConA, SBA and RCA_I, respectively. However WGA-labelled particles up to 64 nm in size reacted with RBC. In RBC, membrane receptors for ConA are found exclusively associated with band 3 (8) and RCA_I reacts with all major glycoproteins (8,9). WGA is known to bind the glycophorin (9). The nature of the membrane components binding SBA is still not established. It would therefore appear that ConA binding sites are close to the lipid bilayer and masked by the glycoprotein chains radiating into the medium since they were capable of binding only the smallest ConA-labelled particles. Since RBC were capable of binding the largest WGA-labelled particles, this confirmed that the oligosaccharides of glycophorin protrude from the cell surface.

Although binding of particles to the cell surface was completely prevented by the presence of inhibitors, only 31% and 28% of bound WGA-Au_{32} and WGA-Au_{50} were desorbed from marked cells in the presence of inhibitor. It is therefore possible that secondary interactions between marker and cell surface such as hydrophobic bonds add to the strength of the binding. Binding constants of lectin markers were determined by Scatchard plot analysis (Table II). The curves for SBA- and RCA-markers were monophasic whereas that for WGA-marker was biphasic as reported previously for the free lectin (12). In all cases binding constants were very high. Contrary to the data obtained with the free lectin, the number of low affinity sites for WGA-Au_{50} was larger than that of the high affinity sites. It is therefore suggested that the high affinity sites are less accessible than the low affinity sites to this marker.

Native RBC were marked with either WGA-Au_5 or WGA-Au_{26}, lysed and the marked ghosts reincubated with one of the two markers (2). The number of particles bound per ghost (surface area 102 μm^2) was counted on 5-6 micrographs (minimum 600 particles counted). A correction was introduced to take into account the covering of WGA-Au_5 particles by the larger particles bound to both faces of the ghost (Table III). These results indicate that new binding sites had become accessible to the markers following lysis of the cells. This effect was not detectable after RBC were fixed for 2 min. in 0.1% glutaraldehyde prior to lysis. In the case where the larger area was occupied by gold particles

Table II : Binding sites of red blood cells for lectins and lectin labelled gold particles (Scatchard plot analysis)

Lectin and lectin markers	Sites/cell	K_{assoc} (M^{-1})
WGA (12)[a]	5×10^6	15×10^6
	2.9×10^6	2.7×10^6
$WGA-Au_{50}$	3'500	20×10^{10}
	12'400	2.7×10^{10}
RCA_I (12)	0.7×10^6	66×10^6
	0.5×10^6	4×10^6
RCA_I-Au_{32}	14'600	0.6×10^{10}
$SBA-Au_{17}$	16'000	5.6×10^9

(a) references for data are given in parenthesis.

Table III : Sequential marking of red blood cells and their ghosts by $WGA-Au_5$ and/or $WGA-Au_{26}$

RBC	Ghosts	Particles bound per ghost (± SD)	
marked by		$WGA-Au_5$	$WGA-Au_{26}$
$WGA-Au_5$		41'800 ± 1'400	
$WGA-Au_{26}$			27'800 ± 3'200
$WGA-Au_5$	$WGA-Au_5$	101'000 ± 14'000	
$WGA-Au_{26}$	$WGA-Au_{26}$		52'900 ± 2'900
$WGA-Au_5$	$WGA-Au_{26}$	34'700 ± 5000	49'600 ± 2'300
$WGA-Au_{26}$	$WGA-Au_5$	49'700 ± 8800	35'100 ± 2'000

(double marking with WGA-Au_{26}), the marker covered only 28% of the surface area. This would indicate that the WGA binding sites are clustered and/or some of them are inaccessible to the particles as suggested above.

Neuraminidase-treated RBC were marked by the indirect protein A gold method (10) by successively incubating the cells with SBA, anti-SBA antibodies and protein A-$Au_{7,20,41}$ (Fig. Ib and d, Fig. IIb). The results were compared to those obtained by the direct method using SBA-$Au_{7,20,41}$ (Fig. Ia and c, Fig. IIa). Control experiments were performed in the presence of 2 mg/ml N-acetyl-D-galactosamine. The number of particles bound was determined by spectrophotometric measurement (1). By the direct method the number of particles bound was 310'000, 45'000 and 2'900 for SBA-$Au_{7,20,41}$, respectively, by the indirect method 400'000, 56'000 and 11'000 for protein A-$Au_{7,20,41}$, respectively. These results also indicate again that the density of marking decreases with increased probe size (see Table I). They also show that with the indirect method the protruding molecules (SBA and antibodies molecules) bound to

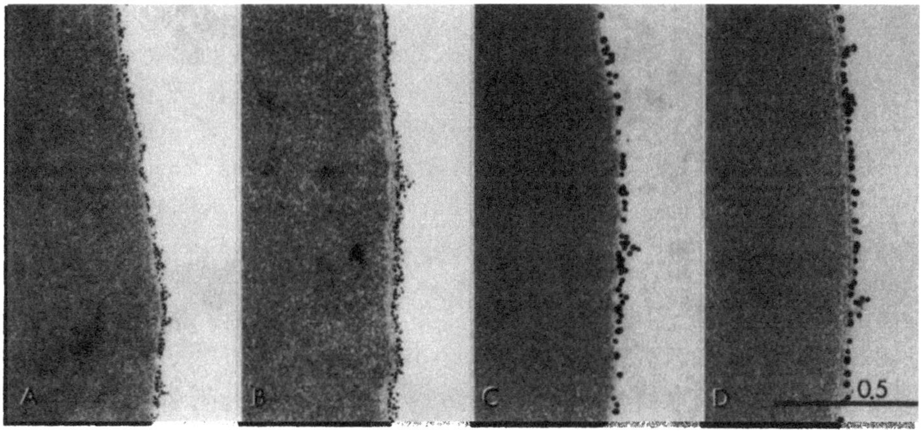

Fig. I. Marking of SBA binding sites on neuraminidase-treated RBC. (a) SBA-Au_7, (b) protein A-Au_7, (c) SBA- Au_{20}, (d) protein A-Au_{20}. The bar represents 0.5 µm.

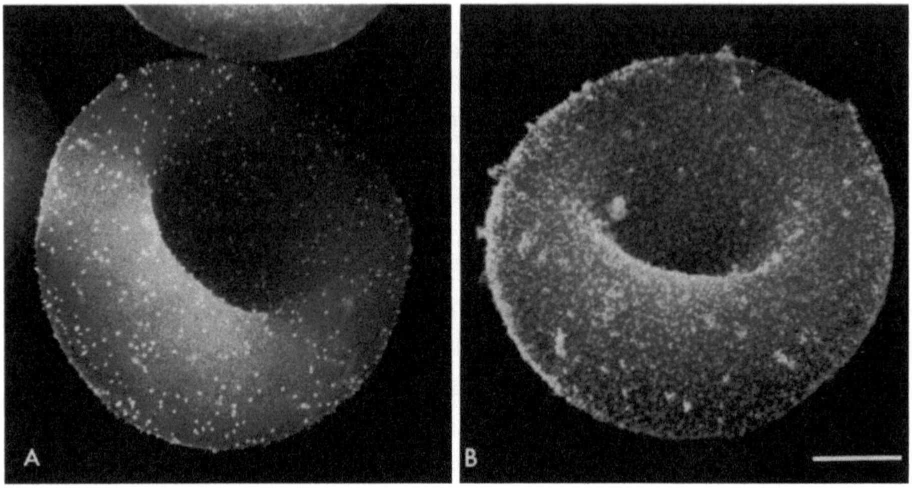

Fig. II. Marking of SBA binding sites on neuraminidase-treated RBC. SEM micrographs. (a) SBA-Au_{41}, (b) protein-Au_{41}. The bar represents 1 μm.

the lectin binding site provide a better stereochemical accessibility to the gold particles especially to the larger probe (compare Fig. IIa with b).

In conclusion, this study demonstrates the importance of varying the size of probes to study the topography of cell membranes by SEM and TEM and that gold particles are well suited for this purpose since they can be prepared in a variety of sizes, are rigid and can be labelled by a tightly bound monolayer of macromolecules (1). Due to steric hindrance, access of lectin-labelled gold markers to receptors depends upon the size of the particles. Therefore no binding reaction does not necessarily mean the absence of cell surface receptors for gold markers. However, the results obtained with gold markers are not necessarily applicable to other markers developed for SEM and TEM since their mode of labelling is different and they may not be as rigid as gold granules. Finally this method permits to gain a better insight not only on the lateral but also on the longitudinal location of cell surface receptors.

Acknowledgements

The authors thank Mrs J. Rosset for the scanning electron micrographs and Mrs M. Beaud for the photographic work.

References

1. Horisberger, M.: in Scanning Electron Microscopy (O. Johari ed.), SEM Inc., AMF O'Hare (Chicago IL 60666, USA), vol. II, p. 9-31, 1981.

2. Horisberger, M., Rosset, J.: J. Histochem. Cytochem. 25, 295 (1977).

3. Horisberger, M., Vonlanthen, M.: J. Microscopy 115, 97 (1979).

4. Schnebli, H.P., Roeder, C., Tarcsay, L.: Exp. Cell Res. 98, 273 (1976).

5. Spurr, A.R.: J. Ultrastr. Res. 26, 31 (1969).

6. Loontiens, F.G., Clegg, R.M., Jovin, T.M.: Biochemistry 16, 159 (1977).

7. Horisberger, M.: Biochim. Biophys. Acta 632, 298 (1980).

8. Tanner, M.J.A., Anstee, D.J.: Biochem. J. 153, 265 (1976).

9. Adair, W.L., Kornfeld, S.: J. Biol. Chem. 249, 4696 (1974).

10. Roth, J., Bendayan, M., Orci, L.: J. Histochem. Cytochem. 26, 1074 (1978).

11. Schnebli, H.P., Bächi, T.: Exp. Cell Res. 91, 175 (1975).

12. Presant, C.A., Kornfeld, S.: J. Biol. Chem. 247, 6937 (1972).

13. Reisner, Y., Lis, H., Sharon, N.: Exp. Cell Res. 97, 445 (1976).

LECTIN BINDING SITES ON SEMITHIN SECTIONS OF EPOXY-EMBEDDED TISSUES

Giuseppe Viale, Patrizia Dell'Orto, Roberto Colombi, Vito De Gennaro, Anna Comi, Guido Coggi

IV Department of Pathology, University of Milan Medical School
Milan, Italy

The identification of sugar residues on tissue sections has proved to be easily accomplished by the use of labeled lectins. In the recent years, several light microscopic investigations have been carried out with fluorescent, peroxidase- or biotin-labeled lectins, in order to define the morphofunctional properties of different tissues and cells, both in normal and pathological conditions (1, 5, 8, 10, 12, 13, 14).

The preservation of the morphological details of tissues is of paramount relevance in this type of investigation, aimed to the precise localization of sugar residues. In this respect , staining of paraffin-embedded material with peroxidase or biotin-labeled lectins has been shown to offer relevant advantages over the staining of frozen tissue sections with FITC-labeled lectins (8, 14).

The use of semithin sections of Epoxy-embedded tissues for light microscopy would result in an even better preservation of morphological details, and in a more precise localization of sugar residues, than those obtained on paraffin-embedded material.

The present study has been undertaken in order to define a staining procedure for the identification of lectin binding sites on semithin sections of tissues routinely fixed and embedded for electron microscopy investigations.

Lectins, Vol. III

Material and methods

Renal biopsies from two rats, killed by cervical dislocation, were fixed with 2.5% Glutaraldehyde in 0.13M phosphate buffer, for 4 hours, at 4°C. A human gastric biopsy, fixed in 2.5% Glutaraldehyde for 2 hours at 4°C, was post-fixed in 2% OsO_4, in 0.13M phosphate buffer for 1 hour at 4°C. After dehydration, the specimens were transferred to propylene oxide, embedded in Epoxy resin (Durcupan, ACM) and allowed to polimerize in an oven at 60°C for 48 hours.
One µ thick sections were floated onto a drop of water on albumin-coated glass slides and left to dry overnight at 37°C.
The tissue sections were treated for 15 min with a saturated solution of potassium hydroxide in absolute ethanol in order to dissolve the resin, according to the method of Lane and Europa (7) with minor modifications, and finally washed in three changes of absolute ethanol for 10 min.
After a washing in PBS for 5 min the tissue sections were incubated with the biotinylated lectins (Vector Lab.) for 30 min at room temperature, in a moist chamber.
CON A, WGA, PNA, were used at a concentration of 0.05 mg/ml and SBA, DBA, UEA_I, RCA_{120} at a concentration of 0.1 mg/ml. Lectins were diluted in 0.01M PBS, pH 7.6, supplemented with 1mM Ca^{++} and 1mM Mg^{++}. Following the incubation with labeled lectins, the sections were washed in buffer and then treated with the avidin-biotinylated-peroxidase complex (Vector Lab.) for 30 min (6).
The peroxidase activity was developed with the 3-3' diaminobenzidine tetrahydrochloride medium of Graham and Karnovsky (4) for 5 min in the darkness, or with 3-amino-9-ethylcarbazole in dimethyl formamide for 3 min, according to Rojas-Espinosa et al. (9).

Results and discussion

Tissues fixed in 2.5% Glutaraldehyde and embedded in Epoxy resin are actually suitable for the identification of sugar residues in light microscopy, with the use of biotin labeled lectins, provided the resin is removed from the tissue sections before staining.

The use of semithin sections of specimens fixed and embedded for electron microscopy results in a more precise localization of sugar residues, because these are sharply stained by labeled lectins and the tissue morphological details are best preserved in Epoxy-embedded sections (figs. 1-6).

The possibility of employing the avidin-biotinylated-peroxidase complex as the second step in the staining procedure ensures high sensitivity and very low background staining, as it has been already observed in immunohistochemical studies (6).

The chromogens currently used for the development of the peroxidase activity on frozen or paraffin-embedded sections are also suitable for Epoxy-embedded tissues. It is noteworthy, nevertheless, that diaminobenzidine cannot be used on osmium postfixed specimens. In fact, its osmiophilia would be responsible for an aspecific binding to the osmium, thus affecting the specificity of the reaction with sugar residues. Osmium itself, on the other hand, oxidizes several sugar residues, hindering the binding of lectin conjugates (13).

Indeed, removal of OsO_4 from tissue sections with periodate, which can be easily performed in immunohistochemistry, is to be avoided in lectins histochemistry, due to the adverse effects of the periodic acid on sugar molecules. As a consequence, if OsO_4 fixed tissues are to be studied, 3-amino-9-ethylcarbazole should be used instead of 3-3' diaminobenzidine for developing peroxidase activity.

In conclusion, specimens routinely processed for electron microscopy are actually suitable not only for traditional carbohydrate histochemistry (PAS, Alcian) (2, 11), but also for lectin histochemistry, in light micro-

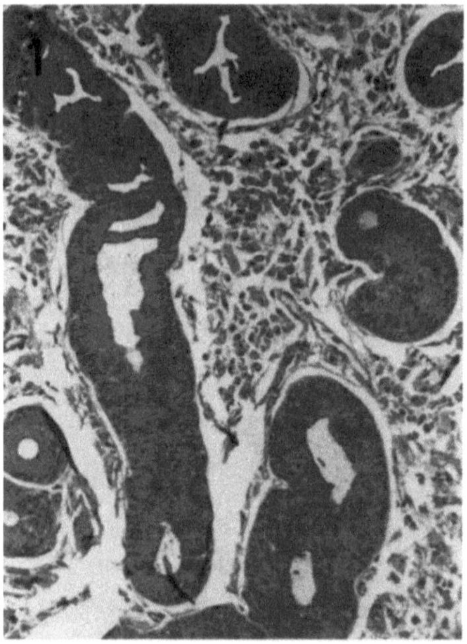

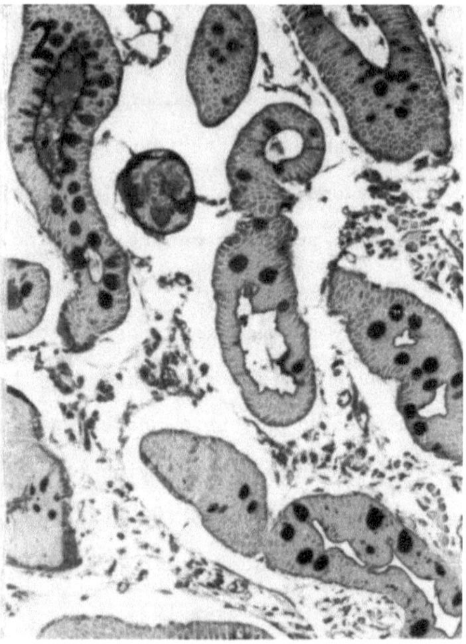

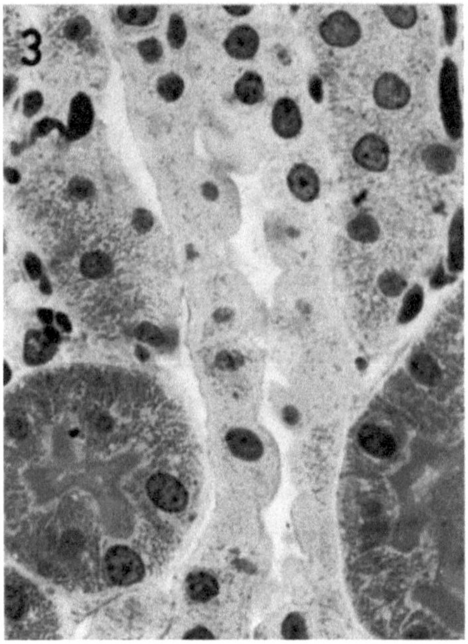

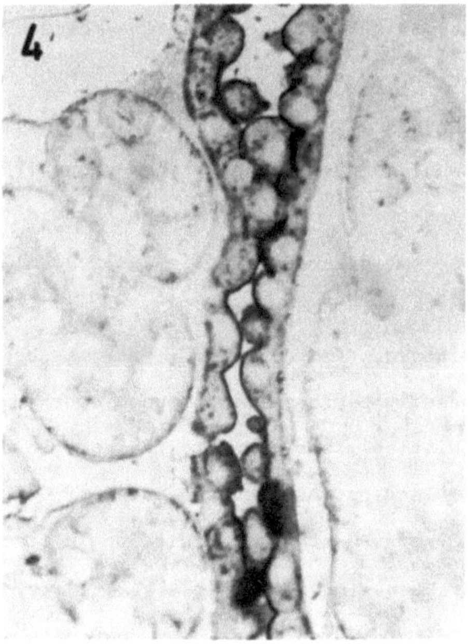

Fig.1: Intestinal metaplasia, toluidine blue, x250.

Fig.2: Goblet cells are strongly stained by PNA, after neuraminidase digestion, x250.

Fig.3: Rat kidney, collecting duct, toluidine blue, x400.

Fig.4: Rat kidney: the luminal border of the epithelial cells strongly reacts with DBA, x400.

scopy. If one considers that tissues processed for electron microscopy can be also used for immunohistochemical studies (3), then the possibility of carrying on histological, histochemical, immunohistochemical and ultra-

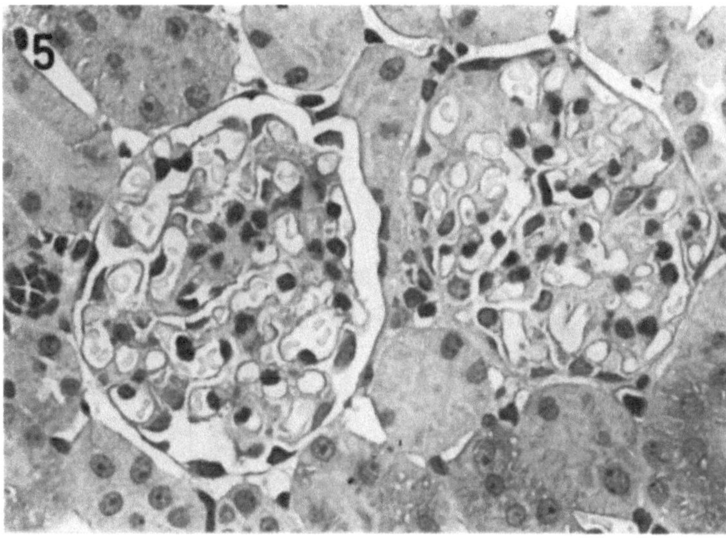

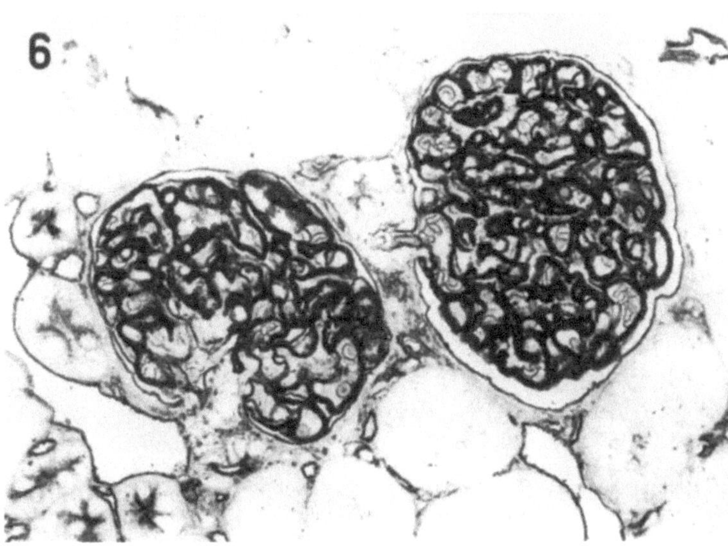

Fig.5: Rat kidney, toluidine blue, x250.
Fig.6: Rat kidney: the glomerular basal membrane reacts with PNA after neuraminidase digestion, x250.

structural studies on adjacent sections of the same block, actually makes Epon-embedded tissues suitable for comprehensive morphofunctional investigations.

References

1. Coggi, G., Dell'Orto, P., Bonoldi, E., Doi, P., Viale, G.: These Proceedings.

2. Colombi, R., Comi, A., Coggi, G.: Proceedings of the IV° National Congress of SIAP, Messina (1981) (in press).

3. Dell'Orto, P., Viale, G., Colombi, R., Braidotti, P., Coggi, G.: J. Histochem. Cytochem. 30, 630-637 (1982).

4. Graham, R.C.Jr., Karnovsky, M.J.: J. Histochem. Cytochem. 14, 291-296 (1966).

5. Howard, D.R., Ferguson, P., Batsakis, J.G.: Cancer 47, 2872-2877 (1981).

6. Hsu, S., Raine, L., Fanger, H.: J. Histochem. Cytochem. 29, 577-580 (1981).

7. Lane, B.P., Europa, D.L.: J. Histochem. Cytochem. 13, 570-582 (1965).

8. Pena, S.D.J., Gordon, B.B., Karpati, G., Carpenter, S.: J. Histochem. Cytochem. 29, 542-546 (1981).

9. Rojas-Espinosa, O., Dannenberg, A.M., Sternberger, L.A., Tsuda, T.: Am. J. Pathol. 74, 1-12 (1974).

10. Rose, M.L., Habeshaw, J.A., Kennedy, R., Sloane, J., Wiltshaw, E., Davies, A.J.S.: Br. J. Cancer 44, 68-74 (1981).

11. Snodgress, A.B., Dorsey, C.H., Bailey, G.W., Dickson, L.G.: Lab. Invest. 26, 329-337 (1972).

12. Stoward, P.J., Spicer, S.S., Miller, R.L.: J. Histochem. Cytochem. 28, 979-990 (1980).

13. Suzuki, S., Tsuyama, S., Suganuma, T., Noboru, Y., Murata, F.: J. Histochem. Cytochem. 29, 946-952 (1981).

14. Watanabe, M., Muramatsu, T., Shirane, H., Ugai, K.: J. Histochem. Cytochem. 29, 779-790 (1981).

COMPARISON OF LECTIN BINDING SITES IN THE KIDNEYS OF DIFFERENT ANIMAL SPECIES.

H. Holthöfer[1], A. Miettinen[2], I. Virtanen[1]

[1]Dept. of Pathology and [2]Dept. of Bacteriology and Immunology, University of Helsinki, Helsinki, Finland.

Introduction

In recent years, lectins have been increasingly used in identifying glycosubstances in tissues (2, 4, 8, 11, 13). In these studies, mainly performed with tissues of laboratory animals, compartmentation of saccharides to some tissues has been revealed (7, 8, 9). There are only a few lectin studies analyzing sugar components in the different tissues of one species (9, 14) and virtually no studies comparing the sugar components of a single organ in various animals. Our previous lectin binding studies with human kidney tissue revealed, that some lectins selectively bind to different parts of the nephron (8), and could be used as markers of these structures. Proteinuria in kidney diseases is thought to be partly due to the loss of the glomerular sialic acids (1, 10), and alterations in other glycocompounds have also been revealed in renal disease (5). As experimental models of renal diseases are widely used (3, 12), we wanted to compare the glycosylation patterns of the nephrons in the normal kidneys of various laboratory animals using fluorochrome coupled lectins with different sugar specificities.

A clear species-specific but not strain-specific compartmentation of saccharides to various parts of the

Lectins, Vol. III

nephron could be demonstrated, although also similarities in the distribution of some lectin receptors were noticed in all animals.

Materials and methods

Tissue samples from rabbit, guinea pig, rat, hen and mouse kidneys were obtained from Department of Bacteriology and Immunology, University of Helsinki. Samples of human kidneys were obtained from normal poles of kidneys removed because of renal carcinoma, as earlier described (8). 10 different mouse strains (C57, Bl/6, A8, CBA, Balb/c, B10Htt, CB20, RGa, AL/N, C3H) and 2 rat strains (Sprague-Dawley, DA) were studied.

The lectins used in this study were Wheat germ agglutinin (WGA; sialic acid, N-acetylglucosamine), Ulex europaeus I (UEA I; α-L-fucose), Peanut agglutinin (PNA; galactose-N-acetylgalactosamine) and Soybean agglutinin (SBA; N-acetylgalactosamine). All FITC- or TRITC- labelled lectins were obtained from Vector laboratories (Burlingame, Pa., USA). Tissue preparation and the controls for staining specificity were performed as previously described (8). In some experiments, frozen tissue sections were pretreated with neuraminidase (Vibrio cholerae neuraminidase, Behringwerke, Marburg, F.R.G) as earlier described (8).

Frozen tissue sections were incubated for 30 min with lectins, washed and mounted in polyvinylalcohol and studied using Zeiss standard microscope equipped with filters for FITC- and TRITC-fluorescence.

Results

The results show, that each fluorochrome coupled lectin presented a species-specific, but not strain-specific binding pattern on frozen sections of kidneys of various laboratory animals. However, some lectins presented uniform binding to the same nephron parts in all animals studied, whereas some others bound to limited parts of the nephrons in some animals. The location of lectin receptors in different parts of the nephrons in the animals studied is summarized in table I.

<u>TABLE I</u> Location of lectin binding sites in different parts of the nephrons of different animal species.

	GLOMERULUS				TUBULUS							
					prox tub				dist tub			
	WGA	PNA	UEA	SBA	WGA	PNA	UEA	SBA	WGA	PNA	UEA	SBA
Rabbit	+[1]	-	-	+[1]	+	-	-	+	+	+	+	+
Guinea pig	+[1]	-	-	+[1]	+	+	-	+	+	-	+	+
Rat	+[1]	-	-	-	+	+	-	+	+	-	-	+
Mouse	+[1]	+[2]	-	-	+	+	-	-	+	+	-	+
Hen	+[1]	+[3]	-	-	+	-	-	+	+	-	-	+
Man	+[1]	+[2]	+[3]	-	+	-	-	+	+	+	-	+

Abbreviations: Prox tub, proximal tubules; dist tub, distal tubules; [1]capillary/vascular walls; [2]Bowmanns capsule; [3]endothelial cells.

WGA presented strong binding to the glomeruli in all the animals studied (fig. 1a,b and c). In tubular areas most animals presented fluorescence in tubular apical aspects. Pretreatment of the tissues with neuraminidase prevented the binding of WGA to glomeruli.

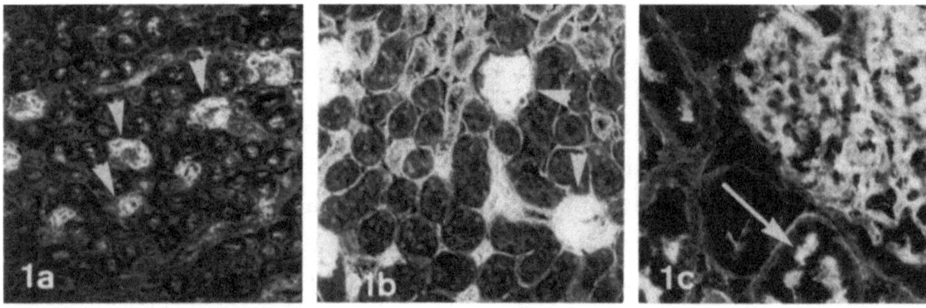

Fig. 1. Hen (a), mouse (b) and rabbit (c) kidney sections stained with TRITC-WGA. A strong binding to the glomerules (arrowheads) is seen. In rabbit kidneys, proximal tubules are visualized also on the apical aspect (arrow).

PNA interestingly bound to the endothelia, but not to other structures in hen kidney (fig. 2a). In human and mouse kidneys PNA reacted with Bowmann's capsules, but other tested animals did not show any glomerular staining. In tubular areas PNA binding sites were found either in proximal (guinea pig, rat) or in distal tubules (rabbit and man, fig 2b,c).

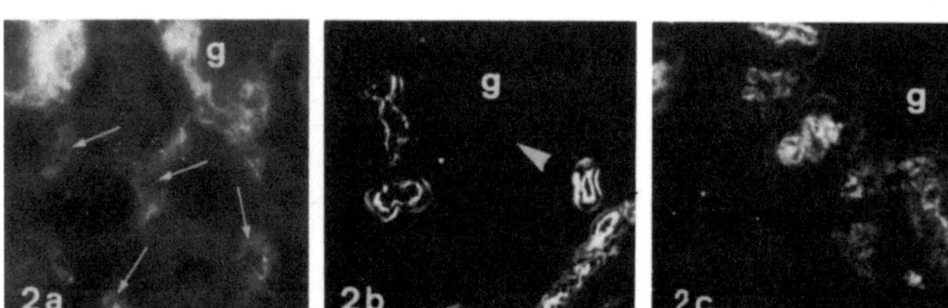

Fig. 2. PNA reactive sites in hen (a), human (b) and guinea pig (c) kidneys. In hen, binding is limited to vessel wall endothelia (arrows, peritubular capillary endothelia). In both human and guinea pig, fluorescence is seen in the distal tubules. (arrowhead in b, Bowmanns capsule), a,c x 280, b x 110.

UEA I, which reacts with endothelial cells in human tissues (9), failed to react with the endothelium, glomerules or proximal tubules in any of the animals studied, but presented a bright fluorescence in the distal tubules of rabbit and guinea pig, fig. 3.

The kidneys of all the animals except mouse presented granular staining of distal tubules when treated with SBA. However, in human and hen kidneys the proximal tubular staining was stronger, whereas in other species preferential staining of distal tubules was seen, fig 4a,b.

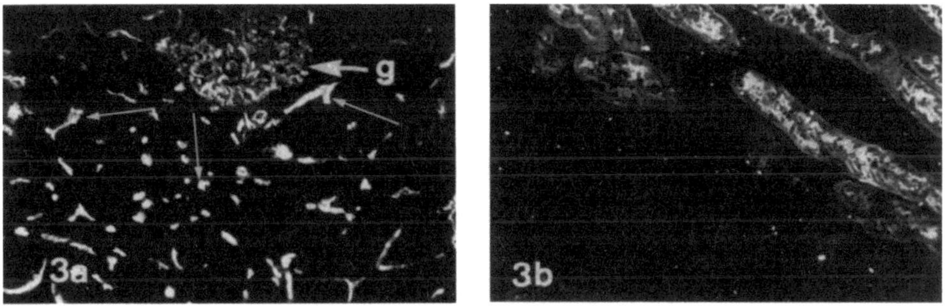

Fig. 3. Human (a) and rabbit (b) kidney sections treated with TRITC-UEA I. Fluorescence located to endothelial cells in human kidney is seen (g, glomerulus; arrows, peritubular capillaries). Only some distal tubules are seen fluorescent in rabbit. a x 110, b x 280.

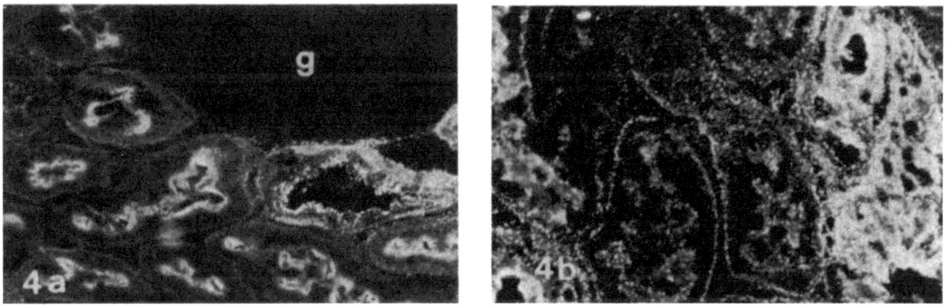

Fig. 4. TRITC-SBA treated frozen sections of rat (a) and rabbit (b) kidneys show a granular fluorescence. In rat kidneys tubular lumina and in rabbit kidneys also the basement membranes are visualized. (g,glomerulus). x 280.

Discussion

The results of this study suggest, that each animal species presents a typical lectin binding pattern on kidney sections, although some lectins seemed to bind regularly to the same structures of the nephrons in all animal kidneys.

The loss of the glomerular WGA receptors after neuraminidase treatment indicates, that sialic acids are the main glomerular binding sites for WGA. The strong binding of WGA to the glomerules in all animals tested suggests, that sialic acids are important to normal glomerular function. Results obtained from experimental models of kidney diseases have indicated, that loss of sialic acids from the polyanionic coat of glomerular capillary walls is often associated with proteinuria (3, 10). WGA binding might thus be used as one indicator for the function the glomerular filtration barrier.

PNA, which on human kidney reacts only with distal tubules and collecting ducts (8), bound only to endothelium in hen kidney. In the kidneys of guinea pig and rat PNA receptors were located in proximal tubules, whereas in rabbit kidney PNA bound only to the distal nephron. After treatment of the tissue sections with neuraminidase, PNA stained brightly glomerules in all animals. This suggests, that a sugar configuration, in which the subterminal galactose containing PNA receptors are coated by sialic acids, is important in the glomerules in a wide variety of animals.

SBA presented granular staining in all the animal kidneys. Moreover, similar to PNA, also SBA presented preferential binding to either proximal or distal tubules. This might reflect a variation in the location of certain functional glycocompouds in the kidneys of different animals.

UEA I, which is specific for some α-L-fucose moieties (6) reacts only with vascular endothelium in human tissues (9). In the animals tested, UEA I did not stain endothelia but bound to some distal tubules in rabbit and guinea pig kidneys. It thus seems, that limited species-spesific sugar configurations may be exposed in vascular endothelia. The function of these glycocompunds is unknown.

Although we studied mainly closely related laboratory animal species, it seems that the glycosylation of nephrons in species-specific, but not strain-specific. Our preliminary results from experimental models of kidney disease (Miettinen et al., unpublished results) indicate, that there may be alterations in kidney glycocompunds demonstrable by lectins. For the expanding use of lectins in tissue studies it is important to note, that lectin binding results obtained with one species cannot be directly applied to another - even closely related - species.

This study was supported by Finnish Medical Research Council, the Finnish Cultural Foundation, The Sigrid Juselius Foundation and the Association of Finnish Life Insurance Companies.

References

1. Blau, E. B., Haas, J. E.: Lab. Invest. 28, 477-481 (1973).
2. Bretton, R., Bariety, J.: J. Histochem. Cytochem. 24, 1093-1100 (1976).
3. Caulfield, J. P., Farquhar, M. G.: Lab. Invest. 39, 505-512 (1978).
4. Freeman, H. J., Lotan, R., Kim, Y. S.: Lab. Invest. 42, 405-412 (1980).
5. Glassock, R. J.: Pathology of basement membranes: alteration in diabetes and other disease states. In: Biology and chemistry of basement membranes, pp. 421-441, Academic Press, New York (1978).

6. Goldstein, I. J., Hayes, C.E.: Adv. Carbohydr. Chem. Biochem. 35, 127-339 (1978).

7. Hatten, M. E., Schahner, M., Sidman, R. L.: Neuroscience 4, 921-935 (1979).

8. Holfhöfer, H., Virtanen, I., Pettersson, E., Törnroth, T., Alfthan, O., Linder, E., Miettinen, A.: Lab. Invest. 45, 391-398 (1981).

9. Holfhöfer, H., Virtanen, I., Kariniemi, A-L., Linder, E., Hormia, M., Miettinen, A.: Lab. Invest. (in press).

10. Latta, H.: Filtration barriers in the glomerular capillary wall. In: Functional ultrastructure of the kidney. pp. 3-29. Academic Press (1980).

11. Maylie-Pfenninger, M-F., Jamieson, J. D.: J. Cell Biol. 86, 96-103 (1981).

12. Michael, A. F., Blau, E. B., Vernier, R. L.: Lab. Invest. 23, 649-657 (1970).

13. Pena, S. D. J., Gordon, B. B., Karpati, G., Carpenter, S.: J. Histochem. Cytochem. 29, 542-546 (1981).

14. Watanabe, M., Muramatsu, T., Shirane, H., Ugai, K.: J. Histochem. Cytochem. 29, 779-790 (1981).

RECEPTORS FOR HELIX POMATIA LECTIN ON NORMAL HUMAN LYMPHOCYTES AND LYMPHOCYTES FROM PATIENTS WITH CHRONIC LYMPHOCYTIC LEUKAEMIA

Günther Hermann, Jürgen Lembke, Hans-Peter Weil
Immunologische Abteilung der Chirurgischen Universitätsklinik,
Robert-Koch-Str. 10, 5000 Köln 41, Germany

Dieter Gerecke
Medizinische Universitätsklinik,
Joseph-Stelzmann-Str. 9, 5000 Köln 41, Germany

Introduction

Chronic lymphocytic leukaemia (CLL) is a well defined clinical entity. The diagnosis is traditionally based on a Giemsa-stained blood-smear showing an absolute increase of lymphocytes of mature appearance. The symptoms of CLL have been staged by Rai et al. (1). CLL-lymphocytes do not differ from normal small lymphocytes by morphology. Immunochemical data about the nature of these lymphocytes come from cell surface studies using agglutination reactions (E-rosettes) as well as immunoglobulin and lectin binding to these lymphocytes. The demonstration of surface immunoglobulin in more than 90% of the cases of CLL classifies this disease as a B-cell disorder. Synthesis and secretion of these immunoglobulins is obviously impaired.

Helix pomatia lectin (Hp) consists of six nearly identical polypeptide chains, each possessing a combining site with specificity for methyl-α-N-acetyl-D-galactosamine. The lectin preferably agglutinates human A-erythrocytes and binds in a specific manner to neuraminidase-treated human T-lymphocytes, a subpopulation of B-lymphocytes, and most B-lymphocytes in human cord blood (2,3). Hp was therefore suggested to be a differentiation marker for B-lymphocytes (3).

In the present investigation, lymphocytes have been qualitatively classified with fluorochrome labeled Hp-lectin and anti human immunoglobulin by immunofluorescence. Hp positive lymphocytes and surface immunoglobulin positive lymphocytes (sIg) were present as well as double labeled cells.

Since most CLL-lymphocytes are Hp-receptor positive (4), peroxidase labeled Hp-lectin was used as a specific probe. By the method indicated below, number and affinity of lectin receptors of normal and leukaemic lymphocytes were calculated following Scatchard (5,6).

Materials and methods

Blood from healthy volunteers (23-67 of age) was used to obtain normal lymphocytes.

CLL-lymphocytes were obtained from patients with chronic lymphocytic leukaemia verified by clinical course and cytology.

Lymphocytes were isolated by Ficoll Paque (Pharmacia Fine Chemicals, Uppsala) density gradient centrifugation as indicated by the manufacturer. Isolated cells are more than 95% mononuclear leukocytes, the viability is more than 95% as measured by trypanblue exclusion.

Neuraminidase-treatment: To unmask receptors for the lectin, 1.25×10^7 lymphocytes were suspended in 1 ml of RPMI 1640. The medium was supplemented with $NaHCO_3$ (2 g/l), and bovine serum albumine (10 mg/ml) (BSA, Behringwerke, Marburg/Lahn), referred to in the following as RPMIs. The resuspended cells were then incubated with 0.005 ml neuraminidase from V.cholerae (specific activity 1 U/ml, Behringwerke, Marburg/Lahn) for 45 min. at 37 ^{o}C, and subsequently washed twice with cold phosphate buffered saline (PBS, Seromed, Munich).

Preparation of Rhodamine-labeled Hp (Hp-TRITC): Hp is labeled with tetra-methyl-rhodamin-isothiocyanat (TRITC, chromatographically pure, Serva, Heidelberg) using a slightly modified method as indicated by Nairn (7). The Hp-TRITC used in these experiments has a Hp-concentration of 0.45 mg/ml and a F/P-ratio of 1.

Fluorescence labeling of cells: Lymphocytes (5 x 10^6/ml) were incubated either with Hp-TRITC (0.05 ml; Hp content 0.1 mg/ml) or with FITC-labeled anti human immunoglobulin (goat-anti-human-immunoglobulin, classes G, A and M; Behringwerke, Marburg/Lahn, 0.05 ml, diluted 1 : 3) for 1 h at 4 °C. After three washings, cells were resuspended in 0.05 ml PBS-glycerol with sodium azide (5 mg/ml). For detachment of immunoglobulins bound by adsorption, cells were preincubated in serum-free medium for at least 45 min. at 37 °C (8). For double labeling studies, the neuraminidase-treated lymphocytes were first incubated with anti-human-immunoglobulin-FITC and then with Hp-TRITC.

Counting of cells: At least 500 cells were counted alternatively by transmitted light and fluorescence under the microscope Zeiss "universal" (filters for FITC-fluorescence: KP 436 nm; LP 520 nm; filters for TRITC-fluorescence: KP 546 nm; LP 590 nm).

Peroxidase-labeled lectin was prepared and purified as described by Weil et al. (5).

Quantitative binding studies: Lymphocytes (5 x 10^5/ml RPMIs) were incubated with increasing amounts of peroxidase-labeled lectin for 6 h at 4 °C.

Measurement of membrane-bound lectin: After incubation, free lectin was removed by washing the cells twice with ice-cold PBS for 5 min. at 1500 x g. The membrane-bound lectin was eluted by a 10 min. incubation at 37 °C with 0.1 ml 0.1 M solution of N-acetyl-D-galactosamine (GalNAc). Then 1 ml of chromogenic substrate solution is added and the cells are spun down for 5 min. at 4000 x g.

Chromogenic substrate: 2 mM solution of 2,2'-azino-di(3-ethylbenzo-thiazolinesulfonate-6); (ABTS, Boehringer, Mannheim) in

Donor	Blood-group	Age	Sex	Hp^+ %	sIg^+ %	K 10^{14} l/mol	r_{max} 10^6
1	A	60	f	54	13°	4,8	12,6
2	A	25	m	61	14°	3,3	11,5
3	A	55	f	55	n.t.	2,8	6,1
4	A	45	m	7o	11°	5,6	3,2
5	A	23	f	59	30^+	1,1	11,3
6	O	67	m	60	32^+	0,9	9,3
7	A	53	f	78	26^+	4,8	5,4
8	n.t.	35	m	68	40^+	1	9,6
9	A	26	m	64	43^+	1,9	4,7
10	A	25	m	74	$1o^o$	3,8	8,1
11	O	39	f	56	12°	1,9	14,3
average		41,2		63,5	$34,2^+$	2,9 (m+f)	8,7
					12°	30,8 (f)	9,94
						2,75(m)	7,o6

Table 1. Experimental data of lymphocytes from 11 healthy donors. Abbreviations: Hp^+; sIg^+; Hp^+-sIg^+: percent of lymphocytes with receptors for Hp-lectin, anti-human-immunoglobulins or both, respectively. K: apparent association constant; r_{max}: maximum number of bound lectin molecules per Hp^+-cell.

0.15 M acetic-acid-sodium-acetat buffer pH 5.0 with BSA (5 mg/ml). After centrifugation, the reaction was started with 0.1 ml 0.05mM H_2O_2 and followed over 2 min. at 405 nm.

Analysis of data: The measured peroxidase activity is proportional to the amount of membrane-bound lectin. From the latter, r = bound lectin molecules per Hp-positive cells and

$$r/A = \frac{\text{bound lectin molecules per cell}}{\text{molar concentration of free lectin}}$$

were calculated. To obtain the so called Scatchard plot, r/A is plotted versus r and should give a linear plot with the equation $r/A = KXr + Kxr_{max}$ (6). K is the apparent association constant and r_{max} the maximum number of membrane bound lectin molecules per cell.

Results

The binding of Hp-TRITC was specifically inhibited by preincubation of the labeled lectin with 0.01 ml of 0.1 M D-GalNAc for 10 min. at 37 °C. Anti-human immunoglobulin bound in a specific manner to lymphocytes since the reaction was inhibited by preincubation of the antiserum with undiluted citrated donor-plasma resulting in precipitation of the anti-immunoglobulin antibodies.

On the average, healthy donors showed 63% Hp-positive lymphocytes versus 72% Hp-positive cells in CLL-patients.

Without preincubation in serum-free medium, 34% of normal lymphocytes had immunoglobulin on their surface. After detachment of adsorbed immunoglobulins by preincubation only 12% of lymphocytes showed membrane-standing immunoglobulins. So about 20% of lymphocytes bound immunoglobulin by adsorption. 64% of CLL-lymphocytes expressed immunoglobulins on their cell surface. 2% double labeled cells - cells with receptors for Hp and anti-human-immunoglobulin - were observed in healthy donors, whereas 47% of leukaemic lymphocytes showed double labeling.

The binding of peroxidase-labeled Hp was inhibited by 0.001 M GalNAc and 0.001 M N-acetyl-D-glucosamine to more than 90%, whereas other monosaccharides only partially inhibited even at high concentrations (0.1 M). At 4 °C incubation, a saturation of peroxidase-labeled lectin binding was reached after 18 h (fig. 1). Under these conditions we observed only nonlinear Scatchard-plots (fig. 2). Therefore, a 6 h incubation period

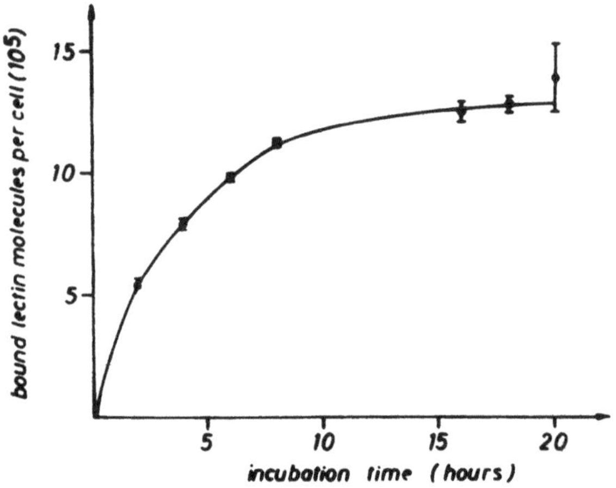

Fig. 1. Kinetics of lectin binding to normal lymphocytes. 5×10^5 neuraminidase-treated normal lymphocytes are incubated with 1.25 µg peroxidase-labeled Hp-lectin at 4 °C.

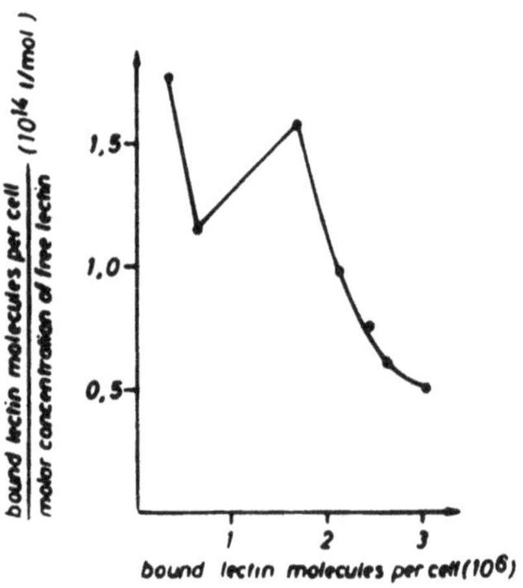

Fig. 2. Binding of peroxidase-labeled lectin to normal human lymphocytes. 2.5×10^5 normal lymphocytes are incubated with increasing amounts (0.4-7.8 µg) of labeled lectin for 18 h at 4 °C. Data are calculated according to Scatchard (6).

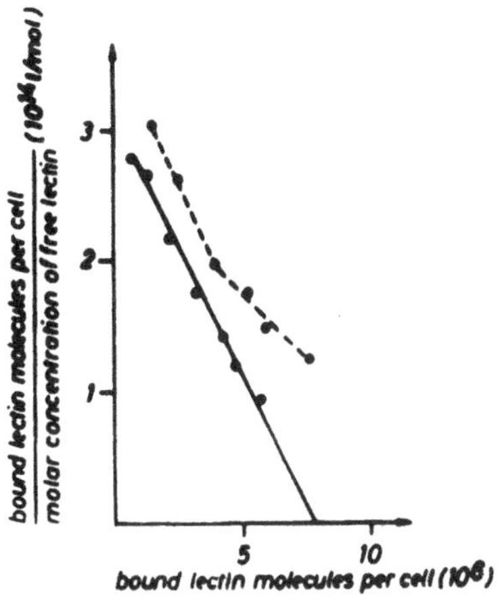

Fig. 3. Binding of peroxidase-labeled lectin to lymphocytes from two healthy donors (no. 10 and no. 11). 5 x 10^5 neuraminidase-treated lymphocytes are incubated with increasing amounts of labeled lectin for 6 h at 4 ^oC. Surface bound lectin is determined, data are calculated according to Scatchard (6).
No. 10 = ●——● No. 11 = ●- - - -●

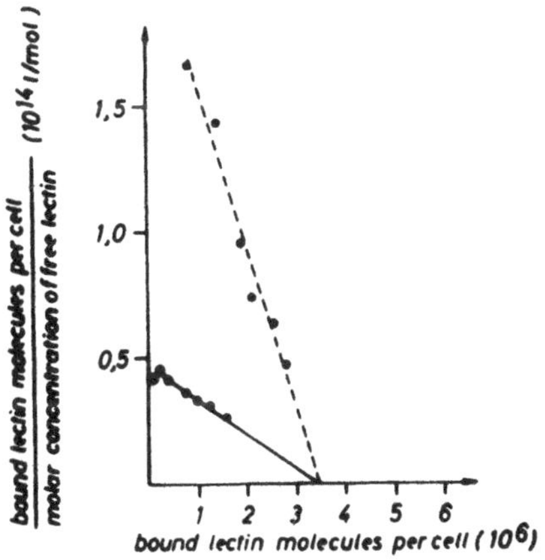

Fig. 4. Binding of peroxidase-labeled lectin to leukaemic lymphocytes. 5 x 10^5 lymphocytes (patients 3 and 8) are incubated with increasing amounts of labeled lectin for 6 h at 4 ^oC. Data are plotted according to Scatchard (6).
No. 3 = ●——● No. 8 = ●- - - -●

Patient	Blood-group	lymphoc. count per mm^3	Hp^+ %	sIg^+ %	Hp^+-sIg^+ %	K 10^{14} 1/mol	r_{max} 10^6
1	A	1300	42	80	34	2,2	4,9
2	O	122000	89	60	60	3,7	3
3	O	35000	75	46	41	1,5	3,2
4	A	48000	55	80	n.t.	0,85	6,9
5	A	23000	80	n.t.	n.t.	1,2	2,9
6	A	44000	64	51	45	n.t.	n.t.
7	n.t.	29000	77	n.t.	n.t.	2,2	3,6
8	A	17000	83	35	35	6,2	3,5
9	O	14000	70	63	3o	1,6	7,7
10	O	65000	84	100	84	1,8	2
average		39800	72	64	47	2,4	4,2

Table 2. Clinical and experimental data from 10 CLL-patients. Abbreviations are the same as in table 1. The staging-scheme of Rai et al. (1) was applied to our patients.

at 4 °C was chosen, since most Scatchard-plots were linear under these conditions (fig. 3 and 4). Lymphocytes of healthy women and men bound, on the average, 8.7 x 10^6 lectin molecules per cell, respectively, whereas the association constant seemed to be the same for both sexes. CLL-lymphocytes bound 4.2 x 10^6 lectin molecules per cell, the association constant being slightly lower compared to normal lymphocytes (2.4 x 10^7 l/mol vs 2.9 x 10^7 l/mol). r_{max} did not seem to be correlated to clinical data as age, duration of disease, stage of disease and lymphocyte count.

Discussion

The qualitative results with fluorescein-labeled indicators showed that 63% of normal neuraminidase-treated lymphocytes are Hp^+. This is in good agreement with the results obtained by other authors (9). In contrast the value of 72% of Hp-TRITC binding leukaemic lymphocytes is lower than the value reported by Hellström et al. (4). This difference may be due to different experimental conditions. We chose only one tenth of the fluorescence-labeled lectin-concentration used by Hellström for the detection of Hp^+-cells in order to decrease agglutination. Also the selection of patients may have been different. Therefore the stage of the disease should be taken into consideration. All examined patients had elevated numbers of immunoglobulin-bearing lymphocytes. In most cases the fluorescence with anti-human-immunoglobulin was weaker than for normal lymphocytes. These findings indicate the B-cell nature of the disease and the functional immaturity of CLL-lymphocytes, confirmed by the lower quantitative binding of Hp-lectin molecules.

It is of importance that we observed only nonlinear Scatchard-plots after 18 h incubation at 4 °C and even some nonlinear plots after 6 h incubation at 4 °C. This is probably an indication for receptor site heterogeneity. One could expect this receptor site heterogeneity from our large value for r_{max}, since it seems improbable that about 10^7 identical glycoprotein-molecules are present on one lymphocyte. The rather high value for r_{max} and the low value for K needs further consideration in comparison with the data of Hammarström (9), who found 1.6×10^6 for r_{max} and $5\text{-}7 \times 10^8$ l/mol for K. The difference may partially be explained by the molecular weight of the employed lectin conjugates. Hp-POD has a somewhat higher molecular weight. Therefore not all receptors with the same affinity may be reached. Another explanation is given by our own experiments (5), showing that the parameters of incubation temperature, incubation time, cell density in the solution and lectin to cell ratio influence widely the obtained and calculated data. Therefore, the quantitative results are only valuable under the strict experimental conditions indicated above.

Leukaemic lymphocytes bind less Hp-molecules than normal cells, whereas the association constant is similar. This suggests that the receptors on both cells are identical. This interpretation is supported by the experiments of Axellsson et al. (lo), who were able to isolate the same Hp-binding molecule from normal and leukaemic lymphocytes. Hp-lectin agglutinates stronger tumor cells from patients with blood-group A than those from patients with other blood-groups (11). Our data (tab.1,tab.2) do not permit conclusions about the influence of blood-groups on normal or CLL-lymphocytes, because most of the donors belong to blood-group A, the number of examined cases is small. But the individual values for K and r_{max} of donors with blood-group O are variable only in usual limits as compared to persons with blood-group A. Our results confirm those of HARDING, CROWTHER and GALLAGHER, who examined the lymphocytes of 4 patients with CLL and 5 volunteers by quantitativ immuno-fluorescence assays and showed that Hp-lectin binds only to desialated cells and that CLL- and other pathologic lymphocyte cell types bind less Hp-lectin than normal peripheral blood-lymphocytes. Double marked lymphocytes are not described by these authors. In conclusion it seems evident that most CLL-lymphocytes belong either to a subgroup of B-lymphocytes characterized by receptors for immunoglobulins and Hp-lectin or that in CLL-lymphocytes a Hp-receptor appears as a transient marker due to abnormal differentiation of these cells.

References

1. Rai, K.R., Sawitsky, A., Cronkite, E.P., Chanama, A.D., Levy, R.N., Pasternak, B.S.: Blood 46, 219-234 (1975).
2. Hammarström, S.: Ann. N.Y. Acad. Sci. 234, 183-197 (1974).
3. Hellström, U., Perlman, P., Robertson, E.S., Hammarström, S.: Scand. J. Immunol. 7, 191-197 (1978).
4. Hellström, U., Mellstedt, H., Perlman, P., Holm, G., Pettersson, D.: Clin. Exp. Immunol. 26, 196-203 (1976).
5. Weil, H.-P., Lempfrid, K.W., Hermann, G.: in: The lectins-biology, biochemistry, clinical biochemistry 2, pp ed. by T.C. Bøg-Hansen, W. de Gruyter, Berlin 1982
6. Scatchard, G.: Ann. N.Y. Acad. Sci. 51, 660-672 (1949).
7. Nairn, R.C.: in: Fluorescent protein tracing, pp 369-371 4th ed., ed. by R.C. Nairn, Churchill Livingstone, Edinburgh, London, New York 1976
8. Winchester, R.J., Fu, S.M.: Scand. J. Immunol. 5, 77-82 (1976).
9. Hammarström, S., Hellström, U., Perlman, P., Dillner, M.L.: J. Exp. Med. 138, 196-203 (1973).
10. Axellsson, B., Kimura, A., Hammarström, S., Wigzell, H., Nillsson, K., Mellstedt, H.: Eur. J. Immunol. 8, 757-764 (1978).
11. Prokop, O., Graffi, A., Hoffmann, F., Schnitzler, St., Dt. Gesundh.-Wesen 23, 1926-1929 (1968)
12. Harding, M., Crowther, D., Gallagher, J., this volume.

LECTIN BINDING PATTERNS OF NORMAL AND LEUKAEMIC LYMPHOID AND MYELOID CELLS

M. Harding, D. Crowther and J.T. Gallagher

Cancer Research Campaign Department of Medical Oncology, Manchester University and Christie Hospital & Holt Radium Institute, Wilmslow Road, Manchester M20 9BX, England

Leukaemia arises predominantly in two cell lineages, and in each the disease may follow an acute or chronic clinical course. Of the lymphoid malignancies, chronic lymphocytic leukaemia (CLL) cells, in the majority of cases, express surface immunoglobulin and are considered to be B lymphocyte disorders (1). Acute lymphoblastic leukaemia (ALL) is classified on the basis of cell surface phenotype, which has important implications for prognosis: the uncommon T (characterised by sheep red cell rosetting) and B cell types have a poor long term survival in both children (9) and adults (16). Common ALL cells which lack surface markers of T or B lymphocytes are defined by the common ALL antigen, a glycoprotein found also on small numbers of normal bone marrow cells (15). Patients with ALL of common or null (i.e. devoid of all surface markers) types have a significantly better prognosis (9,16). By comparison with lymphoid cells, surface phenotyping of the myeloid lineage, both normal and malignant, is limited to the Ia-like antigen (13) (expressed also by immature lymphocytes) and in consequence, there are few cell surface features of prognostic value.

To further characterise alterations in the plasma membrane saccharide composition of leukaemic cells, we have studied quantitatively the cell surface binding of several fluorescein conjugated lectins. Results have been compared with those of their mature counterparts, namely peripheral blood lymphocytes for ALL and CLL and granulocytes for acute myelogenous leukaemia (AML).

Lectins, Vol. III

Materials and Methods

Cells from 11 patients with AML, seven with ALL (one of T cell phenotype) and four with CLL were studied at the time of diagnosis, before therapy was started. A further patient with T cell CLL was included in view of the rarity of this malignant phenotype; although he had been treated for one year, his disease was uncontrolled. Cells from five normal donors were used for comparison.

Normal lymphocytes and leukaemic cells were isolated from peripheral blood by density gradient centrifugation on Ficoll (density 1.077: Flow Laboratories). Monocytes were removed as described by Boyum (8). Granulocytes were separated by the method of Barrett et al (5). An aliquot of each sample was treated with Clostridium perfringens neuraminidase (Sigma Chemicals) 0.05 units/10^7 cells/ml for 30 mins at 37°C in PBS/BSA (phosphate buffered saline, pH 7.2, 0.1% w/v bovine serum albumin). Cell viability was >95% as measured by Trypan blue exclusion. Leukaemic samples rarely included morphologically normal lymphocytes. Those that did were only analysed if there was clear separation of the cell populations on the basis of light scatter, such that the leukaemic cells could be analysed separately. All cell samples were adjusted to 2 x 10^6 cells/ml in PBS/BSA and stored on ice.

Lectins were obtained commercially as affinity purified, fluorescein conjugated preparations, values in parentheses are the molar ratios of fluorescein:protein (F:P). Wheat germ agglutinin, WGA (1.3:1) and Ricinus communis agglutinin, RCA, (6.4:1) from Miles Chemicals Ltd., Helix pomatia agglutinin, HPA (0.56:1) from Polysciences, Glycine max (soybean agglutinin), SBA, (3.9:1) from Vector Laboratories and Wisteria floribunda agglutinin, WFA, (4.5:1) from E-Y Laboratories.

10^5 cells (50 ul of 2 x 10^6 cells/ml) were added to lectin dilutions in 100 ul PBS/BSA. After 30 mins incubation at 4°C the cell surface fluorescence was measured in a Biophysics Cytofluorograf 4800A (Ortho Instruments) interfaced to a Hewlett Packard 9845S computer for data accumulation (7). Mean cell fluorescence was corrected for cell autofluorescence, the F:P ratio of each lectin and the cell surface area calculated from the measured average cell diameter. Specificity of lectin binding was confirmed by >90% reduction in fluorescence of cells incubated with lectin in the presence of an appropriate saccharide inhibitor.

Results are expressed as the relative number of lectin molecules bound per unit surface area as cell diameters varied from 7u for PBL to 17u for some AML cells. Lectin concentrations used are given in the legends to Figs. 1 and 2. They were selected to give >70% saturation of lectin binding sites.

Results

The lectins studied comprised two groups, WGA-F and RCA-F, which bound to untreated cells of all types, although after neuraminidase treatment RCA-F binding increased (results not shown) and WGA-F binding was reduced (Fig.1). In contrast, HPA-F, SBA-F and WFA-F bound significantly only to desialylated cells.

RCA is thought to have an extended carbohydrate binding site (4) and shows the highest affinity for asparagine linked sequences with terminal galactose residues. The interaction of RCA-F with untreated cells is shown in Fig.1. There was a clear distinction between the high density RCA-F binding galactose sites on normal lymphocytes and the lower density on lymphoid leukaemic cells. In the myeloid series normal granulocytes showed significantly more residues recognised by RCA-F per unit surface area than did AML cells. The average binding site density was slightly greater on lymphocytes than granulocytes, but appeared similar between AML and ALL cells.

WGA interacts with both sialic acid residues (6,17) and oligosaccharides of N-acetyl glucosamine (GlcNAc). Fig.1 shows that WGA-F binding to all cell types was reduced by neuraminidase treatment. The percentage reduction was similar for each sample of lymphoid cells. Untreated lymphocytes had a higher density of WGA-F binding saccharides than either ALL or CLL cells. This distinction became more marked after desialylation. In contrast, untreated normal and leukaemic cells of the myeloid lineage showed similar densities of

WGA-F binding sites. However, after neuraminidase treatment, there was a significant reduction in WGA-F binding per unit surface area on AML cells: by comparison WGA-F binding to granulocytes was only slightly reduced by the enzyme treatment (Fig.1). The density of

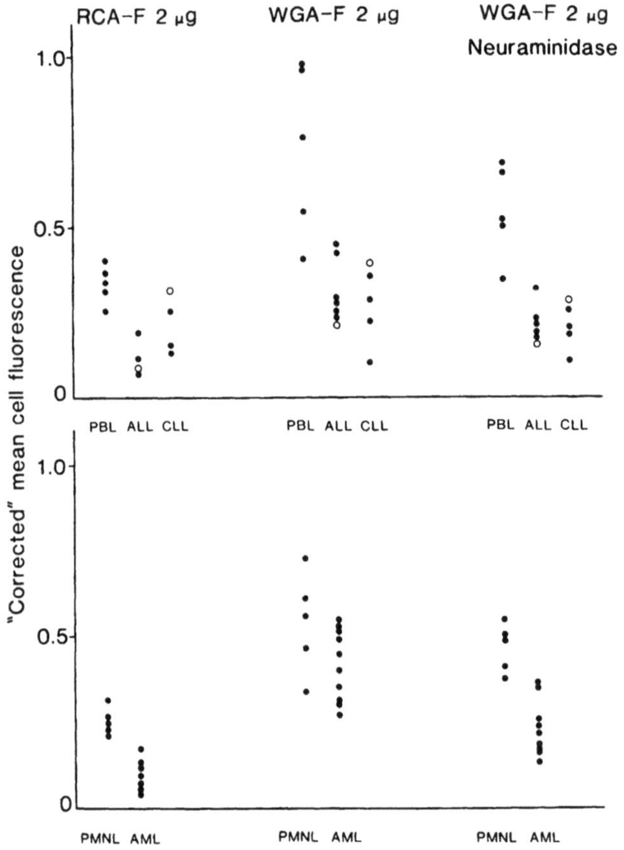

Fig.1. Binding of RCA-F and WGA-F to normal and leukaemic cells. The corrected mean cell surface fluorescence refers to the relative numbers of lectin molecules bound/unit surface area. Open circles are results obtained from malignant lymphocytes of T cell phenotype.
PBL = peripheral blood lymphocytes
PMNL = polymorphonuclear leucocytes ≡ granulocytes (mature myeloid cells)
ALL = acute lymphoblastic leukaemia
AML = acute myeloblastic leukaemia
CLL = chronic lymphocytic leukaemia

neuraminidase resistant WGA-F sites was similar on AML and ALL cells, although with untreated cells the AML population displayed the higher binding site density.

Results for the N-acetyl galactosamine (GalNAc) binding lectins HPA-F, WFA-F and SBA-F to neuraminidase treated cells are shown in Fig.2. Unseparated lymphocytes (mainly of T phenotype) bound more HPA-F molecules than any other cell type.

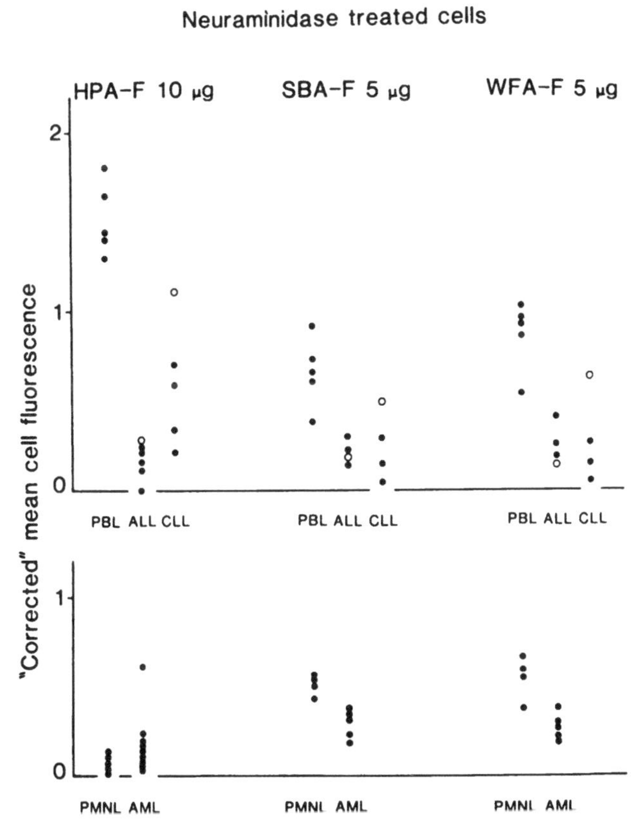

Fig.2. Binding of N-acetyl galactosamine specific lectins to neuramindase treated normal and leukaemic cells.
Experimental details as in text. Abbreviations as for Fig.1.

ALL cells had a binding site density for HPA which was only 10% of the density of HPA sites on normal lymphocytes. Also of note was the higher density of HPA-F binding sites on CLL cells than ALL cells, and the low density of HPA sites on normal and leukaemic myeloid cells. Lymphocytes showed a tenfold greater density of HPA binding sites than granulocytes indicating a significant difference in surface saccharide composition between these two groups of normal peripheral blood cells. However, acute leukaemic cells of lymphoid and myeloid lineages gave low binding of HPA-F.

Desialylated lymphocytes bound similar amounts of both SBA-F and WFA-F although this was only half the amount of HPA-F bound by these cells. In contrast, granulocytes bound four times more WFA-F and SBA-F than HPA-F. The density of GalNAc residues recognised by WFA-F and SBA-F was lower on leukaemic cells than normal cells of the same lineage. Furthermore, although lymphocytes had, on average, more WFA-F and SBA-F sites per unit surface area than granulocytes, acute leukaemia cells of both lineages showed similar binding site densities.

It is of interest that the T ALL cells showed a lower cell surface density of most lectin binding saccharides by comparison with the other ALL samples (Figs. 1 and 2). In contrast, the T CLL cells showed a significantly higher density of all lectin binding sites than the B CLL samples. It is, however, possible that therapy may have influenced the surface saccharide composition of the T CLL cells studied.

Discussion

Our findings showed that within each cell type there was a wide range of surface densities of lectin binding saccharides. However, for each lectin there was usually a clear distinction in surface binding between normal and malignant cells of the same lineage, with a lower density of lectin binding sites on leukaemic cells. As, in general, leukaemic cells were larger than normal cells, numbers of lectin molecules bound were often higher on the malignant populations. (Results not shown).

HPA binds to desialylated mature T lymphocytes, the majority of thymocytes and several T cell leukaemic and lymphoma cell lines (3). The lectin also interacts with CLL cells (12, Hermann et al, this volume), most foetal B lymphocytes (13), a minority of adult B lymphocytes (13) and some non-T leukaemia and lymphoma cell lines (3). It is of interest that in many of these cell types a single glycoprotein, MW 150,000 is responsible for almost all HPA binding (3). Furthermore, the erythroleukaemic K562 cell line has detectable HPA-binding surface glycoprotein MW 150,000 (3), suggesting that this 'HPA-recognised site' may be present on very early haemopoietic cells.

We have shown that unseparated lymphocytes bound significantly more HPA-F than any other cell type studied and that granulocytes were markedly deficient in HPA binding saccharides, as were most acute leukaemic cells irrespective of their lineage. The single AML cell sample with more HPA-F binding than the rest of the group indicates that leukaemic myeloid cells may express an HPA recognised glycoconjugate, either as a persistent constituent normally found only on early haemopoietic cell surfaces, or as a consequence of genetic depression at a later stage of maturation. It would appear that HPA binding glycoconjugates, including GP 150,000, are expressed throughout T lymphocyte maturation, but may be deficient in T cell leukaemias (our T ALL cell data). HPA-binding structures may be a transient feature of B cell differentiation since they are found on CLL cells and normal foetal B lymphocytes, but only weakly expressed on common or null ALL cells, which are postulated to represent malignant transformation of cells in the early stages of B lymphocyte maturation (20). Conversely, the myeloid differentiation pathway appears to be associated with early loss of expression of HPA binding.

Our results indicate at least two types of lectin recognised GalNAc saccharides, in that the HPA site is clearly deficient on granulocytes, whereas these cells bind significant amounts of SBA-F and WFA-F. SBA interacts with the HPA binding GP 150,000 (11), and although this is probably absent on granulocytes (2) with consequent minimal HPA-F binding, SBA-F can clearly interact with other available GalNAc residues. Furthermore, SBA binds to untreated thymocytes with which HPA will interact only after desialylation (19).

We suggest (Gallagher et al, this volume) that the majority of WGA-F binding to untreated cells is mediated through sialyl residues whereas after desialylation, binding occurs predominantly through glycans containing GlcNAc. If this is so, lymphoid leukaemic cells are deficient in both sialic acid and GlcNAc sites per unit surface area by comparison with normal lymphocytes. In contrast, AML cells have a lower density of GlcNAc sites than granulocytes, but an almost equivalent density of sialyl sites (Fig.1). The reduction in WGA-F binding to AML cells following neuraminidase treatment was significantly greater than for any other cell type. This may indicate an increase in surface saccharide sialylation of AML cells. Our observations that leukaemic cells express a low density of RCA-F binding sites and do not bind GalNAc reactive lectins unless treated with neuraminidase, indicates the capacity of these malignant cells to extensively sialylate Gal and GalNAc residues.

Analyses of isolated plasma membrane glycoproteins from desialylated leukaemic cells has revealed similar components on non-T, non-B ALL cells and undifferentiated AML cells (2,10). This is in agreement with our data showing that the surface carbohydrate structure, determined by lectin binding, is remarkably similar on acute leukaemic cells irrespective of their derivation.

Although we have demonstrated significant differences between normal and leukaemic cells, further refinements are required before this technique will be useful diagnostically in distinguishing between the different forms of acute leukaemia.

References

1. Aisenberg, A.C. (1981)
N.Engl.J. Med., 304, 331-335

2. Andersson, L.C., Gahmberg, C.G., Siimes, M.A., Teerenhovi, L. & Vuopio, P. (1979)
Int. J. Cancer, 23, 306-311

3. Axelsson, B., Kimura, A., Hammarstrom, S., Wigzell, H., Nilsson, N. & Mellstedt, H. (1978)
Eur. J.Immunol. 8, 757-764

4. Baenziger, J.U. & Fiete, D. (1979)
J.Biol.Chem., 254, 9795-9799

5. Barrett, S., Garratty, E. & Garratty, G. (1979)
Br. J. Haemat. 43, 575-588

6. Bhavanandan, V.P. & Katlic, A.W. (1979)
J.Biol.Chem. 254, 4000-4008

7. Blackledge, G., Swindell, R., Hodgson, B.W. & Crowther, D.(1980)
Int.J.Bio-Medical Computing, 11, 41-52

8. Boyum, A. (1968)
Scand. J.Clin.Lab.Invest. 21 (Suppl 99), 1.

9. Chessells, J.M., Hardisty, R.M., Rapson, N.T. & Greaves, M.F. (1977) Lancet (ii) 1307-1309

10. Gahmberg, C.G., Andersson, L.C. & Nilsson, K. (1980)
Leuk.Res. 4, 279-286

11. Hammarstrom, S., Murphy, L., Goldstein, I.J. & Etzler, M.E.(1977)
Biochemistry, 16, 2760

12. Hellstrom, J., Mellstedt, H., Perlmann, P., Holm, G. & Pettersson, D. (1976)
Clin.Exp.Immunol. 26, 196-203

13. Hellstrom,U., Perlmann, P., Robersson, E.S. & Hammarstrom,S. (1979) Scand.J.Immunol. 7, 191-197

14. Janossy, G., Francis, G.E., Capellaro, D., Goldstone, A.H. & Greaves, M.F. (1978)
Nature, 276, 176-178

15. Janossy, G., Bollum, F.J., Bradstock, K.F., McMichael, A., Rapson, N. & Greaves, M.F. (1979)
J.Immunol. 123, 1525-1529

16. Lister, T.A., Roberts, M.M., Brearley, R.L., Woodruff, R.K. & Greaves, M.F. (1979)
Cancer Immunol.& Immunother. 6, 227-230

17. Monsigny, M., Roche, A.C., Sene, C., Maget-Dana, R. & Delmotte, F. (1980)
Eur. J.Biochem. 104, 147-153

18. Newman, R.A., Sutherland, R. & Greaves, M.F. (1981)
J.Immunol. 126 (5) 2024-2030

19. Richard, Y., Boumsell, L., Coppin, H., Mishall, Z., Lemerle, J. & Bernard, A. (1981)
J.Immunol. 127, 252-255

20. Vogler, L.B., Crist, W.M., Bockman, D.E., Pearl, E.R., Lawton,A.R. & Cooper, M.D. (1978)
New Engl.J. Med. 298, 872-878

THE BINDING AND STIMULATION OF HUMAN LEUKOCYTES BY THE FIVE POKEWEED LECTINS

M. J. Waxdal, B. F. Haynes, B. J. Fowlkes, C. T. Thomas, III, B. J. Mathieson, and A. Fauci

The National Institute of Allergy and Infectious Diseases, National Institutes of Health, Bethesda, Maryland 20205 USA

For many years crude salt extracts, PWM, of pokeweed (*Phytolacca americana*) have been commercially available and used to stimulate mitosis in lymphoid cell cultures from both mouse and man. Because it is one of the few polyclonal activators of human B cells, PWM is commonly used in assays for human B cell function, and for T cell control of that function (1, 2, 3). It is also apparent that macrophage are important in the activation of human B cells (4).

Previous studies (5) have isolated five separate lectins from salt extracts of *Phytolacca americana* roots. These five glycoproteins have been designated Pa-1 through Pa-5 and have distinct physiochemical properties, different biological activities when assayed with murine lymphoid cells (6, 7, 8, 9, 10), and appear at different times of the year. These lectins are specifically inhibited by oligosaccharides of N-acetyl glucosamine (11, 12) and appear to bind to the same sites on human erythrocytes (5, 11) and murine spleen cells (5). With the exception of Pa-4, all are agglutinins. Highly purified Pa-4 is neither an agglutinin, nor a mitogen for murine cells. However, Pa-4 does compete with the other 4 lectins for the same binding sites on erythrocytes and inhibits agglutination by them (5, 11). Pa-1 appears to exist in solution as a hexamer and dodecamer of 22,000 d subunits (13), which may explain why its avidity for human erythrocytes is about 10-fold that of the other 4 lectins (11), which appear to be monomers.

The effect of the 5 individual lectins has been studied with murine lymphoid cells (6, 7, 8, 9, 10). Published studies with human cells (14, 15) have only included Pa-1 and Pa-2. This communication briefly reports the abilities of the 5 pokeweed lectins to stimulate DNA synthesis and immunoglobulin production in cultures of human lymphoid cells. We also report studies on the binding of these lectins to human lymphoid cells.

Materials and Methods

The pokeweed mitogens, Pa-1 to Pa-5, were isolated as previously described (5) with the following modification. The dialyzed root extracts were chromatographed on DEAE cellulose (10 mM phosphate buffer, pH 8) instead of hydroxyapatite. The major portion of the dissolved material was not retained on the column. After washing, a major fraction was eluted with a phosphate buffer and KCl gradient (0.1 M potassium phosphate, pH 5.3 to 0.2 M potassium phosphate, 0.5 M KCl). The eluted material contained the lectins and was dialyzed and lyophilized. The lectins were then isolated by Sephadex gel permeation chromotography as previously described.

Purified human mononuclear cells were obtained by Ficoll-Hypaque density gradient centrifugation of heparinized veneous blood. DNA synthesis was measured by the incorporation of ^{3}H-thymidine. Immunoglobulin production was determined by the reverse plaque assay which determines the number of immunoglobulin secreting cells (ISC).

For flow microfluormetry, the lectins were labeled with fluorescein isothiocyanate in sodium carbonate buffer. Flow microfluorometry analyses were performed using a FACS II as previously described for lectins (16). Flurorescence intensity and forward light scatter intensity (a measure of viability and cell size) were measured simultaneously on at least 50,000 viable cells. Data capture and processing was performed with a Digital PDP-11/34 computer.

Results and Discussion

The abilities of the five purified pokeweed mitogens to stimulate DNA synthesis and immunoglobulin secretion by cultured human peripheral blood mononuclear cells is presented in Table I. These data were obtained using the lowest maximally effective dose of each lectin. Cultures were assayed each day for 7 days. The maximum responses for both DNA synthesis and immunoglobulin production occurred on day 5 and this is the data presented. The lectins Pa-1, 2, 3, and 5 each stimulated human cells in both assays. Pa-4 did not stimulate either of the responses, although it was tested at a concentration range over 5 orders of magnitude.

Table I

Responses[1] of Human PBL to the Pokeweed Lectins

Lectin	μg/ml	DNA[2]	ISC[3]
PWM[4]	200	85	48
Pa-1	1	34	11
Pa-2	0.1	79	54
Pa-3	0.1	99	55
Pa-4	1	1.5	0
Pa-5	1	36	9

[1] Data are the average values from at least 3 normal patients.

[2] E/C of ^{3}H-thymidine incorporation.

[3] Immunoglobulin secreting cells above background determined by the reverse plaque assay of 1000 cells.

[4] Final dilution of 1 to 5 from commercial material.

When the concentration of each stimulating lectin was increased, the DNA response remained at a plateau level to over 100 µg/ml. However, the immunoglobulin production of these cultures rapidly declined. This suggests that excess binding of lectin to the B cell (see below) may inhibit its differentiation, or that increased lectin levels interfer with the abilities of T cells to provide "help" to the B cells.

Flow microfluorometry studies of the binding of each of these lectins to human peripheral blood mononuclear cells indicated that each of these lectins bound strongly to a subpopulation, comprising about 20% of the cells. Much less binding was observed to the remaining cells until the concentration of the fluorescent lectin was raised well above the required for DNA stimulation and into the region that depressed immunoglobulin secretion. Human peripheral blood mononuclear cells were fractionated into (1) monocytes and macrophages, (2) T cells, (3) B cells and (4) Null cells and retested for lectin binding. In each case, the monocytes fraction (about 20% of the initial cells) readily bound each of the lectins (Table II). Lectin binding to the remaining cell populations was much weaker and only about 10-16% of the cells bound detectable amounts of the lectin at mitogenic concentrations (Table II). At higher concentrations binding increased to 30-40%. This increase could be attributed primarily to the B cells, and correlated with a decrease in the immunoglobulin secretion of these cells.

Table II

Binding of Pokeweed Lectins to Human PBL

	Mitogenic Concentration[1]		Higher[2]
Lectin	Monocytes	Non-Monocytes	Non-Monocytes
	RF	%	%
Pa-1	35[3]	16[4]	40[4]
Pa-2	20	16	41
Pa-3	6	10	32
Pa-4	16	4	9
Pa-5	24	13	-

[1] Lowest lectin concentration giving maximum stimulation, see Table I.

[2] 33 μg/ml for Pa-1, 2, and 3. 333 μg/ml for Pa-4.

[3] 100% of monocytes bind each lectin. RF is the relative fluorescence intensity of the bound lectins. This is an approximation of the amount of lectin bound.

[4] Percent of cells binding detectable amount of lectin.

An exception to these observations is the binding of Pa-4 (Table II). This lectin bound to all monocytes, but to about 4% of other cells at 1 μg/ml (about the limit of accuracy of the assay). When the Pa-4 concentration was raised over 300 fold, the binding to non-monocytes cells increased only to 9% and still was of low intensity. These results suggest that Pa-4 binding may serve as a marker for human monocytes and provide a means for their removal and purification.

References

1. Fauci, A. S. 1979. Human B cell function in a polyclonally induced plaque forming cell system. Cell triggering and immunoregulation. Immunol. Rev. 45:93.

2. Fauci, A. S., G. Whalen, and C. Burch. 1980. Activation of human B lymphocytes. XV. Spontaneously occurring and mitogen-induced indirect anti-sheep red blood cell plaque-forming cells in normal human peripheral blood. J. Immunol. 124:2410.

3. Fauci, A. S., K. R. Pratt, and G. Whalen. 1976. Activation of human B lymphocytes. II. Cellular interactions in the PFC response of human tonsillar and peripheral blood B lymphocytes to polyclonal activation by pokeweed mitogen. J. Immunol. 117:2100.

4. Dimitriu, A., and A. S. Fauci. 1978. Activation of human B lymphocytes. IX. Modulation of antibody production by products of activated macrophages. J. Immunol. 120:1818.

5. Waxdal, M. J. 1974. Isolation, characterization and biological activities of 5 mitogens from pokeweed. Biochemistry 13:3671.

6. Waxdal, M. J., and T. Y. Basham. 1974. The B and T cell stimulatory activities of multiple mitogens from pokeweed. Nature 251:163.

7. Basham, T. Y., and M. J. Waxdal. 1975. The stimulation of immunogloblin production in murine spleen cells by the pokeweed mitogens. J. Immunol. 114:715.

8. Waxdal, M. J. 1975. Differential stimulation of murine T and B cell populations by purified mitogens from pokeweed. In Immune Recognition. Ed. A. S. Rosenthal, Academic Press, NY, p. 85.

9. Waxdal, M. J., Nilsson, S. F., and Basham, T. Y. 1976. Heterogenity of the pokeweed mitogens and the responding lymphocytes. In _The Role of Mitogens in Immunobiology_. Eds. J. J. Oppenheim and D. Rosenstreich, Academic Press, NY, p. 161.

10. Basham, T. Y., S. Toyoshima, F. Finkelman, and M. J. Waxdal. 1981. A series of interleukin molecules which stimulate both murine and human lymphocytes. I. Production by Pa-2 stimulated thymus and thymus derived cells. Cellular Immunology 63:118.

11. Yokoyama, K., T. Terao, and T. Osawa. 1978. Carbohydrate-binding specificity of pokeweed mitogens. Biochim. Biophys. Acta 538:384.

12. Unpublished Results.

13. Yokoyama, K., O. Yano, T. Terao, and T. Osawa. 1976. Purification and biological activities of pokeweed mitogens. Biochim. Biophys. Acta 427:443.

14. Janossy, G., E. Gomez de la Concha, M. J. Waxdal, and T. Platts-Mills. 1976. The effects of purified mitogenic proteins (Pa-1 and Pa-2) from pokeweed on human T and B lymphocytes _in vitro_. Clin. Exp. Immunol. 26:108.

15. Janossy, G., E. Gomez de la Concha, A. Luguetti, M. J. Snajdr, M. J. Waxdal, and T. Platts-Mills, 1977. T cell regulation of immunoglobulin synthesis and proliferation in pokeweed (Pa-1) stimulated human lymphocyte cultures. Scand. J. Immunol. 6:109.

16. Fowlkes, B. J., M. J. Waxdal, S. O. Sharrow, C. A. Thomas, III, R. Asofsky, and B. J. Mathieson. 1980. Differential binding of fluorescein-labeled lectins to mouse thymocytes: Subsets revealed by flow microfluorimetry. J. Immunol. 125:623.

THE USE OF Pa-4 FOR THE IDENTIFICATION AND REMOVAL OF MONOCYTES FROM HUMAN PERIPHERAL BLOOD MONONUCLEAR CELLS

C. Murre and M. J. Waxdal

Laboratory of Immunology, National Institute of Allergy and Infectious Diseases, National Institutes of Health, Bethesda, Maryland 20205 USA

A number of fluorescein isothiocyanate labeled lectins were prepared and assayed for their ability to discriminate between subpopulations of human peripheral blood mononuclear cells. Flow microfluorometric analyses indicated that several of these lectins bound predominantly to human monocytes. The non-agglutinating lectin, Pa-4, isolated from the roots of Phytolacca americana (pokeweed), was found to be the most specific lectin probe for human monocytes. When Pa-4 was covalently bound to an insoluble support, it could be used to remove essentially all monocytes. The bound cells could be recovered by treatment with a specific inhibitor of Pa-4. The results presented here suggest that Pa-4 may provide a useful tool for the identification and separation of human monocytes.

Materials and Methods

Pa-4. This single polypeptide chain, 21,000 dalton glycoprotein was isolated from the roots of Phytolacca americana (pokeweed) by ion exchange and gel permeation chromatography as previously described (1). This lectin, purified to homogeniety, binds to cell surface glycoconjugates, but is neither an agglutinin nor a mitogen. The cellular binding is specifically inhibited by oligomers of N-Acetyl glucosamine (GlcNAc). Pa-4 was labeled with fluorescein isothiocyanate in sodium carbonate buffer to yield a fluorescent reagent for cell staining. Cyanogen bromide activated Sepharose 6MB beads were coupled to Pa-4 in sodium borate buffer (2). Control beads were coupled with bovine serum albumin (BSA).

Cells. Human peripheral blood mononuclear cells were isolated by plateletpheresis of venous blood of healthy donors, followed by centrifugation of the cell pellet on Ficoll-Hypaque. Monocyte depleted cells were obtained by specific agglutination of monocytes with the complex carbohydrate PWCH (3, 4). Monocytes were obtained by disaggregation of the agglutinated complex with α-methyl mannoside and removal of specifically agglutinated T cells by E rosetting (3, 4).

Detection of Monocytes. Non-specific esterase staining of monocytes was performed by a modification of the method described by Tucker et al. (5). Reagents for this stain were obtained from Technicon (Terrytown, New York, USA). Fixed cells were incubated with substrate and dye for 8 minutes at 37°. Microscopic examination of a typical suspension of human cells showed 15-20% dark red to brown staining cells which were scored as esterase positive monocytes. The mouse monoclonal antibody, OKM1, which primarily stains human monocytes was obtained from Ortho Diagnostic systems (Raritan, New Jersey, USA) and used for flow microfluorometric analyses.

Flow Microfluorometry. Low angle forward light scatter (a measure of cell size and viability) and fluorescence intensity analyses were performed using a FACS II (Becton Dickinson, Mountain View, CA, USA) as previously described for lectins (6). Measurements were made on at least 50,000 viable cells in each analysis and the data stored and analysed by computer. The proportion of brightly stained cells, corrected for control values, are expressed as "per cent positive."

Removal of Monocytes from PBL by Binding of Pa-4 Beads. Human mononuclear cells were suspended in PBS or RPMI-1640, containing 5% fetal calf serum at a concentration of 5 x 10^6 cells/ml. One ml of this cell suspension was mixed with 100 µl of a 50% (v/v) suspension of Pa-4 coupled Sepharose 6MB beads. Control experiments were performed using BSA coupled beads. The mixed suspension was centrifuged for 5 min to form a pellet of cells and beads. Most of the supernatant was removed and the pellet was incubated for 30 min at 37°. The cell pellet was then resuspended, centrifuged, and incubated for an additional 15 minutes. The cells which did not adher to the beads were removed by three cycles of suspension and decantation with 1 ml of PBS or Delbucco's PBS. The unbound cells were recovered and assayed for the presence of monocytes by non-specific esterase staining.

The cells which bound to the Pa-4 coupled beads were released by treatment with 1 ml of buffered, Ca^{+2} free, saline containing the Pa-4 inhibitor (GlcNAc)n. This inhibitor was a mixture of oligosaccharides of N-acetylglucosamine and predominately contained $(GlcNAc)_4$. The concentrations stated for this inhibitor refer to the GlcNAc residue. If the experiment had been carried out in Ca^{+2} free media, 10 mM of the inhibitor was sufficient. If Ca^{+2} was present, the eluting solution contained 25 mM of the inhibitor and 2 mM EDTA. The inhibitor-released cells were removed from the beads by repeated decantation and assayed for monocytes by non-specific esterase staining.

Results and Discussion

Flow microfluorometric analyses of human peripheral blood mononuclear cells stained with fluorescene labeled Pa-4 showed a well resolved brightly staining population of cells (Pa-4 positive, Table I). The percent of Pa-4 positive cells correlated very well with the values obtained for OKM1 and non-specific esterase positive cells (Table I). These results suggested that Pa-4 may bind preferentially to human monocytes. Further analyses on purified human monocytes and on monocyte depleted cell populations (Table I), showed that Pa-4 bound specifically to human monocytes and not to other mononuclear cells. The results from experiments on cells from 10 normal patients were similar and consistant, indicating that Pa-4 could be readily used to identify human monocytes.

Although Pa-4 readily bound to human monocytes, it did not agglutinate them, even at high concentrations. This property enhanced the use of fluorescence labeled Pa-4 to stain monocytes, but prevented their easy removal by agglutination. To use the specific-

Table I

Identification of Human Monocytes by Pa-4

Cells	Percent Positive[1,2]		
	Esterase	OKM1	Pa-4
Unfractionated[3]	15	14.5	14
Monocyte Depleted	<0.1	<0.1	<0.1
Monocytes	85	87	86

[1] Values are from a single typical normal patient. Nine other normal patients gave similar results.

[2] Positive cells were scored as described under Materials and Methods.

[3] Human peripheral blood mononuclear cells.

ity of Pa-4 for monocytes in cell separation procedures, several insoluble derivatives of Pa-4 were prepared. One of the most effective was obtained by covalently coupling the lectin to Sepharose 6MB. This procedure resulted in beads which could multivalently bind cells with glycoconjugates recognized by the attached Pa-4. The ability of these beads to remove essentially all monocytes from human peripheral mononuclear cells is presented in Table II. Control experiments, using Sepharose 6MB coupled with bovine serum albumin instead of Pa-4, removed few monocytes (Table II). In these experiments, a critical factor is the effective contact between the Pa-4 derivatized beads and the cells. The double incubation procedure described above must be carefully followed to insure full depletion of monocytes.

Table II

Selective Removal of Monocytes by Pa-4 Derivatized Sepharose Beads

Cell Population	Esterase Positive[1]
Unfractionated	18%
Not bound to BSA beads	16%
Not bound to Pa-4 beads	<1%
Bound and released from Pa-4 beads	69%

[1] Average of 3 experiments with the same preparation of human peripheral blood mononuclear cells.

ed above. The major source of contamination of the specifically bound and released monocytes are cells which settle and lodge between the beads.

The contamination may be kept to a minimum by careful washing of the cell bound beads by repeated suspension and decantation prior to adding the oligomers of GlcNAc.

All these data suggest that Pa-4 interacts specifically and preferentially with human monocytes. In addition to the use of Pa-4 as a marker, this property can be utilized for the removal and enrichment of monocytes from human peripheral blood mononuclear cells.

Acknowledgment

The authors wish to thank S. Trost for technical assistance. S. Sharrow and D. Stephany for the flow microfluorometry analyses and M. Monsigny for the oligomers of GlcNAc.

References

1. Waxdal, M. J. 1974. Biochemistry 13:3671.
2. Murre, C., C. Kieda, and M. J. Waxdal. 1982. Submitted for review.
3. Kieda, C., M. Monsigny, and M. J. Waxdal. 1982. This volume.
4. Kieda, C., and M. J. Waxdal. 1982. Submitted for review.
5. Tucker, S. B., R. V. Pierre, and R. E. Jordan. 1977. J. Immunol. Methods 14:267.
6. Fowlkes, B. J., M. J. Waxdal, S. O. Sharrow, C. A. Thomas, R. Asofsky, and B. J. Mathieson. 1980. J. Immunol. 125:623.

ANALYSIS OF SUBPOPULATIONS OF HUMAN PERIPHERAL BLOOD LEUKOCYTES AND SPLENIC CELLS ISOLATED BY PEANUT AGGLUTININ

B.J. Mathieson[1], B.F. Haynes[2], M.J. Waxdal, B.J. Fowlkes, C.A. Thomas III, and S.O. Sharrow

Laboratory of Microbial Immunity, Laboratory of Clinical Immunology, and Laboratory of Immunology, National Institute of Allergy and Infectious Diseases, and Immunology Branch, National Cancer Institute, National Institutes of Health, Bethesda, Maryland, USA

Introduction

Human mononuclear cells from peripheral blood are comprised of several subpopulations based on cellular morphology and on cell surface characteristics such as the ability to bind heterologous erythrocytes, the ability to bind antigen-antibody complexes or the ability to bind and ingest latex particles. Subpopulations can be shown to selectively respond to different mitogenic stimuli _in vitro_ and to be susceptible to pharmacologic agents _in vivo_ and _in vitro_. The study and dissection of human lymphocyte subpopulations has been recently facilitated by the use of monoclonal antibodies that bind to antigenic determinants expressed on the cell surface. (For review see Haynes, 1; and Reinherz and Schlossman, 2.)

To analyze the apparent differentiation of human peripheral blood lymphocytes (PBL), we have surveyed a panel of lectins that we have previously demonstrated will bind differentially to mouse thymocytes and spleen cells (3, 4). We were interested in comparing the lectin binding patterns to those obtained with monoclonal reagents that are now available from a variety of sources, because this differential binding might give some

Lectins, Vol. III

indication of the relation of the known carbohydrate-binding specificities of the lectins to that of the monoclonal antibodies detecting antigenic determinants on different glycoproteins. Furthermore we were interested in determining whether any of the lectins could detect or delineate specific subsets not recognized by the currently available monoclonals.

A selected panel of 20 lectins was examined for differential binding to human peripheral blood leukocytes and to human splenic cells. This report will be limited to a discussion of the relative binding of lectins from this panel to cells separated by peanut lectin (PNA) agglutination from normal cell donors. We also report the initial characterization and the functional capacity of PNA separated cells measured by the proliferative response when stimulated _in vitro_ with mitogenic lectins.

Materials and Methods

Lectins. The lectins that were used for these analyses included mannose binding lectins from _Pisum sativum_ (Pea lectin), _Conconavalia ensiformis_ (Concanavalin A, Con A), and _Lens culinaris_ (Lentil lectin); galactose binding lectins from _Arachis hypogaea_ (peanut agglutinin, PNA), _Abrus precatorius_ (Abrus lectin), and _Maclura pomifera_ (MPL, osage orange lectin; the N-acetyl galactosamine binding lectin from _Glycine soja_ (Soybean agglutinin, SBA); and several other lectins with more complex carbohydrate specificities that included poly-N-acetyl glucosamine binding lectins from _Phytolacca americana_ (five pokeweed isolectins, Pa-1 to Pa-5), _Solanum tuberosum_, (Potato lectin, Spud-1); the sialic acid binding lectin from _Homarus americanus_ (Lobster lectin); the fucose binding lectins from _Ulex europus_ (Ulex I) and _Tetragonolobus_ (Lotus lectin); _Helix_ pomatia agglutinin (HPA), the agglutinin from Phaseolus vulgaris (PHA) and wheat germ agglutinin (WGA) from Triticum vulgaris. These and other lectins and the sources from which we obtained the lectins used in these studies are tabulated elsewhere (3).

Identification and fractionation of leukocyte populations. Purified mononuclear cell suspensions were obtained by standard Hypaque-Ficoll density centrifugation from heparinized venous blood drawn from normal adults or from peripheral blood suspensions from leukaphareses of normal human volunteers (5,6). Small pieces of cadaver splenic tissue, obtained 24 to 48 hr after death, were obtained from excess material supplied for HLA typing to the tissue typing laboratory, Georgetown University Medical Center, Washington, D.C. Leukocyte suspensions were obtained by teasing cells free of suporting tissue into Hanks' balanced salt solution and by removing red blood cells by ammonium cloride-hypotonic lysis. Further separation of cells to yield sheep erythrocyte rosetting cells, T cells (7); cells binding C3 coated bovine erythrocytes, (complement receptor positive B cells (8); and purified, esterase positive monocytes by counterflow elutriation (9).

PNA agglutination. Peripheral blood or splenic leukocytes, unfractionated by other methods, were agglutinated at high cell concentrations (6-8 x 10^8 cells/ml) with an equal volume of lmg per ml PNA (Boehringer Mannheim, Indianapolis, IN). Agglutinated cells were separated from non-agglutinated cells at unit gravity. These methods were adapted from those of Reisner et al. (10) and are further detailed in Fowlkes et al. these proceedings. It was essential to remove red blood cells before agglutination, because the red cells caused "streaming" in the center of the fetal calf serum gradient which caused contamination of the sedimented, agglutinated fraction with non-agglutinated cells. Because of the low proportion of agglutinated cells, it was necessary to remove the top 15 ml from the discontinuous sedimentation gradient to avoid high losses of the non-agglutinated cells.

Fluorescent-lectin staining and analysis. Fluorescein-conjugated lectins were used at sub-agglutinating concentrations to stain the different populations of human cells as we have reported previously for murine cells (3, and Fowlkes et al., these proceedings). Cells were analyzed by flow microfluorometry on a fluorescence-activated cell sorter (FACS II: Becton Dickinson, Mountain View, CA). Data were collected, stored, processed and displayed as histograms of the fluorescence intensity of the individual cells versus the cell number in a 1024 channel data collection system (11).

Results and Discussion

We have used flow microfluorometry analysis to quantitatively measure the binding of fluorescein-conjugated lectins to human cell populations after separation by 3 non-lectin dependent methods. First, cells were sedimentated by counterflow elutriation. This method largely separates cells on the basis of cell size and can selectively separate monocytes from lymphocytes in leucocyte populations. Peripheral blood cells were also fractionated by separation of human T cells on the basis of sheep erythrocyte binding to form rosettes or by separation of the human B cells on the basis of their ability to form complement receptor dependent rosettes with antibody plus complement coated bovine erythrocytes.

The results of these binding studies can be generalized by the following statements. The highest binding of most of the lectins was seen on monocytes, with B cells having the next highest level of binding and finally with T cells showing the lowest levels of binding. This was true for a number of lectins regardless of their known potential for preferential cellular proliferative stimulation or for the carbohydrate specificity of the lectins used. For example this pattern of monocyte > B cell > T cell was seen with Con A, a T cell mitogen, Pa-2, a B cell mitogen, and PNA which has no mitogenic activity on untreated cells. An example of the differential binding is discussed in the paper by Waxdal et al. (these proceedings), where the pokeweed lectins have been shown to selectively bind to monocytes and apparently indirectly deliver the mitogenic signal to B cells. Crude preparations of the pokeweed mitogens have been shown to stimulate plaque-forming B cells from unfractionated human leukocyte cultures (12). For further discussion of the relation of interacting cellular subsets and mitogenesis also see Stadler et al., these proceedings.

Several reports in the literature have indicated that only a small proportion of peripheral blood leukocytes bind PNA (13-15). These reports have either indicated that the only peripheral blood cells that bind PNA are monocytes (13), or that some T cells may also preferentially bind this lectin (14, 15). It also appears that PNA also binds some T cells from tonsils (14, 15). The binding of fluoresceinated PNA to unfractionated peripheral blood leukocytes, measured by flow microfluorometry, confirms that only a minor proportion of leukocytes bind this lectin. While the previously reported range of binding has been shown to be 2 to 18% (13-19), it can be seen in Fig. 1A, that we have routinely observed a slightly higher level of binding, ranging from 5% to 28%. This simply reflects the increased sensitivity of the method for cell surface analysis by flow microfluorometry.

Only a small proportion of peripheral leukocytes are aggluti-

Table 1. DIFFERENTIAL CELL COUNTS OF PERIPHERAL BLOOD LEUKOCYTES AFTER PNA AGGLUTINATION.

	Lymphocytes	Polymorphs	Monocytes
Unfractionated	65	0	35
PNA^-	90	0	10
PNA^+	25	0	75

Cell morphology was assessed on cytocentrifuge preparations of cells stained with Wright-Giemsa.

nated with PNA, usually less than 10% of the cells. Therefore, we compared the PNA agglutinated (PNA^+) and non-agglutinated (PNA^-) cell populations for relative enrichment of cell subsets as indicated by differential cell counts for morphological cell type observed with stained cytocentrifuge preparations. Table 1 displays the cell counts from one normal cell preparation. The data from the peanut agglutination of a second individual peripheral blood cell preparation are presented later in the comparison with the Pa-2 agglutination in Table 2. The data indicate that there was an enrichment of monocytes (63 to 75%) in the PNA^+ fraction and a relative depletion of monocytes in the PNA^- fraction. Reciprocally, it has been reported that depletion of glass or plastic adherent cells, both of which can be used to remove phagocytic monocytes from cell suspensions, removes PNA staining cells (13, 17).

When the differential cell counts were compared with the flow microfluorometry fluorescence data, it was apparent that only a modest enrichment or depletion was observed for the higher intensity PNA binding cells (see Fig. 1B). At no time did we observe 75% of the cells binding high levels of fluorescent PNA. This would be expected if all monocytes were PNA binding.

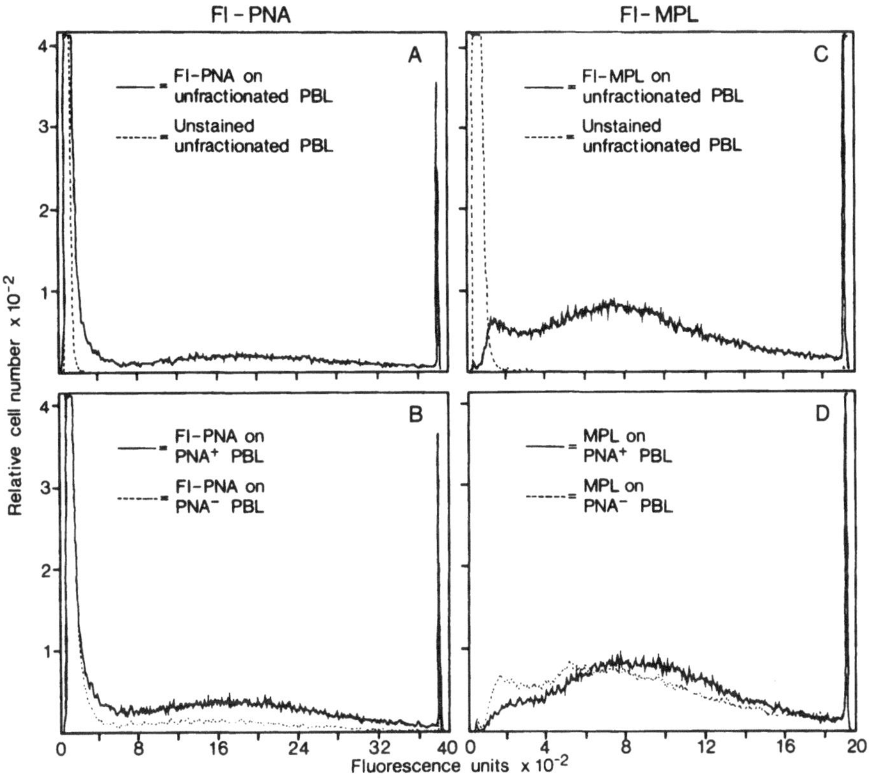

Fig. 1. Fluorescence histograms of fluoresceinated lectin binding to fractionated populations of human peripheral blood leukocytes (PBL). PBL were fractionated with peanut lectin agglutination into PNA^+ and PNA^- populations. The proportion of cells in panel A with greater than 400 fluorescent units of Fl-PNA bound is 23.6%. By comparison, the PNA^+ fraction has 43.8% and the PNA^- fraction has 12.6% of the cells staining above that intensity. MPL bright staining cells can be seen to be enriched in the PNA^+ leukocytes.

Therefore our data are in agreement with that of London et al. (14), who suggested that some non-monocytic, T cells are the PNA^+ cells in peripheral blood leukocyte suspensions.

It was also seen that a large proportion of agglutinated cells did not demonstrate the high level of binding with this lectin

(Fig. 1B). Therefore PNA agglutination was not equivalent to the PNA binding pattern. This result was reproducible for 5 different peripheral blood samples and 3 spleen suspensions, (see also Table 2 below). This lack of correlation has also been observed by others (18).

Because we observed a subpopulation of lymphocytes that were consistently agglutinated with the monocyte enriched fraction, we wished to determine whether there were specific subsets that might be detected by other lectins that would correlate better with the PNA agglutinated fraction. In Fig. 1, the binding pattern of fluoresceinated PNA (Fl-PNA) is compared with the binding pattern of fluoresceinated lectin from Maclura pomifera (Fl-MPL). Although we had observed no clear enrichment of the PNA binding cells, we did observe an enrichment of the MPL binding cells in the PNA^+ fractions (Fig. 1C,D). This lectin has been shown in other species to be a T cell binding lectin (4,20). However, in Fig. 1C,D, it is apparent that MPL is binding with an increased level to more that just T cells in human peripheral blood cell suspensions.

From previous studies, it is clear that different types of cells in suspensions of peripheral blood leukocytes may be binding or agglutinating with PNA. It has also been suggested that PNA can bind to some human T cell subset in tonsils (14, 15) as well as possibly other immature precursor populations of lymphoid and monocytic lineages of cells (10, 13-19, 21-23), (See also, Reisner, these proceedings.) Furthermore, PNA has been shown to bind to some monocytes (14, 15, 17, 19). Ballet et al. (15) have indicated that PNA binding is correlated with peroxidase staining, a characteristic of macrophages. Others have also shown PNA binding to macrophages capable of ingesting latex (14, 19). But there is disagreement about the PNA binding to cells capable of adherence to plastic or glass (15, 19), a method

Table II. COMPARISON OF DIFFERENTIAL CELL COUNTS OF HUMAN PERIPHERAL BLOOD LEUKOCYTES AGGLUTINATED WITH PNA OR Pa-2

Agglutination Phenotype*	Lymphocytes	Polymorphs	Monocytes
PNA^-	88	5	8
PNA^+	26	11	63
$Pa\text{-}2^-$	98	0	2
$Pa\text{-}2^+$	23	10	66

*This phenotype refers only to the characteristic of where the cells were recovered in the agglutination fractions. The starting population for both procedures contained cells in the following proportions 64:5:31.

often used to isolate or remove phagocytic monocytes. With these concerns in mind we compared the cell populations obtained from parallel agglutinations with PNA and Pa-2. Pa-2 was selected because the pokeweed lectins appear to bind more selectively to monocytes. (See Waxdal et al., these proceedings.) Table II shows an example of the data that we obtained. This indicates that while there was no difference between the PNA^+ and $Pa\text{-}2^+$ fractions on the relative purification of monocytes, the $Pa\text{-}2^-$ fraction was slightly more enriched for lymphocytes while being better depleted of monocytes. This was observed both for peripheral blood cells (as seen in Table II), and also for human spleen cells (data not shown).

Finally we tested the possibility of increased T cell functional activity by analysis of the responsiveness to T cell mitogens with the PNA agglutinated fractions. In Table III, it can be seen that the mitogen responsiveness to conventional T cell mitogens Con A or PHA is higher in the PNA^+ fraction although this fraction represents less that 10% of the cells. This could

Table III. PEAK BLASTOGENIC RESPONSE OF PNA FRACTIONATED HUMAN PERIPHERAL BLOOD LEUKOCYTES

	Con A*	PHA*
Unfractionated	64,780 ± 4915	94,999 ± 2114
PNA^-	34,453 ± 5591	53,898 ± 1382
PNA^+	69,053 ± 1499	97,770 ± 3620

*Data are presented as the mean counts per minute of ^{3}H-thmidine uptake by mitogen stimulated cells from 3 normal donors. The PNA^- fraction is different from either the PNA^+ ($p < 0.01$) or the unfractionated cells ($p < 0.02$).

be interpreted in one of several ways, PNA may selectively agglutinate a small subset of T cells that have increased mitogen responsiveness. Alternatively a special subset of monocytes (?peroxidase positive) cells are required for indirect mitogenic stimulation of the T cells and these are enriched in the PNA^+ fraction. When these cells are depleted from the PNA^- fraction, the blastogenic response is reduced. Further experimentation with other fractionation methods may be able to elucidate the two possibilities.

In summary we have shown that lectin agglutination with PNA does not enrich specifically for the higher PNA binding cells. However the PNA agglutinated fraction that is obtained from peripheral blood cells contains the proliferative functional activity. We suspect that this there is a co-isolation of a T cell subset and a monocytic, adherent, antigen-presenting cell that is co- purified by lectin agglutination methodology. This is further supported by the co-isolation of a substantial subpopulation of lymphocytic cells in the PNA^+ fraction and the demonstration that by the minimal criteria of cell morphology and fluorescent lectin analysis, the agglutinated fraction is clearly heterogeneous.

[1]Present address: Basel Institute for Immunology, Grenzacherstrasse 487, CH-4005, Basel, Switzerland. (Founded and supported by F. Hoffman-La Roche & Co., Ltd., Basel, Switzerland.)

[2]Present address: The Division of Rheumatic and Genetic Diseases, Department of Medicine, Duke University School of Medicine, Durham, NC, USA.

Acknowledgements

We would like to thank Howard Mostowski for technical assistance in preparing the cells, Drs. R. Asofsky and A.S. Fauci for the support of their laboratories, Dr. Marion Zatz for arranging the access to human splenic material, Hans-Peter Stahlberger for his helpful preparation of the artwork for this and other papers presented at this meeting, and my colleagues at the Basel Institute for Immunology for reading the manuscript.

References

1. Haynes, B.F.: Immunol. Rev. 57, 127-161 (1981).
2. Reinherz, E.L., and Schlossman, S.F.: Cell 19, 821-827 (1980).
3. Fowlkes, B.J., Waxdal, M.J., Sharrow, S.O., Thomas, C.A., III, Asofsky, R., and Mathieson, B.J.: J. Immunol. 125, 623-630, (1980).
4. Fowlkes, B.J., Mathieson, B.J., Sharrow, S.O., Waxdal, M.J., and Asofsky, R.: Cell. Immunol. In Press (1982).
5. Graw, R.G., Jr., Herzig, G.P., Eisel, R.J., and Perry, S.: Transfusion 11, 94- (1971).
6. Herzig, R.H., Herzig, G.P., Graw, R.G., Jr., Bull, M.I., and Ray, K.K.: N. Engl. J. Med. 296, 701- (1977).
7. Jondal, M., Holm, G., and Wigzell, H.: J. Exp. Med. 136, 207-215 (1972).
8. Haynes, B.F., and Fauci, A.S.: J. Immunol. 121, 559-565 (1978).
9. Contreras, T.J., Jemionek, J.F., Stevenson, H.C., Hartwig, V.M., and Fauci, A.S.: Cell. Immunol. 54, 215-229 (1980).
10. Reisner, Y., Linker-Israeli, M., and Sharon, N.: Cell. Immunol. 25, 129-134 (1976).

11. Mathieson, B.J., Sharrow, S.O., Campbell, P.S., and Asofsky, R.: Nature 277, 478-480 (1979).

12. Fauci, A.S., and Pratt, K.R.: J.Exp. Med. 144, 674-684 (1976).

13. Reisner, Y., Biniaminov, M., Rosenthal, E., Sharon. N., and Ramot, B.: Proc. Natl. Acad. Sci. (USA) 76, 447-451 (1979).

14. London, J., Perrot, J.Y., Berrih, S., Laroche, L., and Niaudet, P.: Scand. J. Immunol. 9, 451-459 (1979).

15. Ballet, J.-J., Fellous, M., Sharon, N., Reisner, Y., and Agrapart, M.: Scand. J. Immunol. 11, 555-560 (1980).

16. Levin, S., Russell, E.C., Blanchard, D., McWilliams, N.B., Maurer, H.M., and Mohanakumar, T.: Blood 55, 37-39 (1980).

17. Galili, U., Galili, N., Or, R., and Polliack, A.: Clin. Exp. Immunol. 43, 311-318 (1981).

18. Maccario, R., Ferrari, F.A., Siena, S., Vitiello, M.A., Martini, A., Siccardi, A.G., and Ugazio, A.G.: Thymus 2, 329-337 (1981).

19. Elfenbein, G.J., Bellis, M.B., and Santos, G.W.: Transpl. Proc. 13, 273-277 (1981).

20. Jones, J.M., and Feldman, J.D.: J. Immunol. 111,1765-1770 (1973).

21. Haimovitz, A., Fuks, Z., Galili, N., and Treves, A.J.: J. Reticuloendothelial Soc. 31, 187-192 (1982).

22. Richard, Y., Boumsell, L., Coppin, H., Mishall, Z., Lemerle, J., and Bernard, A.: J. Immunol. 127, 252-255 (1981).

23. Osmond, D.G., Saveriano, N., Drinnan, M., Santer, V., Rahal, M.D., Owen, J.J.T., and Rijnbeek, A.-M.: Dev. in Immunol. 15, 103-110 (1981).

DETECTION AND CHARACTERIZATION OF MURINE THYMOCYTE SUBSETS DELINEATED BY LECTINS AND FLOW MICROFLUOROMETRY

B.J. Fowlkes[1], B.J. Mathieson[2], and M.J. Waxdal

Laboratory of Microbial Immunity and Laboratory of Immunology, National Institute of Allergy and Infectious Diseases, National Institutes of Health,
Bethesda, Maryland, USA

The mouse thymus has been of interest to immunologists because cellular differentiation in this organ leads to the production of immunocompetent T lymphocytes. The steps involved in the differentiation and maturation of thymocytes have yet to be elucidated. It is well accepted that at least two easily distinguished cell populations exist in the thymus (1). One population has been referred to as immunoincompetent cortical thymocytes. This population bears a high concentration of Thy-1 and low concentration of H-2 antigen. It is sensitive to cortisone and is agglutinated with peanut lectin (PNA). A second population of immunocompetent medullary thymocytes, about 5-10% of normal thymocytes, bears a low concentration of Thy-1 and high concentration of H-2 antigen, is resistant to cortisone and is not agglutinated with PNA (1-3). Likewise, two subsets exist with respect to Lyt-2 antigen. About 90% of normal thymocytes express Lyt-2, 10% do not (4). Whether cortical thymocytes give rise directly to the immunocompetent medullary cells is a matter of current controversy. To elucidate this maturation sequence, differentiation and phenotypic markers for identifying component cell types, as well as separation techniques for isolating subsets, will be extremely important.

Lectins, Vol. III

In this respect, peanut lectin (PNA) has been extremely useful as a means of characterizing and enriching thymocyte subpopulations (3,5-13). In the present report, we indicate that lectins other than PNA may be valuable for marking and isolating functionally important thymocytes. Using fluorescein labeled lectins and flow microfluorometry we find that a number of lectins are able to delineate one of the same thymocyte subpopulations delineated with PNA. We also demonstrate the value of flow microfluorometry in developing and monitoring separation methods, and we discuss new techniques for isolating thymocyte subpopulations using lectins.

Materials and Methods

Mice. C57BL/6 (B6) and BALB/cAnN (BALB/c) mice (6-8) weeks of age were obtained from the Small Animal Section, Veterinary Resources Branch, Division of Research Services, National Institutes of Health.

Lectins. The source, references, and nominal specificity of lectins used have been reported previously (14). Lectins from *Arachis hypogaea* (PNA), *Ricinus communis* (Ricin), *Abrus precatorius* (Abrin and Abrus lectin), *Sophora japonica* (SJ-II), *Maclura pomifera* (MPL), *Glycine soja* (Soybean agglutinin; SBA), *Phytolacca americana* (Pa-2 and Pa-3), and *Triticum vulgaris* (Wheat germ agglutinin; WGA) were used in this work. Lectins were labeled with fluorescein isothiocyanate as described previously (14).

Cell preparation and fluorescence-staining. Thymocytes were prepared as single cell suspensions from carefully disected thymic lobes and stained with fluorescent lectin as previously described (14).

Flow microfluorometry of lectin binding. Analyses of fluorescent lectin binding to cells were performed using a fluorescence-activated cell sorter (FACS II: Becton Dickinson, Mountain View, California) and data were stored and processed in a PDP 11/40 computer (Digital Equipment Corporation, Maynard, Mass.) (15,16). Fluorescence data were collected only from viable cells as detected by forward light scatter intensity (1-15°). For each histogram, 50,000 viable cells were analyzed.

Cell separation using differential agglutination by PNA. Thymocytes were separated into agglutinated (PNA^{+}) and non-agglutinated (PNA^{-}) cells either by modified methods for unit gravity separation (7) initially described by Reisner et al. (3) or by a modified method using Ficoll (see below). Thymocytes were prepared suspended at 6-8 x 10^{8} cells/ml in separation medium (Hanks' balanced salt solution containing 0.1% bovine serum albumin). An equal volume of PNA (1mg/ml in Hanks') (Boehringer Mannheim, Indianapolis, Ind.) was added and incubated 15 min at ambient temperature.

For unit gravity (1g) sedimentation, the suspension was layered on a fetal calf serum discontinuous gradient consisting of 30ml of 20% fetal calf serum in separation medium, overlaid with 3ml of separation medium in a 40ml disposable conical centrifuge tube. The best fractionations were attained if the interface between layers was disturbed gently while layering cells. Agglutinated cells were allowed to settle for 45 min after which the top 10ml was removed with a Pasteur pipet. This fraction was designated the top or PNA^{-} fraction and contained mostly unagglutinated cells. The bottom 3ml contained the agglutinated cells and was designated the bottom or PNA^{+} fraction. The middle fraction contained very few cells and was discarded. The top fraction was washed by centrifugation (10 min, 200 x g, at 4°C) with 30ml of separation medium. To disperse aggregates and remove bound PNA, cells in both fractions were resuspended in 0.2 M galactose solution (adjusted to 305-310 mOsM with 10x Hanks' balanced salt solution) + 10% fetal calf serum). Warming the galactose solution to 37°C facilitated disaggregation. After 15 min, 25°C, cells were washed 3 times by centrifugation (see above) to remove any remaining PNA, galactose, or serum. This washing procedure was extremely important since all three substances could inhibit the binding of many of the lectins used in subsequent steps.

For fractionation with Ficoll, the PNA agglutination procedure was the same but thymocytes were layered on 8ml of Ficoll Paque (Pharmacia Fine Chemicals, Piscataway, N.J.) in round bottom centrifuge tubes (Falcon 2057) and centrifuged at 250 x g for 9 min at 25°C. After centrifugation, the suspended lymphocyte layer was found to contain unagglutinated cells and was designated the top or PNA^- fraction. The cell pellet, which contained agglutinated cells, was designated the bottom or PNA^+ fraction. Both fractions were washed free of Ficoll, aggregates were dispersed, and PNA was removed with the galactose treatment described above.

Results and Discussion

In previous studies (14), we have shown by flow microfluorometry that many fluorescent labeled lectins bind differentially to subsets of mouse thymocytes. Those studies demonstrated that binding was carbohydrate specific since it could be inhibited by appropriate sugars. However the differential binding by various lectins was not related to the nominal specificity of the lectin nor to cell size. Those results indicated that flow microfluorometry could be a quick and reliable means of monitoring purity of cells separated using lectins.

PNA has had a wide application in the separation of lymphocyte subsets (see Reisner et al., these proceedings). In contrast to in vivo cortisone treatment (an alternate method for isolating medullary thymocytes), this method is advantageous because no major subpopulation of thymocytes is eliminated. Although several studies have made use of the differential binding and agglutination property of PNA to separate two subpopulations of thymocytes for use in functional studies (3,5,8,11-13), most investigators employing this separation method have failed to report the purity of separated populations. Our results indicate that flow microfluorometry analysis of fluoresceinated PNA binding provides an excellent assay for the purity of PNA fractionated thymocyte subsets. Figure 1 demonstrates the use of such analyses with B6 thymocytes agglutinated and fractionated with PNA. While the two subpopulations obtained after lectin agglutination have been referred to as PNA^- and PNA^+ thymocytes, it should be pointed out that there are virtually no unstained cells in the unfractionated population or in any of the fractions with Fl-PNA (Fig. 1a-f).

Once PNA fractionated cells are washed with galactose to remove bound PNA, fluoresceinated PNA can be used to stain the cells (Fig. 1b,c,e,f). Staining the fractions of cells obtained after PNA agglutination, unit gravity sedimentation, and subsequent galactose elution reveals that the duller staining cells of the unfractionated thymocytes correspond to PNA^- cells. These analyses of fractionated B6 thymocytes reveal a 4-5-fold enrichment of PNA dull staining cells in the top, PNA^- fraction and a depletion of about one-half of PNA dull staining cells in the bottom, PNA^+ fraction. These results are highly reproducible when using the same strain of mouse. However, when BALB/c and C3H/HeN thymocytes are fractionated by the same methods, higher purity is seen in the top fraction for PNA dull cells (B.J. Mathieson, and B.J. Fowlkes, unpublished observations).

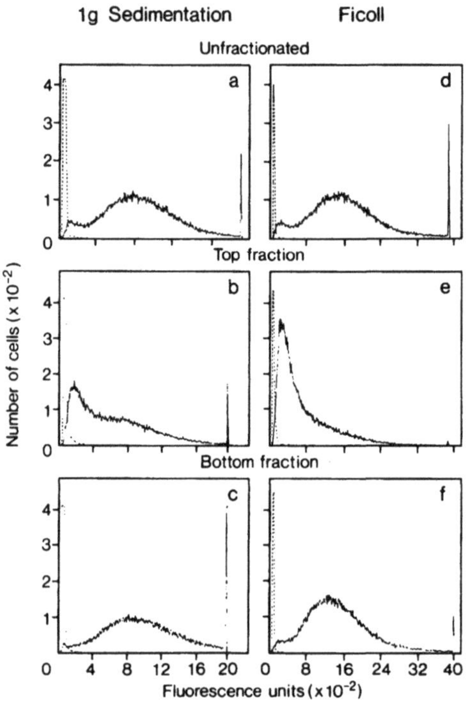

Fig. 1. Fluorescence histograms obtained with fluoresceinated PNA on fractions of B6 thymocytes (——). Fractionated thymocytes were obtained after PNA agglutination and fractionation by unit gravity sedimentation with discontinuous fetal calf serum (a,b,c) or Ficoll (d,e,f) as described in Materials and Methods. The top fractions contain non-agglutinated cells (PNA^-), whereas the bottom fractions contain mostly agglutinated cells (PNA^+). All profiles are superimposed with the fluorescence histogram obtained on unstained control thymocytes (----).

Other methods of separating the PNA^+ from PNA^- (dull) cells are currently being investigated. Figure 1d-f shows a fractionation of thymocytes using Ficoll. Compared to the cells from the unit gravity sedimentation, the top, PNA^- fraction shows a better (6-7-fold) enrichment of PNA dull staining cells, while the bottom, PNA^+ fraction shows less depletion of dull staining cells. These data reveal that Ficoll separation is superior to the simple sedimentation method for enrichment of PNA dull staining cells (top fraction), whereas sedimentation at unit gravity over fetal calf serum is better for depletion of these cells from the PNA^+ bottom fraction.

The quantitation of fluorescence, which is possible with flow microfluorometry analysis, demonstrates that neither the unit gravity (Fig. 1b,c) nor Ficoll (Fig. 1e,f) separation technique totally depletes PNA dull staining cells from the PNA^+ fraction.

This phenomenon of repeated inability to remove a small proportion of PNA dull cells may be due to "trapping" of single cells in larger aggregates of PNA^+ cells during the unit gravity sedimentation or centrifugation step. However, the agglutination method has been used because it has the advantage of yielding two functionally distinct subpopulations (3,8,11-13) with minimal manipulation and minimal loss of either subpopulation. In addition, both methods of separation of PNA agglutinated cells are rapid, convenient and can be used to prepare the large numbers of cells often needed in assays of lymphocyte function with a high level of viability and a high recovery of the total starting cell numbers.

It can be concluded from the analyses using fluoresceinated lectin, that while PNA agglutination and separation methods may enrich for PNA dull and/or bright staining cells, the resultant populations often are not pure. Thus, flow microfluorometry could provide a rapid analysis for subpopulation composition to aid in the development of better cell fractionation methods. These analyses would be useful also for comparison of functional data from different laboratories.

The size, phenotype, and functional properties of PNA fractionated thymocytes have been characterized in several studies (3,5,7-9,11-13,17-18). We had earlier shown that a group of lectins, which included several galactose or N-acetyl galactosamine binding lectins and the pokeweed lectins, also indicated a subset of thymocytes with lower binding of the fluoresceinated lectins (14). Because the proportion of dull binding cells was similar to that seen with fluoresceinated PNA, we wished to determine the correlation of the subsets detected by these several lectins. Therefore, we compared the relative enrichment of thymic subsets detected by other lectins, after PNA agglutination and separation of normal thymocytes.

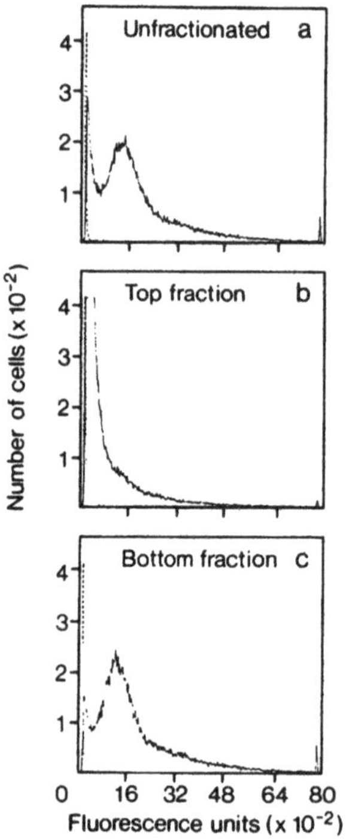

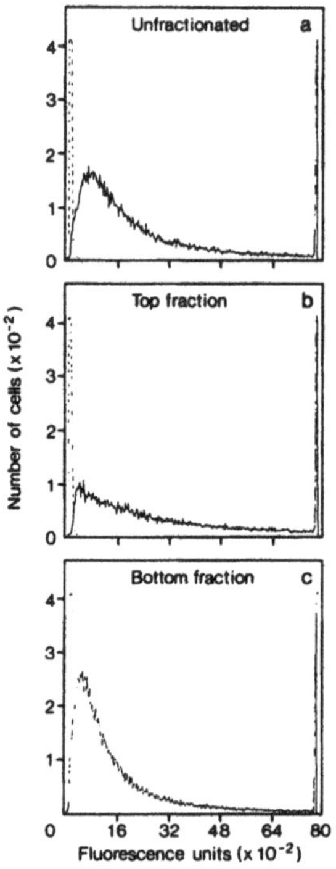

Fig. 2. (left). Fluorescence histograms obtained with fluorescent pokeweed lectin (Pa-2). Fig. 3. (right). Fluorescence histograms obtained with fluoresceinated WGA. Fractions of B6 thymocytes after binding with fluoresceinated lectins (——), obtained after PNA (unlabeled) agglutination and fractionation by unit gravity sedimentation (a,b,c) as described in Materials and Methods. These profiles are superimposed with fluorescence profiles obtained on unstained control thymocytes (----).

Fluoresceinated lectins were used to stain unfractionated normal thymocytes and these staining patterns were compared with PNA$^+$ or PNA$^-$ thymocyte populations fractionated at unit gravity. Figure 2 shows staining of PNA fractionated cells (after galactose elution of PNA) with fluorescent Pa-2. The results indicate that dull staining thymocytes with this pokeweed isolectin correspond to the PNA$^-$ (dull staining) thymocytes (compare Fig. 2b with Fig. 1b). Likewise, using other fluoresceinated lectins, it was found that Ricin, Pa-3, SBA, MPL, Abrin, and SJ-II dull staining thymocytes are also enriched in the PNA^- thymocyte fraction. In dual parameter analyses using fluorescence and forward angle light scatter (to measure size), the dull staining subset seen with these lectins is comprised of medium sized cells. See Fowlkes et al. (14) for an example of this type of analysis, data not shown here. Correlation of the relative intensity of the staining patterns was also seen with some of these same fluoresceinated lectins on human monocyte fractions (Waxdal et al., and Mathieson et al., these proceedings).

These findings indicate that a number of lectins are able to delineate a minor population of dull or unstained thymocytes which correlates with the same population delineated by PNA. This conclusion has been supported with results from cortisone resistant (medullary) thymocytes. Previously we demonstrated that cortisone treatment enriched for PNA dull staining thymocytes (14). The residual cells after in vivo cortisone treatment were of medium size, as determined by the light scatter parameter, also indicating that they were medullary thymocytes. When cortisone resistant thymocytes were stained with MPL, Pa-2, and Ricin, and the flow microfluorometry histograms compared with fluorescence histograms obtained from thymocytes of untreated mice, it was found that cortisone resistant thymocytes were enriched for the dullest staining cells (14). Therefore, these lectins, like PNA, were able to distinguish a difference between medullary thymocytes and the bulk of thymocytes which reside in cortical areas of the thymus.

With WGA, (wheat germ agglutinin) a contrasting result was found that indicated that WGA was unlike the previous group of lectins in its differential binding pattern on thymocytes. Figure 3 shows staining of PNA fractionated cells stained with

WGA. The WGA dull staining cells are enriched in the PNA agglutinated fraction, while the WGA bright staining cells were enriched in the PNA nonagglutinated fraction. Since WGA has been shown to bind sialic acid residues (M. Monsigny, personal communication), the data that indicates a higher level of WGA binding on the PNA^- thymocytes would support the view of Reisner et al. (3, and these proceedings) that some PNA receptors of the "mature" population are masked by sialic acid residues. This data is further corroborated by that of Favero et al. (these proceedings) and data from fetal ontogeny (see Drakeford et al., these proceedings).

It was of particular interest to note that the enrichment of the lectin dull staining cells with PNA, Pa-2 and several of the other lectins was also correlated with earlier findings from our laboratory that indicated that PNA non-agglutinated cells have an enriched proportion of $Lyt2^-$ cells (7). The contrasting enrichment of WGA bright staining cells that is shown here in the PNA^- fraction is correlated with the enrichment of thymocytes with increased expression of Lytl antigen expression that was also reported earlier by Betel et al. (7). Since the correlation is not precise, we do not feel that these Ly antigens, which are glycoproteins, account for the observed differentials in flourescent lectin binding. But these glycoproteins may have acceptors that contribute to this phenomenon. It will be interesting to see if there is any correlation with the molecular characterization of the PNA binding molecules on the cell surface with the already reported characterization for the Ly antigens. (See Favero et al., these proceedings for data on the proteins detected with cell surface binding of PNA).

Flow microlfuorometry analysis and foward angle light scatter of cortisone resistant thymocytes and of the fractions obtained by PNA agglutination both indicate that medium sized normal thymocytes (10% of the population) stain weakly with PNA and correspond to the PNA nonagglutinable cells. The results using selected fluoresceinated lectins on PNA fractionated thymocytes indicate that many other lectins may also identify differentially expressed markers on medullary and/or cortical thymocytes. Furthermore these subsets do not appear to be identical to each other. It is obvious, therefore, that lectins other than PNA may be able to further fractionate functional thymocyte subpopulations.

These studies reveal that lectins can be a valuable tool both for marking and isolating functionally important cell populations and may be very useful tools in elucidating molecular mechanisms involved in cell recognition, interaction, and differentiation.

[1]This work was done in partial fulfillment of the requirements for the Ph.D. degree, George Washington University, Washington, D.C.

[2]Present address: Basel Institute for Immunology, Basel, Switzerland.

Acknowledgements

The authors are very grateful to Mrs. Virginia Shaw and Ms. Margrit Bühler for expert assistance in the preparation of the manuscript. We also thank Dr. Richard Asofsky for his advice, guidance and support throughout this work.

References

1. Raff, M.C.: Nature New Biol. 229,182- (1971).
2. Shortman, K., Cerottini, J.-C., and Brunner, K.T.: Eur. J. Immunol. 2,313- (1972).
3. Reisner, Y., Linker-Israeli, M., and Sharon, N.: Cell. Immunol. 25,129-134 (1976).
4. Mathieson, B.J., Sharrow, S.O., Campbell, P.S., and Asofsky, R.: Nature 277,478-480 (1979).
5. Irlé, C., Piguet, P.-F., and Vassalli, P.: J. Exp. Med. 148,32-45 (1978).
6. London, J., Berrih, S., and Bach, J.-F.: J. Immunol. 121,438-443 (1978).
7. Betel, I., Mathieson, B.J., Sharrow, S.O., and Asofsky, R.: J. Immunol. 124,2209-2217 (1980).
8. Kruisbeek, A.M., and Astaldi, G.C.B.: J. Immunol. 123,984- (1979).
9. Roelants, G.E., London, J., Mayor-Withey, K.S., and Serrano, B.: Eur. J. Immunol. 9,139-145 (1979).
10. Waxdal, M.J., Fowlkes, B.J., Sharrow, S.O., and Mathieson, B.J.: Protides of Biological Fluids, Colloquium 27, edited by Peeters, H., Pergamon Press, Oxford, pp.525-528 (1980).
11. Mathieson, B.J., Mage, M., Betel, I., and Sharrow, S.O.: Protides of Biological Fluids, Colloquium 27, edited by Peeters, H., Pergamon Press, Oxford, pp.535-538 (1980).
12. Cooley, M.A., and Schmitt-Verhulst, A.-M.: J. Immunol. 123,2328- (1979).
13. Oppenheim, J.J., Northoff, N., Greenhill, A., Mathieson, B.J., Smith, K., and Gillis, S.: Biochemistry of Lymphokines, edited by De Weck, A., Kristensen, F., and Landy, M., Academic Press, New York, pp.399-404 (1980).
14. Fowlkes, B.J., Waxdal, M.J., Sharrow, S.O., Thomas, C.A. III, Asofsky, R., and Mathieson, B.J.: J. Immunol. 125, 623-630 (1980).
15. Loken, M.R., and Herzenberg, L.A.: Ann. N.Y. Acad. Sci. 254,163-171 (1975).
16. Miller, M.H., Powell, J.I., Sharrow, S.O., and Schultz, A.R.: Rev. Scient. Instrum. 49,1137-1142 (1978).
17. Umiel, T., Linker-Israeli, M., Itzchaki, M., Trainin, N., Reisner, Y., and Sharon N.: Cell. Immunol. 37,134- (1978).
18. London, J., Petit, M., Horton, M.: Protides of Biological Fluids, Colloquium 27, edited by Peeters, H., Pergamon Press, Oxford, pp.545-549 (1980).

THE LATE APPEARANCE OF A PNA "DULL" SUBSET IN THE MOUSE THYMUS DEVELOPMENT

Dorothy Drakeford, William M. Leiserson, and Susan O. Sharrow

Laboratory of Microbial Immunity, National Institute of Allergy and Infectious Diseases, and Immunology Branch, National Cancer Institute, National Institutes of Health,
Bethesda, Maryland

Bonnie J. Mathieson

The Basel Institute for Immunology,
Basel, Switzerland

Previous studies of peanut agglutinin (PNA) binding to lymphoid cells have shown that mouse thymocytes can be separated into two fractions by agglutination; cells that were agglutinated by the peanut lectin (PNA^+) and those that were not agglutinated (PNA^-) (1). Using visual fluorescence, the appearance of PNA^+ and PNA^- subsets of lymphocytes was detected in fetal ontogeny from day 11 of gestation (2). Since a subset of thymocytes bearing the Lyt2 antigen ($Lyt2^+$) and the subset which was PNA^+ were highly correlated among adult murine thymocytes (3), we wished to determine further their correlation during ontogeny in the thymus.

Materials and Methods

Mice. B6 congenic mice for Lyt antigens or a subline of BALB/c mice, partly congenic for the nu (nude) gene, were used for the initial studies. These mice were maintained at Cor Bel Laboratories, Rockville, MD.

Reagents. Monoclonal anti-Lyt2.2 was the generous gift of Dr. Ulrich Hämmerling, Sloan Kettering Institute for Cancer Research, New York, N.Y. (4). The preparation and fluorescein labelling of the Fab fragment of goat anti-mouse immunoglobulin (FlGAMIg) used to counterstain anti-Lyt2.2 has been reported previously (5,6). PNA was purchased from Boehringer Mannheim Biochemicals, Indianapolis, Indiana. Fluorescein conjugation was carried out using the celite/fluorescein isothiocyanate technique (7). Fluoresceinated-PNA (Fl-PNA) was diluted for use to 50 mg/ml with phosphate buffered saline containing 0.1% NaN_3.

Cell preparation. Mice were killed by cervical dislocation. Each thymus was extracted gently with fine forceps. Fetal thymi were pooled; each neonatal thymus was individually teased in cold Medium 199. The thymocyte suspensions were transferred into centrifuge tubes and allowed to settle 2-5 min to remove tissue debris and cell clumps. Each cell suspension was transferred to a second tube and washed by centrifugation (7 min, 200 x g at $4^{o}C$). Cells were counted using a hemacytometer and adjusted to a concentration of 2×10^7 cells/ml in Hanks' balanced salt solution + 0.1% bovine serum albumin + 0.1% NaN_3, and thereafter handled in this medium.

Immunofluorescence. Twenty microliters of fluorescent PNA or 10 µl of anti-Lyt2.2 were added to tubes containing 50 µl (10^6 cells) of the cell suspension. The cells were incubated for 20 min at $4^{o}C$ with the lectin or for 45 min at $4^{o}C$ with anti-Lyt2.2 with intermittent agitation. After incubation the cells were washed twice with 2 ml of medium by centrifugation. Cells treated with anti-Lyt2.2 were counterstained with 10 µl of fluorescent goat anti-mouse immunoglobulin and incubated for 45 min as described above. Next, cells were washed twice and resuspended in 0.3 ml medium for flow microfluorometry.

Control thymocytes for the PNA stained cells were prepared in the same manner, but not incubated with fluorescent PNA. Control thymocytes for the anti-Lyt antibody treated cells were prepared as described above but only incubated for 45 min with fluorescent goat anti-mouse immunoglobulin.

Detection of anti-Lyt2.2 and PNA binding. Immunofluorescence was measured by flow microfluorometry (8,9) on a fluorescence-activated cell sorter (FACS II; Becton Dickinson, Mountain View, California) coupled to a PDP 11/34 computer. The data configuration and means of data processing have been described (10). Data are presented as histograms of cell frequency versus fluorescence intensity. For convenience, these histograms will be referred to as fluorescence profiles. For comparison with other systems, fluorescence intensity is measured in arbitrary fluorescence units (FU) relative to a gain 16 amplification on the FACS II machine. Splenic B cells stained for surface immunolgobulin have a mean fluorescence intensity of approximately 6000 units on this scale.

Results

Fluorescence distribution of PNA and anti-Lyt2 binding to thymocytes. The percentage of cells binding PNA was determined from the intersection of the fluorescence profile of fluorescent-PNA stained cells with that of the fluorescence profile of control unstained cells. PNA^+ "bright" cells were defined as the subpopulation with > 400 fluorescence units when analyzed by flow microfluorometry and PNA^+ "dull" cells as those cells that had < 400 units of fluorescence but which had stained above the level of control cells. Lyt2 phenotype was simply determined as positive or negative although quantitative differences were noted and indicated in the data below.

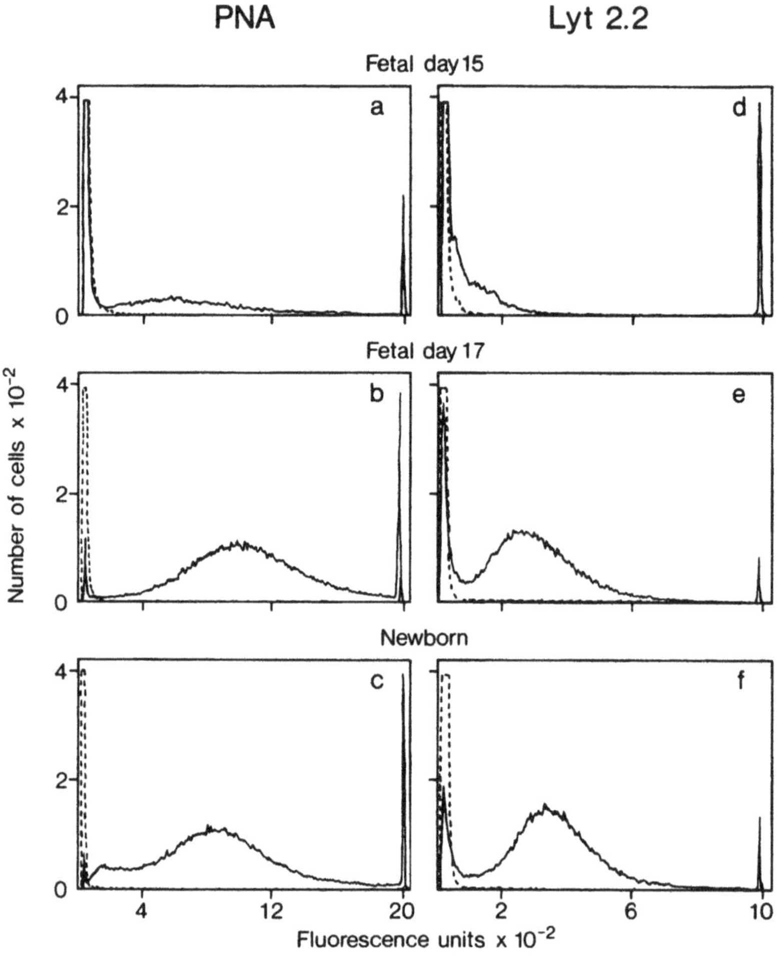

Fig. 1. Fluorescence histograms of thymocytes treated with fluoresceinated PNA or anti-Lyt2 + fluoresceinated goat anti-mouse immunoglobulin are shown above. Control thymocyte profiles (----) are superimposed on the stained thymocyte profiles (——). At each time point in the ontogeny, the same cell suspension was treated with the lectin and in a second tube, with the antibody for direct comparison. Controls were either unstained thymocytes for the fluoresceinated lectin, or cells incubated only with the fluoresceinated anti-mouse immunoglobulin for the anti-Lyt2 controls.

Table 1 Thymocyte Subsets Determined by Lyt2 and PNA Binding

Age	PNA^- %	PNA^+ "Dull" %	PNA^+ "Bright" %	$Lyt2^+$ %
Fetal day 15[1]	69.2	7.6[2]	23.2	35.3
Fetal day 17[1]	1.8	3.8[2]	94.4	80.5
Newborn	1.3	12.3	86.4	87.8
3 day	2.4	10.7	86.9	90.4
5 day	1.0	12.2	86.6	90.1
10 day	< 1	13.2	86.6	89.1
12 day	< 1	8.9	90.7	91.4
13 day	< 1	13.1	86.6	88.3
18 day	< 1	10.0	89.7	92.2
22 day	< 1	14.2	85.6	87.6

[1] Fetal age was determined from detection of a vaginal mating plug (plug date = 0). Fetal thymocytes were pooled from 7 embryos.

[2] These fluorescence profiles did not show a separate peak of "dull" cells like that seen in thymocytes of newborn or older mice, e.g. see Fig. 1. The numbers here represent the integration of the same area between the unstained control and the 400 fluorescence unit point used to deliberate bright versus dull cells on the PNA profiles of adult thymocytes.

The analysis of PNA binding by flow microfluorometry at day 15 of gestation revealed 2 distinct subpopulations. The first of these was a relatively homogeneous fluorescence distribution of PNA^+ staining thymocytes (Fig. 1). A second larger subset of PNA^- cells (with no detectable fluorescence above background) was also seen early in gestation (day 15). By day 17 of fetal development, this PNA^- population of thymocytes had nearly disappeared (Table 1, Fig. 1): at least 94% of the thymocytes were PNA^+ "bright" (see Table 1) with virtually no PNA^+ "dull" thymocytes (Fig. 1) visible in this BALB/c strain. In C57BL/6 mice, this pattern persists till sometime after birth.

From newborn to young adult animals, the PNA binding for PNA^+ "bright" cells is consistent with earlier results reported by Reisner et al. (1). Using this lectin, they showed about 90% of the cells in the normal adult thymus are PNA^+ for agglutination. Only a very small percentage of the PNA^- cells are still detectable with the appearance of the PNA^+ dull subset. This PNA^+ dull subset has been correlated with non-agglutinating thymocytes (7, and Fowlkes et al. these proceedings).

In Figure 1, PNA binding is compared with anti-Lyt2 binding on thymocytes. Previous studies with C57BL/6 mice showed that $Lyt2^+$ cells are first seen at day 16-17 of gestation (10); however, in this BALB/c strain about 35% of the total thymic population is $Lyt2^+$ at low levels on day 15 of gestation. The proportion of $Lyt2^+$ cells increases from 80% at day 17 of gestation to 90% after birth. In contrast, the proportion of PNA^+ bright cells decreases from close to 95% to 85-90% during the same time period.

Discussion

It has been demonstrated by other investigators that differential PNA binding leads to the detection of two thymic subsets (1,3,12,13). Our data indicate that an essentially homogeneous population of PNA^+ thymocytes exists before birth and that there is a later development of a PNA "dull" subset of cells.

Roelants et al. have indicated that approximately 19% of newborn thymocytes are unstained by PNA (2). This contrasts with our results obtained by flow microfluorometry analysis because at this time point fewer than 2% of cells are PNA^-, i.e. they fall below the autofluorescence background of unstained cells. Perhaps this discrepancy could be explained by the higher sensitivity of our method for analyzing fluorescence intensity since the earlier data from Roelants et al. was obtained by visual assessment. Also, the "dull" staining population, which is seen at this time, is probably equivalent to the PNA^- population described by others in adult mice (1,12). Finally, the discrepancy may be attributed to PNA binding differences in mouse strains. We also have noted differences between B6 and BALB/c mice elsewhere (Fowlkes et al. these proceedings).

London et al. (12) have suggested that peanut lectin could be used as a marker for pre-T cells in mice . For several reasons, that earlier data might be interpreted differently. First, Rose et al. (14) recently have shown PNA "bright" staining cells in germinal centers. Second, others (15,16) have claimed that PNA is a marker for early pre-B cells. While these two studies extend the repertoire of PNA^+ cells, they do not rule out the possibility that PNA binding may be, in addition, a marker for pre-T cells. However, many pre-T cells already in the fetal thymus at day 15 do not express the PNA acceptors.

It was suggested earlier that a subpopulation of agglutinated (PNA^+) $Lyt2^+$ cells could be the precursors of non-agglutinated (PNA^-) $Lyt2^+$ cells (3). The data presented are compatible with with the concept that during maturation PNA^+ cells convert to non-agglutinated, PNA^+ "dull" cells by attachment of sialic acid to the majority of PNA binding sites (1, also see Reisner et al., and Favero et al., these proceedings).

Thus, the relation between the $Lyt2^+$ subset of cells and PNA^+ "bright" cells which is reasonably concordant in the adult animal is highly discrepant in fetal animals. While there is clear evidence of two thymic subsets by analysis of Lyt2 antigen expression, the PNA binding is seen on virtually all of the cells at day 17 of gestation. These data therefore suggest a reduction in PNA binding on a maturing subset of thymic cells. However maturation of PNA^+ dull cells within the thymus from a separate lineage of PNA^- cells is still possible.

Acknowledgements

The authors are grateful to Mrs. Virginia Shaw and Ms. Margrit Bühler for their assistance in preparation of the manuscript. Special thanks go to B.J. Fowlkes for the preparation of fluorescent reagents, to Dr. Ronald H. Schwartz for providing the mice, to Dr. Richard Asofsky for his guidance and encouragement, and to Dr. H. von Boehmer for review of the manuscript.

REFERENCES

1. Reisner, Y., Linker-Israeli, M., and Sharon, N.: Cell. Immunol. 25,129-134 (1976).
2. Roelants, G.E., London, J., Mayor-Withey, K.S., and Serrano, B.: Eur. J. Immunol. 9,139-145 (1979).
3. Betel, I., Mathieson, B.J., Sharrow, S.O., and Asofsky, R.: J. Immunol. 124,2209-2217 (1980).
4. Hämmerling, G.J., Hämmerling, U., and Flaherty, L.: J. Exp. Med. 150,108-116 (1979).
5. Morse, H.C. III, Neiders, M.E., Lieberman, R., Lawton, A.R. III, and Asofsky, R.: J. Immunol. 118,1682-1689 (1977).
6. Mathieson, B.J., Campbell, P.S., Potter, M., and Asofsky, R.: J. Exp. Med. 147,1267-1279 (1978).
7. Fowlkes, B.J., Waxdal, M.J., Sharrow, S.O., Thomas, C.A., Asofsky, R, and Mathieson, B.J.: J. Immunol. 125,623-630 (1980).
8. Loken, M.R., and Herzenberg L.A.: Ann. N.Y. Acad. Sci. 254,163-171 (1975).
9. Mathieson, B.J., Sharrow, S.O., Campbell, P.S., and Asofsky, R.: Nature 277,478-480 (1979).
10. Miller, M.H., Powell, J.I., Sharrow, S.O., and Schultz, A.R.: Rev. Sci. Instrum. 49,1137-1142 (1978).
11. Mathieson, B.J., Sharrow, S.O., Rosenberg, Y., and Hämmerling, U.: Nature 289,179-181 (1981).
12. London, J., Berrih, S., and Bach, J.F.: J. Immunol. 121,438-443 (1978).
13. Irlé, C., Piguet, P.-F., and Vassalli, P.: J. Exp. Med. 148,32-45 (1978).
14. Rose, J.L., Berbeck, M.S.C., Wallis, V.J., Forrester, J.A., and Davies, A.J.S.: Nature (London) 284,364-366 (1980).
15. Newman, R.A., and Boss, M.A.: Immunology 40,193-200 (1980).
16. Osmond, D.G., Saveriano, N., Drinnan, M., Santer, V., Rahal, M.D., Owen, J.J.T., and Rijnbeek, A.-M.: Dev. Immunol. 15,103-110 (1981).

CHARACTERIZATION OF PEANUT AGGLUTININ RECEPTORS AS POSSIBLE EXPRESSION SITES OF SIALIC ACID DURING THE MATURATION OF MOUSE THYMOCYTES

Jean Favero, Jean-Claude Bonnafous, Jacques Dornand, Richard L. O'Brien* and Jean-Claude Mani
Laboratoire de Biochimie des Membranes, ER CNRS 228, ENSCM, 8 rue de l'Ecole Normale, F-34075 Montpellier (France), and
* Department of Pathology, USC School of Medicine, 2025 Zonal Avenue, Los Angeles Ca 90033 (USA)

INTRODUCTION

As it has been exposed by Reisner in the preceding paper, peanut agglutinin (PNA) has been widely used for the fractionation of mouse and human thymocytes (1,2). The major subpopulation of thymus cells (85-90%) binds PNA (PNA^+) and can be isolated by agglutination with the lectin. This subset which reacts poorly, if at all, in assays of immune functions, has been shown to be identical with the cortical *immature* cells. The non-agglutinated cells (PNA^-, 10-15%) have been described as the medullar hydrocortisone-resistant *mature* thymocytes. The good correlation between maturation stage and PNA binding has been confirmed by several independent studies for the majority of immune cells from T lineage (3,4).
PNA binds specifically to the D-galactosyl-β(1→3)-D-N-acetyl-galactosamine sequence (5). On cell membranes this sequence is generally terminated by sialic acid, and it has been hypothesized (see preceding paper) that the free galactosyl groups of PNA receptors present on immature thymocytes could be masked by sialic acid residues during lymphocyte differentiation (6,7); this attachment of sialic acid residues to the peanut lectin receptors may be an important step in the maturation of thymic cells.
The purpose of the work presented here is to characterize the

Lectins, Vol. III

PNA receptors on murine thymocytes and to provide some evidence that these receptors are expression sites of sialic acid during the maturation of thymic cells. PNA-binding cell surface molecules have been isolated from radiolabelled thymocytes by immunoprecipitation with rabbit anti-PNA serum and *Staphylococcus aureus* protein A which binds to Fc fragments of G class immunoglobulins. The receptors have been characterized in sodium dodecylsulfate-polyacrylamide gel electrophoresis (SDS-PAGE) followed by fluorography.

MATERIALS AND METHODS

In all experiments 3 week old male Swiss mice were used. Thymuses were removed, minced in Hank's balanced salt solution and passed through a stainless steel mesh to give single cell suspension; when necessary erythrocytes were selectively killed by 0.14 M ammonium chloride. Surface membrane integrity was checked using trypan blue exclusion test. Mature and immature thymocyte populations were isolated by agglutination with PNA (1,8). The efficiency of the separation was assessed by the respective 5'nucleotidase and adenosine deaminase activities of the two subsets, according to Dornand *et al* (8,9).
Mouse spleen cells, freed from erythrocytes, were prepared in Hank's medium. Highly purified T lymphocytes, unable of undergoing mitogen-induced proliferation without Ia^+ adherent cell reconstitution, were prepared by passage of spleen lymphocytes through both nylon wool and Sephadex G-10 columns (10).
PNA^+ and PNA^- cells, unseparated thymocytes or purified spleen T cells (2×10^7 cells per ml) were labelled with tritium, either on their terminal D-galactosyl residues using the galactose oxidase/tritiated sodium borohydride method (0.5 U galactose oxidase for 30 min at room temperature, then 20 µl aliquot tritiated borohydride, 1 mCi for 15 min at 4°C), or on their sialic acid residues using sodium periodate (1 mM) as oxidizing agent (11). In some experiments cells were treated with neura-

minidase (0.04 U per ml) prior to oxidation, in order to remove the sialic acid moieties. ^{125}I labeling was performed as described in (12).

Immunoprecipitation of PNA receptors was performed following the method described by Kessler (13). Unseparated labelled thymocytes were treated with PNA under the same conditions as reported for the agglutination process. After extensive washing, the cells were solubilized by a 15 min incubation at 4°C in PBS (phosphate buffer saline) pH 7.4, containing 1% detergent (NP-40), 0.1% bovine serum albumin and 2 mM phenylmethylsulfonyl fluoride, then centrifuged at 10,000g for 5 min (10^8 cells per ml). The supernatant (NP-40 supernatant) was recovered. *Staphylococcus aureus* (Cowan I) was washed three times with a solution of the same detergent and incubated 1 h at 4°C with rabbit anti-PNA immunoglobulin G fraction (50 μl per 200 μl *Staphylococcus* initial suspension) or with normal rabbit serum, then extensively washed with the same buffer. The NP-40 supernatant was incubated (60 min at 4°C) with washed *Staphylococcus aureus* not coated with immunoglobulins in order to remove non-specifically bound components. After centrifugation the supernatant was incubated (60 min at 4°C) with immunoglobulin-treated *Staphylococcus aureus*. The pellet was washed three times with PBS. The samples (cells and immunoprecipitates) were solubilized in electrophoretic sample buffer containing sodium dodecylsulfate and 2-mercaptoethanol, then kept for 5 min in a boiling water bath. SDS-PAGE was performed according to Laemmli (14). The gels were stained with Commassie brilliant blue, destained with 10% acetic acid in water, dried and processed for fluorography as described in (15).

RESULTS AND DISCUSSION

SDS-PAGE Pattern of Tritium-Labelled Unseparated Murine Thymocytes.

When unseparated thymocytes were labelled with tritiated borohydride alone (Fig 1a), one weak band (molecular weight 46 kd) could be determined by fluorography of SDS-PAGE. When thymocytes were labelled with galactose oxidase/tritiated borohydride, an intense labelled band (apparent molecular weight 175 kd) was determined, in agreement with the results of Gahmberg *et al* (16) and Hoessli *et al* (17) (170 kd), but other weaker bands could also be detected (Fig 1b). Because galactose oxidase does not enter the cell (16) and oxidizes terminal galactosyl and N-acetyl-galactosaminyl moieties to the corresponding C_6 aldehydes, these bands represent cell surface molecules having free galactosyl or N-acetyl-galactosaminyl groups.

Determination of PNA Receptors Isolated by Immunoprecipitation

Unseparated labelled thymocytes were treated with PNA and solubilized in PBS containing NP-40 and bovine serum albumin. After centrifugation the supernatant was incubated with untreated *Staphylococcus aureus* , in order to remove components interacting non-specifically with the bacteria. After centrifugation the supernatant was divided into two equal volumes and immunoprecipitated with *Staphylococcus aureus* coated with either rabbit anti-PNA immunoglobulins G or normal rabbit immunoglobulins G. When normal rabbit serum was used, low levels of labelled material was recovered, as compared to the experiments in which rabbit anti-PNA was used (350 and 1760 cpm/5 µl immunoprecipitate aliquot, respectively). Figure 1c shows that no detectable band was obtained when normal rabbit serum was used, whereas seven bands (apparent molecular weights 180, 175, 130, 115, 65, 26, 23 kd) could be characterized by rabbit anti-PNA immunoprecipitation from tritium-labelled thymocytes (Fig 1d). Among these bands the most intense one represents the major labelled surface molecule which binds PNA. The 175 kd band proved to be the major immunoprecipitated band obtained from ^{125}I-labelled cells (Fig 1e). Because PNA binds only to immature thymic cells the seven bands obtained in SDS-PAGE by immunoprecipitation from unseparated labelled cells represent the PNA receptors of immature thymocytes. The labelled bands from whole cells which are not immunoprecipitated probably represent cell surface molecules possessing free galactosyl groups with too low affinity for the lectin.

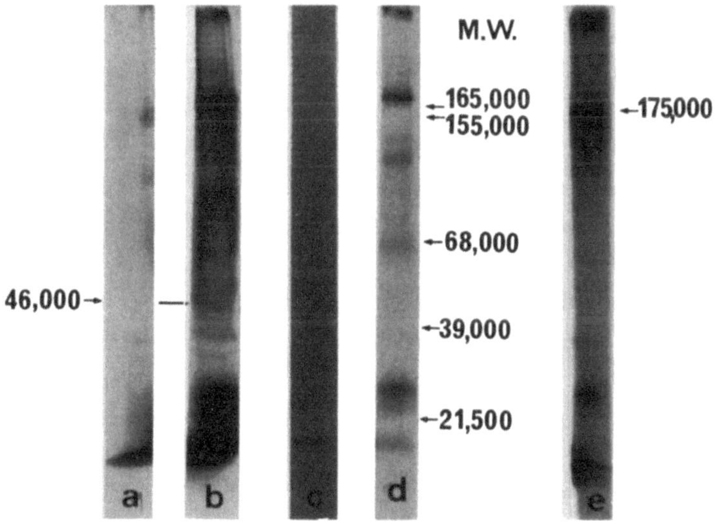

Figure 1. SDS-PAGE of cell surface molecules from labelled unseparated thymocytes and corresponding immunoprecipitates: nonspecific labeling with tritiated borohydride without enzymatic treatment (a); unseparated thymocytes labelled by galactose oxidase/tritiated borohydride (b); immunoprecipitate from galactose oxidase/tritiated borohydride labelled thymocytes using *Staphylococcus aureus* coated with either normal rabbit immunoglobulins G (c) or rabbit anti-PNA immunoglobulins G (d); immunoprecipitate from ^{125}I-labelled thymocytes (e).

Specific Labeling of PNA^+, PNA^- Thymocytes and Purified Spleen T cells

We have labelled PNA^+, PNA^- thymocytes and nylon wool/G-10 purified spleen T lymphocytes with galactose oxidase/tritiated borohydride (Fig 2a,b,c). Electrophoretic bands representing cell surface molecules which have incorporated tritium were only obtained from PNA^+ cells. This result supports the hypothesis that only immature population of thymus-derived lymphocytes bears, on their surface, molecules with free galactosyl moieties. This is conflicting with the results of Sharon and Reisner (18 and preceding paper) who have shown that both mature and immature thymocytes have receptors for soybean agglutinin (SBA), a lectin which is specific for N-acetyl-galactosamine, and to a lesser extent for D-galactose. Actually we think that a large majority of free galactosyl groups are masked in PNA^- mature cells, but a few molecules which are not PNA receptors possess free terminal galactosyl groups able to bind SBA. These SBA receptors on PNA^- cells probably represent a small number of molecules not detected by our labeling method.
When PNA^+ and PNA^- subsets were pretreated with neuraminidase before oxidation/reduction labeling, bands corresponding to PNA receptors were observed in both types of cells (Fig 2d,e). This result confirms the hypothesis that PNA receptors are covered with sialic acid during lymphocyte maturation and is in agreement with our observation that the PNA^- fraction can be agglutinated by PNA if pretreated with neuraminidase. Sialic acid residues have been labelled on PNA^+ and PNA^- thymocytes, using sodium periodate as oxidizing agent (Fig 2f,g). Numerous labelled surface molecules were detected in both populations and actually no noticeable difference could be evidenced in their electrophoretic pattern, but it is difficult to compare the intensity of the bands. Some of these bands have the same electrophoretic mobility as the PNA receptors, which implies that PNA receptors on immature cell surface possess both free galactosyl groups and sialyl residues, and thus are not completely

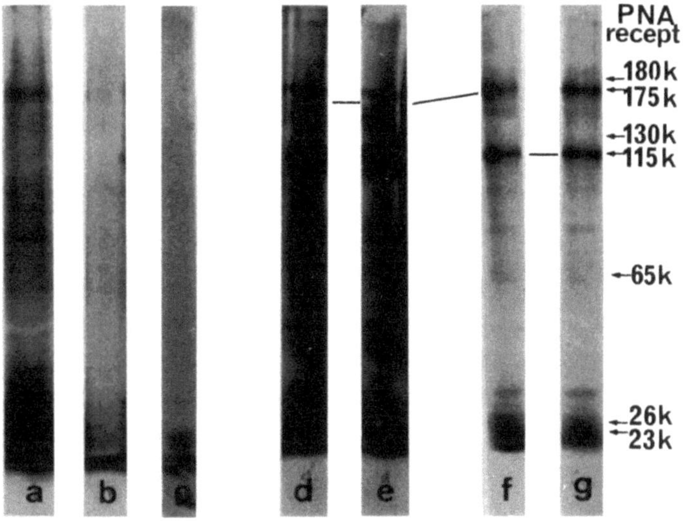

Figure 2. SDS-PAGE of cell surface molecules from labelled PNA$^+$, PNA$^-$ thymocytes and nylon wool/G-10 purified spleen T cells: galactose oxidase/tritiated borohydride labelled PNA$^+$ (a) or PNA$^-$ (b) thymocytes, or splenic T cells (c); neuraminidase/galactose oxidase/tritiated borohydride labelled PNA$^+$ (d) or PNA$^-$ (e) cells; periodate/tritiated borohydride labelled PNA$^+$ (f) or PNA$^-$ (g) cells.

desialylated. Unfortunately we failed to obtain good immunoprecipitates from periodate/tritiated borohydride labelled PNA$^+$ or PNA$^-$ cells in order to ascertain this result.

CONCLUSION

Mature and immature thymocyte subpopulations can be separated by agglutination with PNA through specific receptors present on immature cell surface. The PNA receptors have been immunopreci-

pitated from labelled thymocytes and characterized by SDS-PAGE and fluorography.
The immunoprecipitation has not been performed according to the standard method using lectin-particles conjugates to fractionate detergent-extracts from surface-labelled cells; in this method the detergent solubilization may reveal cryptic receptors normally inaccessible to the lectin and thus cause qualitative and quantitative changes in the molecules which bind PNA on the intact cell. In the immunoprecipitation system we used, PNA is able to bind cell surface molecules on intact cells, under physiological conditions, i.e. before NP-40 solubilization.
Among the seven immunoprecipitated bands corresponding to PNA receptors of immature thymocyte membranes, the 175 kd one represents the major cell surface component which binds PNA. A similar band was obtained by Jilg *et al* (19) by immunoprecipitation from rat thymocytes by anti-thymocyte immunoglobulins, and has widely been described as a thymocyte membrane component (16,17). The doublet 175-180 kd probably represents the two molecules (170-180 kd) described by Hoessli *et al* (17); these molecules have already been defined as glycoproteins common to immature and differentiated T lymphocytes which acquire additional sialic acid residues during the course of differentiation. This is in agreement with our results which show that these molecules loose the ability to be oxidized by galactose oxidase during the maturation. The 180 kd surface membrane component has been defined in the litterature as a molecule antigenically linked to a 190 kd one which is present on differentiated T lymphocytes and absent on thymocytes (20), these two mocules are though to be identical to the rat leukocyte common antigen (21,22) and the T-200 glycoprotein described by Trowbridge *et al* (23,24). The 26 kd band certainly represents the Thy-1 antigen, the molecular weight of which is known to be in the 25-30 kd range. Thy-1 has its sialic acid content increased during T cell differentiation (6). The other immunoprecipitated molecules were not identified to electrophoretic bands already described by others.

The characterization of PNA binding sites on immature thymocytes can be of great importance as specific markers of differentiaton. These receptors are probably covered with sialic acid during maturation, which may confer to the T lymphocyte its negative charge which has been shown to be essential for the normal circulation of T lymphocytes in the body (25).

ACKNOWLEDGEMENTS

This work received financial support from the Centre National de la Recherche Scientifique and the Fondation pour la Recherche Médicale Française. The authors acknowledge skilful technical assistance of N. Bernad and R. Larguier.

REFERENCES

1. Reisner, Y., Linker-Israeli, M., Sharon, N.: Cell. Immunol. 25, 129-134 (1976).
2. Reisner, Y., Biniaminov, M., Rosenthal, E., Sharon, N., Ramot, B.: Proc. Natl. Acad. Sci. USA 76, 447-451 (1979).
3. Irle, J., Piquet, P.F., Vassali, P.: J. Exp. Med. 148, 32-45 (1978).
4. London, J., Berrih, S., Bach,J.F.: J. Immunol. 121, 438-443 (1978).
5. Lotan, R., Skutelsky, E., Danon, D., Sharon, N.: J. Biol. Chem. 250, 8518-9523 (1975).
6. Hoessli, D.C., Bron, C., Pink, R.L.: Nature 283, 576-578 (1980).
7. Despont, J.P., Abel, C.A., Grey, M.M.: Cell. Immunol. 17, 487-494 (1975).
8. Dornand, J., Bonnafous, J.C., Mani, J.C.: FEBS Lett. 118, 225-228 (1980).
9. Dornand, J., Bonnafous, J.C., Favero, J., Mani, J.C.: Biochem. Med. (1982 in press).
10. Phillips, M.L., Parker, J.W., Frelinger, J.A., O'Brien, R. L.: J. Immunol. 124, 2700-2707 (1980).
11. Gahmberg, C.G., Andersson, L.C.: J. Biol. Chem. 252, 5888-5894 (1977).

12. Salisbury, J.G., Graham, J.M.: Biochem. J. 194, 351-355 (1981).

13. Kessler, S.W.: J. Immunol. 115, 1617-1624 (1975).

14. Laemmli, U.K.: Nature 227, 680-685 (1970).

15. Bonner, W.H., Laskey, R.A.: Eur. J. Immunol. 46, 83-88 (1974).

16. Gahmberg, C.G., Häyry, P., Andersson, L.C.: J. Cell Biol. 68, 642-653 (1976).

17. Hoessli, D.C., Vassali, P., Pink, J.R.L.: Eur. J. Immunol. 10, 814-821 (1980).

18. Sharon, N., Reisner, Y.: Molecular Mechanisms of Biological Recognition (Balaban, M., ed.) Elsevier, Amsterdam 1979, pp. 95-106.

19. Jilg, W., Hannig, K.: Hoppe Seyler's Z. Physiol. Chem. 362, 1475-1485 (1981).

20. Hoessli, D.C., Vassali, P.: J. Immunol. 125, 1758-1763 (1980).

21. Standring, R., Mc Master, W.R., Sunderland, C.A., Williams, A.F.: Eur. J. Immunol. 8, 832-839 (1978).

22. Sunderland, C.A., Mc Master, W.R., Williams, A.F.: Eur. J. Immunol. 9, 155-159 (1979).

23. Trowbridge, I.S., Mazauskas, C.: Eur. J. Immunol. 6, 557-562 (1976).

24. Trowbridge, I.S.: J. Exp. Med. 148, 313-323 (1978).

25. Woodruff, J.J., Gesner, B.M.: J. Exp. Med. 129, 551-563 (1969).

EFFECT OF 32 PURIFIED ANIMAL AND PLANT LECTINS ON HUMAN T LYMPHOCYTES.

Federico Licastro, Mariella Chiricolo, Luigi Barbieri, Fiorenzo Stirpe

Istitute of General Pathology, University of Bologna, 40126 Bologna, Italy.

Anna Falasca, Carlo Alfonso Rossi

Istitute of Biochemistry, University of Bologna, 40126 Bologna, Italy.

Claudio Franceschi

Chair of Immunology, Istitute of General Pathology, University of Padova, 35100 Padova, Italy.

Introduction

In plant and animal tissues proteins or glycoproteins have been found which have various biological activities, such as cell agglutination, mitogenic activity and, sometimes, toxic activity for intact cells and/or in cell-free systems. Such substances are called lectins owing to their major biochemical property, i.e. their binding specificity for carbohydrate groups. The aim of the present study was to ascertain if a single lectin could have more than one of these properties and exert different biological effects. For these reasons thirty-two purified animal and plant lectins were studied regarding to their mitogenic and/or toxic activity on purified human T lymphocytes. On the basis of the present study a classification of the lectins in three major groups will be proposed, i.e. those with mitogenic activity, those with toxic activity and those which do not exert either of these effects.
Furthermore a comparison between the results of this study and those obtained in a previous investigation (1) shows that a single lectin can exert different and apparently contradictory

biological effects. In fact some lectins appear to be mitogenic on intact lymphocytes and capable of inhibiting protein synthesis in a cell-free system. These observations allow some speculations about the interactions between lectin and their specific receptors in order to produce a biologically relevant effect.

Materials and methods

Lectins. The lectins used in the present study were obtained from different sources, as indicated in Table 1.

TABLE 1. Sources of the lectins used in the present study.

Lectins	Sources
Crotalaria juncea;	Dr. B. Ersson
Sarothamnus scoparius, Ulex europaeus;	Dr. L.G. Gürtler
Arachis hypogea, Coregonus lavaretus maraena, Datura stramonium, Dolichos biflorus, Erythrina indica, Glycine max, Helix pomatia, Ononis spinosa, Perca fluviatilis, Phaseolus coccineus, Phaseolus lunatus, Pisum sativum, Robinia pseudoacacia, Rutilus rutilus Vicia cracca;	Dr. J. Kocourek
Wistaria floribunda;	Dr. T. Kurokawa
Vicia villosa;	Dr. H. Wigzell
Abrus precatorius (agglutinin), Abrus precatorius (Abrin C), Adenia digitata (modeccin 4B), Adenia digitata (modeccin 6B), Hura crepitans (latex), Hura crepitans (seeds), Momordica charantia, Ricinus communis (R60), Ricinus communis (R120), Vicia sativa;	Our laboratories
Bandeiraea simplicifolia, Triticum vulgare	Commercial preparations

Human lymphocytes. 30-40 ml of heparinized venous blood were collected from a total of 10 blood donors. The peripheral blood lymphocytes were separated by density gradient centrifugation through Ficoll-Hypaque (2).

Enriched T lymphocytes were obtained by incubating lymphocytes with 2-aminoethylisothiouronium bromide hydrobromide (AET)-treated sheep red blood cells (SRBC), according to Pellegrino et al. (1975) (3) with minor modifications as described elsewhere (4). Briefly 2.5 x 10^6/ml lymphocytes were incubated with AET-SRBC (0.5%) for 15 min at 37°C in RPMI 1640 plus 10% of foetal calf serum. After incubation in an ice bath for 10 min, the rosetted cells (T lymphocytes) were separated from non-rosetted cells (B lymphocytes) by a Ficoll-Hypaque gradient. Preparation of T lymphocytes were 90% pure as judged by re-rosetting.

Effect of lectins. 0.1 ml of cell suspension (1 x 10^5 T lymphocytes) in complete medium (RPMI 1640, containing 2mM glutamine, 100 U/ml penicillin, 100 μg/ml streptomicin and 10% inactivated pooled human AB serum) was distributed in triplicate in microplate wells (Falcon Plastics, Los Angeles) and 0.1 ml of complete medium containing 2 μg/ml of PHA (PHA-P, Difco Detroit) or 0.01,0.1,1,10 and 100 μg/ml of the 32 lectins tested was added. T lymphocyte cultures were incubated in a humidified atmosphere of 5% CO_2 and 95% air for 72 hours. Methyl-^{3}H-thymidine (^{3}H-TdR, the Radiochemical Center, Amersham, spec. act. 5 Ci/mmol) was added for the last 6 hours (0.5 μCi/well).

The inhibitory experiments with Dolichos biflorus, Perca fluviatilis and Momordica charantia lectins were performed by adding each lectin to 1 x 10^5 T lymphocytes stimulated with PHA or ConA (10 μg/ml, ConA 3x crist.,Miles, Lausanne) as described above. Inhibitory lectins were added at the initiation of the culture period in appropriate amounts to reach final concentrations of 0.01,0.1,1,10 and 100 μg/ml. At the end of the incubation period the T lymphocytes were harvested and washed on glass fiber filters with the aid of a multiple cell culture harvester (Skatron, Norway) and ^{3}H-TdR incorporation (expressed as cpm) was measured by liquid scintillation counting as previously described (5).

Data evaluation. The results obtained in cell cultures were expressed as Stimulation Index (S.I.). S.I. was calculated as the ratio between cpm of stimulated and unstimulated cultures:

$$S.I. = \frac{\text{cpm of cultures stimulated with lectin}}{\text{cpm of cultures without lectin}}$$

Results

To test the mitogenic and/or the toxic activity of the lectins included in the present study more than ten experiments were performed. In each experiment a negative (T lymphocytes without mitogen) and a positive control (T lymphocytes plus PHA) were always included. In the different experiments the ^{3}H-TdR incorporation ranged from 200 to 4,000 cpm in negative controls (mean 1,120 cpm) and from 44,880 to 232,015 cmp (mean 99,840 cpm) in positive PHA-stimulated control cultures.

TABLE 2. Lectins which do not affect the ^{3}H-TdR incorporation of human purified T lymphocytes.

	Lectin concentrations (μg/ml)				
	0.01	0.1	1	10	100
Arachis hypogaea	1.13 [a]	1.11	0.73	0.58	0.84
Coregonus lavaretus M.	0.68	1.12	0.97	0.68	0.22
Helix pomatia	0.47	0.85	0.57	0.46	0.33
Pisum sativum	0.66	0.51	0.43	0.52	0.49
Sarothamnus scoparius	0.81	0.57	0.79	0.82	1.27
Triticum vulgare	1.01	0.68	1.01	1.00	0.88
Ulex europaeus	0.74	0.50	0.29	0.34	0.65
Vicia villosa	0.53	0.20	0.34	0.30	0.38

(a) Results are expressed as stimulation index (see Mat. and Met.)

Table 2 shows the results obtained with a group of lectins which do not affect the ^{3}H-TdR incorporation of the T lymphocytes. Although using a wide range of lectin concentrations (0.01-100 µg/ml) the S.I. values are close to 1.
Twelve lectins were found which caused a ^{3}H-TdR incorporation significantly higher than that observed in control unstimulated cultures.
The results of these experiments are shown in Table 3.

TABLE 3. Lectins which have mitogenic effect on human purified T lymphocytes.

	Lectin concentrations (µg/ml)				
	0.01	0.1	1	10	100
A)					
Crotalaria juncea	0.55[a]	54.91	189.87	32.77	7.63
Glycine max	0.85	0.78	0.69	0.81	4.15
Hura crepitans (latex)	2.84	45.10	132.59	157.57	15.67
Hura crepitans (seeds)	2.74	51.00	103.73	98.24	1.70
Phaseolus lunatus	0.56	0.76	1.25	16.86	2.97
Vicia sativa	0.69	0.68	1.67	11.47	64.15
B)					
Datura stramonium	0.77	1.02	1.92	2.41	0.11
Erythrina indica	0.78	4.98	22.59	9.69	0.10
Ononis spinosa	1.26	1.66	6.24	52.36	0.55
Phaseolus coccineus	1.58	74.08	278.37	132.125	0.80
Robinia pseudoacacia	2.97	48.74	105.78	17.13	0.34
Rutilus rutilus	1.59	4.28	14.04	2.75	0.04

(a) Results are expressed as stimulation index (see Materials and Methods); A = lectins which are mitogenic even at the highest concentrations tested; B = lectins which bring about a decreased ^{3}H-TdR incorporation at higher concentrations, after a peak of optimal activity.

The mitogenic power of many lectins (*Crotalaria juncea*, *Hura crepitans* from latex, *Hura crepitans* from seeds, *Vicia sativa*, *Ononis spinosa*, *Phaseolus coccineus* and *Robinia pseudoacacia*) is quite strong. The mitogenic activity of other lectins such as *Glycine max* and *Datura stramonium* is weak, while other lectins such as *Erythrina indica*, *Phaseolus lunatus* and *Rutilus rutilus* have, in these experimental condition, an intermediate mitogenic power.
These mitogenic lectins have been further divided into two subgroups. In the first lectins are mitogenic even at the highest concentrations tested, while in the second lectins bring about a decreased incorporation of ^{3}H-TdR at higher concentrations, after a peak of optimal activity.
Many lectins caused a decreased ^{3}H-TdR incorporation below the level observed in control unstimulated cultures. They are listed in Table 4.

TABLE 4. Lectins with toxic effect on human purified T lymphocytes.

	Lectin concentrations (µg/ml)				
	0.01	0.1	1	10	100
Abrus precatorius (agglutinin)	2.13[a]	---	0.62	0.50	0.27
Abrus precatorius (Abrin C)	0.02	0.02	0.01	0.02	0.03
Adenia digitata (modeccin 4B)	0.02	0.02	0.02	0.02	0.01
Adenia digitata(modeccin 6B)	0.02	0.02	0.04	0.01	0.01
Dolichos biflorus	0.38	0.36	0.36	0.18	0.13
Momordica charantia	0.93	0.39	0.24	0.22	---
Perca fluviatilis	0.55	0.49	0.28	0.08	0.07
Ricinus communis (ricin 60)	0.02	0.02	0.02	0.01	0.02
Ricinus communis (ricin 120)	0.04	0.11	0.06	0.03	0.06
Vicia cracca	0.54	0.50	0.39	0.09	0.10
Wistaria floribunda	---	1.04	0.65	0.49	0.13

(a) Results are expressed as stimulation index (see Materials and methods).

Some of these lectins such as ricin, abrin and modeccin are well known inhibitors of protein synthesis (6-8).
The toxic activity of three of these lectins has been confirmed by their addition to T lymphocytes previously stimulated by PHA and ConA (Fig. 1). In this case the inhibition of ^{3}H-TdR incorporation is dose-dependent and Momordica charantia lectin appears to be the strongest inhibitor.
The results obtained with Bandeiraea simplicifòlia (data not shown) were not clear enough to allow an assignment to one of three aforementioned group of lectins.

Discussion

The main results of the present study may be summarized as follows.

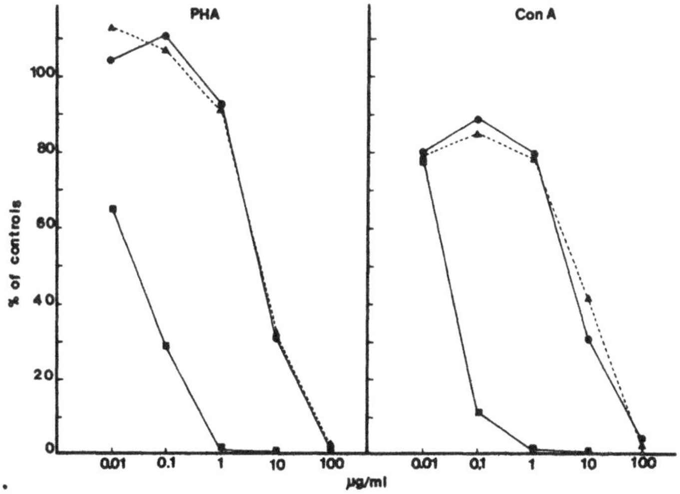

Fig. 1.
Effect of Dolichos biflorus (●—●), Perca fluviatilis (▲--▲) and Momordica charantia (■—■) lectins on the ^{3}H-TdR incorporation of PHA and ConA stimulated lymphocytes.
Results are given as per cent of control values.

First, as far as we know, the mitogenic activity for human T lymphocytes of Datura stramonium, Erythrina indica, Hura crepitans (from latex), Ononis spinosa and Rutilus rutilus lectins is described for the first time. The mitogenic activity of other lectins has been previously described (9,10).

It is interesting to note that two of the most potent mitogenic lectins derive from the same plant, i.e. Hura crepitans. Preliminary data suggest that the two lectins which can be purified from the seeds and from the latex respectively are different although very similar molecules. (Barbieri et al., unpublished results). Particularly interesting seems also to be the lectin from Rutilus rutilus which is of animal origin and has also toxic properties in a cell-free system (as will be later discussed). Further studies are needed to ascertain whether the mitogenic activity of some of these lectins is restricted to the T lymphocytes.

Many mechanisms have been proposed to account for the bell-shaped responses observed with mitogenic lectins (11). Taking into account the dose-response curves the mitogenic lectins have been divided in two subgroups, a first one in which the ^{3}H-TdR incorporation is always higher than that observed in unstimulated control cultures, even at the highest concentration used, and a second subgroup in which high lectin concentrations cause a ^{3}H-TdR incorporation lower than that of control cultures. This subdivision is proposed here as a working hypothesis to ascertain whether it can help in finding differences among the lectins of the two subgroups in terms of cell binding and toxicity, preferential stimulation of lymphocyte subpopulations with suppressor activity etc.

Second, we observed that the lectins with the highest inhibitory power on in vitro protein synthesis, such as abrin, ricin and modeccin, are also highly toxic for intact T lymphocytes. However apparent contradictions between the toxic activity in a cell-free system and on intact lymphocytes were also observed. In fact the lectins from Crotalaria juncea, Hura crepitans (latex), Hura crepitans (seeds) and Rutilus rutilus are all mitogenic (and in the case of the first three lectins among the strongest mitogens) for T lymphocytes and still they exert a strong inhibitory effect on protein synthesis in a cell-free system (1, and unpublished results). This strongly suggests that these lectins bind to cell receptors but either any or very few molecules are able to enter within the cells or that, even if they are able of entering, they are rapidly inactivated. The mechanisms and the structural requirements which determine and control the penetration of a lectin into a cell and the subsequent fate within the cell (enzimatic cleavage, breakdown of disulphide bridges etc.) seem to be critical as far as the overall biological effect (stimulation, killing etc.) on the cell is concerned.

Third, a group of lectins which did not affect the ^{3}H-TdR incorporation in intact lymphocytes was identified. It is interesting to note that, apart from _Coregonus lavaretus maraena_ lectin, which has a weak inhibitory activity, all other lectins of this group are also inactive as protein synthesis inhibitors in a cell-free system (1).

However, there are also lectins, such as those from _Dolichos biflorus_ and _Perca fluviatilis_, which exert a toxic effect on the spontaneous blastogenesis of T cells and totally inhibit the ^{3}H-TdR incorporation of PHA or ConA stimulated T lymphocytes at concentrations (100 μg/ml) that do not affect at all protein synthesis (1).

Thus, the mechanism by which these two lectins exert a toxic effect on intact lymphocytes is different from that of the majority of the other toxic lectins, i.e. ribosome inactivation and consequently inhibition of protein synthesis (6).

In conclusion our study suggest that different and apparently opposite properties such as the capacity to stimulate cell mitosis and to inhibit protein synthesis may belong to the same molecule and stress the importance of the experimental system used to assess the lectin activity. Moreover we hope that our approach i.e. the study of the effects exerted by lectins on a homogeneous population of intact purified human T cells, together with the study of the effects exerted by the same lectins on protein synthesis in a cell-free system, may help in the understanding of the structural requirements and the type of cell interactions that control binding, penetration, activation or inactivation, and consequently the final biological effect exerted by each lectin.

This work was supported by grants n. 81.01371.96 and number 81.01469.96 from Consiglio Nazionale delle Ricerche to prof. C. Franceschi and F. Stirpe, respectively, within the "Progetto Finalizzato sul controllo della Crescita Neoplastica", and by grant n. 81.00340.04 to Prof. C. A. Rossi.

References

1. Barbieri, L., Lorenzoni, E., Stirpe, F.: Biochem. J. 182, 633-635 (1979).
2. Böyum, A.: Scand. J. Clin. Lab. Invest. 21, Suppl. 97, 77-89 (1968).
3. Pellegrino, M.A., Ferrone, S., Dierich, M.I., Reisfeld, R.A.: Clin. Immunol. Immunopathol. 3, 324-333 (1975).
4. Licastro, F., Franceschi, C., Chiricolo, M., Battelli, M.G., Tabacchi, P., Cenci, M., Barboni, F., Pallenzona, D.: Carcinogenesis 3, 45-48 (1982).
5. Franceschi, C., Licastro, F., Paolucci, P., Masi, M., Cavicchi, S. and Zannotti, M.: J. Ment. Defic. Res. 22, 179-191 (1978).
6. Olsnes, S., Pihl, A.: In Cuatrecasas, P. (Ed.), the Specificity and Action of Animal, Bacterial and Plant Toxins 129-173, London: Chapman and Hall (1977).
7. Olsnes, S., Haylett, T., Refsnes, K.: J. Biol. Chem. 253, 5069-5073 (1978).
8. Gasperi-Campani, A., Barbieri, L., Lorenzoni, E., Montanaro, L., Sperti, S., Bonetti, E., Stirpe, F.: Biochem. J. 174, 491-496 (1978).
9. Falasca, A., Franceschi, C., Rossi, C.A., Stirpe, F.: Biochim. Biophys. Acta 577, 71-81 (1979).
10. Falasca, A., Franceschi, C., Rossi, C.A., Stirpe, F.: Biochim. Biophys. Acta 632, 95-105 (1980).
11. Hume, D.A., Weidemann, M.J.: Mitogenic Lymphocyte Transformation, Elsevier, North Holland Biomedical Press, Amsterdam. New York. Oxford 1980.

LECTINS AS PROBES OF BLOOD PLATELET ACTIVITY

Kenneth J. Clemetson*, John L. McGregor*, and Ernst F. Lüscher*

*Theodor Kocher Institute, University of Berne, P.O. Box 99, CH-3000 Berne 9, Switzerland.

**INSERM Unité 63, 22 ave du Doyen Lépine, F-69500 Bron and Faculté de Médicine, Alexis Carrel, rue Guillaume Paradin, F-6900 Lyon, France.

Blood platelets form an ideal test-system for studying the effect of lectins in inducing cellular activity and in inhibiting its induction by physiological stimulators. In vitro, they show three major activation stages: a) Shape change, where the resting discoidal platelet is transformed to a spherical form, which rapidly develops long pseudopods; b) the platelets become "sticky" and aggregate; c) the contents of storage organelles, many of which are also stimulators, are released. This latter stage may produce an enhanced aggregation, the so-called second wave. In addition, in vivo, platelets can adhere to exposed subendothelium, a process in part dependent on a plasma glycoprotein, the von Willebrand factor.

These processes can be either stimulated or inhibited by various lectins and several studies have been made, either on the effects of lectins on platelet activity (1-3) or to find out which lectins bind to which platelet membrane glycoproteins (4-7). However, little work has been donc on correlating the binding of a lectin to a particular glycoprotein with the biological effect produced.

Lectins, Vol. III

Considerable details are known of the role of some of the major platelet glycoproteins in the functions detailed above. These come from studies on platelets with genetically absent glycoproteins (Bernard-Soulier syndrome, GPIb, and Glanzmann's thrombasthenia, GPIIb and IIIa) (8-11) and studies with antibodies directed against specific glycoproteins (12-15). The major sialic acid containing glycoprotein (GPIb) is known to be the von Willebrand factor receptor and to contain a thrombin binding site, while GPIIb and IIIa form Ca^{2+}- dependent complexes and probably constitute the platelet aggregation receptors, possibly via fibrinogen binding. Glycoprotein V is the only thrombin-cleavable glycoprotein and is indispensable for thrombin-induced activation (16).

Methods and Materials

Platelets were isolated from citrate-treated blood collected for the Central Laboratory of the Blood Transfusion Service of the Swiss Red Cross and washed as previously described (17). Platelet membranes were prepared from washed platelets by sonication and differential centrifugation as previously described (6). In order to avoid degradation of membrane glycoproteins by endogenous Ca^{2+}-activated proteases, leupeptin (100 µg/ml, Sigma) was added before sonication.

Lectin-affinity chromatography was carried out with platelet membranes solubilized in 1% sodium deoxycholate on lectins bound to Sepharose 4B either prepared as previously described (4,6) or purchased from Pharmacia (Uppsala, Sweden) or Medac (Hamburg, F.R.G.).

Two-dimensional gel electrophoresis was carried out either by the method of Wang and Richards (18) (unreduced, reduced) or by the method of O'Farrell (19). Lectin binding components were identified by treating the fixed, washed gels with ^{125}I-labelled lectins as previously described (20).

Aggregation studies. The effects of lectins in inducing platelet agglutination/activation or in inhibiting platelet activation induced by physiological agents were studied by adding different concentrations of lectin to 200 µl of platelets ($5x10^8$/ml). When inhibition of other agents was tested, the agent was added 5 sec after the lectin. The effects of lectins on platelets which had been desialated by treatment with neuraminidase (Behringwerke) (6) were also checked.

Results and Discussion

The effects of various lectins on platelets have been studied by several groups (1 - 3) but the glycoprotein to which these lectins bind have not been so well characterized and little attempt has been made to correlate binding of lectins with the effects produced. In Table I such a correlation is attempted and the major glycoproteins bound and effects on platelets are shown for those lectins which have been studied in the most detail. In part, the effects on platelets are drawn from previously work (1-3) complemented by our own studies, reported here, while the results for the glycoproteins bound come principally from our studies with lectin-affinity chromatography (4,6) and binding of ^{125}I-labelled lectins to platelet glycoproteins separated by two-dimensional polyacrylamide gel electrophoresis (results reported here). Although the various lectins produce a wide range of effects on platelets it is nevertheless possible to draw certain broad conclusions and to see that the lectins fall roughly into two classes. These are firstly, those lectins which induce aggregation but only at relatively high concentrations and which are not strong inducers of the platelet release re-

TABLE I

Lectin	Major GP bound	Direct effect	Inhibition of
Lens culinaris	Ib, IIb, IIIa (4)	Aggregation (2)	Second wave†
Concanavalin A	IIIa, IIb	Shape change, release (2)	Aggregation†
Wheat germ agglutinin	Ia, Ib, IIIb (4)	Strong aggregation and release (3)	vWf induced aggregation
Ricinus communis lectin I*	Ia, Ib, IIIb (V)	Aggregation and release (1,2)	-
Soy bean agglutinin*	None,glycolipid?	No shape change but Aggregation (1)	-
Peanut agglutinin*	Ib (5,6)	Strong aggregation and release (1)	vWf induced aggregation
Maclura aurantiaca	Ib, high mol.wt. component (5)	Strong aggregation and release	vWf induced aggregation
Phaseolus vulgaris	Ib, IIa, IIIb	Aggregation + release (1)	Aggregation†

* With platelets treated with neuraminidase

† Induced by ADP, epinephrine or arachidonic acid

action and secondly, lectins which induce strong aggregation and release at low concentrations. In inhibition studies the first group tends to inhibit aggregation by a wide range of agents but not by ristocetin/von Willebrand factor while the second group inhibits ristocetin/von Willebrand factor induced aggregation strongly but has little effect on aggregation by other agents. The main members of the first group are Lens culinaris lectin, concanavalin A and Phaseolus vulgaris lectin while wheat germ agglutinin (with normal platelets) and peanut agglutinin and Ricinus communis lectin (with desialated platelets) are the principal lectins of the second group.

Fig. 1 shows the results obtained with Lens culinaris lectin as representative of the first group while Fig. 2 shows those obtained with peanut lectin as representative of the second. It is of course not yet possible to exclude that lectins bind also to glycolipids and thus affect platelet function. Soybean agglutinin may be a lectin which falls in this category. In addition, because most of the lectins bind to more than one glycoprotein there is some overlapping in the effects which they produce. None of the lectins tested had any effect on thrombin induced release of storage granules though lectins of the first group inhibit aggregation. This despite the fact that several lectins including wheat germ agglutinin and Lens culinaris lectin bind to glycoprotein V (the thrombin-cleavable glycoprotein) albeit relatively weakly. The results obtained can be interpreted as follows; those lectins which bind strongly to GPIb such as wheat germ agglutinin or, after neuraminidase treatment, peanut agglutinin, strongly induce both aggregation and release and inhibit vWf-induced aggregation if they can be used at a high enough concentration. Lectins which bind to GPIIb and/or GPIIIa such as Lens culinaris lectin or concanavalin A induce aggregation, although larger amounts are required, but are generally only weak inducers of release. They inhibit aggregation or second-wave of aggregation induced by agents such as ADP, collagen, adrenalin or thrombin. The relationship between the glycoproteins bound and the effect produced is in good agreement with results obtained using monospecific or monoclonal antibodies directed against specific glycoproteins (12-15). There are some differences in effect which can be explained in terms of the binding sites on the glycoprotein for lectins compared to antibodies. Thus, the inhibition of von Willebrand factor-induced platelet aggregation by lectins binding GPIb supports earlier suggestions (21) that it is the outer,

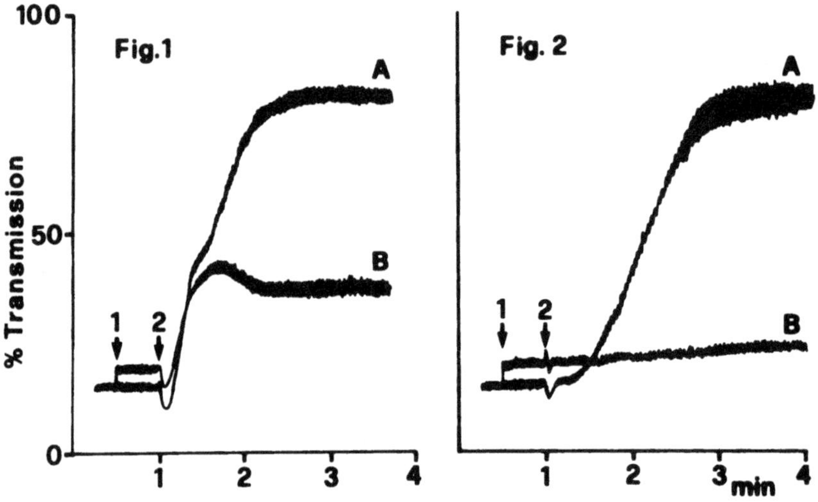

Fig. 1: Effect of <u>Lens culinaris</u> lectin on ADP/fibrinogen-induced platelet aggregation ($x10^8$ platelets/ml, 4μM fibrinogen). **A:** Aggregation curve with 5μM ADP added at arrow 2. **B:** Aggregation curve with <u>Lens culinaris</u> lectin (final concentration (200μg/ml added at arrow 1, followed by 5μM ADP at arrow 2.

Fig. 2: Effect of peanut lectin on ristocetin-induced aggregation of neuraminidase treated platelets resuspended in plasma. **A:** Aggregation curve with ristocetin (final concentration, 1.5 mg/ml) added at arrow 2. **B:** Aggregation curve with peanut agglutinin (20μg/ml, final concentration) added at arrow 1 followed by ristocetin as in A at arrow 2.

carbohydrate-rich part of GPIb which is involved in von Willebrand factor binding. The effects of <u>Lens culinaris</u> lectin and concanavalin A in inhibiting platelet aggregation by a variety of agents (excluding von Willebrand factor) gives further weight to the idea that this agglutination is mediated by complex formation between GPIIb and IIIa, producing fibrinogen binding sites on the platelet surface (22). The dimeric fibrinogen molecules may then form bridges between the platelets causing aggregation. The lectins may act either by blocking the fibrinogen binding sites directly or by preventing complex formation between GPIIb and IIIa. It is not clear why with Lens culinaris lectin only the second wave is affected but similar results have been observed with antibodies directed against GPIIb/IIIa. Possibly two different mechanisms are involved whereby only one i.e. the aggregation produced by fibrinogen and ADP released from the platelets, can be inhibited. In conclusion, it is now clear that many of the effects of lectins on platelets can be explained in terms of the glycoproteins to which they bind but that not all of the problems have yet been solved.

Acknowledgements

This work was supported by grants from the Swiss National Science Foundation and INSERM CRL 817016.

References

1. Patscheke, H., Brossmer, R., and Wörner, P. Biochem. Biophys. Res. Commun., 75, 200-206, (1977).
2. Patscheke, H. and Wörner, P. Thromb. Res. 11, 391-402, (1977).
3. Greenberg, J.H. and Jamieson, G.A. Biochem. Biophys. Acta, 345, 231-242, (1974).
4. Clemetson, K.J., Pfueller, S.L., Lüscher, E.F. and Jenkins, C.S.P. Biochim. Biophys. Acta, 464, 493-508, (1977).
5. Judson, P.A. and Anstee. D. In: Proc. of the 27th Meeting, Protides of the Biological Fluids, H. Peeters (Ed.) Pergamon, Oxford, New York 1980 pp. 871-874.
6. Clemetson, K.J., Naim, H.Y. and Lüscher, E.F. Proc. Natl. Acad. Sci. USA, 78, 2712-2716, (1981).
7. Hagen, I. and Gogstad, G. In: Lectins, Biology, Bio chemistry, Clinical Biochemistry, T.C. Bøg-Hansen (Ed) Vol 1, de Gruyter, Berlin, New York 1981 pp 347-357.
8. Jenkins, C.S.P., Phillips, D.R., Clemetson, K.J., Meyer D., Larrieu, M.-J. and Lüscher, E.F. J. Clin. Invest. 57, 112-125, (1976).
9. Hagen, I., Nurden, A., Bjerrum, O.J., Solum, N.O. and

Caen, J.P. J. Clin. Invest. 65, 722-731, (1980).

10. Nurden, A.T. and Caen, J.P. Semin. Haematol. 16, 234-250, (1979).

11. Phillips, D.R. and Poh Agin, P. J. Clin. Invest. 60, 535-545, (1977).

12. Ali-Briggs, E.F., Clemetson, K.J. and Jenkins, C.S.P. Br. J. Haematol. 48, 305-318, (1981).

13. Jenkins, C.S.P., Ali-Briggs, E.F. and Clemetson, K.J. Br. J. Haematol. 49, 439-447, (1981).

14. Ruan, C., Tobelem, G., McMichael, A.H., Drouet, L., Legrand, Y., Degos, L., Kieffer, N., Lee, H. and Caen, J.P. Br. J. Haematol. 49, 511-519, (1981).

15. Kornecki, E., Lee, H., Tuszynski, G.P. and Niewiarowski, S. Fed. Proc. 41, 528 (1982) abstr.

16. Berndt, M.C. and Phillips, D.R. J. Biol. Chem. 256, 59-65, (1981).

17. Massini, P. and Lüscher, E.F. Biochim. Biophys. Acta 372, 109-121, (1974).

18. Wang, K. and Richards, F.M. J. Biol. Chem. 252, 8005-8018, (1974).

19. O'Farrell, P.H. J. Biol. Chem. 250, 4007-4021, (1975).

20. McGregor, J.L., Clemetson, K.J., James, E., Capitanio, A., Greenland, T., Lüscher, E.F. and Dechavanne, M. Eur. J. Biochem. 116, 379-388, (1981).

21. Jamieson, G.A., Okumura, T. and Hasitz, M. Thromb. Haemostas. 42, 1673-1678, (1979).

22. Nachman, R.L. and Leung, L.L.K. J. Clin. Invest. 69, 263-269, (1982).

SIALIC ACID AS A CELL SURFACE BINDING SITE FOR WHEAT GERM AGGLUTININ

John T. Gallagher*, Margaret Harding* and Robert E. Dale#
*Cancer Research Campaign Department of Medical Oncology,
#Paterson Laboratories,
Christie Hospital & Holt Radium Institute, and Manchester University,
Wilmslow Road, Manchester M20 9BX, England.

The combining site of wheat germ agglutinin (WGA) consists of three identical sub-sites each complementary to N-acetylglucosamine -GlcNAc-: the free sugar is a poor inhibitor of lectin binding to glycoconjugates but the di- and trisaccharide forms (chitobiose and chitotriose) are potent inhibitors suggesting that high affinity binding is associated with ligands that interact simultaneously with two or more of the available subsites (1,2). In glycoproteins terminal GlcNAc residues in asparagine-linked glycans are not reactive with WGA although the lectin will precipitate with the isolated trisaccharide man-GlcNAc-GlcNAc which forms the linkage sequence to asparagine (3,7). In native glycoproteins the proximity of this sequence to the core protein may impair recognition by WGA. The interaction of WGA with cell surfaces is inhibited by treatment of cells with neuraminidase suggesting that sialic acid may be an important sugar in the binding process. (5,6,9) When used as a saccharide inhibitor of WGA, sialic acid has very low activity. (3) However, multiple interactions with sialyl residues may be favoured at cell surfaces because of the location of this sugar at the terminal non-reducing end of carbohydrate chains. In an attempt to elucidate the roles of sialic acid and GlcNAc we have carried out kinetic and inhibition studies on the influence of neuraminidase on WGA binding to the surfaces of K562 cells.

Materials and Methods

Cells used were an erythroleukaemic cell line, K562, derived from a patient with chronic myeloid leukaemia in blast crisis. (8) The binding of F-WGA (Miles Chemical Co., molar ratio, fluorescein :

Lectins, Vol. III

protein 1.3:1) was determined as described previously (4). Briefly, varying concentrations of F-WGA were added to K562 cells (10^5 cells in 150 ul) and incubated at 4°C for 30 minutes. Cell surface-associated lectin was then measured by flow cytometry. Sugar inhibitors (GlcNAc and chitobiose) were mixed with the lectin before addition of cells. In kinetic analyses lectin binding to cells was effectively stopped by diluting the incubation tenfold with cold phosphate buffered saline (PBS) (pH 7.2). Competitive interaction between F-WGA and the native form of the lectin was carried out by first incubating the cells for varying times with the fluorescein-derivative and then adding a two to tenfold excess of the unconjugated form. Further details may be found in the figure legends. Neuraminidase (Type IV - from Sigma Chemical Co.) treatment was carried out by incubating 10^7 cells with 0.05 units of enzyme in a total volume of 1 ml of PBS. Maximum decrease in F-WGA binding capacity was achieved under these conditions.

Results

Specificity of interaction of WGA with the K562 cell surface was established by the complete inhibition of lectin binding in the presence of chitobiose and the significant but less complete inhibition by GlcNAc. (Fig.1)

Distinctive binding curves were obtained when comparing control and neuraminidase-treated cells: in the former binding approached a plateau between 1 ug and 5 ug of lectin, whereas in the treated cells binding increased quite sharply over this concentration range (Fig.2). The degree of inhibition of binding to treated cells was dependent upon the lectin concentration, 64% inhibition at 0.1 ug declining to 39% at 5 ug.

Kinetic analysis indicated that the rate of WGA binding was considerably more rapid to untreated than to neuraminidase-treated cells (Fig.3). In the treated cells fast and slow phases of the binding process were observed with each phase representing about 50% of the binding curve. In the control cells, however, 70% of binding was complete at 1 min (fast phase) and the slow phase was quantitatively less significant (Fig.3). Further complexity was evidenced when binding was examined over very short time increments of 5 secs (Fig.4). In the control population, 60% binding site occupancy was achieved within 5 secs whereas the corresponding fast phase in enzyme-treated cells led to only 25% occupancy. We would emphasise that different lectin concentrations were used in the kinetic experiments (Figs. 3-5). 1 ug was chosen for the untreated control population as this gave virtual saturation of binding sites preventing selective analysis of sites occupied at low lectin concentration. 5 ug wås used for the neuraminidase-treated cells because this was the highest lectin concentration we could employ without causing significant agglutination. The latter was not a 'saturating' concentration so some potential sites would remain unoccupied.

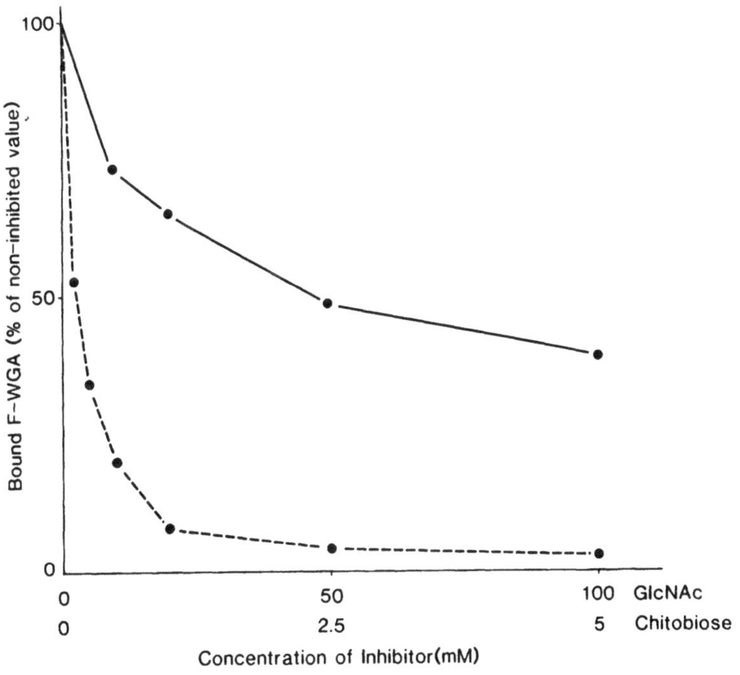

Fig.1. Saccharide inhibition of F-WGA binding to K562 cells

GlcNAc (———) and Di,N-acetylchitobiose (-----) were mixed with the lectin (1ug/150ul) before addition of K562 cells. For each saccharide concentration incubations with cells were for 30 mins at 4°C. Cell fluorescence was measured by flow cytometry.

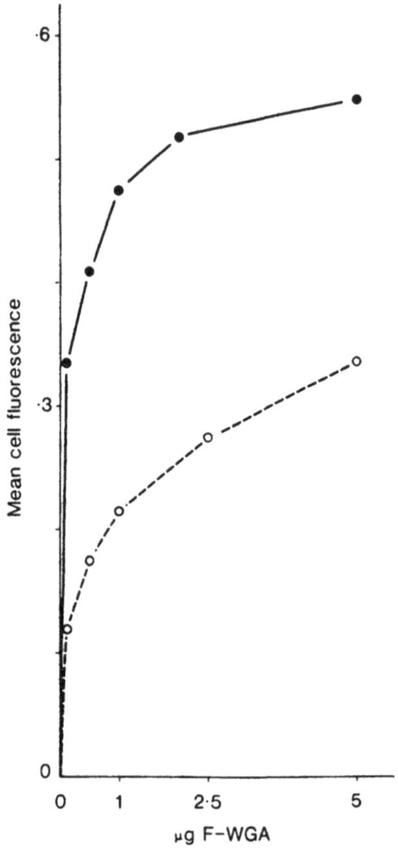

Fig.2. <u>Influence of neuraminidase on F-WGA binding</u>

Neuraminidase-treated(-----) and untreated cells (———) were incubated with the stated range of F-WGA concentrations for 30 mins at 4°C. 10^5 cells were mixed with the appropriate lectin concentration in a total volume of 150 ul of PBS.

When the fluorescein-conjugated WGA (1 ug) was incubated with K562 cells for 30 mins and then a tenfold excess of native WGA added, no decrease in cell fluorescence, and hence no equilibrium between bound and free forms of the lectin was observed (result not shown). However, in view of the complex binding kinetics of WGA we decided to explore the possibility that irreversibility of binding may be time dependent, i.e. that a certain residence time of the lectin at the cell surface is necessary for irreversible binding. Cell suspensions were therefore incubated with F-WGA for various times and the binding process interrupted by addition of an excess of native WGA. The native lectin should both compete with F-WGA for unoccupied binding sites and establish an equilibrium with bound F-WGA if lectin binding

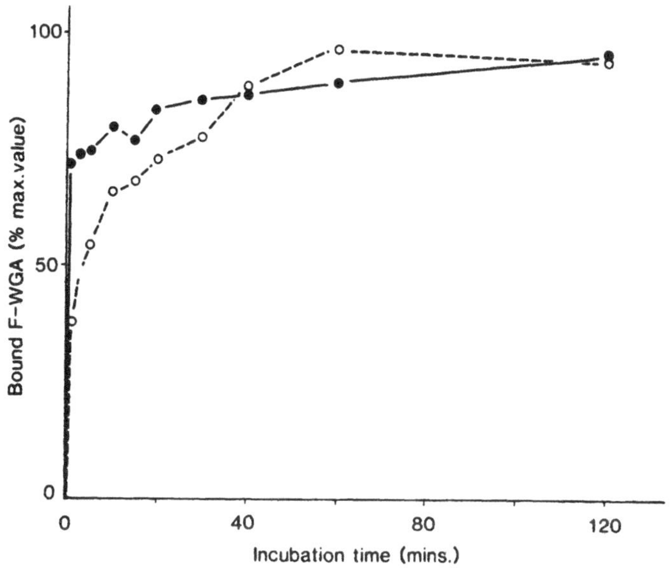

Fig.3. Rate of binding of F-WGA

F-WGA was added to untreated (———) and neuraminidase-treated (-----) cells at a lectin concentration of 1 ug/150 ul and 5 ug/150 ul respectively. At the times indicated F-WGA binding was stopped by addition of 10 volumes of PBS.

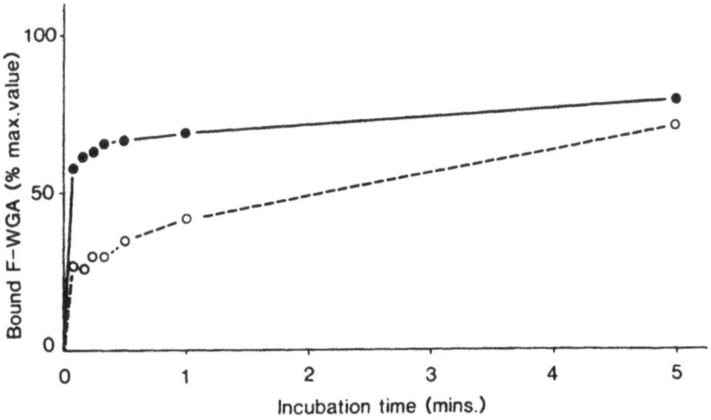

Fig.4. Rate of binding of F-WGA

Conditions as in Fig.3 except much shorter time increments were used.

is reversible. The results showed that in contrast to the data in Fig.4, the binding curves for untreated and neuraminidase-treated cells were almost identical (Fig.5). At the 5 sec. time point, the concentration of bound F-WGA on the untreated cells (Fig.5) was only about 25% of the value observed when F-WGA binding was effectively stopped by dilution (Fig.4). In contrast, the binding curves for neuraminidase-treated cells were almost identical, irrespective of whether the lectin binding process was interrupted by dilution (Fig.4) or excess native lectin (Fig.5).

Discussion

Neuraminidase treatment clearly decreases F-WGA binding to K562 cells (Fig.2) indicating an important role for sialic acid in the cell surface reactivity of this lectin. These results are in accord with findings of other workers (see Introduction). Sialyl residues may be of particular significance at low lectin concentrations since the neuraminidase-inhibition effect was more pronounced at 0.1 ug than 5 ug of lectin (Fig.2). However, interpretation of F-WGA binding data is complex. The lectin recognises both sialic acid and GlcNAc residues and it is conceivable that removal of sialyl groups may

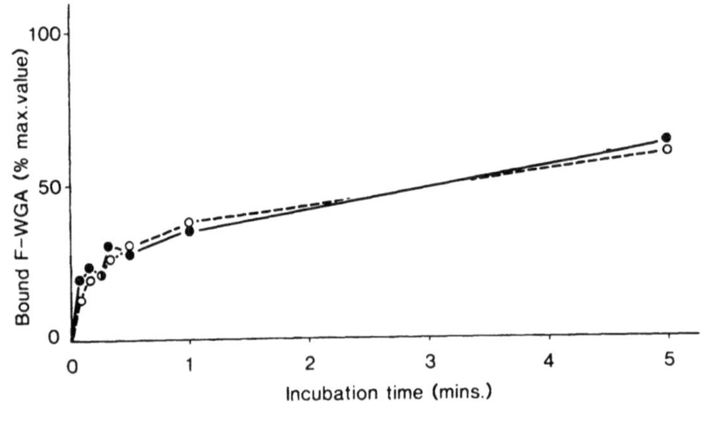

Fig.5. Competitive inhibition of F-WGA binding

F-WGA was added to untreated (———) and neuraminidase-treated (-----) cells using lectin concentrations of 1 ug/150 ul and 5 ug/150 ul respectively. At the specified times, (identical to those in Fig.4) 10 ug of native lectin was added to the incubation mixture and cell fluorescence determined immediately.

expose GlcNAc sites not normally available to the lectin. Changes in F-WGA binding induced by neuraminidase may thus comprise a balance between loss of sialyl sites and accompanying expression of cryptic sites. Sialic acid is a terminal sugar on saccharide chains and it could be preferentially localised at the periphery of the surface-associated glycocalyx. These sites may interact readily at low lectin concentrations. Accessibility and rate of interaction of internal sites, presumed to be mainly GlcNAc and resistant to neuraminidase, may be influenced by steric factors and the rate of lectin diffusion through the pericellular matrix. Additionally, the binding of WGA to sialyl residues may impede WGA binding to GlcNAc. We suggest that the fast component of WGA binding, most notable in untreated cells, involves an interaction between the lectin and sialic acid. The slower component, quantitatively more important in the neuraminidase-treated cells, represents WGA recognition of sites more proximal to the membrane surface and consisting principally of GlcNAc residues.

Interruption of F-WGA interaction with cell surfaces by addition of excess native lectin yielded similar net rates of lectin binding to treated and untreated cells (Fig.5) and this contrasted with the conspicuous differences observed in these two cell populations when binding rates were determined after dilution rather than addition of native lectin (Figs. 3 & 4). These results indicate that in the primary stage of the binding process to untreated cells F-WGA interacts mainly with sialic acid residues. Initially, occupancy of these sites is reversible so that when an excess of native lectin is added, cell fluorescence declines as a new equilibrium is established between bound and free forms. However, with prolongation of the contact time, 'lectin-binding site' complexes undergo a progressive, acquired irreversibility and addition of native lectin after a pre-incubation of 30 min. with F-WGA did not influence cell fluorescence values. In contrast, binding to neuraminidase-treated cells is, in the main, irreversible, so that binding curves are almost identical irrespective of the method of interrupting the binding process (cf.Figs.4 and 5).

In conclusion, our results point to considerable heterogeneity in F-WGA binding sites at the K562 cell surface. Dissection of the binding process by kinetic analysis, competitive inhibition and neuraminidase-induced inhibition indicates that sialyl residues at the periphery of the glycocalyx are the most rapidly-recognised, quantitatively predominant sites and that following desialation a group of relatively inaccessible, slow-reacting GlcNAc sites are exposed which appear less readily saturable than the sialyl sites. By careful choice of the experimental conditions (short incubation time, low lectin concentration) it seems possible that WGA may be used as a selective probe of cell surface sialic acid.

References

1. Adair, W.L. & Kornfeld, S. (1974)
J.Biol.Chem. 249, 4696-4704
2. Allen, A.K., Neuberger, A. & Sharon, N. (1973)
Biochem.J. 131, 155-162
3. Bhavanandan, V.P. & Katlic, A.W. (1979)
J.Biol.Chem. 254, 4000-40C8
4. Blackledge, G., Gallagher, J.T., Swindell, R. & Crowther, D.(1980)
Flow Cytometry IV, 222-226
5. Briles, E.B., Li, E. & Kornfeld, S. (1977)
J.Biol.Chem. 252, 1107-1116
6. Cuatracasas, P. (1973)
Biochemistry, 12, 1312-1323
7. Goldstein, I.J., Hammastrom, S. & Sundblad, G. (1975)
Biochim. Biophys. Acta 405, 53-61
8. Lozzio, C.B. & Lozzio, B.B. (1975)
Blood, 45, 321-334
9. Stanley, P. & Carver, J.P. (1977)
Proc.Natl.Acad.Sci. U.S., 74, 5056-5059

INTERACTION OF HUMAN YOUNG AND OLD ERYTHROCYTES WITH WGA. A FLUORESCENCE STUDY.

Dominique Bladier°, Marie-Ange Deugnier°° and Michel Caron°
°Laboratoire de Biochimie et Immunochimie (Pr P.Cornillot)
°°Laboratoire de Microscopie Quantitative (Pr J.-C.Bisconte)
U.E.R. Biomedicale, 74, rue Marcel Cachin-93012 BOBIGNY, France

Introduction

During its lifespan in the circulation, the macromolecular components of the red blodd cell (RBC) are submitted to progressive modifications leading to the alteration of its physico-chemical and functionnal properties and ultimately resulting in its removal by the macrophages. Erythrocyte aging, among other events, gives rise to the homogeneous loss of membrane fragments (1) and to modifications of some membrane dynamic properties (2). Changes at the membrane level have an even higher importance since it is the membrane which interacts with the extracellular environment. The role attributed to the surface glycoconjugates in such interactions led us to examine these structures and their properties in relation with the age of the cells.
The aim of the study, reported here, was thus to investigate the interaction of human RBC with rhodamine labelled WGA, looking for possible differences between young and old cells.

Material and methods

<u>Cell labelling</u>. Young and old human O erythrocytes, obtained from healthy donors of the blood bank (Centre Departemental de transfusion Sanguine, Bobigny, France) were separated accor-

ding to Murphy (3).Rinsed RBC were incubated with the rhodamine wheat germ agglutinin conjugate (WGA-T,IBF) for 30 min at 20°C.A subagglutinating concentration,i.e. 9.6µg of WGA-T per 10^8 cells,was used for all experiments.Cells were washed twice with 0.9% NaCl solution before being resuspended in 1mL of phosphate buffered saline solution (PBS).Fluorescein diacetate (FDA) was used as a freely diffusing intracellular probe, at 6 µg per 10^8RBC (4).The labelling was performed for 10 min at 37°C and followed by 3 whashings with PBS.The labelled samples were kept at 4°C until the analysis.

Fluorescence measurements.Suspensions of erythrocytes were mounted on a glass slide covered with a coverslip and examined under a microscope equipped with a laser illumination.The basic instrument has been previously described (5) and detail of the computer control is given in (6).Briefly,the laser beam (Spectra Physics,λemission=514.5 nm) is focused through the microscope objective (x125) to a 3.5 µm diameter spot, which allows cell by cell measurements.Light pulses of preselected duration (500 msec) and intensity (30µW) are delivered using an accousto-optical modulator (SORO) monitored by a PDP 11-34 (DEC) computer.The fluorescence emission is detected by a photomultiplier (Leitz,MPV_2) also connected to the computer.The fluorescence intensity values,stored for further processing,are proportionnal to the number of fluorescent sites per illuminated cell location.

Data analysis.Two types of analysis have been made:
i)determination of the mean ($\pm$sd) fluorescence intensity of a given cell population.In this case,one measurement per cell was performed and about 100 cells were examined.A histogram can be plotted with these 100 values.
ii)determination of the reproducibility of the measurement on a single cell.In this case,5 successive measurements were performed per cell and about 20 cells per sample were assayed

It should be pointed out,that before each of the 5 measurements the chosen cell was repositionned at the center of the microscopic field,where the laser beam is focused,thus resulting in the eventuality that the 5 locations,successively examined,are not exactly superimposable.The results are expressed as follows:for each cell,the differences between the 5 measured values of fluorescence intensity were determined 2 by 2.Ten numbers are thus obtained.The set of numbers,calculated for the 20 assayed cells,are then distributed,into classes of 1 arbitrary unit wide,on a histogram.

Results

1)Mean fluorescence intensity of the RBC populations.It was measured:i)on young and old WGA-T-labelled RBC.The results obtained from 5 samples are listed in table 1.It appears that the mean values found for the old RBC populations are significantly higher than that found for the young ones.A typical his togram of fluorescence distribution is shown in figure 1 B. ii)on old FDA-labelled RBC,figure 1 A.Very similar fluorescence distributions were obtained for both WGA-T and FDA labelled populations,figure 1 A and B.

2)Reproducibility of the measurements on a single cell.It was tested on both FDA and WGA-T labelled RBC.Typical histograms are presented in figure 1 A' and B' respectively.These histograms are very different,suggesting that the reproducibility of the measurements on a single cell is better on FDA labelled RBC than on WGA-T labelled RBC.

Discussion

The results obtained in this study of WGA interaction with

TABLE 1

Mean value of fluorescence intensity (arbitrary units) of WGA-T labelled old and young RBC.

Sample N°	Old Cells m	sd	n	Young Cells m	sd	n
1	51	13.5	100	41	9	102
2	59	10	104	52	10	99
3	61	12	100	50	8.5	100
4	65	14	100	58	10	100
5	61	14	101	47	10	101

n=number of measurements per sample.
The mean value of autofluorescence was less than 11 a.u.In all cases, the difference between fluorescence intensity of young and old RBC are significant (p 0.001)

young and old human RBC reveal a double heterogeneity of WGA binding:on a single cell,and within a same cell population.

i)<u>Heterogeneity on a single cell:</u>it was established by comparing the measurements obtained on FDA(a freely diffusing intracellular probe)and WGA-T labelled RBC.It can be seen that the reproducibility of the measurements on the FDA labelled RBC is rather good,fig.1 A'.This means that whatever each of the 5 cell locations successively illuminated,the corresponding measurements are very close together.This is true for each of the 20 analysed cells.The distribution of the FDA probe can thus be considered to be homogeneous within the cell.On the contrary,the wide histogram exhibited by WGA-T labelled RBC,fig.1 B' reveals that the measurements,obtained for esch of the 5 cell locations successively illuminated,are different.The distribution of the lectin,at such subagglutinating concentration,can thus be considered to be heterogeneous at the surface of the cell,due to the presence of clusters.Whether these clusters are induced by the lectin interaction,or preexist,is currently under study.It cannot be excluded that differences in some

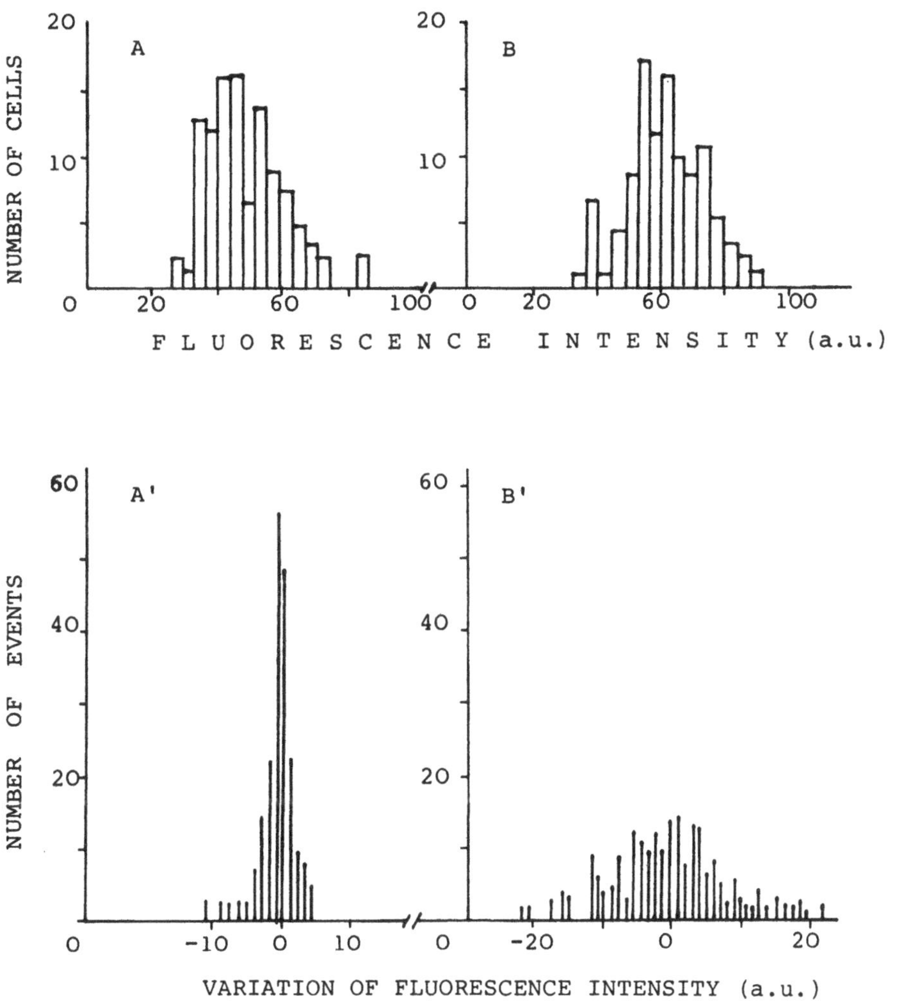

Fig. 1 Fluorescence of FDA and WGA-T labelled RBC:

Above: Histograms of fluorescence distribution, one measurement per cell (100 cells per sample), classes of 4 a.u. wide. A = FDA labelled RBC; B = WGA-T labelled RBC.

Below: Reproducibility of the measurements: the difference of fluorescence intensity between the 5 successive measurements per cell (20 cells per sample are distributed into classes of 1 a.u. wide). A' = FDA labelled RBC; B' = WGA-T labelled RBC.

dynamic properties of the cell membrane with aging(2)could also participate in the mechanism of cluster formation.

ii)heterogeneity within a same cell population:old RBC populations are characterized by both higher mean values of fluorescence intensity and wider fluorescence distribution.As WGA binding to cells is known to be biphasic,due to the presence of high and low affinity binding sites(7)this difference may be due either to a higher density of WGA binding sites on old RBC or to modifications in the balance of the number of high and low affinity binding sites,or to variations of the affinities. In the interaction between the cell surface and WGA,the non reducing terminal residues (NeuNac) play a more important role rather than the usually internal residues (GlcNac)(8).In the case of RBC,NeuNac content is known to decrease with aging (9) but its surface densities on young and old RBC are the same(10) The higher fluorescence of old RBC may thus not be attributed to a higher density of NeuNac residues on these cells.On an other hand,we have established (data not shown) that increasing the WGA concentration results in wider histograms of fluorescence distribution,reflecting the heterogeneity of low affinity binding sites.The wider distribution of fluorescence of old RBC may thus be related to a higher density of low affinity binding sites on their surface.

In conclusion,these results reveal,whatever the hypotheses, that old RBC bind more WGA than do young ones and thus that the surface glycoconjugates distribution and/or properties are modified with aging.This is of interest in the hypothesis of the participation of such structures in the interaction of old RBC with their environment and particularly with macrophages.

References

1. Weed,R.I.,Reed,C.F.:Am.J.Med.41,681-698 (1966).

2. Bartosz,G.,Szabo,G.,Szöllosi,J.,Damjanovich,S.:Mech.Ageing Dev.16,265-274 (1981).

3. Murphy,J.R.:J.Lab.Clin.Med.:82,334 (1973).

4. Rotman,B:in Fluorescence Techniques in Cell Biology,Thaer A.A. and Sernetz,M.Eds,Springer Verlag,255-258 (1973).

5. Deugnier,M.-A:,Albe,X.,Bisconte,J.-C.:Biochimie 63,873-876 (1981).

6. Albe,X.,Deugnier,M.-A.,Caron,M.,Bisconte,J.-C.:to be published in Comp.Biomed.Res.

7. Adair,W.L.,Kornfeld,S.:J.Biol.Chem. 249,4696-4704 (1974).

8. Bhavanandan,V.P.,Katlic,A.W.:J.Biol.Chem. 254,4000-4008 (1979).

9. Greenwalt T.J.,Steane,E.A.:Br.J.Haematol.25,207-215 (1973)

10. Seaman,G.V.F.,Knox,R.J.,Nordt,F.J.,Regan,D.H.:Blood 50, 1001-1011,1977.

LABELING OF RAT SPERMATOGENIC CELLS WITH FLUORESCEIN ISOTHIOCYANATE CONJUGATED LECTINS

Raija Malmi, Karl-Ove Söderström, Kristian Karjalainen
Departments of Anatomy and Pathology, University of Turku
SF-20520 Turku 52, Finland

Lectins are proteins which are characterized by the differences in their affinities for particular carbohydrate moieties. They are valuable probes for investigating cell surfaces, useful for following cell differentiation or cell transformation and for elucidating the organization of cell membrane carbohydrate containing antigens.
Most studies concerning the binding of lectins to male germinal cells have been made with spermatozoa. The first experiments investigating the reaction of lectins with mammalian spermatozoa showed that soybean extracts interacted with bull sperm to produce predominantly tail-to-tail agglutination of the sperm (1). Human spermatozoa from donors of blood group A have been specifically agglutinated by a lectin from the snail Helix Pomatia, while phytohemagglutinin, ricin and wax bean lectin agglutinated all sperm tested (2). The quantification of lectin binding sites on the sperm surface has indicated that spermatozoa do not differ greatly from somatic cells in the total number of lectin receptor sites per cell (3, 4, 5).
Preliminary studies on lectin binding to mouse spermatogenic cells suggest that there are significant differences in the binding of various lectins to isolated spermatogenic cell fractions of mouse testes (6). Pachytene spermatocytes, for example, bind concanavalin A to a lesser extent than any of the

Lectins, Vol. III

other proteins studied. By the time of spermatid differentiation Con A receptors have increased in number and predominate on the surface membrane (6) .

Materials and methods

The testis from three months old Sprague-Dawley rats was fixed in Bouin´s fluid, embedded in paraffin and 5 um thick sections were cut. After deparaffinization the tissue sections were incubated in a moist chamber under a drop of lectin solution at room temperature. The following fluorescein isothiocyanate conjugated (FITC) lectins (Vector Laboratories, Burlingame, CA) were used in the experiment: Concanavalin A (Con A), Dolichos biflorus (DBA), Glycine maximum (SBA), Arachis hypogaea (PNA) and Triticum vulgaris (WGA). The lectin conjugates were used at the concentration of 100 ug/ml in phosphate buffered saline (PBS), pH 7,4. After one hour´s incubation the sections were washed three times in PBS and once in destillated water, dried and mounted with Fluoromount.

The specificity of the staining obtained was tested by preincubating the lectin in a solution of the appropriate inhibitory sugar: alfa-methyl-mannoside for Con A, N-acetyl-D-galactosamine for DBA, D-galactose for PNA and SBA, and N-acetylglucosamine for WGA. The concentration of the inhibitor was 0.1 M, except alfa-methylmannoside, the concentration of which was 0.3 M.

The tissue sections were examined with a Leitz epifluorescence microscope equipped with the appropriate filter combination for FITC and phase contrast optics, in order to be able to determine the stage of the seminiferous epithelium (7).

Results

The results show that there are considerable differences in the staining pattern of various spermatogenic cells between different fluorochrome coupled lectins. PNA gave a bright acrosomal fluorescence which was most intense in early spermatids but decreased in late spermatids in the course of spermatid differentiation (Fig.1). WGA also presented a selective staining of the acrosomes. Binding of WGA to acrosomes was weak in early spermatids but increased gradually during the spermatid differentiation, the acrosomal staining being most bright in late spermatids (Fig. 2 - 3).
The labeling of germinal cells with Con A differed considerably from the staining pattern obtained with WGA and PNA. Con A presented no particular staining of the acrosomes but stained the cytoplasm of all spermatogenic cells as well as some nuclear structures, showing no differences in the intensity of this cytoplasmic fluorescence between various germinal cell types (Fig.4). DBA and SBA did not exhibit any specific labeling of spermatogenic cells in the used lectin concentration.

Discussion

The particular binding of PNA and WGA to the acrosomes of the spermatids indicates that specific lectin binding sugar residues are located in this region of the spermatid. The acrosome is a modified lysosome like structure originating in the Golgi complex and is known to contain complex carbohydrates. Many of the acrosomal enzymes are glycoproteins, e.g. hyaluronidase (8). Little is known about the formation of the polysaccharides in the acrosome, and it has been suggested that their function is related to the capacitation of the spermatozoa and acrosome activation may be associated with these acrosomal polysaccharides (9). The results of this study indicate that carbohydra-

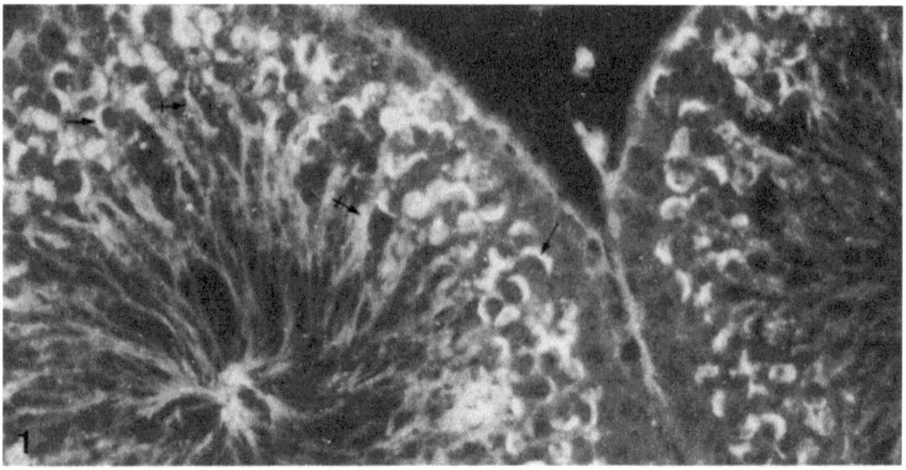

Fig.1. PNA staining of testicular tissue. The picture shows to the left a seminiferous tubule of stage VII. The acrosome of the developing spermatids is intensely stained. The acrosome of the step 7 spermatids (arrow) is more brightly stained than that of later developmental steps (crossed arrow). Magnification 700x.

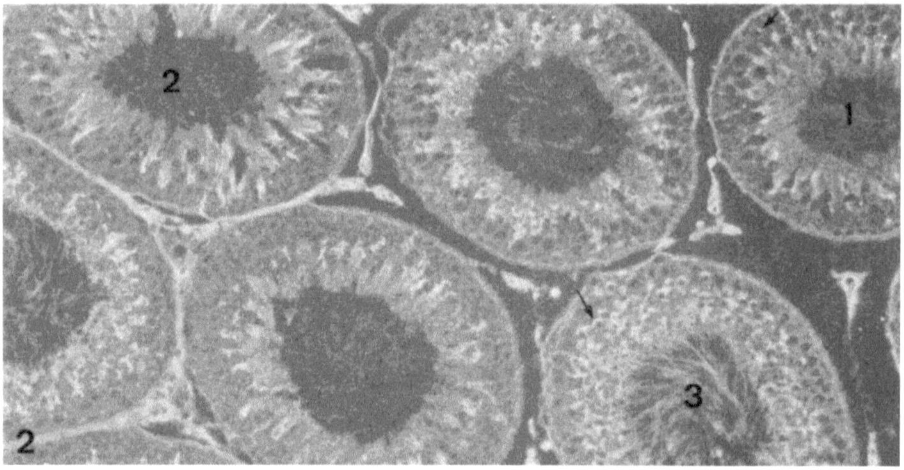

Fig.2. WGA staining of testicular tissue. In stage II (tubule 1) the acrosomes of the early spermatids are stained (arrow) as well as the acrosomes of later stages. In stage V (tubule 2) the late spermatids descend towards the basal lamina and the acrosome of the step 5 spermatid remains visible as a little spot. In stage VII (tubule 3) the acrosome spreads upon the nuclear surface, being visible as a cup shaped structure (arrow). Magnification 250x.

Fig.3. A higher magnification of a seminiferous tubule of stage III stained with WGA. The acrosomes of both early (arrow) and late spermatids (crossed arrow) give a bright fluorescence. Magnification 700x.

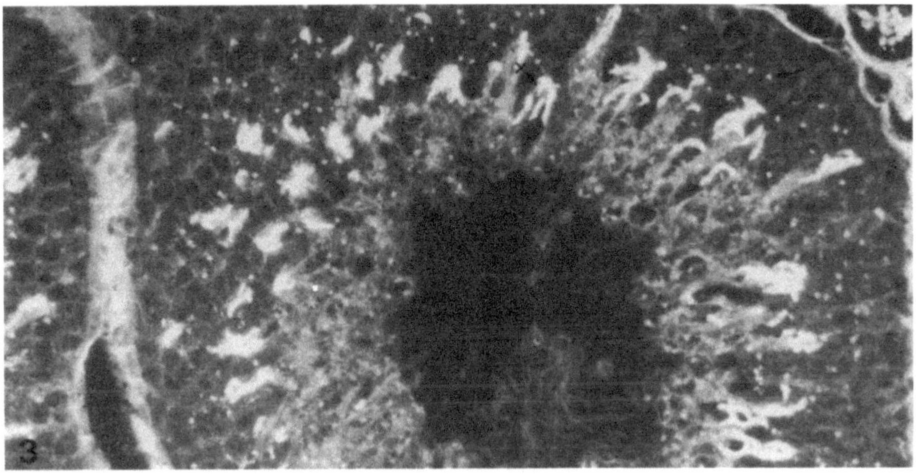

Fig.4. The staining of testicular tissue with Con A. Con A did not specifically bind to the acrosome of the spermatids but stained the cytoplasm of all spermatogenic cells, showing no differences in the intensity of this cytoplasmic fluorescence between different cell types. Magnification 700x.

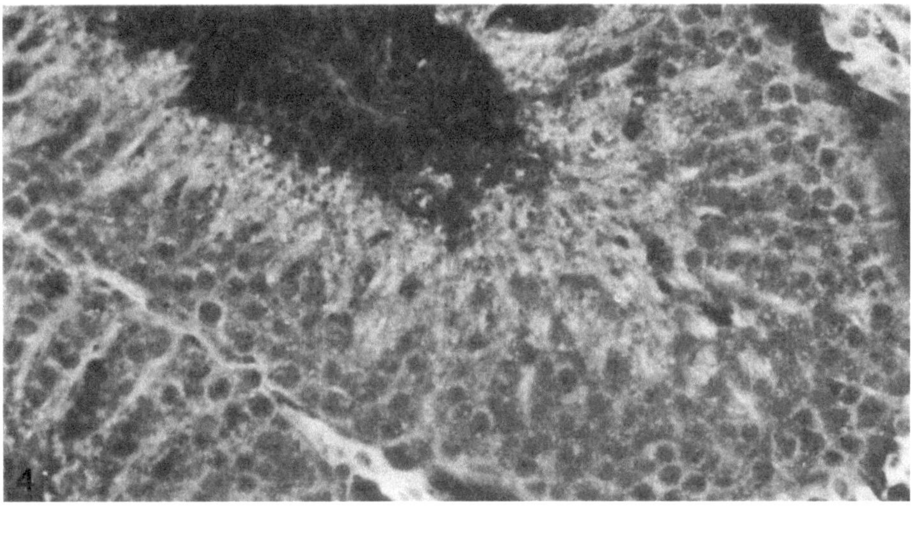

te containing compounds are formed already in the beginning of spermatid differentiation at the same time as the acrosome starts to be formed in the early spermatids of stage I of spermatogenesis (7). The decrease in the intensity of PNA staining in late spermatids possibly reflects the fact that specific PNA receptors are not synthesized any more in late spermatids but are formed in earlier developmental stages, thereafter being preserved within the acrosome. On the contrary the increasing intensity in the binding of WGA to the acrosomes in the course of spermiogenesis suggests that receptor molecules for this lectin are being formed continuosly during spermatid differentiation.

It has previously been reported that there are Con A receptors on the cell surface of isolated mouse spermatogenic cells (3, 10). In our work we were able to demonstrate that in histologic sections Con A, unlike other lectins, stained the cytoplasm of all spermatogenic cell types as well as some nuclear structures. However, we could not notice any difference in the intensity of this cytoplasmic staining between various cell types.

References

1. Kashiwabara, T., Tanaka, R., Matsumoto, T. : Nature 207, 831-832 (1965).

2. Uhlenbruck, G., Herrman, W.P. : Vox sanguinis 23, 444-451 (1972).

3. Edelman, G.M., Millette, C. F. : Proc. Natl. Acad. Sci. Usa 68, 2436-2440 (1971).

4. Nicolson, G. L., Yanagimachi, R. : Science 177, 276-279 (1972).

5. Millette, C. F. : Doctoral thesis, The Rockefeller University, New York (1975).

6. Millette, C. F. : Immunobiology of Gametes, Johnson, M. H., Edidin, M., editors, Cambridge University Press, London 50-71 (1977).

7. Leblond, C. P., Clermont, Y. : Ann. N. Y. Acad. Sci. 55, 548-573 (1952).

8. Gould, S. F., Bernstein, M. H. : Differentiation 3, 123-135 (1975).

9. Gould, S. F., Bernstein, M. H. : Techniques of Human Andrology, Hafez, E. S. E., editor, Elsevier / North Holland, Biomedical Press, 189-207 (1977).

10. Millette, C. F., Spear, P. G., Gall, W. E., Edelman, G.M.: J. Cell. Biol. 58 : 662-675 (1973).

AFFINITY OF TEN INSOLUBILIZED LECTINS TOWARDS VARIOUS GLYCOPEPTIDES WITH THE N-GLYCOSYLAMINE LINKAGE AND RELATED OLIGOSACCHARIDES

H. Debray, A. Pierce-Crétel, G. Spik and J. MONTREUIL
Laboratoire de Chimie Biologique, Université des Sciences et Techniques de Lille I et Laboratoire Associé au CNRS n° 217, 59655 Villeneuve d'Ascq Cédex, France

Insolubilized on various supports, lectins are now widely used to fractionate soluble or membrane glycoproteins of diverse origins (1) as well as glycopeptides obtained from soluble or membrane-bound glycoproteins (2, 3).

For some lectins and in particular for Con A, the structural basis for the interaction of glycopeptides with the N-glycosylamine linkage and related oligosaccharides is now well defined (4, 6), but for other lectins, these structural requirements are still poorly understood.

In a previous study (7), we used oligosaccharides or glycopeptides related to glycoproteins with the N-glycosylamine linkage (8) to precise the specificities of various lectins, by inhibition of agglutination of human red blood cells. These results are here completed by studying the affinity of the insolubilized lectins towards these oligosaccharides or glycopeptides.

Materials and Methods

Lectins.- All the lectins used in this study (*Lens culinaris* agglutinin : LCA, *Vicia faba* agglutinin : VFA, *Pisum sativum* agglutinin : PSA, *Ricinus communis* agglutinins I and II : RCA_I

Lectins, Vol. III

and RCA_{II}, wheat germ agglutinin : WGA, *Lotus tetragonolobus* agglutinin : LTA, *Ulex europeus* agglutinin I : UEA_{I}), except Concanavalin A which was purchased from Pharmindustrie, Clichy, France, have been purified by affinity chromatography as described (7). *Griffonia simplicifolia* agglutinin II : GSA_{II} was purified from seeds of *Griffonia simplicifolia* on a Sepharose-ovomucoid column (unpublished results).

Insolubilization of lectins on Sepharose 4B.- Lectins were coupled to Sepharose 4B activated according to the procedure of March *et al.* (9), at a concentration of 2 mg of lectin per ml of gel.

Origin of glycopeptides and oligosaccharides.- Oligosaccharides, native or after digestion with various exoglycosidases (Structures 2, 3, 4, 8, 9, 11, 12, 13, 14, 20, 21, 25, 27, 28, 34, 35, 36, 38 shown in fig. 1 to 4) were isolated by Dr. G. Strecker from the urines of patients with various lysosomal diseases (10). Glycopeptides, native or after digestion with various exoglycosidases, isolated from human serotransferrin (Structures 1, 7, 19), from human lactotransferrin (Structures 5, 6, 10, 22), from hen ovotransferrin (Structure 33), from bovine lactotransferrin (Structure 37), from Cohn's fraction IV (Structure 29) and from human milk IgA (Structures 15, 16, 17, 18) were gifts from Dr. G. Spik. Glycopeptides 26 and 27, isolated from human α_1-acid glycoprotein, were obtained from Dr. B. Fournet. Glycopeptide 30 was isolated from calf fetuin and the hybrid type glycopeptide (Structure 39) isolated from hen ovalbumin was a gift from Dr. A. Kobata. All these structures are related to glycoproteins with the N-glycosidic linkage (8).

Affinity chromatography of oligosaccharides or glycopeptides on lectin-Sepharose column.- Glycopeptides, N-[^{14}C]acetylated according to Koide *et al.* (11) or related oligosaccharides, reduced at the terminal N-acetylglucosamine residue with tritiated sodium borohydride, were applied on the columns of lectin-Sepharose (10 x 1 cm) equilibrated in phosphate buffered saline pH 7.4 or in 0.005 M sodium acetate buffer pH 5.2 containing 0.1 M NaCl and $CaCl_2$, $MgCl_2$, $MnCl_2$ (1 mM each) in the case of insolubilized Con A. Elution was carried out first with the above buffers and then with buffer containing 0.01 M or 0.3 M α-methyl glucoside in the case of Con A-Sepharose or 0.15 M sugar for the other insolubilized lectins.

Results and Discussion

Four elution profiles can be obtained with insolubilized lectins as previously described (6) :

i) The non-retained profile (FNR) reflects a lack of interaction between the lectin and the saccharide.

ii) The retarded profile (FR) is obtained by elution with the starting buffer and reflects a weak affinity between the lectin and the saccharide. The elution volume can be more or less important (FR^-, FR^+, FR^{++}).

iii) The sharp elution profile (FE), obtained with the starting buffer containing the competiting sugar, reflects a higher affinity between the lectin and the saccharide.

iv) The "tightly bound" profile is obtained also with the starting buffer containing the competiting sugar. This broad, trailing peak expresses a very great affinity between the saccharide and the lectin.

Lectins with a specificity for α-D-mannose or α-D-glucose.- The results obtained with four lectins (Con A, LCA, VFA and PSA), considered to be identical in terms of α-D-mannose or α-D-glucose specificity were previously reported (6) and are

	GLYCAN STRUCTURE
1	α-NeuAc-(2→6)-β-Gal-(1→4)-β-GlcNAc-(1→2)-α-Man-(1→3)\\ β-Man-(1→4)-β-GlcNAc-(1→4)-β-GlcNAc-(1→4)-Asn α-NeuAc-(2→6)-β-Gal-(1→4)-β-GlcNAc-(1→2)-α-Man-(1→6)/
2	α-NeuAc-(2→6)-β-Gal-(1→4)-β-GlcNAc-(1→2)-α-Man-(1→3)\\ β-Man-(1→4)-GlcNAc α-NeuAc-(2→6)-β-Gal-(1→4)-β-GlcNAc-(1→2)-α-Man-(1→6)/
3	α-NeuAc-(2→3)-β-Gal-(1→4)-β-GlcNAc-(1→2)-α-Man-(1→3)\\ β-Man-(1→4)-GlcNAc α-NeuAc-(2→3)-β-Gal-(1→4)-β-GlcNAc-(1→2)-α-Man-(1→6)/
4	α-NeuAc-(2→3)-β-Gal-(1→4)-β-GlcNAc-(1→2)-α-Man-(1→3)\\ β-Man-(1→4)-GlcNAc α-NeuAc-(2→6)-β-Gal-(1→4)-β-GlcNAc-(1→2)-α-Man-(1→6)/
5	α-NeuAc-(2→6)-β-Gal-(1→4)-β-GlcNAc-(1→2)-α-Man-(1→3)\\ β-Man-(1→4)-β-GlcNAc-(1→4)-β-GlcNAc-(1→4)-Asn α-NeuAc-(2→6)-β-Gal-(1→4)-β-GlcNAc-(1→2)-α-Man-(1→6)/ | 1,6 α-Fuc
6	α-NeuAc-(2→6)-β-Gal-(1→4)-β-GlcNAc-(1→2)-α-Man-(1→3)\\ β-Man-(1→4)-β-GlcNAc-(1→4)-β-GlcNAc-(1→4)-Asn β-Gal-(1→4)-β-GlcNAc-(1→2)-α-Man-(1→6)/ | 1,6 α-Fuc | 1,3 α-Fuc
7	β-Gal-(1→4)-β-GlcNAc-(1→2)-α-Man-(1→3)\\ β-Man-(1→4)-β-GlcNAc-(1→4)-β-GlcNAc-(1→4)-Asn β-Gal-(1→4)-β-GlcNAc-(1→2)-α-Man-(1→6)/
8	β-Gal-(1→4)-β-GlcNAc-(1→2)-α-Man-(1→3)\\ β-Man-(1→4)-GlcNAc β-Gal-(1→4)-β-GlcNAc-(1→2)-α-Man-(1→6)/
9	β-Gal-(1→4)-β-GlcNAc-(1→2)-α-Man-(1→3)\\ β-GlcNAc-(1→4)-β-Man-(1→4)-GlcNAc β-Gal-(1→4)-β-GlcNAc-(1→2)-α-Man-(1→6)/
10	β-Gal-(1→4)-β-GlcNAc-(1→2)-α-Man-(1→3)\\ β-Man-(1→4)-β-GlcNAc-(1→4)-β-GlcNAc-(1→4)-Asn β-Gal-(1→4)-β-GlcNAc-(1→2)-α-Man-(1→6)/ | 1,6 α-Fuc
11	α-NeuAc-(2→3)-β-Gal-(1→4)-β-GlcNAc-(1→2)-α-Man-(1→3)\\ β-Man-(1→4)-GlcNAc β-Gal-(1→4)-β-GlcNAc-(1→2)-α-Man-(1→6)/

Fig. 1.- Interaction of insolubilized lectins N-glycosylamine linkage.

Con A	LCA	VFA	PSA	RCA_I	RCA_{II}	GSA_{II}	WGA	UEA_I	LTA
FE 10mM αMG	FNR	FNR	FNR	FR^-	FR^+	FNR	FNR	FNR	FNR
FE 10mM αMG	FNR	FNR	FNR	FNR	FR^-	FNR	FNR	FNR	FNR
ND	ND	ND	ND	FNR	FNR	FNR	FNR	FNR	FNR
ND	ND	ND	ND	FNR	FR^-	FNR	FNR	FNR	FNR
FE 10mM αMG	FE	FE	FE	FR^-	FR^+	FNR	FNR	FNR	FNR
FE 10mM αMG	FE	FE	FE	FR	FE	FNR	FNR	FNR	FNR
FE 10mM MG	FNR	FNR	FNR	FR	FR^+	FNR	FNR	FNR	FNR
FE 10mM αMG	FNR	FNR	FNR	FR	FE	FNR	FNR	FNR	FNR
FNR	FNR	FNR	FNR	FNR	FR^-	FNR	FNR	FNR	FNR
FE 10mM αMG	FE	FE	FE	FR	FE	FNR	FNR	FNR	FNR
FE 10mM αMG	FNR	FNR	FNR	FNR	FNR	FNR	FNR	FNR	FNR

with glycans from glycoproteins with the

	GLYCAN STRUCTURE
12	α-NeuAc-(2→6)-β-Gal-(1→4)-β-GlcNAc-(1→2)-α-Man-(1→3)\ β-Man-(1→4)-GlcNAc α-Man-(1→6)/
13	α-NeuAc-(2→3)-β-Gal-(1→4)-β-GlcNAc-(1→2)-α-Man-(1→3)-β-Man-(1→4)-GlcNAc
14	α-NeuAc-(2→6)-β-Gal-(1→4)-β-GlcNAc-(1→2)-α-Man-(1→3)-β-Man-(1→4)-GlcNAc
15	-β-GlcNAc-(1→2)-α-Man-(1→3)\ β-Gal-(1→4) < β-Man-(1→4)-β-GlcNAc-(1→4)-β-GlcNAc-(1→4)-Asn β-GlcNAc-(1→2)-α-Man-(1→6)/
16	-β-GlcNAc-(1→2)-α-Man-(1→3)\ β-Gal-(1→4) < β-GlcNAc-(1→4)-β-Man-(1→4)-β-GlcNAc-(1→4)-β-GlcNAc-(1→4)-Asn -β-GlcNAc-(1→2)-α-Man-(1→6)/
17	-β-GlcNAc-(1→2)-α-Man-(1→3)\ β-Gal-(1→4) < β-Man-(1→4)-β-GlcNAc-(1→4)-β-GlcNAc-(1→4)-Asn -β-GlcNAc-(1→2)-α-Man-(1→6)/ 1,6 α-Fuc
18	-β-GlcNAc-(1→2)-α-Man-(1→3)\ β-Gal-(1→4) < β-GlcNAc-(1→4)-β-Man-(1→4)-β-GlcNAc-(1→4)-β-GlcNAc-(1→4)-Asn -β-GlcNAc-(1→2)-α-Man-(1→6)/ 1,6 α-Fuc
19	β-GlcNAc-(1→2)-α-Man-(1→3)\ β-Man-(1→4)-β-GlcNAc-(1→4)-β-GlcNAc-(1→4)-Asn β-GlcNAc-(1→2)-α-Man-(1→6)/
20	β-GlcNAc-(1→2)-α-Man-(1→3)\ β-Man-(1→4)-GlcNAc β-GlcNAc-(1→2)-α-Man-(1→6)/
21	β-GlcNAc-(1→2)-α-Man-(1→3)\ β-GlcNAc-(1→4)-β-Man-(1→4)-GlcNAc β-GlcNAc-(1→2)-α-Man-(1→6)/
22	β-GlcNAc-(1→2)-α-Man-(1→3)\ β-Man-(1→4)-β-GlcNAc-(1→4)-β-GlcNAc-(1→4)-Asn β-GlcNAc-(1→2)-α-Man-(1→6)/ 1,6 α-Fuc
23	α-Man-(1→6)-β-Man-(1→4)-β-GlcNAc-(1→4)-β-GlcNAc-(1→4)-Asn 1,6 α-Fuc

Fig. 2.- Interaction of insolubilized lectins N-glycosylamine linkage.

Con A	LCA	VFA	PSA	RCA_I	RCA_{II}	GSA_{II}	WGA	UEA_I	LTA
FE 10mM αMG	FNR	FNR	FNR	FNR	FNR	FNR	FNR	FNR	FNR
ND	ND	ND	ND	FNR	FNR	FNR	FNR	FNR	FNR
ND	ND	ND	ND	FNR	FNR	FNR	FNR	FNR	FNR
FE 10mM αMG	FNR	FNR	FNR	FR	FE	FNR	FR^-	FNR	FNR
FE 10mM αMG	FNR	FNR	FNR	FR^-	FR	FNR	FR	FNR	FNR
FE 10mM αMG	FE	FE	FE	FR	FE	FNR	FR^-	FNR	FNR
FE 10mM αMG	FE	FE	FE	FR^-	FR	FNR	FNR	FNR	FNR
FTB	FR	FNR	FNR	FNR	FNR	FR	FR^-	FNR	FNR
FTB	FNR	FNR	FNR	FNR	FNR	FR	FNR	FNR	FNR
FR	FNR	FNR	FNR	FNR	FNR	FR^-	FNR	FNR	FNR
FTB	FTB	FTB	FTB	FNR	FNR	FR	FNR	FNR	FNR
FR	FNR	FNR	FNR	FNR	FNR	FNR	FNR	FNR	FNR

with glycans from glycoproteins with the

here extended. For Con A, bi-antennary structures of the N-acetyllactosaminic type possessing an additional β-(1 → 4)-linked N-acetylglucosamine on the β-linked mannose residue can be either retarded (Structure 21) or weakly retained (Structures 16 and 18) on the condition that at least one β-GlcNAc-(1 → 2)-mannose sequence is unmasked on the oligosaccharide or on the glycopeptide.

Moreover, in the case of this oligosaccharide and these glycopeptides, Con A-Sepharose possesses a higher affinity for the glycopeptides than for the related oligosaccharides, which could be explained by the fact that the glycan-amino acid linkage leads to structures more rigid than those of the corresponding oligosaccharide.

The affinity for Con A-Sepharose towards an hybrid type glycopeptide from hen ovalbumin (Structure 39) is also interesting : the insolubilized Con A possesses a relatively weak affinity for this structure which can be eluted from the column with 10 mM α-methyl glucoside, but it must be noted that the elution volume is larger than that obtained with bi-antennary N-acetyllactosaminic type glycopeptide (Structure 1) which is also eluted with 10 mM αMG. However, the affinity of Con A-Sepharose is weaker for this hybrid-type glycopeptide than for an oligomannosidic type glycopeptide (Structure 37) which is eluted from the column with 0.3 M α-methyl glucoside. The same hybrid-type glycopeptide is not retained on LCA, VFA or PSA-Sepharose columns.

As previously reported (6, 7), these three insolubilized lectins recognize the pentasaccharidic core substituted by two β-(1 → 2)-N-acetylglucosamine residues and possessing an α-L-fucose residue at the C-6 position on the N-acetylglucosamine residue involved in the N-glycosylamine bond.

These results on the specificities of LCA and PSA were recently confirmed and extended by Kornfeld *et al.* (12), who have shown that LCA or PSA can also interact with tri-antennary glycopeptides possessing an α-L-fucose residue substituting in C-6 position the N-acetylglucosamine residue involved in the N-glycosylamine linkage, but an important condition for this interaction to take place is that this third branch substitutes a core α-mannose residue in the C-6 position. If this substitution is at the C-4 position, LCA or PSA do not have any affinity for these new tri-antennary structures.

Our results show also that, in contrast to Con A, LCA, VFA or PSA can interact with bi-antennary structures possessing an additional β-(1 → 4)-linked N-acetylglucosamine on the β-linked mannose residue, but on the condition that these structures possess an α-L-fucose residue in C-6 position on the N-acetylglucosamine residue involved in the N-glycosylamine bond (Structure 18).

Lectins with a specificity for galactose.- *Ricinus communis* agglutinins (RCA_{I} et RCA_{II}) are very useful for the fractionation of N-acetyllactosaminic type glycopeptides or oligosaccharides. The affinity between the lectins and the saccharide increases with the number of N-acetyllactosamine sequences present in the carbohydrate structure (Structure 7 compared to Structure 31). Substitution of these N-acetyllactosamine sequences by sialic acid residues at the C-6 position of galactose decreases the affinity of the lectins for the saccharide (Structure 5 compared to Structure 10). Moreover, substitution of the β-mannose residue by an additional β-(1 → 4)-N-acetylglucosamine residue decreases the affinity of the lectins for these structures as compared to the unsubstituted ones (Structure 9 compared to structure 8).

Lectins with a specificity for N-acetylglucosamine

Wheat germ agglutinin.- Our results agree with those previously published by Yamamoto *et al.* (13) and show that insolubilized WGA presents a stronger affinity for hybrid-type

	GLYCAN STRUCTURE
24	α-Man-(1→6)-β-Man-(1→4)-β-GlcNAc-(1→4)-β-GlcNAc-(1→4)-Asn
25	α-Man-(1→3)\ β-Man-(1→4)-GlcNAc α-Man-(1→6)/
26	β-Gal-(1→4)-β-GlcNAc-(1→4)\ β-Gal-(1→4)-β-GlcNAc-(1→2)-α-Man-(1→3)\ β-Man-(1→4)-β-GlcNAc-(1→4)-β-GlcNAc-(1→4)-Asn β-Gal-(1→4)-β-GlcNAc-(1→2)-α-Man-(1→6)/
27	β-Gal-(1→4)-β-GlcNAc-(1→4)\ β-Gal-(1→4)-β-GlcNAc-(1→2)-α-Man-(1→3)\ β-Man-(1→4)-GlcNAc β-Gal-(1→4)-β-GlcNAc-(1→2)-α-Man-(1→6)/
28	β-Gal-(1→4)-β-GlcNAc-(1→2)-α-Man-(1→3)\ β-Man-(1→4)-GlcNAc β-Gal-(1→4)-β-GlcNAc-(1→2)-α-Man-(1→6)/ β-Gal-(1→4)-β-GlcNAc-(1→6)/
29	α-NeuAc-(2→3)-β-Gal-(1→4)-β-GlcNAc-(1→4)\ α-NeuAc-(2→6)-β-Gal-(1→4)-β-GlcNAc-(1→2)-α-Man-(1→3)\ β-Man-(1→4)-β-GlcNAc-(1→4)-β-GlcNAc-(1→4)-Asn α-NeuAc-(2→6)-β-Gal-(1→4)-β-GlcNAc-(1→2)-α-Man-(1→6)/
30	α-NeuAc-(2→3)-β-Gal-(1→4)-β-GlcNAc-(1→4)\ α-NeuAc-(2→6)-β-Gal-(1→4)-β-GlcNAc-(1→2)-α-Man-(1→3)\ β-Man-(1→4)-β-GlcNAc-(1→4)-β-GlcNAc-(1→4)-Asn α-NeuAc-(2→3)-β-Gal-(1→4)-β-GlcNAc-(1→2)-α-Man-(1→6)/
31	β-Gal-(1→4)-β-GlcNAc-(1→4)\ β-Gal-(1→4)-β-GlcNAc-(1→2)-α-Man-(1→3)\ β-Man-(1→4)-β-GlcNAc-(1→4)-β-GlcNAc-(1→4)-Asn β-Gal-(1→4)-β-GlcNAc-(1→2)-α-Man-(1→6)/ ↑ β-Gal-(1→4)-β-GlcNAc-(1→6)/ α-Fuc
32	β-GlcNAc-(1→4)\ β-GlcNAc-(1→2)-α-Man-(1→3)\ β-GlcNAc-(1→4)-β-Man-(1→4)-β-GlcNAc-(1→4)-β-GlcNAc-(1→4)-Asn β-GlcNAc-(1→2)-α-Man-(1→6)/

Fig. 3.- Interaction of insolubilized lectins
N-glycosylamine linkage.

Con A	LCA	VFA	PSA	RCA_I	RCA_{II}	GSA_{II}	WGA	UEA_I	LTA
FR	FNR	FNR	FNR	FNR	FNR	FNR	FNR	FNR	FNR
FE	FNR	FNR	FNR	FNR	FNR	FNR	FNR	FNR	FNR
FNR	FNR	FNR	FNR	FR^{++}	FE	FNR	FNR	FNR	FNR
FNR	FNR	FNR	FNR	FR	FE	FNR	FNR	FNR	FNR
FNR	FNR	FNR	FNR	FR	FE	FNR	FNR	FNR	FNR
FNR	FNR	FNR	FNR	ND	FR	FNR	FNR	FNR	FNR
FNR	FNR	FNR	FNR	ND	FE^{++}	FNR	FNR	FNR	FNR
FNR	FNR	FNR	FNR	FR	FE	FNR	FNR	FNR	FNR
FNR	FNR	FNR	FNR	FNR	FNR	FR	FNR	FNR	FNR

with glycans from glycoproteins with the

	GLYCAN STRUCTURE
33	β-GlcNAc-(1→4) β-GlcNAc-(1→2)-α-Man-(1→3) β-Man-(1→4)-β-GlcNAc-(1→4)-β-GlcNAc-(1→4)-Asn β-GlcNAc-(1→2)-α-Man-(1→6) β-GlcNAc-(1→6) α-Fuc
34	β-GlcNAc-(1→2)-α-Man-(1→3) β-Man-(1→4)-GlcNAc β-GlcNAc-(1→2)-α-Man-(1→6) β-GlcNAc-(1→6)
35	β-Gal-(1→4)-β-GlcNAc-(1→4) β-Gal-(1→4)-β-GlcNAc-(1→2)-α-Man-(1→3) β-Man-(1→4)-β-GlcNAc β-Gal-(1→4)-β-GlcNAc-(1→2)-α-Man-(1→6) β-Gal-(1→4)-β-GlcNAc-(1→6)
36	β-GlcNAc-(1→4) β-GlcNAc-(1→2)-α-Man-(1→3) β-Man-(1→4)-β-GlcNAc β-GlcNAc-(1→2)-α-Man-(1→6) β-GlcNAc-(1→6)
37	α-Man-(1→2)-α-Man-(1→2)-α-Man-(1→3) β-Man-(1→4)-β-GlcNAc-(1→4)-β-GlcNAc-(1→4)-Asn α-Man-(1→3)-α-Man-(1→6) 1,6 α-Man
38	β-GlcNAc-(1→4) β-GlcNAc-(1→2)-α-Man-(1→3) β-Man-(1→4)-GlcNAc β-GlcNAc-(1→2)-α-Man-(1→6)
39	β-Gal-(1→4)-β-GlcNAc-(1→4) β-GlcNAc-(1→2)-α-Man-(1→3) β-GlcNAc-(1→4)-β-Man-(1→4)-β-GlcNAc-(1→4)-β-GlcNAc-(1→4)-Asn α-Man-(1→6)-α-Man-(1→6) α-Man-(1→3)

Fig. 4.- Interaction of insolubilized lectins N-glycosylamine linkage.

Con A	LCA	VFA	PSA	RCA_I	RCA_{II}	GSA_{II}	WGA	UEA_I	LTA
FNR	FNR	FNR	FNR	FNR	FNR	FR^+	FNR	FNR	FNR
FNR	FNR	FNR	FNR	FNR	FNR	FR^-	FNR	FNR	FNR
FNR	FNR	FNR	FNR	FR	FE	FNR	FNR	FNR	FNR
FNR	FNR	FNR	FNR	FNR	FNR	FE	FNR	FNR	FNR
FE 0.3M αMG	FR	FNR	FNR	FNR	FNR	FNR	FNR	FNR	FNR
FNR	FNR	FNR	FNR	FNR	FNR	FR^{++}	FNR	FNR	FNR
FE 10mM αMG	FNR	FNR	FNR	FNR	FNR	FNR	FE	FNR	FNR

with glycans from glycoproteins with the

glycopeptides such as those found in hen ovalbumin (Structure 39). Structure 16 which possesses also a β-N-acetylglucosamine residue linked to C-4 of the β-linked mannose residue but belongs to the N-acetyllactosaminic type family, presents a weaker affinity for WGA, and structure 18 which just differs from structure 16 by the presence of an α-L-fucose residue in C-6 position on the N-acetylglucosamine residue involved in the N-glycosylamine bond is no more recognized by insolubilized WGA. The lectin does not present any affinity for oligomannosidic-type glycopeptides (Structure 37). As found also by Yamamoto *et al.* (13), the N-N'-diacetylchitobiose moiety has to be intact, for, oligosaccharide 21 which possesses the β-N-acetylglucosamine residue in C-4 position on the β-mannose residue, but where the β-GlcNAc-(1 →)-Asn sequence is lacking is not retained on WGA-Sepharose. It can also be noted that sialylated glycopeptides (Structures 1, 29, 30) or sialylated oligosaccharides (Structure 2) are not recognized by insolubilized WGA. These interactions involve a charge effect and depend on a mechanism called by Monsigny *et al.* (14) "avidity effect" : only glycopeptides with a high density of terminal non-reducing N-acetylglucosamine or N-acetylneuraminic acid residues interact with WGA.

Griffonia simplicifolia agglutinin.- This insolubilized lectin represents a very useful tool for the fractionation of glycopeptides or oligosaccharides with unmasked N-acetylglucosamine residues in a terminal non-reducing position and the affinity of the lectin increases with the number of such residues : glycopeptide 19 is more retarded on the column than glycopeptide 33. But, at least two of such N-acetylglucosamine residues are necessary for the interaction to take place, for glycopeptides 15, 16, 17, 18 or 39 with only one unmasked N-acetylglucosamine residue are not retained on the column. In the case of structures with two terminal non-reducing N-acetylglucosamine, GSA_{II} presents a greater affinity for the glycopeptide than for the related oligosaccharide (Glycopeptide 19 compared to oligosaccharide 20 or 21). This can also be explained by the fact that the glycan-amino acid linkage leads to structures more rigid than those of the oligosaccharides.

Lectins with a specificity for L-fucose.- *Lotus tetragonolobus* agglutinin or *Ulex europeus* agglutinin I have no affinity for glycopeptides or oligosaccharides of the N-acetyllactosaminic type, even if these structures possess one or two fucose residues α-(1 → 3) or α-(1 → 6)-linked to N-acetylglucosamine residues as in glycopeptide 5 or 6.

As previously described by Pereira *et al.* (15), the specificity of these two lectins seems to be directly towards the α-(1 → 2)-linked fucose residues found in blood-group H active substances.

Conclusions

This work completes and extends our previous results obtained by inhibition of the agglutination of human-red blood cells induced by lectins (7). These results confirm that lectins, considered "identical" in terms of monosaccharide specificity, possess the ability to recognize fine differences in more complex structures and this can be used to fractionate mixtures of N-glycopeptides of various origins.

Acknowledgements

This work was supported in part by the Centre National de la Recherche Scientifique (Laboratoire Associé n° 217) and by the Délégation Générale à la Recherche Scientifique et Technique (Grant 79-7-0669). We thank Drs G. Strecker, B. Fournet and A. Kobata for their generous gifts of oligosaccharides and glycopeptides.

References

1. Lotan, R., Nicolson, G.L. : Biochim. Biophys. Acta 559, 329-376 (1979).

2. Finne, J., Krusius, T., Järnefelt, J.: 27th International Congress of Pure and Applied Chemistry (ed. A. Varmavuori) Pergamon Press, Oxford-New York, pp. 147-159, 1980.

3. Dall'Olio, F., Serafini-Cessi, F., Scannavini, M., Campadelli-Fiume, G. : These proceedings.

4. Ogata, S., Muramatsu, T., Kobata, A. : J. Biochem. 78, 687-696 (1975).

5. Krusius, T., Finne, J., Rauvala, H. : FEBS-Letters 71, 117-120 (1976).

6. Debray, H., Montreuil, J. : Lectins-Biology, Biochemistry, Clinical Biochemistry, Vol. 1, pp. 221-230, Walter de Gruyter, Berlin-New York (1981).

7. Debray, H., Decout, D., Strecker, G., Spik, G., Montreuil, J. : Eur. J. Biochem. 117, 41-55 (1981).

8. Montreuil, J. : Adv. Carbohyd. Chem. Biochem. 37, 157-223 (1980).

9. March, S.C., Parikh, I., Cuatrecasas, P. : Anal. Biochem. 60, 149-152 (1974).

10. Strecker, G., Montreuil, J. : Biochimie 61, 1199-1246 (1979).

11. Koide, N., Muramatsu, T. : J. Biol. Chem. 249, 4897-4904 (1974).

12. Kornfeld, K., Reitman, M.L., Kornfeld, R. : J. Biol. Chem. 256, 6633-6640 (1981).

13. Yamamoto, K., Tsuji, T., Matsumoto, I., Osawa, T. : Biochemistry 20, 5894-5899 (1981).

14. Monsigny, M., Roche, A.C., Sene, C., Maget-Dana, R., Delmotte, F. : Eur. J. Biochem. 104, 147-153 (1980).

15. Pereira, M.E.A., Kisailus, E.C., Gruezo, F., Kabat, E.A. : Arch. Biochem. Biophys. 185, 108-115 (1978).

HERPESVIRUS GLYCOPROTEINS FROM WILD-TYPE AND RICIN-RESISTANT BHK CELLS. CHARACTERIZATION OF GLYCOPEPTIDES BY CON A-SEPHAROSE

F. Dall'Olio, F. Serafini-Cessi
Istituto di Patologia Generale, Università di Bologna, Italy

M. Scannavini and G. Campadelli-Fiume
Istituto di Microbiologia e Virologia, Bologna, Italy

Herpes simplex virus-1 (HSV-1) specifies the synthesis of several antigenically distinct classes of glycoproteins (1, 2), which mainly contain N-linked glycans. In the infected cells both the mature forms gB, gC, gD and gE, and their partially glycosylated precursors (gA-pgB, pgC, pgD and pgE) accumulate. Sialylated complex-type glycans are present in the mature forms whereas mannose-rich glycans predominate in the precursors (3) In a preceding paper (4) we reported that the synthesis of HSV-1 glycoproteins is altered in a mutant cell line (Ric^R14) defective in the N-acetylglucosaminyltransferase (5), which is required for the processing of mannose-rich to complex-type glycans. Mature forms of HSV-1 glycoproteins were absent in Ric^R14 cells whereas underglycosylated proteins accumulated. In the present study the detailed structure of glycans synthesized in HSV-1-infected Ric^R14 and parent BHK cells were examined using a combined technique of ion-exchange chromatography and Con A-Sepharose binding.

Materials and Methods

Radiolabeling of infected cells.
Ric^{R}14 (5) and BHK cells in 25 cm^2 Falcon flasks were infected with HSV-1(MP) at an imput multiplicity of 10 PFU/cell. The cells were labeled between 5 and 18 h after infection with minimum essential medium containing one-half the usual concentration of glucose, 1% foetal calf serum and 4 µC/ml of ^{14}C-glucosamine.

Polyacrylamide gel electrophoresis of HSV glycoproteins.
Cells in Tris/HCl, pH 7, were solubilised with 2% SDS and 5% β-mercaptoethanol, and heat denatured for 2 min. Samples were run in a slab gel containing 8.5% acrylamide and analyzed by autoradiography of the dried slab (3).

Digestion of glycoproteins with endo-β-N-acetylglucosaminidase H.
Portions of sonicated cells were dialyzed against 0.1 M citrate buffer, pH 5.5. 300 µl aliquots of the suspensions (protein concentration 1.5 mg/ml) were incubated at 37° C with 0.025 U of endo-β-N-acetylglucosaminidase H (Seikagaku Kogio Co., Tokio). At the indicated times 50 µl aliquots were withdrawn and proteins precipitated with 50 µl of 12.5% TCA. The release of acid-soluble labeled glycans was measured on 50 µl of the supernatant.

Separation and analysis of glycopeptides.
Infected labeled cells were harvested in 0.1 M phosphate buffer, pH 7.9, and digested with pronase (3). Pronase-digests were chromatographed on 1 x 75 cm column of Bio-Gel P-10. The fractions corresponding to elution position of acidic and neutral glycopeptide markers (from 4800 to 1500 daltons) were

pooled. A portion (2 ml) of this pool was loaded on a DEAE-Sephacel column (1 x 15 cm) equilibrated with 0.05 M acetic acid. After a washing with H_2O (10 ml), a salt gradient (50 ml 0.05 M acetic acid → 50 ml 0.05 M acetic acid and 0.5 M NaCl) was applied to the column. Fractions of 1 ml were collected and the radioactivity of 0.1 ml was measured. The central fractions of each peak were pooled. One portion (1 ml) was used for neuraminidase treatment as previously described (3) and another portion (2 ml) for affinity chromatography on Con A--Sepharose. The column (0.5 x 10 cm) was washed with Tris/HCl, pH 7.4, containing 0.1 M NaCl and 0.02% NaN_3, $CaCl_2$, $MgCl_2$ and $MnCl_2$; it was subsequently eluted with 5 mM α-methyl-mannoside and then with 200 mM α-methyl-mannoside in the same buffer.

Results

Glycoprotein analysis.

Fig. 1 shows the pattern of electrophoretically separated glycoproteins synthesized in HSV-1(MP)-infected Ric^R14 and BHK

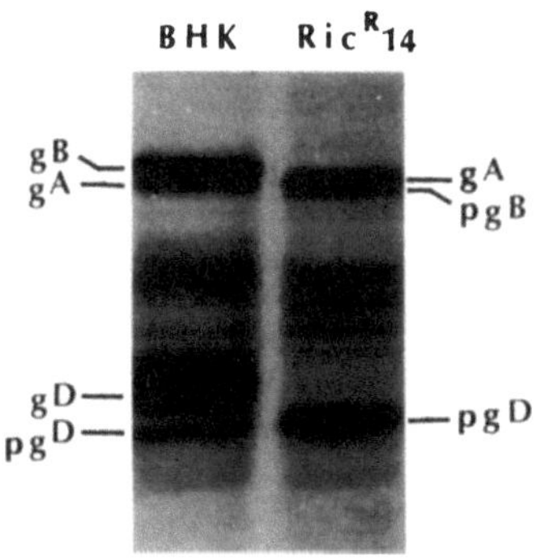

Fig. 1. SDS-PAGE of glycoproteins from HSV-1(MP)-infected Ric^R14 and BHK cells labeled with ^{14}C-glucosamine

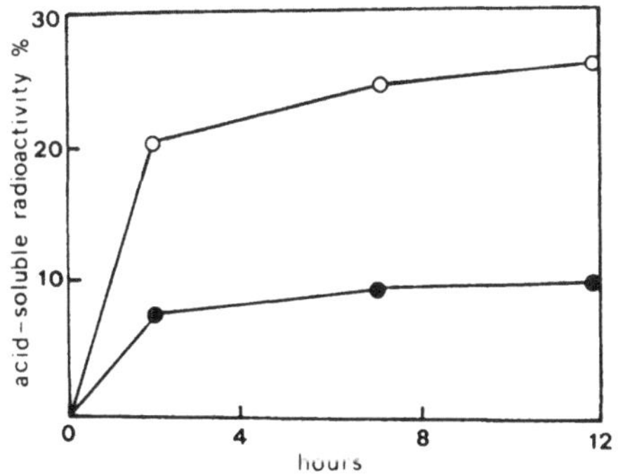

Fig. 2. Acid-soluble glycans released by endo-β-N-acetylglucosaminidase H from infected-Ric^R14 (O), and BHK (●)

cells. In the mutant cells only the partially glycosylated precursors (pgD, pgB and gA) accumulated, whereas the fully glycosylated forms (gB and gD) were absent. Since previous observations (3) indicate that precursors of HSV-glycoproteins, namely gA and pgD contain glycans of the mannosyl-rich type, they should be sensitive to endo-β-N-acetylglucosaminidase H. This enzyme cleaves between the two N-acetylglucosamine residues in high-mannose glycans (four to nine residues) (6).
Fig. 2 shows the results of a typical digestion experiment of ^{14}C-glucosamine-labeled HSV glycoproteins from BHK and $Ric^{R}14$ cells. The amount of acid-soluble glycans released by the enzyme from glycoproteins of mutant cells was significantly higher than that from wild-type cells. This result confirms that mannosyl-rich glycans predominate in precursors glycoproteins of mutant cells and suggests that glycan processing is arrested before the action of N-acetylglucosaminyltransferase I.

Glycopeptide characterization.
Both neutral (N) and acidic (A) pronase-glycopeptides obtained by preparative Bio-Gel filtration (free from undigested material and low molecular weight molecules) were analyzed on DEAE-Sephacel (Fig. 3). Five peaks N, AIa, AIb, AII and AIII were separated. N was not absorbed by the column and emerged with the neutral mannosyl-rich glycopeptide marker. The most acidic glycopeptide AIII was eluted together with a triantennary marker (fetuin glycopeptide) and AII with a diantennary glycopeptide (prepared from human transferrin). In the elution position of slightly acidic glycopeptides two distinct peaks were obtained, AIa and AIb. In the BHK sample all peaks were present and AIII was the most represented. In the $Ric^{R}14$ sample N represented the major peak; among the acidic glycopeptides only AIa and

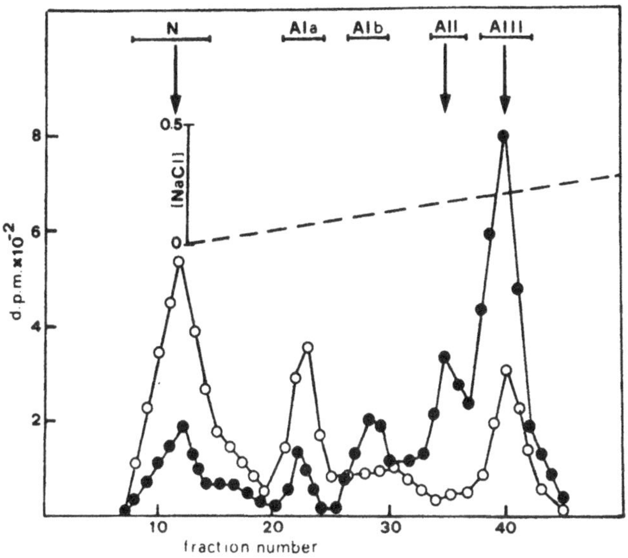

Fig. 3. DEAE-Sephacel chromatography of pronase-glycopeptides from ^{14}C-glucosamine-labeled HSV-infected BHK (●) and Ric^{R}14 (o) cells. The arrows represent, from left to right, the elution positions of ovalbumin-, human transferrin- and fetuin-glycopeptide. The horizontal bars indicate the fractions of each peak which were pooled and chromatographed on Con A-Sepharose.

AIII were present but the latter was small in amount. After neuraminidase treatment the acidic glycopeptides were all split in two peaks when rechromatographed on DEAE-Sephacel: one emerged as desialylated glycopeptide (unretained) and the other one emerged as free sialic acid. This shows that the negative charge responsible for absorption to the DEAE column was due to sialic acid. The amount of radioactivity cleaved by neuraminidase as sialic acid from AIII, AII, AIb and AIa was 40%, 25%, 13% and 10%, respectively. These results reinforce the indication that the structures of AIII, AII and AI (a and b) are, respectively, those of triantennary, biantennary and monoantennary glycans. Since the processing of mannosyl-rich to complex-type glycans changes the affinity of glycopeptides to ConA (7),

Table. Comparison of glycopeptides from HSV-1(MP)-infected BHK and Ric^{R}14 cells separated with DEAE-Sephacel chromatography.

Glyco-peptide	% ^{14}C-glucosamine incorporated		NANAase suscep-tibility	Elution from Con A-Sepharose by Me-Man (mM)	Proposed structure
	BHK	Ric^{R}14			
N	16	51	-	200	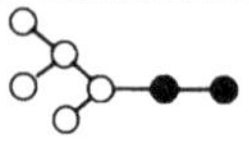
AIa	5	20	+	200	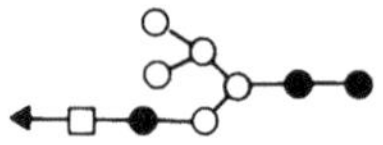
AIb	18	8	+	5	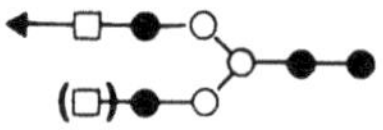
AII	20	2	+	5	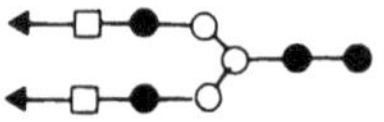
AIII	40	18	+	unbound	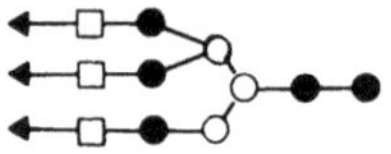

●N-Acetylglucosamine, ○Mannose, □Galactose, ▲sialic acid.

the five glycopeptides were further characterized by Con A-affinity chromatography using two different concentrations of α-methyl-D-mannoside (Me-Man) as specific eluent. N, AIa, AIb and AII were all retained by Con A-Sepharose; N, and AIa were eluted by 200 mM Me-Man, whereas AIb and AII were eluted at the lower concentration (5 mM Me-Man); AIII from both kinds of cells was not retained by the column. The table summarizes the experimental data and shows the relative distribution of the glycopeptides in each cell type. On the basis of the results a putative structure of the glycopeptides is proposed.

Discussion

The combined technique of ion-exchange separation and Con A--affinity chromatography was a good approach for the structural analysis of pronase-glycopeptides from HSV-infected cells. Neuraminidase treatment and DEAE-Sephacel chromatography provided informations on the number of sialic acid residues which accounts for the antennary structure of complex-type glycans, while the binding to Con A-Sepharose correlated with processing of the oligomannosyl core. It was shown that in N-linked glycans the presence of the oligomannosyl moiety at the non-reducing terminal position results in a strong binding to Con A and that any addition of sugars to the mannosyl core causes a progressive decrease in the association constants of the binding between glycopeptide and lectin (7).

The results reported here indicated that (i) AIII was not bound by Con A-gel as expected for triantennary glycans (8, 9); (ii) diantennary AII was retained by Con A-Sepharose and was eluted with a low concentration of α-methyl-mannoside; (iii) AIb was eluted by Con A-Sepharose like a diantennary glycans; after neuraminidase treatment the amount of label emerging by DEAE-Sephacel in the elution position of sialic acid was half of that released with the same treatment from AII. All together these results suggest that AIb has a diantennary structure in which only one branch carries the terminal sialic acid; (iv) AIa was strongly retained by Con A column like neutral mannosyl rich glycans (N), suggesting that it still retains the mannosyl cluster which is usually trimmed during the assembly of complex-type oligosaccharides. On the other hand its behaviour on DEAE-Sephacel indicated the presence of terminal sialic acid, in that before neuraminidase digestion it behaved as acidic structure (bound to the DEAE column) but after digestion the single radiolabeled peak was split in two fractions, the major one was neutral (unbound to the DEAE) whereas the other one coeluted with unlabeled sialic acid. When AIa was subjected to high resolution gel filtration it emerged in the elution position of the glycopeptides remarkably larger than oligo-mannosyl-rich glycans (five mannose residues) (unpublished results). For these reasons we suggest that AIa has a hybrid--type structure similar to that present in Rous sarcoma virus glycoproteins. Hunt and Wright(10) showed in this virus the presence of a hybrid-type glycan with an oligomannosyl core ($Man_5GlcNAc_2$-ASN) charateristic of neutral structures, plus "branch" sugars (NeuNAc-Gal-GlcNAc) charateristic of complex acidic structures.

The reduction of complex-type glycans in Ric^R14 cells is consistent with the glycosyltransferase deficiencies of these mutant cells and shows the relevance of polybranched glycans to the binding of ricin by the cell surface.

This investigation was supported by grants from Consiglio Nazionale delle Ricerche and Ministero della Pubblica Istruzione.

References

1. Spear, P. G.: J. Virol. 17, 991-1008 (1976);
2. Baucke, R. B., Spear, P. G.: J. Virol. 32, 779-789 (1979);
3. Serafini-Cessi, F., Campadelli-Fiume, G.: Arch. Virol. 70, 331-343 (1981);
4. Campadelli-Fiume, G., Poletti, L., Dall'Olio, F., Serafini-Cessi, F.: J. Virol. 43, (1982) (in press);

5. Vischer, P., Hughes, R. C.: Eur. J. Biochem. 117, 275-284 (1981);

6. Tarentino, A. L., Maley, F.: J. Biol. Chem. 249, 811-816 (1974);

7. Baezinger, J. U., Fiete, D.: J. Biol. Chem. 254, 2400-2407 (1979);

8. Krusius, T., Finne, J., Rauvala, H.: FEBS Lett. 71, 117-120 (1976);

9. Debray, H., Montreuil, J.: Biochimie, 60, 697-704 (1978);

10. Hunt, L. A., Wright, S. E.: J. Virol. 39, 646-650 (1981).

CON-A-REACTIVITY OF SERUM AND LEUCOCYTE α_1-ANTITRYPSIN. ELECTROPHORETIC HETEROGENEOUS α_1-ANTITRYPSINS STUDIED BY CROSSED AFFINITY IMMUNOELECTROPHORESIS.

Mette M. Andersen, Susanne Noack
The Finsen Laboratory, The Finsen Institute, Copenhagen, Denmark.

Introduction

In a previous study (1) we demonstrated that human leucocytes contain α_1-antitrypsin (α_1AT) (also called α_1-protease inhibitor). We also showed (2) that the human promyelocytic leukemic cell line, HL-60 synthesizes α_1-antitrypsin. Leucocyte and HL-60 α_1AT's have similar precipitate morphologies in crossed immunoelectrophoresis both being very heterogeneous (2). Analysis of these α_1AT's by traditional affinity immunoelectrophoresis (Bøg-Hansen et al. (6)) does not give precise information about the various subpopulations involved (2). In order to study more closely the molecular variants present in these α_1AT's and to compare them with more defined α_1AT's the present study was undertaken. We also studied α_1AT from α_1AT-deficient sera in order to compare this with normal serum α_1AT.

Materials and Methods

Concanavalin A (conA) (Batch no GL 23289) was from Pharmacia Fine Chemicals, Sweden; Porcine pancreatic elastase (type III) was from Sigma, U.S.A., and Trypsin 1:250, "Difco" certified from Difco Lab., Detroit, U.S.A. Five normal human sera were obtained from volunteers and 8 α_1AT-deficient sera from individuals with genotype ZZ. The normal sera were used to prepare a normal serum pool. Serum α_1AT was purified by agarose electrophoresis followed by chromatography on Blue Sepharose. This α_1AT, adjusted to 1 mg/ml, was mixed with trypsin or elastase (both 1 mg/ml) in 1:1 volume ratios ½ h 37 ^{0}C, and run immediately thereafter. Rabbit anti-α_1AT and anti-prealbumin were from DAKOPATTS, Copenhagen, Denmark.

Cells: Human leucocytes were obtained as described (1) and HL-60 cells, propagated in RPMI 1640 with 10% FCS and 50 µg/ml of penicillin and streptomycin respectively as described (2).

Affinity immunoelectrophoresis. This was performed either as described by Bøg-Hansen et al. (6) or as a two-step procedure (Fig. 1): Firstly the antigen(s) were electrophorized in a free agarose gel electrophoresis (i.e. a first dimension electrophoresis without conA). The respective sections of gel containing fractions of separated antigen(s) were cut out and transferred to secondary plates for affinity immunoelectrophoresis. The width of gel sections mostly used was 5 mm in order to minimize the width of the applied antigen. The gel sections were placed at the cathodic edge of the secondary plate and on the remaining space the conA-containing gel was cast. Prealbumin, used as internal standard, was added into a well cut out in the border between the antigen containing gel slice and the conA gel. First dimension electrophoresis was performed followed by normal second dimension electrophoresis into antibody containing gel.

Both in the traditional and the two-step technique the conA concentration in the gel was standardized at 140 µg/cm^2 throughout. All conA-reactivity fractions were defined by their retardation coefficient (R = Lo/Lr-1) (7). α-methyl-D-mannoside (MM) 5 % was included as specific displacer in all second dimension gels. Technical details otherwise as in (6).

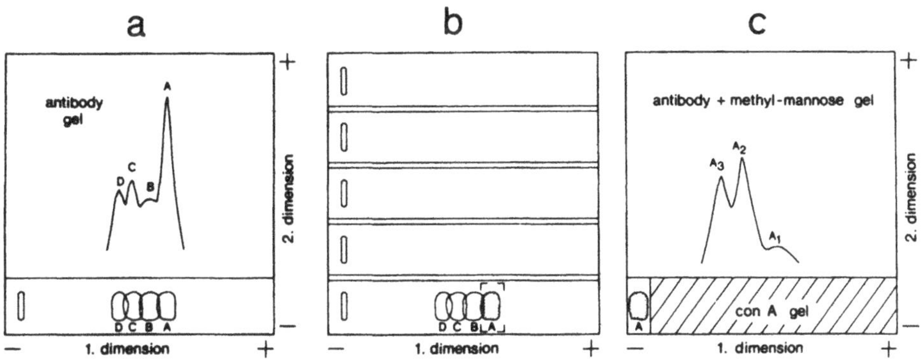

Fig. 1: Two-step crossed affinity immunoelectrophoresis (schematic drawing). a the controlplate, illustrating the position of the heterogeneous molecular variants of a protein after electrophoretic separation (A, B, C and D) and the precipitate pattern of these molecular variants after second dimension electrophoresis. b the first step of this technique with electrophoretic separation of multiple protein samples in parallel (controls are run next to test samples). c the second step, in which a molecular variant (A) has been transferred to a secondary plate where first and second dimension electrophoresis has been performed.

Results

1. α_1AT from normal serum and α_1AT-deficient serum. If normal serum was applied in low amounts, eg 0,3 µl,either directly or diluted 30 µl (1:100) in traditional affinity immunoelectrophoresis, the α_1AT was divided in three conA-reactive fractions, fig. 2 a. The retardation coefficients (R +/- $S_{\bar{x}}$) of these fractions were the same in all individual normal sera ($R(S_1)$ = 0.62 +/- 0.02, $R(S_2)$ = 1.48 +/- 0.04, $R(S_3)$ = 2.39 +/- 0.07). If however serum was applied in higher amounts, e.g. 3 µl, either directly or diluted 30 µl (1:10), a fourth fraction appeared. The fourth fraction was precipitated in the first dimension gel since it was not eluted if specific displacer was not present. The retardation of this fraction was not well defined since it depended on the length of the general conA-affinity precipitate, which differed in individual sera. Both the relative size and the retardation of this fraction depended

on the technical details of the experiments.

The conA-reactivity of α_1AT from α_1AT-deficient sera was similar to that of normal serum α_1AT. Since the concentration was less than 10 % of normal values these sera were run most convenient in 3 µl-amounts producing 4 fractions (fig. 2 b). The retardation coefficients of the 3 genuine fractions and their relative sizes were the same as in normal serum.

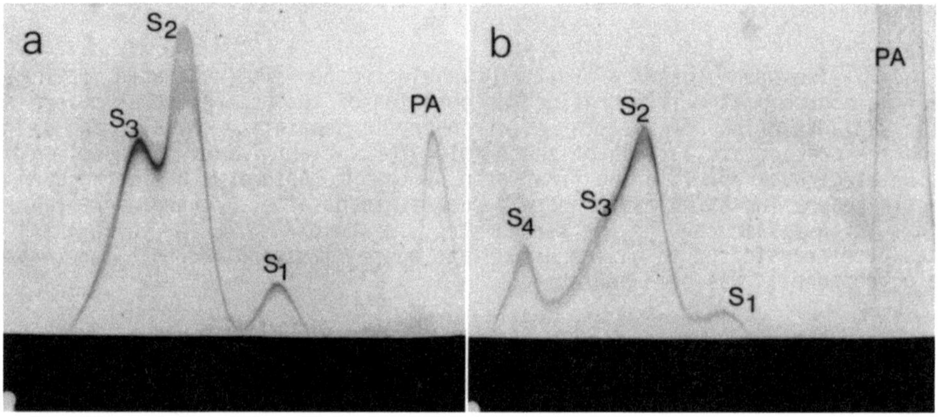

Fig. 2: Crossed affinity immunoelectrophoresis of 0,3 µl normal serum (a) and 3 µl α_1AT-deficient serum (b). Both plates contained 0,36 µl/cm^2 anti-α_1AT plus in (a) 0,06 µl/cm^2, and in (b) 0,14 µl/cm^2 anti-prealbumin. A total of three fractions (S_1 to S_3) was seen in normal serum and four fractions (S_1 to S_4) in α_1AT-deficient serum (see text).

When normal serum α_1AT was run in the two-step technique being divided into three serial sections of 5 mm gel it was seen that the relative sizes of the three conA-fractions (S_1, S_2 and S_3) varied systematically within the electrophorized α_1AT (fig. 3). The most anodic gelslice (containing α_1AT corresponding to the front precipitate leg) contained relatively more α_1AT(S_1) and S_2 than the most cathodic gelslice (α_1AT corresponding to the hind leg) where almost no S_1 was present and S_3 dominated. The intermediate gelslice contained more total α_1AT, but the relative distribution of S_1, S_2 and S_3 was intermediate to the other gelslices (not shown in figure). This demonstrated that S_1 has

the highest and S_3 the lowest electrophoretic mobility among the three fractions.
Similar results were obtained with α_1AT from α_1AT-deficient sera.

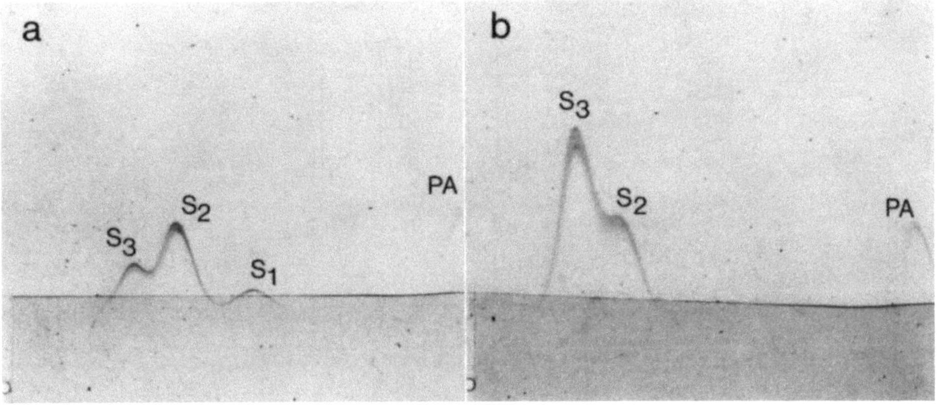

Fig. 3: Crossed affinity immunoelectrophoresis (the two-step technique) of serum α_1AT. (a) and (b) illustrate the distribution of conA-reactive fractions (S_1, S_2 and S_3) in the anodic and cathodic fractions, respectively. In the first step 3 µl normal serum was applied. The second step was performed with 3 gelsections of 5 mm each. Antibody: 0,14 µl/cm^2 anti-α_1AT and 0,06 µl/cm^2 anti-prealbumin.

2. α_1AT-elastase and α_1AT-trypsin reaction products. α_1AT reacted with elastase produced two precipitate peaks in crossed immunoelectrophoresis, one with similar mobility as serum α_1AT and one with β-mobility. α_1AT reacted with trypsin produced three peaks, one with mobility as serum α_1AT, one with α_2-mobility and one with β-mobility.

Serial sections of 5 mm through the entire length of the α_1AT-protease patterns were performed. The results were as follows: The α_1AT having mobility corresponding to serum α_1AT had the same three conA-reactive fractions as serum α_1AT.
The α_1AT corresponding to the α_1AT-elastase or the two α_1AT-trypsin fractions of slower mobility had different and less defined conA-reactivity. Only one (or two badly separated) fraction(s) could be seen with these α_1AT's and their precipitates were often double contoured. The retardation coefficient varied from 1,19 to 2,20.

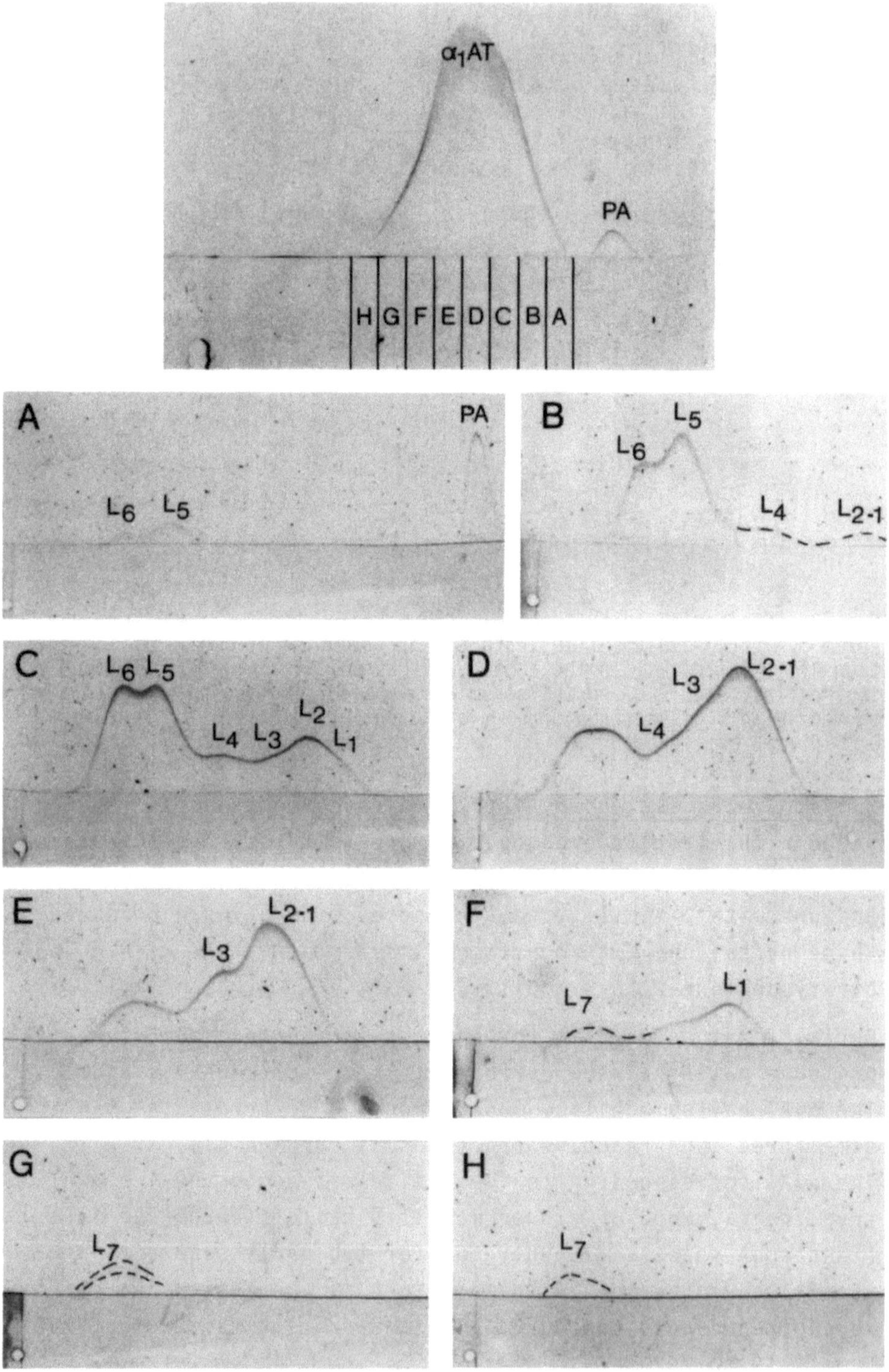
α1AT
PA
H G F E D C B A
A
PA
L6 L5
B
L5
L6
L4
L2-1
C
L6 L5
L4 L3 L2 L1
D
L2-1
L3
L4
E
L2-1
L3
F
L1
L7
G
L7
H
L7

3. HL-60 cells. This human promyelocytic cell line was grown in fetal calf serum for more than one year in our lab. In all samples taken α_1AT was present, proving it to be synthesized by these cells. The conA-reactivity of this α_1AT was heterogeneous, all 5 mm gel sections containing one or two very wide precipitates with strong reactivity towards conA. The precipitate area of HL-60 α_1AT was smaller after conA-affinity electrophoresis if specific displacer was omitted despite no visible affinity precipitate was formed in the first dimension gel. In the most anodic gelslices a fraction having retardation coefficient similar to S_1 was present, as well as traces of a fraction with a lower conA-reactivity. Whether S_2 and S_3 were present could not be determined due to the wide precipitates.

4. Leucocytes. A total of 7 different conA-reactive fractions could be distinguished (fig. 4). Up to 5 fractions were present within a single 5 mm gel section (fig. 4C). All 3 serum-α_1AT fractions were found ($L_4 = S_1$, $L_5 = S_2$ and $L_6 = S_3$) and they were present in the anodic part of the leucocyte-α_1AT (fig. 4 A-C). The most cathodic part of the α_1AT pattern contained α_1AT reacting similarly to that of the α_1AT-proteases, with rather undefined retardation coefficient and a double contoured precipitate (fig. 4 F-H). In the middle of the leucocyte α_1AT pattern three fractions with zero or very low conA-reactivity were found ($R(L_1) = 0 \pm 0{,}00$, $R(L_2) = 0{,}21 \pm 0{,}01$, $R(L_3) = 0{,}45 \pm 0{,}05$), fig. 4B-F.

Fig. 4: Crossed affinity immunoelectrophoresis (the two-step technique) of leucocyte α_1AT. The upper picture illustrates the control plate (the 5 mm gel sections transferred from the corresponding experimental plate are marked A-H). Plates A-H illustrate the various conA-reactive fractions within each gel slice. Antigen: 3×10^6 leucocytes. Antibody: control plate 0,43 $\mu l/cm^2$ and A-H 0,14 $\mu l/cm^2$ anti-α_1AT. All plates 0,06 µl anti-prealbumin. The prealbumin precipitate (PA) is illustrated on plate A but has been cut off from plates B-H.

Discussion

Serum α_1AT was found to be divided into three distinct molecular variants by the present methods. Similar results have very recently been obtained by Bayard et al. (5). These workers moreover provided evidence for uniformity of glycans within each molecular variant. They found fraction S_1 having triantennary, fraction S_2 having biantennary and fraction S_3 biantennary with intercalated N-acetylglucosamine residue carbohydrate chains. Our results showed that fraction S_1 had higher electrophoretic mobility than the other fractions. This could be due to higher content of sialic acid on this fraction of α_1AT, since sialic acid would both provide increased electrophoretic mobility and decreased conA-reactivity.

The fourth fraction (S_4) appearing when serum was applied in high amounts most probably represent α_1AT unspecificically entrapped, since it varied in individual sera according to the size of the general conA-affinity precipitate. A fourth fraction of serum α_1AT was never seen with the two-step technique, probably because the major glycoproteins producing the conA-affinity precipitate (e.g. α_2-macroglobulin) are excluded by this procedure.

The presence of such unspecific entrapment could lead to errously conclusions eg. in the comparisom of normal serum with α_1AT-deficient sera if the latter are run in higher concentration (fig. 2a, b). In their study of serum α_1AT Hinnerfeldt et al.(10) found 4 conA-reactive fractions. Since their conA concentration was the same as ours, the fourth fraction can be explained on basis of the amount of serum applied (1 µl). However, they moreover found a small fraction (0,93 % of the total α_1AT) non-reactive to conA. This we could not reproduce.

The conA-reactivity pattern of α_1AT after reaction with protease was the same whether elastase or trypsin was used. Some α_1AT had unchanged conA-reactivity compared to serum α_1AT. This could be due to surplus of free α_1AT despite theoretical higher molar ratios of enzymes. The remaining α_1AT had changed conA-reactivity compared to serum α_1AT indicating that both enzymes interacted with the carbohydrate residues of α_1AT. In agreement with this Baumstark et al. (4) found net loss of 8 N-acetylglucosamine and 10 hexose residues per mole α_1AT after interaction with elastase.

Although trypsin and elastase most probably reacts at different sites of α_1AT (Feste and Gann (8)) both enzymes may cleave a peptide of similar size from α_1AT. Banda et al. (3) found that macrophage elastase removed a 4000-5000 peptide from α_1AT and trypsin cleares a peptide of MW about 4000 from α_1AT (37 residues (Carell et al. (7)). The interaction of either enzyme with α_1AT could in this way have indirect effect on carbohydrate residues, linked to these peptides.

α_1AT synthesized by HL-60 cells exhibited heterogeneous conA-reactivity. This could be due to either multiple steps of synthetic α_1AT products being present or continuous synthesis of α_1AT with more glycan types than present in serum α_1AT. The strong reactivity of most of this α_1AT with conA could be due to lack of sialic acid residues on the glycans.

Leucocyte α_1AT differed from that of the other α_1AT's studied. Three fractions were similar to serum α_1AT and one fraction was similar to that of α_1AT reacted with protease. The remaining three fractions with zero or very low conA-reactivity could be in complex with specific leucocyte components or be break down products of α_1AT devoid of carbohydrate residues. Thirdly if leucocytes do synthesize α_1AT, as suggested by us (2) and others (11, 12), this α_1AT could be stored at different synthetic steps compared to HL-60 α_1AT.

Conclusion.

1. α_1AT from α_1AT-deficient sera contained the same three conA-reactive fractions as normal serum α_1AT (S_1, S_2 and S_3).

2. Seven individual molecular variants of α_1AT could be distinguished in human leucocytes on basis of their conA-reactivity. Three of the variants were similar to the variants of α_1AT in serum as determined by their retardation coefficients. One variant was similar to α_1AT reacted with trypsin or elastase. The three remaining molecular variants in leucocytes had zero or very low conA-reactivity.

3. The α_1AT synthesized by HL-60 cells reacted heterogeneously with conA. One variant was similar to S_1, but for the remaining no definite retardation coefficients could be given.

Acknowledgements

This study was supported by the Danish Medical Research Council (grant no. 12-1054).

References

1. Andersen, M.M. Scand. J. Immunol. 15, 399-407 (1982).
2. Andersen, M.M. (submitted to Scand. J. clin. lab. Invest.).
3. Banda, M.J., Clark, E.J., Werb, Z. J. Exp. Med. 152, 1563-1570 (1980).
4. Baumstark, J.S., Babin, D.R., Lee, C.T., Allen, R.C. Biochim. Biophys. Acta 628, 293-3o2 (1980).
5. Bayard, B., Kerckaert, J.-P., Laine, A., Hayem, A. Eur. J. Biochem. 124, 371-376 (1982).
6. Bøg-Hansen, T.C., Bjerrum, O., Ramlau, J. Scand. J. Immunol. 4, suppl. 2, 141-147 (1975).
7. Bøg-Hansen, T.C., Prahl, P., Løwenstein, H. J. Immunol. Methods 22, 293-307 (1978).
8. Carrell, R.W., Boswell, D.R., Grennan, S.O., Owen, M.C. Biochem. Biophys. Res. Comm. 93, 402-399 (1980).
9. Feste, A., Gann, J.C. J. Biol. Chem. 256, 6374-6380 (1981).
10. Hinnerfeldt, F., Albrecsen, J.M., Bøg-Hansen, T.C. Electrophoresis '82, ed. Statachos, W. de Gruyter, Berlin, (in press).
11. Ikuta, T., Okubo, H., Kudo, J., Ishibashi, H., Inoue, T. Biochem. Biophys. Res. Comm. 104, 1509-1516 (1982).
12. Wilson, G.B., Walker, J.H., Watkins, J.H., Wolbroch, D. Proc. Soc. Exp. Biol. Med. 164, 105-114 (1980).

SEMINAL ACID PHOSPHATASE.

Purification by lectin affinity chromatography.

Mineo Iwasa and Kaoru Sagisaka
Department of Legal Medicine Gifu University
School of Medicine, Gifu, 500 Japan

Acid phosphatase (AcP,E.C.3.1.3.2.)from human prostate gland has been known to be a phosphaoric monoester hydrolase (5) which is used for identification of seminal stains in forensic practice (3). AcP is widely distributed in animals, plants and microorganisms (7,10). As for human body, AcPs from prostate and seminal plasma have the same specificity of substrate but a few differences in the pattern of polyacrylamide electrophoresis were pointed out (6). On the reactivity with lectins, human urinary AcP was demonstrated to have different activity to Con A from serum AcP (4). On the other hand, some plant AcPs identified as glycoproteins were inhibited or activated by Con A (8,9). From these results lectins have been used to isolate glycoproteins in affinity chromatography. In this paper, we report the reaction of AcP from seminal plasma with 5 different lectins and purification of AcP by affinity chromatography with *Canavalia gladiata* DC lectin.

Materials and Methods

Preparation of lectins. Ground seeds of *Phaseolus coccineus* L., *Vigna Catiang* Endl. varis Sinensis, *Cytisus scoparius* Link., *Wistaris brachbotrys* Sieb et Zucc.and *Canavalia gladiata* DC (Can G) were extracted with 10 volumes of saline at room temperature for 60 min. The suspensions were centrifuged and the supernatants were defatted with an equal volume of ethyl ether. Purification of Can G was performed by affinity chromatography (1). The defatted extract was precipitated with 60 % saturated $(NH_4)_2SO_4$ and precipitate was applied onto a Sephadex G-200 column (Fig.1). Elution was carried out stepwise with 1.0 M NaCl and 1.0 M NaCl containing 0.1 M α-methyl-D-mannoside as specific displacer. AcP activity of Can G was eluted as a single peak with maximum activity in tube number 55. Inhibition tests revealed that only the specifically displaced fraction had lectin activity and reacted with AcP of seminal plasma. The affinity purified lectin was coupled to agarose by the cyanogen bromide method (2).

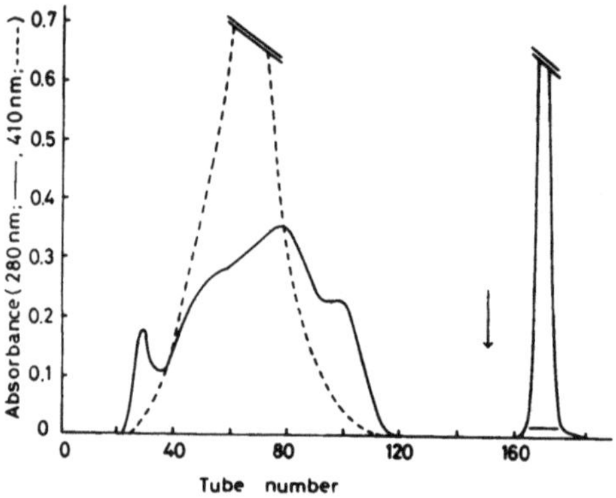

Fig.1. Affinity chromatography of $(NH_4)_2SO_4$ fraction of Can G lectin on Sephadex G-200. The dotted line is the activity of Can G AcP. The arrow indicates addition of α-methyl-D-mannoside. The fraction indicated with the black bar contained Can G lectin with agglutinin activity (the second peak).

Seminal plasma samples were centrifuged at 3000 rpm for 20 min, and the supernatants were separated. Seminal plasma diluted with four volumes of saline was applied onto a Sephadex G-200 column. Lyophilized lectins were dissolved in seminal plasma of which AcP activity was beforehand adjusted to an appropriate level and incubated at 4 °C for 12 hr. After centrifugation, AcP activity of the supernatants was assayed using 20 mM p-nitrophenyl phosphate as a substrate.

Isolation of AcP by affinity chromatography. Pooled seminal plasma diluted with 9 volumes of saline was mixed with half volume of the Can G lectin-coupled agarose gel, followed by incubation at 4°C overnight. After washing with saline, AcP was eluted with a linear 0 to 0.1 M α-methyl-D-mannoside gradient.

Crossed immunoelectrophoresis. One percent agarose gel in glycine-Veronal-Tris buffer pH 8.9 (27.5 g glycine, 22.4 g Veronal, 48.9 g Tris and 2.0 g calucium lactate in 1000 ml of water) was used. Electrophoresis in the first dimension was run with 10 mA for 60 min. The second run was performed in gel containing 15 % rabbit anti-seminal plasma (see below) with 5 mA for 12 hr. After electrophoresis, the gel plate was washed and stained with 0.2 g α-naphtyl phosphate and 0.4 g diazonium o-dianisidine in 100 ml of 0.2 M acetate buffer pH 5.0.

Preparation of rabbit antisera. Three rabbits were injected intramuscularly with 1 ml of pooled seminal plasma once a week for 7 weeks. At initial two injections emulsion with Freund's complete adjuvant was immunized. Can G-coupled gels were mixed with seminal plasma and washed sufficiently with saline, followed by homogenization. Another three rabbits were immunized with the homogenate by the same way.

Results

Effects of crude lectin on seminal AcP activity. Compared to an AcP activity of pooled seminal plasma of about 100,000 μM ρ-nitrophenol/hr/mg protein, the blank AcP activity of the crude Can G was 95 μM/hr/mg protein. Crude lectins of Can G, *Phaseolus coccineus* L., *Vigna Catiang* Endl. varis Sinensis, *Cytisus scoparius* Link., *Wistaris brachbotrys* Sieb et Zucc were added to seminal plasma with AcP activity adjusted to 250 X 10^3 (ρ-nitrophenol μM/hr). After incubation for 12 hr, the supernatants of each mixture were assayed for total AcP activity. As shown in Fig.2, some lectins activated the seminal AcP (*Phaseolus coccineus* L.,50 %; *Vigna Catiang* Endl. varis Sinensis, 35 %; *Cytisus scoparius* Link., 31 %; *Wistaris brachbotrys* Sieb et Zucc., 15 %). Of the five lectin extracts tested, only Can G lectin reduced the activity of AcP. When the lectins of less than 5 mg/ml were added, AcP activity of seminal plasma changed scarcely. However, adding of the lectins of concentration of 10 mg/ml or more resulted in increasing or reducing of the AcP activity.

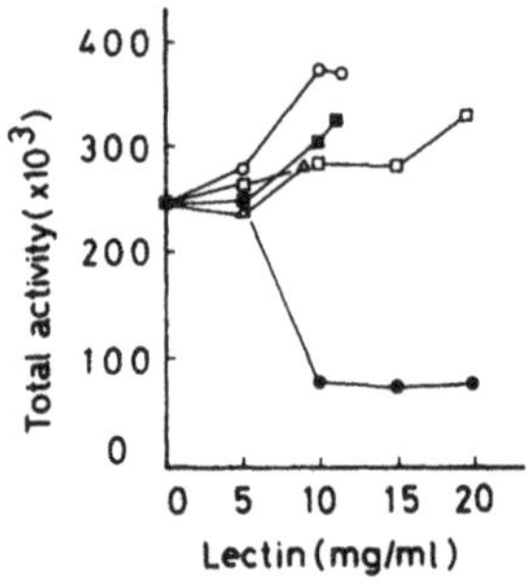

Fig.2. Inhibition of AcP activity with various lectins. O-O, *Phaseolus coccineus* L.; □-□, *Vigna Catiang* Endl. varis Sinensis; ■-■, *Cytisus scoparius* Link.; △-△, *Wistaris brachbotrys* Sieb et Zucc.; ●-●, *Canavalia gladiata* DC.

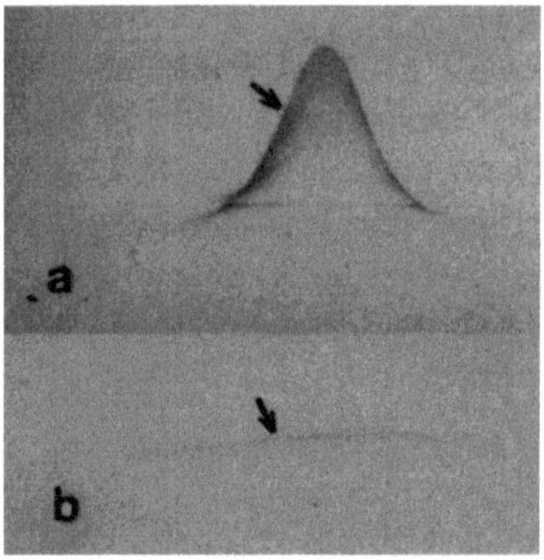

Fig.3. Crossed immunoelectrophoretic patterns of seminal plasma (a) and the supernatant of seminal plasma precipitated with Can G lectin (b). The gel plates were stained for AcP.

Crossed immunoelectrophoresis was performed of the supernatants and the height of the AcP peak was remarkably reduced with Can G lectin(Fig.3). When gel plates were stained with Coomassie brilliant blue, about 15 precipitation peaks were observed and the peaks other than AcP were not affected by the lectins.

Fractionation of seminal plasma. Three protein peaks were observed by gel filtration of seminal plasma on Sephadex G-200. The AcP assay revealed that some activity was eluted with the first peak and most of the activity with second peak. By addition of pure lectin to the active fractions, we found that the activity of both fractions was some what reduced (Fig.4). The rate of reduction was 3.6 % at 1.5 mg/ml of lectin added and

13.7 % at 2.0 mg/ml or more.

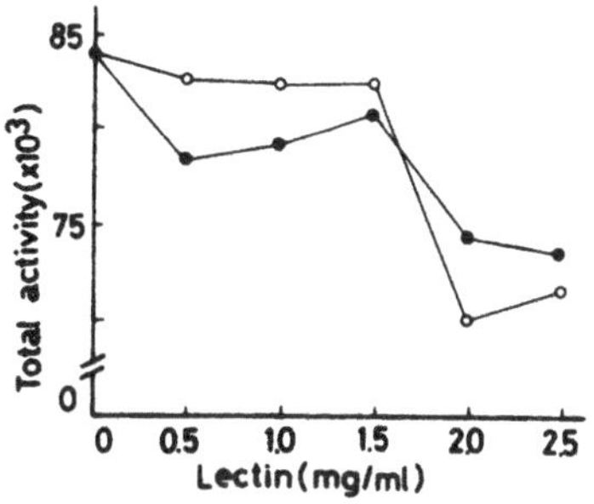

Fig.4. Inhibition of the two fractions possessing AcP activity separated by Sephadex G-200 gel filtration with purified Can G lectin. o—o ,Fraction 1; ●—●,Fraction 2.

Isolation of AcP by affinity chromatography. By elution with a linear 0-0.1 M α-methyl-D-mannoside gradient, the AcP fraction was recovered with the eluent at low concentration (Fig.5). The specific activity of AcP was elevated (54 fold) by this method. Rabbits immunized with AcP-coupled gels produced potent antibody to AcP. A week activity to one of the seminal proteins was absorbed with serum.

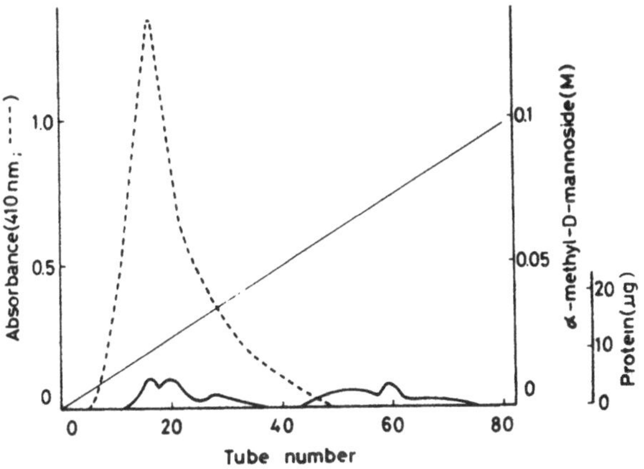

Fig.5. An affinity chromatographic pattern of seminal plasma on Can G lectin-coupled agarose gel with gradient elution. After washing with 0.85 % NaCl, AcP (dotted line) was eluted with 0.1 M α-methyl -D-mannoside linear gradient.

Discussion

We wanted to identify a lectin which could bind and inhibit seminal plasma acid phosphatase (AcP) because we wanted to purify AcP. The reason for our interest in AcP is that AcP activity from prostatic gland has been used for identification of seminal stains in forensic science (3). Of five bean lectins examined, four lectins activated and only Can G lectin inhibited AcP of seminal plasma. The crossed immunoelectrophoresis with antibody against seminal plasma revealed that AcP was precipitated from incubation mixture with Can G lectin. Bøg-Hansen and Brogren (4) demonstrated by crossed immuno-affino-electrophoresis that there were two type enzymes possessing one and two or more binding sites to Con A. Although the analytical procedure of the present study was exactly same as the affinoelectrophoresis, our results suggested that AcP of seminal plasma had one binding site to Can G lectin per molecule. As the lectin was proved to be specific for α-mehtyl-D-mannoside and mannose by hemagglutination in this experiment, we conclude that seminal plasma AcP is a glycoprotein with mannose in the carbohydrate part.

Two active fractions separated by Sephadex G-200 gel filtration were inhibited with Can G lectin to the similar extent. It seemed that the two isozymes of seminal plasma AcP had the similar binding activity to Can G.

Van Etten and Saini (11) reported a procedure for purification of human prostatic AcP in which Con A Sepharose was employed as absorbent. By this procedure, AcP was prepared with a relative activity of 16.3 which is relatively low as compared with the value (54 fold) of seminal plasma AcP. It meant that Can G lectin have a great affinity for AcP of seminal plasma.

It is concluded that Can G lectin is much useful as a tool to isolate AcP of seminal plasma and to prepare anti-AcP for forensic reseachers who investigate seminal stains concerning sexual crimes.

References

1. Agrawal,B.L. & Goldstein,I.J.: Biochem.J.,96,23-25 (1965)
2. Axén,R.,Porath,J. & Ernback,S.: Nature(Lond),214,1302-1304 (1967)
3. Baxter,S.J.: Medicine Sci.Low,13,155-165 (1973)
4. Bøg-Hansen,T.C. & Brogren,C.-H.: Scand.J.Immunol.,4,Suppl.,2,135-139 (1975)

5. Jutisx,M. & De La Llosa,P. (1972) in Gottshalk,A.(ed.) Glycoproteins,Elsevier Publishing Co.,Amsterdam, pp. 1019-1068

6. Lam,W.K.W.,Yam,L.T.,Wilbur,H.J.,Taft,E. & Li,C.Y.:Clin. chem.,25,1285-1289 (1979)

7. Leithoff,H. & Kurizias,G.A.:Dtsch.Z.ges.gerichtl.Med.,46, 260-261 (1957)

8. Lorenc-Kubis,I. & Bøg-Hansen,T.C. (1981) in Bøg-Hansen,T.C. (ed.) Lectin-Biology,Biochemistry,Clinical Biochemistry,Vol. 1, Walter de Gruyter, Berlin·New York, pp.157-167

9. Lorec-Kubis,I.,Morawiecka,B.,Wieczorek,E.,Wisniowska,J., Wierzba,E.,Ferens,M. & Bøg-Hansen,T.C. in Bøg-Hansen,T.C. (ed.) Lectin-Biology,Biochemistry,Clinical Biochemistry, Vol.1, Walter de Gruyter, Berlin·New York, pp. 169-178

10. Newmark,M.Z. & Wenger,B.S.: Arch.Biochem.Biophys.,89,110-117 (1960)

11. Van Etten,R.L. & Saini,M.: Biochem.biophys.Acta,484,487-492 (1977)

ISOLATION AND CHARACTERIZATION OF POA PRATENSIS SEED ACID PHOSPHATASES INTERACTING WITH CON A

I. Lorenc-Kubis
Institute of Biochemistry, University of Wroclaw, Poland and the Protein Laboratory, University of Copenhagen, Denmark.

The ability of lectins to interact with glycoproteins has been successfully employed to purify some enzymes as revieved in (1), to fractionate glycopeptides (2-4), to study the carbohydrate structure of glycoproteins (5-7, 27) and the modulation of enzyme activity (8, 20). In crude extracts of grass seeds we have shown two groups of acid phosphatases which differ in their binding to con A. One of the groups exhibits a high affinity to con A, whereas the other group does not bind (8). In the present paper I describe that the non-binding acid phosphatase developed a binding capacity towards con A after purification.

Materials and methods.

Ground seeds were extracted with 0.1 M acetate buffer, pH 5,1 as described in (9). Acid phosphatase activity was measured spectrophotometrically (10). Protein was determined by the method of Lowry with bovine serum albumin as standard (11). Affinity chromatography was performed with con A-Sepharose (Pharmacia Fine Chemicals) from where the non-binding proteins were washed out by the acetate buffer, pH 5,6. Gel filtration was performed on Biogel P-100 and Biogel P-300. Hydrophobic interaction chromatography on phenyl-Sepharose (Pharmacia Fine Chemicals) was performed according to (12). Extraction and chromatography was performed reproducibly about ten times. Quantitative immunoelectrophoresis including modifications with lectins was carried out as described in (6, 13, 14, 21). Antibodies were purified rabbit antiserum (15) raised against protein extracts of Poa pratensis and against partially purified acid phosphatase from Poa pratensis seeds. Acid phosphatase activity was developed histochemically (16). Treatment with proteolytic enzymes: trypsin, α-and β-chymotrypsin, pronase and thermolysin was performed according to (17). Treatment with glycosidases as described in (18). Neutral sugars were detected by phenol-sulphuric acid method (19).

Results

Acid phosphatases from Poa pratensis seeds were separated by affinity chromatography on con A-Sepharose into two fractions: the non-binding and the con A-binding fraction (16). The con A non-binding acid phosphatase (AcPase A) was further purified by gel filtration in Bio-gel P-100 (step 1). The enzyme activity was eluted from the column as a symmetric, sharp peak localized in the first protein peak (Fig. 1). Tubes 62-72 (Fig. 1) were pooled and subjected to a second affinity chromatography in con A-Sepharose (step 2). It was found that after gel filtration more than 80% of the total activity of AcPase A was retained on con A-Sepharose. The enzyme liberated from the column with 10% α-methyl-D-mannopyranoside (pH 5.6) was further fractionated by gel filtration on Bio-gel P-300 (step 3) and by hydrophobic interaction chromatography on phenyl-Sepharose (step 4). Active fractions were eluted from the phenyl-Sepharose with 0.8 M ammonium sulphate, pH 8.7 (AcPase A_1), and with Tris-glycine buffer (AcPase A_2) pH 8.7. The specific activity of AcPase A_1 and AcPase A_2 was 6.0 u/mg and 10.3 u/mg of protein, respectively, as compared to an activity of 0.28 u/mg of the con A non-binding fraction (step 1). The purified enzymes which contained 40% and 20% of neutral carbohydrate

The in vitro activity of AcPase A_1 and A_2 was measured in presence of con A, WGA and SBA but I did not find any activity change, in contrast to what was found for other enzymes (16, 20). In crossed immunoelectrophoresis (Fig. 2) I found that both fractions precipitated with con A, but did not react with

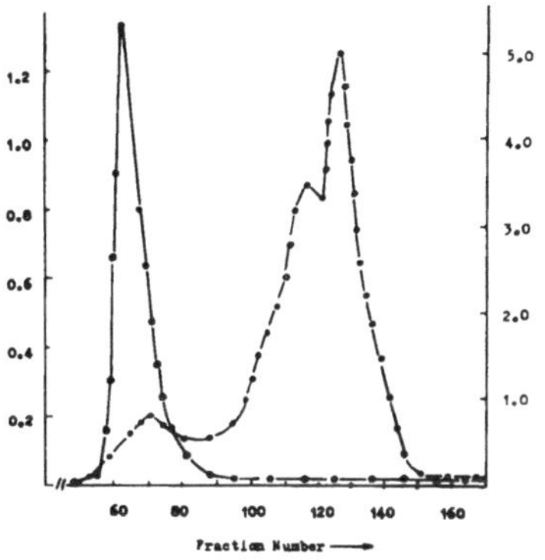

Fig. 1. Elution profile of acid-phosphatase A (AcPase A) from gelfiltration on Biogel P-100.
The abscissa gives fraction numbers.
Left ordinate is absorbance at 280 nm.
Right ordinate is phosphatase activity.

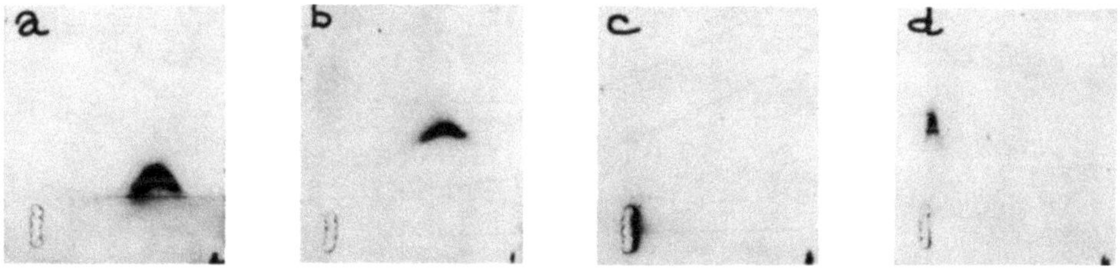

Fig. 2. Lectin affinity immunoelectrophoresis of AcPase A_2 (20 µg).
a: Normal crossed immunoelectrophoresis of AcPase A_2.
b: Crossed immunoelectrophoresis with an intermediate gel with 0.02% α-methyl-D-Mannopyranoside.
c: As b, except con A (50 µg/cm^2) in the first dimension gel.
d: As b, except con A in the first dimension gel and 0.02% of sugar in the intermediate gel.

The plates were stained for acid phosphatase activity (16).

neither WGA nor SBA. In crossed immunoelectrophoresis with con A in the first dimension gel and with an intermediate gel (Fig. 2b) with different amounts of α-methyl-D-mannopyranoside (0.02 - 3.0%), I found that AcPase A_1 was released from the affinity precipitate (Fig. 2c) in the presence of 0.2% sugar, while AcPase A_2 required a 10-times lower concentration for release. After treatment of both enzymes with glycosidases (24 h) they were still active when analyzed spectrophotometrically and by crossed immunoelectrophoresis. The glycosidase treatments were also monitored by crossed immunoelectrophoresis with con A in the first dimension gel. I found that the α-L-fucosidase-treated, the α-glucosidase-treated and the β-galactosidase-treated AcPase A_1 and AcPase A_2 were completely precipitated with free con A to form an affinity precipitate in the first dimension gel (as Fig. 2c). When the enzymes were treated with α-mannosidase, β-glucosidase or mollusk β-glucuronidase only a small part of the activity was found in the affinity precipitate in the first dimension gel with the major part as an immunoprecipitate (as Fig. 2b) in the second dimension gel.

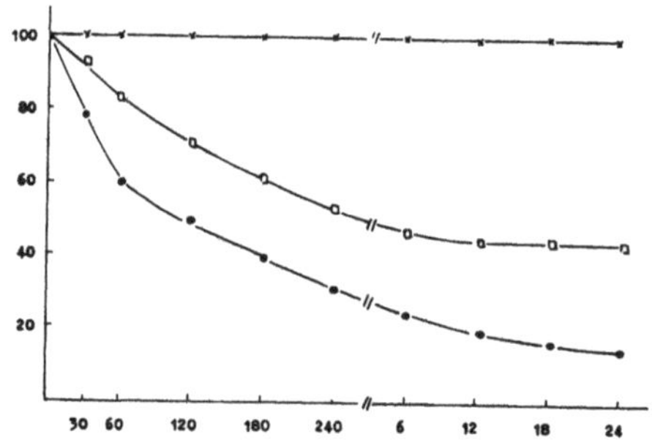

Fig. 3. Activity of acid phosphatase AcPase A_2 treated with thermolysin in presence and absence of con A.
AcPase A_2 with or without con A (upper straight curve).
AcPase A_2 with thermolysin (lower curve).
AcPase A_2 with con A and thermolysin (intermediate curve).
Lectin-AcPase A_2 ratio was 10.1; AcPase A_2-thermolysin ratio was 2:1. Con A was added 10 min prior to addition of thermolysin. Abscissa is the incubation time in min/hr.
Ordinate is the measured enzyme activity.

AcPase A_1 and AcPase A_2 treated with trypsin and chymotrypsin were still 100% active, but lost about 60-80% activity after incubation with pronase and thermolysin. In the presence of con A both enzymes were more resistant to pronase and thermolysin digestion (Fig. 3).

Discussion

Lectin affinity immunoelectrophoretic studies of Poa pratensis acid phosphatases have proved con A to be the best lectin amongst con A, SBA, WGA, LCA and pea lectin for affinity chromatography of these enzymes (16). Using affinity chromatography with con A-Sepharose, I separated grass acid phosphatases into two fractions: the con A non-binding (AcPase A), and the con A-binding (AcPase B) (9). Here I have shown that the non--binding enzyme (AcPase A) develop the ability to bind to con A and that it may be further separated into two fractions, AcPase A_1 and AcPase A_2. The change in the affinity of AcPase A to con A may indicate that the enzyme exists as aggregates or complexes with other proteins or glycoproteins without mannose or glucose units exposed. After sieving chromatography the con A-binding units became exposed.

Hydrophobic interaction chromatography on phenyl-Sepharose separated acid Phosphatase A into two enzyme fractions: AcPase A_1 and AcPase A_2 which differ in their content of sugar.

Specific displacer (α-methyl-D-mannopyranoside) in the intermediate gel in crossed immunoelectrophoresis with free con A in the first dimension gel allowed to see the differences between the enzyme fractions in the dissociation of the affinity precipitate in the first dimension gel.

Both AcPase A_1 and AcPase A_2 were still active after treatment with glycosidases. Studies with several different enzymes

(22) have shown that enzymatic activity was affected only slightly when the carbohydrate residues or glycoproteins were oxidized, or the activity was essentially unchanged after removal of carbohydrate residues (23, 24).

The degradation of the carbohydrate part may be studied by testing the affinity to con A (25, 26). The affinity of AcPase A_1 and AcPase A_2 to con A after treatment with glycosidases was analyzed by lectin affinity immunoelectrophoresis with free con A in the first dimension gel. The changes in the affinity precipitate in the first dimension gel of both AcPase A_1 and AcPase A_2 and the appearance of an immunoprecipitate in the second dimension gel indicated that these enzymes were hydrolyzed by α-mannosidase, β-glucosidase and mollusk β-glucuronidase (an enzyme which manifest hydrolytic activity towards α-mannosidase (24)). On the other hand, α-glucosidase and β-galactosidase did not change the affinity of the enzymes to con A.

Similar to other glycoproteins grass acid phosphatases (8), AcPase A_1 and AcPase A_2 lost their activity after prolonged treatment with pronase and thermolysin. In the presence of con A both enzymes were more resistant to proteolytic digestion.

Depressed susceptibility of con A AcPase complex to proteolytic digestion may indicate a possible physiological role for lectins in plants as a protector against proteolysis during the biological processes.

Acknowledgements. Professor dr. hab. Bronislawa Morawiecka and dr. T.C.Bøg-Hansen are thanked for discussions, advice and support. The Danish Ministery of Education is thanked for a scholarship. Lectins were partly provided by the Danish Research Council through a grant to T.C.Bøg-Hansen. The Polish Academy of Sciences supported the study by grants no. MR II 1.2 and R 1.9.

References

1. Dunaley, J.T. Mol.Cell.Biochem. 21, 43-63 (1978)
2. Javaid, J.I., Hof, H., Brunngraber, E.G. Biochim,Bioph. Acta 404, 74-82 (1975)
3. Krusius, T. FEBS Lett. 66, 86-89 (1976)
4. Ogata, S., Muramatsu, T., Kobata, A. J.Biochem. 78, 687-696 (1975)
5. Bøg-Hansen, T.C., Brogren, C.H. Scand.J.Immunol. 4. Suppl. 2, 135-139 (1975)
6. Bøg-Hansen, T.C., Bjerrum, O.J., Ramlau, J. Scand.J. Immunol. 4, Suppl. 2, 141-147 (1975)
7. Kornfeld, R., Ferris, C. J.Biol.Chem. 250, 2614-2619 (1975)
8. Lorenc-Kubis, I., Morawiecka, B., Bøg-Hansen, T.C. These proceedings, Vol. 2, 509-515 (1982)
9. Lorenc-Kubic, I., Morawiecka, B. Soc.Bot.Polon. 42, 337-369 (1973)
10. Yoshida, H., Tamiya, N. J.Biochem. 69, 525-534 (1971)
11. Lowry, O.H., Resenbrough, Farr, A.L., Randall, R.J. J. Biol.Chem. 193, 265-275 (1951)
12. Bjerrum, O.J., Blirup-Jensen, S., Larsen, P. Prot.Biol. Fluids. 27, 775-780 (1980)
13. Weeke, B. Scand.J.Immunol. 2, Suppl. 1, 47-56 (1973)
14. Svendsen, P.J. Scand.J.Immunol. 2, Suppl. 1, 69-77 (1973)
15. Harboe, N., Ingild, A. Scand.J.Immunol. 2, Suppl. 1, 161-164 (197)
16. Lorenc-Kubis, I., Bøg-Hansen, T.C. These proceedings, Vol. 1, 157-167 (1981)
17. Bjerrum, O.J., Bøg-Hansen, T.C. Scand.J.Immunol. 4, Suppl. 2, 89-99 (1975)
18. Kawasaki, T., Ashwell, G. J.Biol.Chem. 251, 5292-5299 (1976)
19. Dubois, M., Gilles, K.A., Hamilton, J.K., Rebers, P.A., Smith, E. Anal.Chem. 28, 350-356 (1956)
20. Lorenc-Kubis, I., Morawiecka, B., Wieczorek, E., Wisniowska, J., Wierzba, E., Ferens, M., Bøg-Hansen, T.C. These proceedings , Vol. 1, 169-178 (1981)
21. Salier, J.P., Faye, L., Vergaine, D., Martin, J.P. Electrophoresis 1, 193-197 (1980)
22. Lee, T., Hager, L.P. Fed.Proc. 29, 599- (1970)

23. Pazur, J.H., Simpson, D.L., Knull, H.R. Biochem.Biophys. Res.Commun. 36, 394- (1969)

24. Scocca, J., Lee, Y.C. J.Biol.Chem. 254, 4852-4863 (1979)

25. Kornfeld, R., Fenis, C. J.Biol.Chem. 250, 2614-2619 (1975)

26. Onishi, H.R., Tkacz, J.S., Lampen, J.O. J.Biol.Chem. 254, 11943-11952 (1978)

27. Iwasa, M., Sagisaka, K. These proceedings.

WGA AFFINITY IMMUNOELECTROPHORESIS OF GLYCOPHORIN FROM HUMAN ERYTHROCYTE MEMBRANES. CAPACITY MEASUREMENTS OF WGA GEL AND BINDING CONSTANT DETERMINATION.

Niels H.H. Heegaard[1], Ulla Christensen[2] and Ole J. Bjerrum
[1]The Protein Laboratory and [2]Chemical Laboratory IV, H.C. Ørsted Institute, University of Copenhagen, Denmark.

Lectin affinity immunoelectrophoresis of soluble glycoproteins has been used to identify lectins in complex mixtures, to identify and characterize glycoproteins including investigations of microheterogeneity, to determine affinity constants, and as a qualitative method to predict the outcome of affinity column experiments (2,3,5-10,12,16). The principle has not yet been used extensively to investigate membrane glycoproteins and their affinity constants have not been determined using this technique.

In this communication we present our studies on the binding of wheat germ agglutinin (WGA) to glycophorin from human erythrocyte membranes both in crude extracts obtained by solubilization of membranes with Triton X-100 (2) and after purification (15). The binding of Triton extracts was studied using immobilized lectin. The dissociation constant for both Triton extracts and purified glycophorin was calculated using free lectin after the method of Bøg-Hansen and Takeo (9). In addition we tried to correlate the value of the dissociation constant obtained in this way to the value determined by direct binding experiments using a batchwise separation of glycoprotein and immobilized WGA. Finally, we elaborated a simple table-top immunoelectrophoretic technique to estimate the binding capacity of an immobilized lectin matrix.

Lectins, Vol. III

Materials and methods

WGA, Sepharose CL 6B and CNBr-activated Sepharose 4 B was obtained from Pharmacia Fine Chemicals. WGA was coupled to Sepharose 4B in a concentration of 5 mg/ml gel as described by the manufacturer. N-acetyl-β-D-glucosamine was from Sigma and agarose type HSA from Litex (Glostrup, Denmark). Kodak delivered lithiumdiiodosalicylate, BDH Triton X-100 (polyoxyethylene (9-10) p-t-octylphenol, scintillation grade) and human albumin came from Behringwerke. Rabbit immunoglobulins against human erythrocyte membranes were from Dakopatts (Copenhagen, Denmark) and used without absorption for the experiments with free WGA in the 1st dimension. Otherwise, it was absorbed with human albumin before use (4). Human erythrocyte membranes were prepared and solubilized according to (2). Final protein concentration: 2 mg/ml. Glycophorin was prepared from lithiumdiiodosalicylate membrane extracts according to Marchesi (15). The concentration of glycophorin in water was determined by dry weight measurements. The apparent molecular weight of glycophorin including bound detergent was determined by gel filtration on a calibrated 80 cm x 1.2 cm^2 Sepharose CL 6B column in presence of 0.5 % Triton X-100, 0.038 M Tris and 0.1 M glycine, pH 8.7 (4) and was found to be 490,000. A minor fraction appeared with a molecular weight of 3 mill.

Immunoelectrophoreses were all performed in 1 % (w/v) agarose gel containing 0,5 % (v/v) Triton X-100, 0.038 M Tris and 0.1 M glycine, pH 8.7 (2). Crossed affinity immunoelectrophoresis with free WGA (9) and immobilized WGA in an intermediate gel (5) and 1st dimension gel (8,16) was performed as described. Free WGA was applied in the range of 1-30 $\mu g/cm^2$. To liberate protein bound to the matrix in 1st dimension 5 % (w/v) N-acetyl-D-glucosamine was incorporated in an intermediate gel of the 2nd dimension electrophoresis. Except otherwise stated, the 1st dimension electrophoresis was performed at 10 V/cm and lasted for 1 h., and the 2nd dimension electrophoresis ran overnight at

2.5 V/cm. The plates were stained with Coomassie Brilliant Blue and in some cases also for acetylcholinesterase (2). Glycophorin concentrations were determined by rocket immunoelectrophoresis at 2.5 V/cm (1) using standard curves (0.1 to 0.012 µg glycophorin) at an anti-membrane antibody content of 2.5 $\mu l/cm^2$. Immunoelectrophoretic determination of the binding capacity of the WGA matrix for glycophorin was performed using the principle of line absorption immunoelectrophoresis of Krøll and Andersen (14). A quantitative aspect was introduced by measuring the rocket-shaped deflections of the line preciputate of glycophorin which passed wells containing WGA matrix (applied from a slurry where the gel bed was diluted 1:3 in detergent containing electrophoresis buffer. Afterwards the wells were sealed with a drop of agarose). Batchwise binding of glycophorin to immobilized WGA was performed as follows: Glycophorin at a concentration of 6.7 µg/ml in 5 % (v/v) Triton X-100, 0.1 M glycine and 0.038 M Tris, pH 8.7, was mixed with varying amounts of WGA-Sepharose 4B washed in the same buffer (20-100 µl) in a total volume of 500 µl. After 15 min of equilibration the samples were centrifuged for 10 min at 1500 x g at 22°C, whereafter the glycophorin concentration of the supernatants was determined by rocket immunoelectrophoresis (1). Determination of the binding capacity on columns were performed in 1 ml tuberculin plastic syringes containing 150 µl WGA matrix or 150 µl Sepharose 4B. The columns were loaded with a solution of glycophorin (67 µg/ml) and 30 µl fractions were collected by hand. The amount of retained glycophorin was determined by rocket immunoelectrophoresis.

Results and discussion

<u>Binding to immobilized WGA.</u> The existence of glycoproteins binding to WGA in a Triton X-100 extract of human erythrocyte membranes appears from Fig. 1 where a crossed affinity immunoelectrophoresis with WGA-Sepharose in an intermediate gel is

shown. Earlier the various precipitates were characterized in detail (2,4) and the presence of 3 WGA binding glycoproteins: Glycophorin, band 3 complex, and antigen no. 25 were verified. Apparently neither glycophorin nor complexed band 3 protein in Fig. 1 are retained completely in the intermediate lectin gel. However, the glycophorin precipitate disappeared with less protein load while complexed band 3 protein still showed a fractional binding. The relation between the gel matrix area below the precipitate and that protein load where the peak just has disappeared will give a weak idea of the binding capacity of the WGA matrix.

To detect possible antigens with less affinity to WGA-Sepharose, the matrix was incorporated in the 1st dimension gel (Fig. 2) (8,16). Interaction under these conditions is seen as a cathodic shift of the position of the precipitate compared with thw
shift of the position of the precipitate compared with the con-

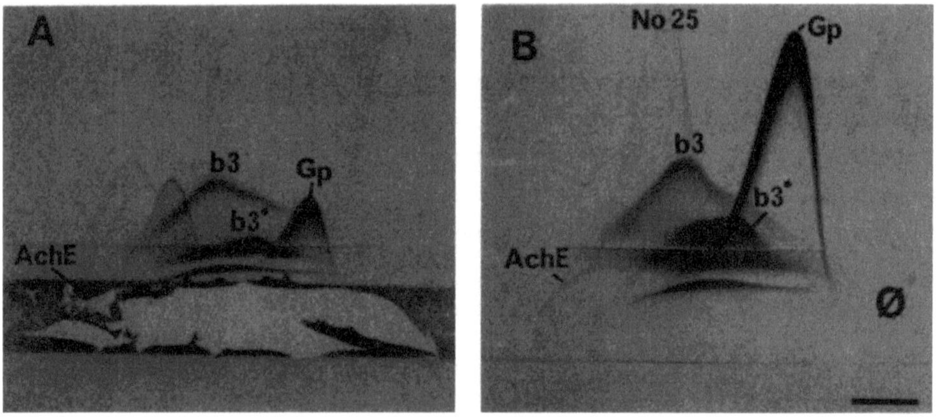

Fig. 1. Crossed affinity immunoelectrophoresis of 20 µg Triton X-100 solubilized human erythrocyte membrane proteins. In (A) the intermediate gel contains WGA-Sepharose (50 $\mu g/cm^2$) and in (B) it is blank (Ø). Nearly all the glycophorin (Gp) is retained on the matrix, and with less load the precipitate completely disappears. Other glycoproteins showing binding in this set up are no. 25 (complete), and band 3 containing complex (b3*) (partial), whereas non-complexed band 3 (b3) and acetylcholinesterase (AchE) do not bind. The upper gel contains anti-membrane antibodies (7 $\mu l/cm^2$). The bar is = 1 cm.

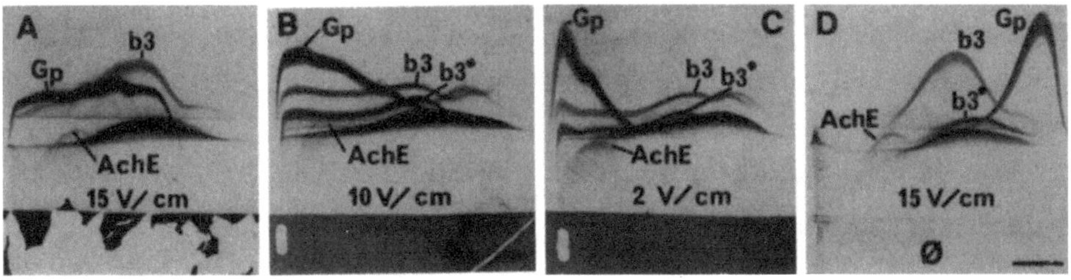

Fig. 2. Crossed affinity immunoelectrophoresis of 20 µg Triton solubilized erythrocyte membrane proteins. In (A,B,C) the 1st dimension contains WGA-Sepharose (120-180 µg/cm^2) and in (D) it is blank (Ø). The intermediate gels contain N-acetyl-D-glucosamine (5 % (w/v)). The voltage applied in 1st dimension is indicated. Otherwise conditions and designations are as in Fig. 1.

trol (8,16). Incorporation of the lectin specific sugar in an intermediate gel of the 2nd dimension electrophoresis ensures liberation of antigens from the lectin matrix. The precipitation patterns were found to be highly dependent on the applied voltage in the 1st dimension electrophoresis (compare Figs. 2A,2B, and 2C) indicating that a fast migration interferes with the protein-lectin interaction. The most pronounced retardation is obtained at 2 V/cm. Besides glycophorin, complexed band 3 protein and antigen no. 25, also acetylcholinesterase and free band 3 protein were retarded. These observations are in agreement with preparative column experiments and prove the higher sensitivity when having the matrix in the 1st dimension electrophoresis.

Concerning the application of affinity immunoelectrophoresis for prediction of the outcome of a preparative column experiment (3,5) it is a drawback that the binding capacity of the matrix cannot be directly determined. This is done by the experimental set up shown in Fig. 3. By absorption to WGA matrix in wells placed anodically to an intermediate gel containing glycophorin, deflections appear in the line precipitate as a

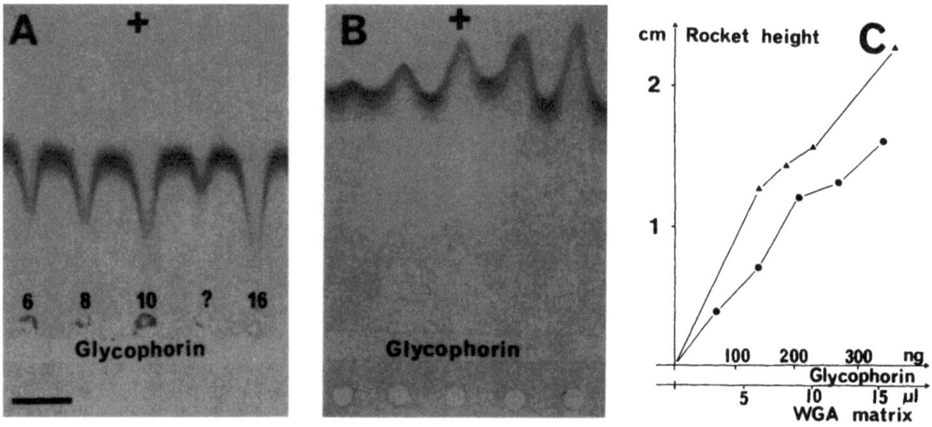

Fig. 3. Table-top immunoelectrophoretic determination of the binding capacity of WGA-Sepharose 4B for glycophorin. Two line immunoelectrophoreses of purified glycophorin are performed. In (A) glycophorin passes wells containing the WGA matrix in the amounts indicated (µl). In (B) rockets are formed by applying glycophorin (Gp) in wells below the intermediate gel. The anti-membrane antibody content of the gels were 3 µl/cm^2. In (C) the relation between the reversed rockets of (A) (▲——▲) and the normal rockets of (B) (●——●) is drawn.

kind of "reversed" rockets (Fig. 3A). Application of increasing amounts of WGA matrix gives the quantitative relation (Fig. 3C). A similar correlation can be obtained by measuring the standard rockets with increasing concentrations of glycophorin (Figs. 3B and 3C). By equalization of the rocket heights above 1 cm it was calculated that 1 ml gel bed of WGA matrix bound 3.3 µg glycophorin. Column experiments on the same matrix gave a value of 2.7 µg per ml gel, thus proving the predictability of the technique.

Dissociation constant. To characterize the binding of glycophorin to WGA further we estimated the dissociation constant (K) of the complex between WGA and glycophorin. K was calculated both for the purified glycophorin and for glycophorin in a crude Triton X-100 extract of membranes using the electrophoretic method of Bøg-Hansen and Takeo (9). Fig. 4 shows the re-

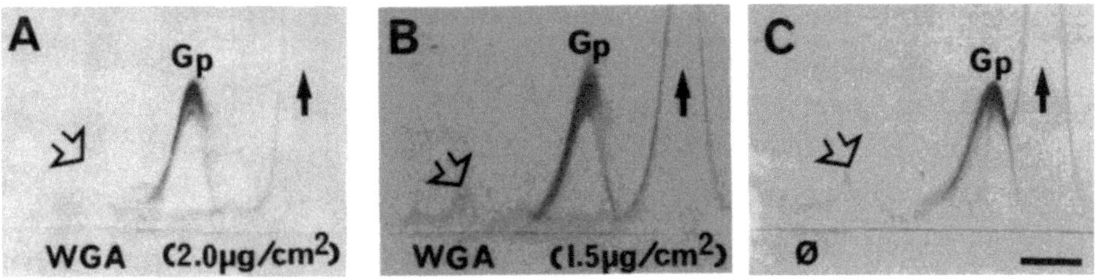

Fig. 4. Crossed affinity immunoelectrophoresis of 1 μg purified glycophorin mixed with 0.5 μg human albumin. The 1st dimension gel in (A) and (B) contains free WGA in the amounts indicated. In (C) the gel is blank (Ø). The anti-membrane antibody content of the 2nd dimension gels was 6 μl/cm^2. Filled arrows indicate the median for the albumin peak and open arrows some albumin impurities from the buffer vessels.

tardation of purified glycophorin. The migrations are expressed as the first dimension migration distances from the front edge of the application well to the projection of the median of the immunoprecipitate. In our experiments all migration distances were expressed relative to that of albumin which acted as an internal standard since in this connection it is a non reactive protein. Our results for glycophorin of both sources are plotted in Fig. 5. As it was not possible to separate them into two groups the apparent dissociation constant is calculated

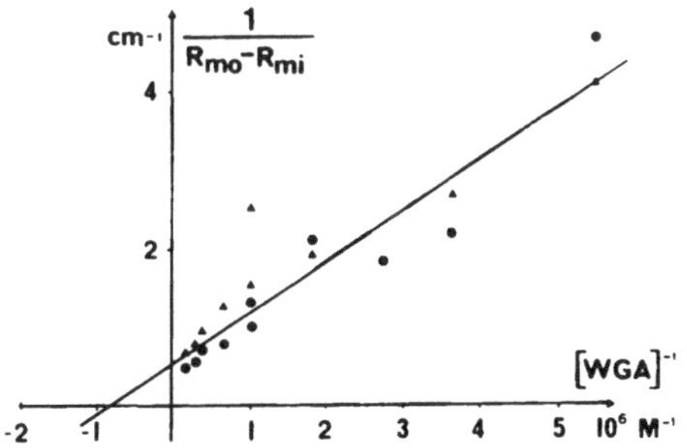

Fig. 5. The Bøg-Hansen - Takeo plot for calculation of the dissociation constant for the complex formation of WGA and glycophorin. (●—●): Purified glycophorin, (▲—▲): Glycophorin in a Triton extract of human erythrocyte membranes.

from the line including all experimental points: K_{app} = $1.28\ 10^{-6}$ M (least squares method, $R^2 = 0.93$). The published dissociation constants obtained in this way have never been correlated to values measured by other methods, so we found it desirable to verify the constant for the glycophorin-WGA complex by other means. In general it is difficult to determine binding constants between two macromolecules. Equilibrium experimentse.g. with gel filtration technique (13) or by elution analysis affinity chromatography (11) both required more material than we had acces to. Using the advantage of having the lectin in immobilized form we derived an expression for the apparent dissociation constant K_{app} based on measurements of the free glycophorin concentration after simple mixing with the WGA matrix. On the assumption of only one type of binding of a ligand (glycophorin) to an acceptor (WGA) forming the complex we have at equilibrium:

$$K_{app} = \frac{b\ a}{c}, \qquad (1)$$

where a, b, and c are the equlibrium concentrations of free acceptor sites, free ligand, and formed complex, respectively.

The total concentration B of glycophorin is:

$B = b + c = c(1 + b/c)$, the total concentration of WGA (2)

is: $A = a + c$ (3)

At conditions with the same B but different concentrations of A (A_1 and A_2), where

$$b_1/(B - b_1) = b_1/c_1 = p \qquad (4)$$

$$b_2/(B - b_2) = b_2/c_2 = r \qquad (5)$$

it follows from Eqns. 2 and 3 that:

$$s = A_1/A_2 = (a_1 + c_1)/(a_2 + c_2) = \frac{a_1 + B/(1 + p)}{a_2 + B/(1 + r)} \qquad (6)$$

and from Eqn. 1 that:

$$K_{app} = p\ a_1 = r\ a_2 \Leftrightarrow a_2 = (p/r)\ a_1 \qquad (7)$$

Eqns. 6 and 7 then give:

$$K_{app} = B\ \frac{1/(1 + p) - s/(1 + r)}{s\ p/r - 1}, \qquad (8)$$

which means that Kapp can be calculated from a chosen ratio of acceptor concentrations, the total ligand concentration B, and the measured quantities after equilibrium b_1/c_1 and b_2/c_2. Addition of increasing amounts of WGA matrix in a 5-fold range at a glycophorin concentration of 1.3 10^{-8} M resulted in a successive reduction of the rocket heights to about 60 % of the original value corresponding to 1.0 10^{-8} M. However at the actual concentrations the tip of the rockets were not well-defined (blurred). When the concentrations of free glycophorin within this uncertainty were inserted in Eqn. 1 K_{app} values ranging from 10^{-6} to 10^{-8} were found but also negative values could be obtained. In addition the molar concentrations were relatively uncertain due to the heterogeneity in the molecular weight of glycophorin. Thus it can be concluded that the derived equations demand very precise measurements, i.e. well-defined rockets and a molecular homogenous ligand. As the present system cannot fulfil these requirements the range given above should only be considered a rough delineation of the true K value. At least no disagreement between these data and the values obtained using free lectin is seen.

We thank Miss Pia H. Olsen for skilful technical help, dr. S. Blirup-Jensen for advice concerning the dry weight measurements, and Harboefonden, Lundbeck fonden, and Danish Medical Research Council (grant no. 12-3228) for financial support.

References

1. Axelsen,N.H., Krøll,J., and Weeke,B. (eds.) A manual of quantitative immunoelectrophoresis. Universitetsforlaget, Oslo (1973).

2. Bjerrum,O.J., and Bøg-Hansen,T.C. in Maddy, A.H. (ed.) Biochemical analysis of Membranes. Chapman and Hall, London, 378-426, (1976).
3. Bjerrum,O.J., Ramlau,J., Bock,E., and Bøg-Hansen,T.C. in Jacobs,S.J. and Cuatrecasas,P. (eds.) Membrane receptors: Methods for Purification and Characterization. Chapman and Hall, London, 116-156, (1980).
4. Bjerrum,O.J., Bjerrum,P.J., Larsen,K.P., Norrild,B., and Bhakdi,S. in Bjerrum,O.J. (ed.) Electroimmunochemical Analysis of Membrane Proteins. Elsevier, Amsterdam. (In press), (1983).
5. Bøg-Hansen,T.C.: Anal.Biochem. 56, 480-488 (1973).
6. Bøg-Hansen,T.C., Bjerrum,O.J., and Ramlau,J.: Scand.J. Immunol. 4, suppl. 2, 141-147 (1975).
7. Bøg-Hansen,T.C. in Scouten,W.H. (ed.) Solid phase Biochemistry - Analytical and Synthetic Aspects. Wiley and Sons, New York. (in press), (1982).
8. Bøg-Hansen,T.C.: Prot.Biol.Fluids 27, 659-664 (1980).
9. Bøg-Hansen,T.C., and Takeo,K.: J.Electrophoresis 1, 67-71 (1980).
10. Bøg-Hansen,T.C., Jensen,P., Hinnerfeldt,F., and Takeo,K., in Bøg-Hansen,T.C. (ed.) Lectins - Biology, Biochemistry, Clinical Biochemistry, vol. 1, Wlater de Gruyter, Berlin, New York, 241-258 (1981).
11. Dunn,B.M., and Chaiken,I.M.: Biochemistry 14, 11, 2343-2349 (1975).
12. Horejsí,V., and Kocourek,J.: Methods Enzymol. 34, 178-181 (1974).
13. Hummel,J.P., and Dreyer,W.J.: Biochim.Biophys.Acta 63, 530-532 (1962).
14. Krøll,J., and Andersen,M.M.: J.Immunol.Methods 9, 141-146 (1975).
15. Marchesi,V.T.: Methods Enzymol. 28, 252-254 (1972).
16. Plesner,T., and Bjerrum,O.J.: Scand.J.Immunol. 11, 341-351 (1980).

DISSOCIATION CONSTANTS FOR CONCANAVALIN A AT VARIOUS PH VALUES STUDIED BY AFFINITY ELECTROPHORESIS

An Approach to Standardization of Affinity Electrophoresis for Concanavalin A - Carbohydrate Interaction.

Kazusuke Takeo, Masanori Fujimoto and Akira Kuwahara
Department of Biochemistry, Yamaguchi University School of Medicine, 755-Ube, Japan

Disc gel electrophoresis of concanavalin A (Con A) has been carried out in an acidic medium(1, 2), mainly using the discontinuous buffer system described by Reisfeld et al.(3). Using this buffer system, we (4) studied the interactions between Con A and various polysaccharides and oligosaccharides. The results obtained were highly reproducible. Using this method we (5) calculated the dissociation constants and the thermodynamic parameters of those Con A - carbohydrate interactions.

Tichá et al. (6) already reported association constants of Con A to non-cross-linked mannosyl polyacrylamide using discontinuous buffer systems at acidic pH values. While a discontinuous buffer system is superior to a continuous buffer system to gain a sharp band, the choice of pH value is quite limited. Furthermore, it is impossible to perform electrophoresis at any constant concentration of metal ions. Thus, we have developed a continuous buffer system for disc gel electrophoresis over broad pH range. Using this buffer system, we have investigated the effects of pH, metal ion concentrations, and duration of electrophoresis on the interaction between Con A and dextran. From the results obtained, we have recommended a standard affinity electrophoresis for Con A - carbohydrate interactions.

Method

Electrophoresis was carried out with a continuous buffer system at 20°C with a thermostatic electrophoresis apparatus(5). Stock buffer solutions were prepared mixing 2M of the stock acid and base solutions and H_2O as seen in

Table 1. Stock buffer solutions for the continuous buffer system.

pH	HCl (2M)	HAc (2M)	β-alanine (2M)	Imidazole (2M)	Tris (2M)	Diethanolamine (2M)	H_2O
	ml	ml	ml	ml	ml	ml	ml
3.5	100		200				300
3.7	100		250				250
3.8	100		300				200
4.3		300			100		200
4.9		150			100		350
6.8		100		150			350
7.1		100		200			300
7.9		100			150		350
8.5		100			300		200
9.3		100				300	200

Table 1. The stock buffer solutions were diluted 20 times to the final concentration and used as the working buffer solution for the electrode vessels and the electrophoresis gel buffers. $Ca(Ac)_2$ and $Mn(Ac)_2$ were added to the working buffer solutions.

Electrophoresis gels were prepared 5 cm in height using a 6.0% polyacrylamide gel for Con A and a 7.5% polyacrylamide gel for s-Con A by photopolymerization. For determination of the dissociation constants(K_d), one set of electrophoresis was run with 12 tubes, duplicated gels of 6 different concentrations of dextran (0, 0,05, 0,10, 0,15, 0,20, and 0,25%).

The Con A (1 mg per ml) solution of 20 mM NaCl was diluted with the working buffer solution containing 1 mM $Ca(Ac)_2$ and $Mn(Ac)_2$, 8% sucrose and methylene blue and 25 µl (1.25 µg) was applied directly on the top of the electrophoresis gel. A succinyl concanavalin A (s-Con A) solution was prepared similarly except adding BPB instead of methylene blue.

The electrophoresis was carried out with varying currents at 3, 2, 1, and 0.5 mA per tube until methylene blue or BPB has migrated 4.5 cm from the origin in the electro-

Fig. 1. Affinity electrophoresis patterns of Con A at various pH values.

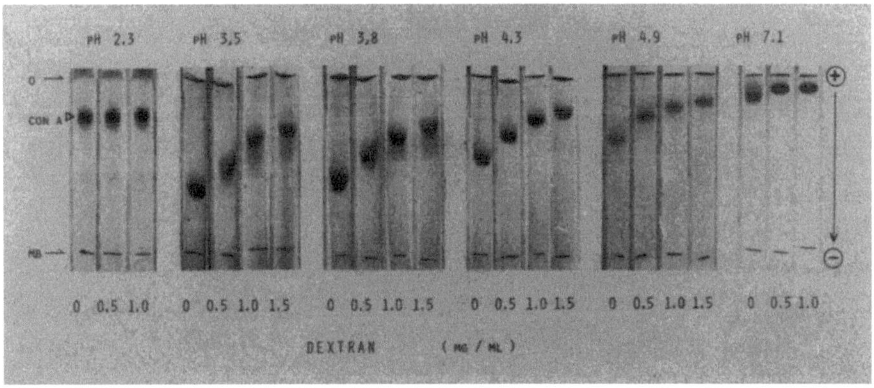

phoresis gels. Electrophoresis at pH 2.3 was performed using a discontinuous buffer system according to the method described by Shepherd and Gurley (7). Protein fractions were visualized by Coomassie BBR-250-staining. The relative migration distance is the ratio of the distance of migration of the Con A or s-Con A to that of the dye.

Results and Discussion

1) Electrophoresis patterns of Con A at various pH values.

In Fig. 1, the affinity electrophoresis patterns obtained in the presence of 0.1 mM of the metal ions at a current of 2 mA per tube were demonstrated. Below pH 3.8, electrophoresis patterns of Con A became diffuse and in the presence of dextran, a new Con A fraction having a low affinity appeared just ahead of the main band. At pH 2.3, Con A lost all affinity to dextran. At pH 4.3 and 4.9, Con A migrated as a sharp band in the presence of dextran. In the absence of dextran a faint tailing pattern appeared in the gel at pH 4.9. At pH 7.1, Con A migrated slowly, but still has a strong affinity to dextran. Those results coincided well with those reported by Tichá et al.(6).

2) Electrophoresis patterns of s-Con A at various pH values.

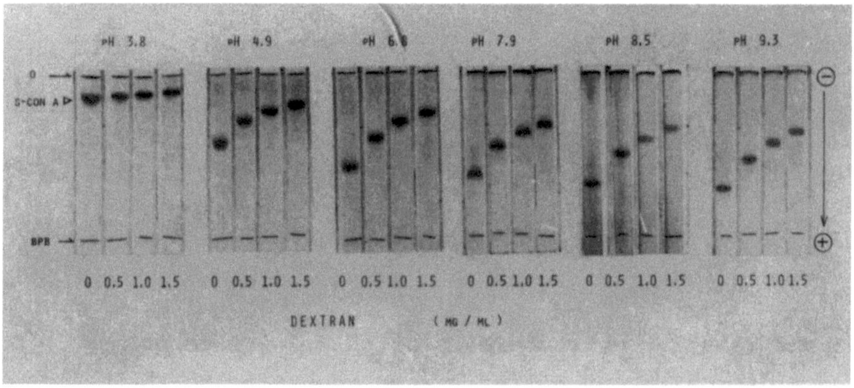

The pI value of s-Con A was nearly pH 3. As seen in Fig. 2, at pH 3.8, it migrated very slowly showing only slight retardation in the gel containing dextran. At higher pH values its mobility was increased and at pH 7.9 or 8.5, the relative migration distance in the gel containing no dextran was approximately 0.7. At the neutral pH values, s-Con A migrated as a sharp band.

3) Dissociation constants of Con A and s-Con A at various pH values.

Dissociation constants(K_d) of Con A and s-Con A were

calculated in the presence of 0.1 mM of Ca^{2+} and Mn^{2+} ions according to the following equation (Eq. I)(8):

$$\frac{1}{Rm_i} = \frac{1}{Rm_o} \left(1 + \frac{c}{K_d} \right) \text{.......(I)},$$

where Rm_o and Rm_i are the relative migration distances of Con A or s-Con A in the absence or the presence of dextran. c was the concentration of dextran in the electrophoresis gels. In Fig. 3, the K_d - pH pattern is illustrated. The molar concentration of dextran was calculated based on the assumption that 1 mg/ml of dextran is equivalent to 0.404 mM(9). K_d values for Con A were practically constant thr-

Fig. 3. K_d - pH curves for Con A and s-Con A.

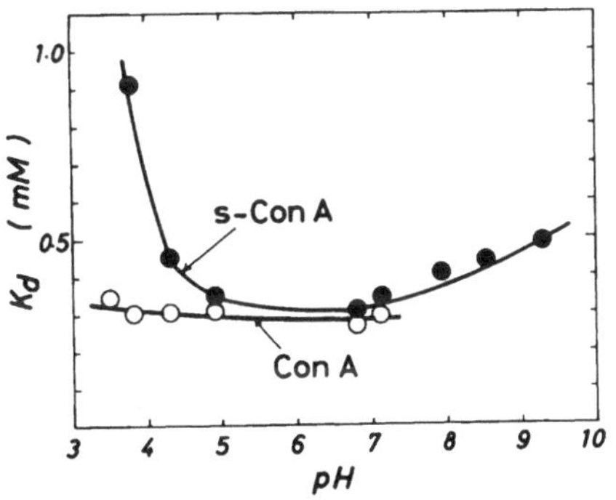

oughout the pH range from 3.5 to 7.1. K_d values for s-Con A also remained constant in the pH range from 4.9 to 7.1. They were practically equal to those for Con A. When pH values were outside this pH range, K_d values increased progressively.

At neutral pH, Con A exists as a tetrameric structure, whereas at pH values below 5.5, a dimeric structure predominates(10). On the other hand, s-Con A exists as a dimeric structure throughout both neutral and slightly acidic pH ranges(11). From the K_d values obtained, it can be concluded that either the transformation of Con A from a tetrameric to a dimeric structure or succinylation of Con A has no effect on its binding affinity to carbohydrate.

4) Effect of metal ions on the Con A - carbohydrate interactions.

Fig. 4A shows K_d values of Con A against the variation in the length for electrophoresis obtained at 4 different electric currents of 0.5, 1.0, 2.0, and 3.0 mA per tube. Electrophoresis was carried out at pH 3.8 containing 0, 0.1, and 2 mM Ca^{2+} and Mn^{2+}. The K_d - duration curves obtained at 0.1 mM and 2 mM of the metal ions were almost identical.

Fig. 4. Effects of duration of electrophoresis and metal ions on K_d values of Con A and s-Con A.

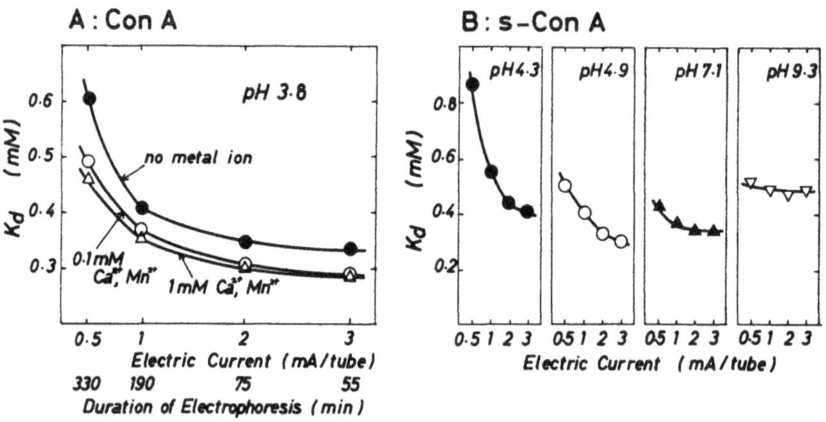

Within 1.5 hours of electrophoresis or at currents over 2 mA per tube, K_d values remained almost constant. When electrophoresis was further prolonged with low currents (0.5 or 1 mA per tube), K_d values increased. On the other hand, when electrophoresis was carried out in the absence of the metal ions, K_d values at any duration of electrophoresis became larger than those obtained in the presence of the metal ions and they were progressively increased when the duration of electrophoresis was prolonged. The parallel results were obtained at pH 4.9. Thus, the prolonged elec-

trophoresis accelerated the loss of the metal ions from Con A molecules resulting in an increase in K_d values. The larger K_d values for Con A to dextran, reported previsouly using Reisfeld's system(3), (0.55 mM)(5) indicate the dissociation of the metal ions from Con A molecules based on the stacking mechanism in the discontinuous buffer system. Denaturation of dextran molecules in the gel during the prolonged exposure at acidic pH could be another reason for the increase in K_d values.

Fig. 4B shows the K_d - duration curves for s-Con A at pH 4.3, 4.9, 7.1 and 9.3 in the presence of 0.1 mM of Ca^{2+} and Mn^{2+}. While at pH 4.3 and 4.9, K_d values were sharply increased during prolonged electrophoresis over 4 hours, at pH 7.1 or pH 9.3, K_d values of s-Con A remained almost constant for durations of electrophoresis over 5 hours. At neutral and slightly alkaline pH values, the affinity of the metal ions to s-Con A appeared to be very high.

The recommeded condition of affinity electrophoresis for Con A.

From the results obtained the following conditions can be recommended for the affinity electrophoresis of Con A or s-Con A.

For Con A: A continuous buffer system at pH 4.3 containing 0.1 mM $Ca(Ac)_2$ and $Mn(Ac)_2$ with a 6.0% polyacrylamide gel using a current of 2 mA per tube. For s-Con A: A continuous buffer system at pH 7.1 or 8.5 containing 0.1 mM $Ca(Ac)_2$ and $Mn(Ac)_2$ with a 7.5% polyacrylamide gel using a current of 2mA per tube.

References

1. Olson, M. O. J. and Liener, I. E.: Biochemistry 6, 105-111 (1967).
2. Hayes, C. H. and Goldstein, I. J.: J. Biol. Chem. 249,

1904-1917 (1974).

3. Reisfeld, R. A., Lewis, U. J., and Williams, D. E.: Nature 201, 281-283 (1962).

4. Takeo, K., Fujimoto, M., Suzuno, R., and Kuwahara, A.: The Physico-Chem. Biol. 22, 139-144 (1978).

5. Takeo, K.: Lectins--Biol. Biochem. Clin. Biochem. 2, 583-594 (1982).

6. Tichà, H. M., Horejsi, V., and Kocourek, J.: Biochim. Biophys. Acta 583, 103-109 (1979).

7. Shepherd, G. R. and Gurley, L. R.: Anal. Biochem. 14, 356-363 (1966).

8. Takeo, K. and Nakamura, S.: Arch. Biochem. Biophys. 153, 1-7 (1972).

9. Takeo, K. and Kabat, E. A.: J. Immunol. 121, 2305-2310 (1978).

10. Kalb, A. J. and Lustig, A.: Biochim. Biophys. Acta 168, 366-367 (1968).

11. Gunther, G. R., Wang, J. L., Yahara, I., Cunningham, B. C., and Edelman, G. M.: Proc. Natl. Acad. Sci. USA 70, 1012-1016 (1973).

RAPID METHODS FOR THE DEMONSTRATION OF SUGAR RESIDUES IN TISSUE EXTRACTS, FLUIDS AND LECTINS IN PLANT EXTRACTS

J.Kohn

Department of Biochemistry, University of Surrey, Guildford
and Ludwig Institute for Cancer Research (London Branch),
Royal Marsden Hospital, Sutton, Surrey, U.K.

J. Raymond

Royal Marsden Hospital, Sutton, Surrey, U.K.

A.Voller and P.Turp

Nuffield Institute of Comparative Medicine, The Zoological
Society of London, Regent's Park, London, U.K.

The purpose of this communication is to discuss methods designed for the demonstration of sugar residues in tissue and cell extracts and their identification based on their affinity for lectins. The techniques described here are based on precipitation reactions on solid substrates (1-4), affinity fixation (5,6) and a novel approach that introduces the enzyme linked immunosorbent assay, ELISA, principle (7) to the study of lectin-sugar interaction.

These techniques are relatively simple and rapid and they have been shown to be useful in preliminary studies of the sugar composition of various biological materials such as body fluids and tissue or cell extracts.

They are also very helpful in monitoring the separation products obtained from affinity columns. Previous studies have shown that these methods can be used in studies involving serum glycoproteins (4,5,8).

Materials and Methods

Commercially available lectins (Miles, Pharmacia) were used, as welll as those made available from the Chester Beatty Research Institute Kindness of Dr. A. Forester). Conjugated lectins were obtained from MEDAC, Hamburg, kindly presented by dr. Mai.

Tissue and cell extracts were prepared by conventional techniques using mechanical and ultrasonic homogenisers followed by high speed centrifugation and/or filtration in order to obtain clear, non-viscous materials. pH was adjusted by addition of appropriate buffers and if necessary ionic constituents such as Ca, Mn and Mg have been added.

Plant extracts were prepared in a similar manner sometimes if the lectin was heat stable with heating. This proved to be a useful isolation and purification procedure, by removing certain inhibitory or contaminating substances, e.g. haemolysing agents. Albumin, ethyl ether, polyvinyl-polypyrrolidone and caffeine have been used to remove tannins and phenolic substances. The pH was adjusted to neutrality and Ca, Mn and Mg have been added to the extractins solutions if required.

Standards and reference materials. In our initial experiments pooled human serum has been used as a source of glycoproteins and for the construction of a calibration curve (5,8). In later studies, substances of known sugar composition such as fetuin, thyroglobulin have also been employed. We are now engaged in selecting the most suitable reference materials which may not only assist in detecting and measuring the sugar residues present but also provide useful indication as to their structure. Known positive and negative controls are introduced into each series of experiments.

Countercurrent electrophoresis is performed predominantly on cellulose acetate (1,3,8). As a rule sheets of cellulose acetate 150 and 75 mm wide are used. The procedure is economical and allows multiple applications thereby ensuring comparable conditions in the electrophoretic run. The samples studied and the lectin solutions are placed in 'wells' pressed into wet, buffer-impregnated cellulose acetate by means of a rounded rod using a perspex template to provide the desired application pattern. The wells may be circular or in the shape of short troughs. The buffer impregnated cellulose acetate is lightly blotted and placed in position on the bridge of the electrophoretic chamber.

The sample volumes applied to the wells (2-5 μl, usually 3 μl) can be adjusted depending on the sample concentration, the lectin and type of material studies. The concentration of the lectin solution was between 0.1 and 1% depending on the degree of affinity and the type of the lectin used for the glycoconjugates studied. At the beginning of the experiment, one of the samples, the more 'precious' one, is placed in a central position and the other at both the cathode and anode ends at a distance of 8-12 mm from each other. Using 0.5 M diethyl barbiturate buffer, pH 8.6, at a constant current setting of 0.5 mA/cm (approximately 180-200 V) the electrophoretic run takes 40 min. After electrophoresis the cellulose acetate is washed in 3% NaCl in water containing a trace of Tween 80 for about one hour and then stained in a weak Nigrosin solution (0.01%) in 5% acetic acid. Destaining is performed in 5% acetic acid or in tap water. The dried cellulose acetate is inspected against a strong source of light for the presence of precipitation lines, which become clearly visible after destaining (Fig. 1). The technical details reported here have been applied in our studies but they are not necessarily the most suitable ones for all purposes and have rather been quoted as an example. The use of peroxidaseconjugated lectins has been shown to increase the sensitivity of this method.

Affinity fixation technique (affinity electrophoresis).

This method is based on the same principle as the immunofixation technique. Cellulose acetate proved to be a very useful substrate for the technique (10) and has been used for the demonstration of impurities in lectin preparatior (Fig. 2). Affinity fixation has also been used for studying the interaction of lectins with electrophoretically separated glycoconjugates.

The addition of polyethylene glycol 6000 to the lectin solutions at a final concentration of about 3% enhances the sensitivity of all techniques based on precipitation reactions.

Solid phase lectin sandwich enzymoassay for the measurement of lectin binding glycoconjugates. This technique is a variant of ELISA (7) and is based on lectin coated microplates which capture the required analyte which is then measured by an enzyme labelled lectin followed by the enzyme substrate. The test conditions were as follows: Polystyrene microplates (Dynatech Labs) were coated with 200 µl per well of the lectin at 10 µg/ml in carbonate-bicarbonate buffer (0.05 M pH 9.6). The plates were incubated at 4°C overnight and were then washed 3 times in phosphate buffered saline (1.5 M pH 7.4) containing 0.5 ml Tween 20 per litre (PBS Tween). The plates were then rinsed and dried and stored, in foild bags containing silica gel at 4°C until used.

Conjugates. The lectins were conjugated with horse-radish peroxidase by the one-step glutaraldehyde method (11). They were stored in the concentrated form at 4°C. The optimal dilution for use of each conjugate was determined by titration using a reference analyte. The enzyme substrate was orthophenylene-diamine solution. This was prepared shortly before use as follows: 40 mg orthophenylene-diamine was added to 100 ml citrate phosphate (0.1 M pH 5.0) and 40 µl 30% H_2O_2 added immediately prior to use.

The test:

1. 200 µl test sample was diluted 1/200 with PBS Tween and added to duplicate wells incubated 2 h at room temperature. Plates were washed 3 times with PBS Tween.
2. 200 µl lectin conjugate was added to each well and incubated for 3 h at room temperature. The plates were washed 3 times in PBS Tween.
3. 200 µl of fresh substrate solution was added to each well and incubated for 30 min at room temperature. Enzyme activity was stopped by the addition of 50 µl 2.5 M HCl to each

well. Absorbance of each well was read at 492 nm and a calibration curve prepared using a suitable standard, in our experiments a human normal serum pool (Fig. 3). The results are expressed in arbitrary units.

Results and Discussion

The techniques described yielded equally satisfactory results with all the pure lectins tested: con A, ricin, abrin, Arachis hypogea agglutinin, soyabean agglutinin, Phaseolus vulgaris agglutinin, wheat germ agglutinin, and phytohaemagglutinin.

So far the main use of the technique presented was in demonstra-

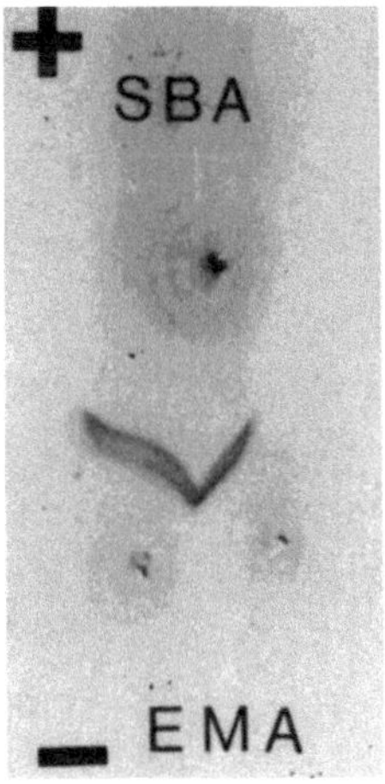

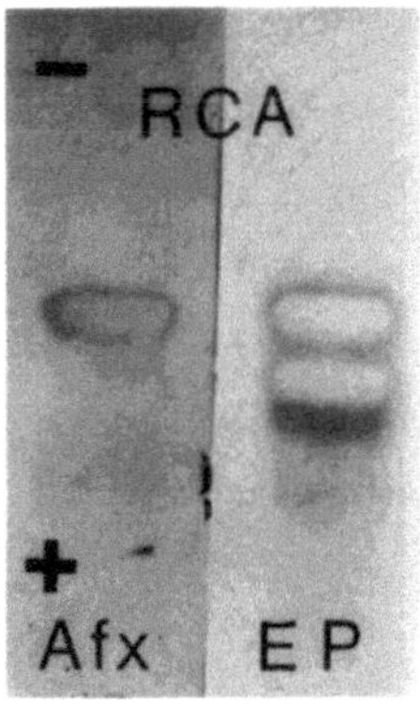

Figure 1. Demonstration of sugar residues, in this case N-acetyl galactosamine in a tissue antigen, using countercurrent electrophoresis on cellulose acetate. 0.05M barbitone buffer pH 8.6, 40 mins. run on constant 0.4 mA/cm width. Nigrosin stain. EMA - Epithelial Membrane Antigen (isolated from human milk). SBA - Soya Bean Agglutinin. Note confluence of precipitation lines between the identical samples differing only by volume applied.

Figure 2. Demonstration of impurities in a lectin (RCA) preparation by affinity (lectin) fixation on cellulose a etate. EP-electrophoresis only. AFX - affinity fixation. Human serum impregnated cellulose acetate as a source of glycoconjugates placed on top of electrophoretic pattern for 10 mins. This is followed by washing and Nigrosin staining (9). Note four bands on EP but only one on AFX, indicating presence of lectin. Central clearing zone in band due to lectin excess.

tion of the presence and type of sugar structures in biological materials and in checking results of specific deglycosylation procedures intended to ascertain the nature of some membrane antigens. Satisfactory results have been obtained in identifying the sugar residues present in the epithelial membrane antigen isolated from human milk and of considerable interest in immunocytochemistry (Fig.4). The results obtined were checked against conventional techniques used for the study of glycoprotein structures and good agreement has been obtained. The advantage of using lectins is rapidity and simplicity of the procedures.

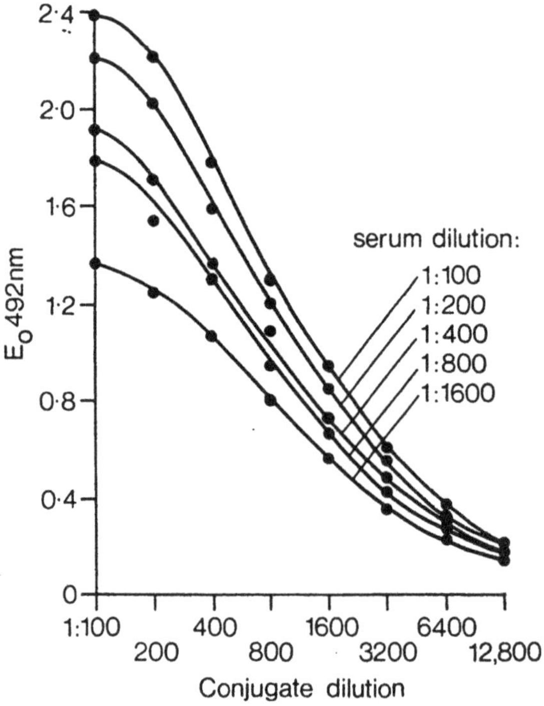

Figure 3. Solid phase lectin sandwich enzymoassay for the demonstration and quantititation of lectin binding substances. Pooled human serum was used as a source of glycoproteins (the reference analyte) for the construction of the titration curve as an example of the procedure. Note the high dilution of sample (serum) and lectin (ricin-peroxidase conjugate) used.

Using the countercurrent electrophoresis technique on cellulose actetate the presence of lectin has been demonstrated in a variety of plant extracts, e.g. Cotoneaster horizontalis (Fig. 5).

As a model system, the sugar affinities of con A have been studied by inhibition techniques based on countercurrent electrophoresis (Fig. 6) and the results were confirmed by conventional haemagglutination procedures.

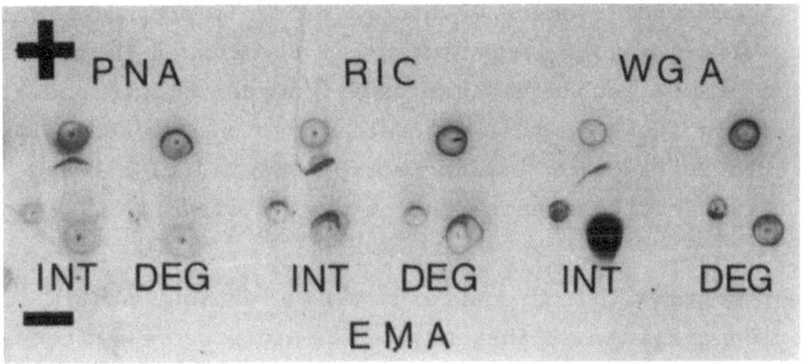

Figure 4. Demonstration of effect of deglycosylation on tissue extract, in this case epithelial membrane antigen - EMA (from human milk). Countercurrent electrophoresis on cellulose acetate. INT - intact molecules. Lectin on anode side, EMA on cathode. Different sample volumes to produce variable reactant ratios to reduce prozone phenomenon. Note ppt. line only with larger sample.

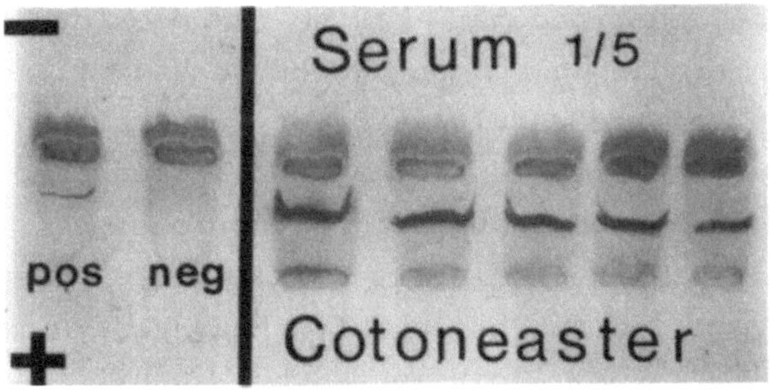

Figure 5. Detection of lectin in crude plant extract - Cotoneaster horizontalis. Increasing dilutions - neat to 1/16 on anode side. Human serum as source of glycoconjugates of cathode. Neg - negative control: diluted serum albumin. Pos - control: Con A - human serum. Inhibition studies showed a galactose affinity.

The main advantage of the countercurrent electrophoretic technique is speed, simplicity and increased sensitivity compared with the more commonly used double diffusion techniques in which the reactant molecules spread out by slow passive diffusion. In countercurrent electrophoresis the molecules are pushed towards each other simultaneously by the electrophoretic 'propulsion' (9). However, it must be stressed that the precipitation techniques are less sensitive than the immunocytochemical methods or the ELISA techniques. Higher concentration of sample material and lectin are required for countercurrent electrophoresis (of the order of 1 mg/ml as compared with 1 µg/ml when immunocytochemical or ELISA techniques are used). Other electrophoretic techniques are shown in this proceedings volume, lectin blotting (12), lectin interaction immunoelectrophoresis (13-15) and lectin affinity electrophoresis (16), a review of which will appear as (17).

Negative results with the countercurrent technique should not be taken for conclusive, as they may be due to prozone effects (glycoconjugate or lectin excess), insufficient concentration of the reactants or low degree of affinity. Another possible problem is that in some instances both reactants may have the same electrophoretic mobility and may therefore not be able to meet each other during the electrophoretic run. Therefore it may be advisable to perform experiments at different pH values.

The ELISA method described here has so far been used for the assay of lectin binding glycoconjugates in body fluids, such as serum or plasma and the results are very encouraging. As in previous studies in which nephelometry has been used it was found that levels of lectin binding glycoproteins in serum, closely correlated with the levels of the acute phase proteins in serum (5,8). It is suggested that these assays may be particularly useful in animal experiments when specific antisera may not be readily available. Preliminary studies in experimental animals with adjuvant arthritis and their response to treatment appear to be promising. Satisfactory results were also obtained

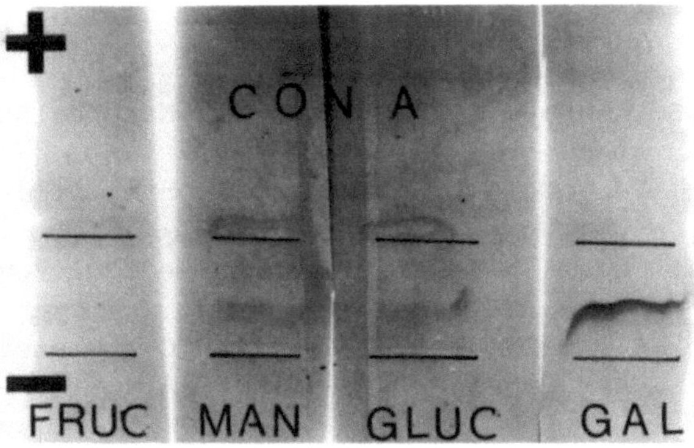

Figure 6. Inhibition studies using countercurrent electrophoresis technique. Human serum as source of glycoconjugates. Con A + sugars in cathodic troughs. Con A in anodic troughs. Control - Con A to which galactose has been added - note strong precipitation line as galactose does not inhibit Con A activity. No precipitation line where inhibiting sugars added to Con A.

in detecting and followed the course of inflammatory and infective diseases in domestic animals (8). Studies are in progress regarding the use of the lectin techniques in sick wild animals in the zoo. Countercurrent electrophoresis, affinity fixation and ELISA in reverse could be used for the demonstration and identification of lectins in crude plant extracts (Fig. 5).

The procedures described could be useful for:

1. Detection of and preliminary characterisation studies on complex carbohydrates.
2. As preliminaries to affinity column studies to indicate the type of lectin to be used for a particular application.
3. Quantitative assay of substances with a specific affinity can be performed easily using the ELISA technique.
4. Inhibition studies using simple oligosaccharides and countercurrent electrophoresis or ELISA techniques.
5. Detection and characterisation of plant and animal lectins.
6. Study of animal and plant taxonomy.
7. Assay and study of patterns of serum glycoproteins in animals.

References

1. Nakamura,S., Tanaka,K., Murakawa,S.: Nature 188 144-145 (1980).
2. Goldstein,I.J., So,L.L.: Biochem.Biophys. 111, 407-414(1965)
3. Kohn, J.: Chromatographic and Electrophoretic Techniques, Vol. 2 (4th ed.), William Heinemann Medical Books Ltd., London (1976) pp. 90-137.
4. Uhlenbruck,G., Baldo,B.A., Schwick,H.G., Chaterjee,B.P., Horejsi,V., Krajhanzl,A., Kocourek,J.: J.Clin.Chem.Clin. Biochem. 16, 19-23 (1978).
5. Warren,C., Whicher,J.T., Kohn,J,: J.Immunol.Methods 32, 141-150 (1980).
6. Horejsi,V.: Analytical Biochem. 112, 1-8 (1981).
7. Voller,A., Bidwell,D.E., Bartlett,A.: The Enzyme Linked Immunosorbent Assay, Dynatec, Europe (1979).
8. Kohn.J., Whicher,J.T., Warren,C., O'Kelly,T.: FEBS Lett. 109, 257-261 (1980).
9. Kohn,J., Kahan,M.J.: J.Immunol.Methods 11, 303-309 (1976).
10. Kohn,J., Riches,P.G.: J.Immunol.Methods 20, 325-331 (1978).
11. Avrameas,S.: Immunocytochemistry 6, 43 (1969).
12. Groeger, B. et al. These proceedings, pp. 179
13. Andersen, M.M., Noack, S. These proceedings, pp. 361
14. Lorenc-Kubis, I. These proceedings, pp. 379
15. Heegaard, N.P.P. et al. These proceedings, pp. 387
16. Takeo, K. These proceedings, pp. 394
17. Bøg-Hansen, T.C., in SOLID PHASE BIOCHEMISTRY. ANALYTICAL AND SYNTHETIC ASPECTS. Ed. by W.H. SCOUTEN, J. Wiley and sons, Inc. New York, in press, 1982.

HAEMADSORPTION LECTIN TEST (HALT) : A SIMPLE, SENSITIVE AND FAST SOLID PHASE METHOD FOR STUDYING LECTINS.

J.A. KINT, Pediatric Department, University Hospital,
Rijksuniversiteit Gent
B-9000 Gent, Belgium

Inhibition of agglutination is up to now the most simple way when investigating lectins. With this technique, lectins may be a useful tool in the study of glycolipids and glycoproteins. However, agglutination of red cells has a number of drawbacks which makes it often not very suitable.
The high adsorptive capacity of plastics for protein is well known. Such adsorption has been used extensively during recent years for different kinds of enzyme immunoassays.
With the present study, a new technique is offered which is based on the adsorption of lectins to plastic.

Materials and methods

Lectins were obtained from IBF through LKB Belgium. Normal microtest plates with U-form wells from several suppliers were used. All chemicals were of analytical grade (Merck, Darmstadt, FRG). Sephadex G-75 and Sepharose 4-B were obtained from Pharmacia (Uppsala, Sweden). Human red blood cells were trypsinized according to Lis and Sharon (3). They were stored at 4° C at a concentration of 50 % in PBS. Immediately before use, they were diluted to a final concentration of 0,25 %. The purification of Con A from jack bean meal (1) and of RCA from castor beans (2) was carried out according to well-established methods, using as affinity adsorbent either G75 for Con A or Sepharose 4B for the ricinus lectin. The coating buffer consisted of Tris 0,1 M pH = 7,4 supplemented with $MnCl_2$ 10^{-3} M and sodium azide 0,1 %. The HALT was carried out in two phases : a coating phase and an incubation phase.

A. Coating the plates with lectin

The lectin to be applied was dissolved in coating buffer at a concentration of at least 1 microgram per millilitre. The wells were coated with 200 µl of the lectin solution for at least 15 min. at room temperature or overnight at 4° C. The latter incubation resulted in a somewhat higher sensitivity. Plates can be stored at 4° C for at least two months without any significant change in sensitivity.

B. Incubation with cells

Immediately before the final incubation, the coating solution was discarded and the plate was washed 3 times under running tap water and shaken until dry. The wells were then filled with 150 µl of a solution containing 0,2 % of trypsinized human erythrocytes in the coating buffer, together with the component to be studied. The latter might consist either of a dissociating monosaccharide or a glycoconjugate - glycoprotein or glycolipid - whose sugar structure had to be investigated. The cells were than allowed to attain equilibrium. After an incubation period of one hour at 37° C or three hours at room temperature, the plate could be studied : a negative result was recognizable as a solid button of cells at the bottom of the well, a positive test showed a thin pink film of erythrocytes. Plates could be stored in this stage of the test for several days at room temperature without any visible change in the result. One precaution, however, should be taken : by pressing a sheet of parafilm on top of the plate, drying out is prevented.

Results

I developed a simple and sensitive method for studying lectins. It is based on the adsorption of lectin to plastic in a microtiter plate.
Eight different lectins were tested : Con A, RCA, WGA, UeA, PWM, SBA, PNA and HPA (Figure 1). They were adsorbed very well and kept their cell binding property after adsorption except PWM and PNA which did not bind cells. The HPA showed a positive HALT reaction only when blood group A cells were applied. This is in line with the specificity of HPA which is indeed a blood group A specific lectin (Figure 1).
About 100 nanograms per well are sufficient to provoke a positive HALT response. The sugar specificity was tested with twelve different mono- and disaccharides. The same sugars which are known to inhibit cell agglutination, prevented the adsorption of the cells in the coated well.
The process of adsorption was fairly independent of the pH of the coating solution. Any buffer between pH = 5,0 and pH = 9,0 was satisfactory. However, in contrast to immunoglobulins, most lectins are not very stable in alkaline solution. Therefore it was decided to use a tris buffer at pH = 7,5 as a general coating fluid. Ca^{+2} and Mn^{+2} ions were added because some lectins need them for optimal sugar binding.
The adsorption of cells in the coated well was markedly enhanced at higher temperature : at 37° C the rate of cell adsorption was about twice as great as at room temperature and more cells could be adsorbed resulting in a sharper distinction between positive and negative results.

Sensitivity of the test

The sensitivity of HALT is somewhat higher than the common agglutination test : whereas for most lectins, the minimum concentration for agglutination lies somewhere between 1 µg per ml and 3 µg per ml, the lowest concentration for HALT lies between 0,1 µg per ml and 1,0 µg per ml.

Figure 1 : Haemadsorption of 8 lectins at 11 different coating concentrations. A twofold dilution of each lectin was made, starting with 25 µg/ml in well A and terminating with 0,39 µg/ml in well G.

1 = Con A	5 = P.W.M.	A = 25 µg/ml	E = 1,56 µg/ml
2 = R.C.A.	6 = S.B.A.	B = 12,5 µg/ml	F = 0,78 µg/ml
3 = W.G.A.	7 = P.N.A.	C = 6,25 µg/ml	G = 0,39 µg/ml
4 = U.e.A.	8 = H.P.A.	D = 3,12 µg/ml	H = BUFFER ONLY

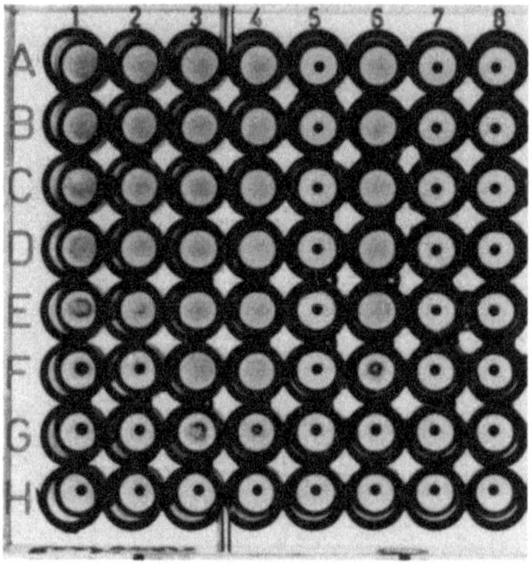

BLOOD GROUP O
CELLS

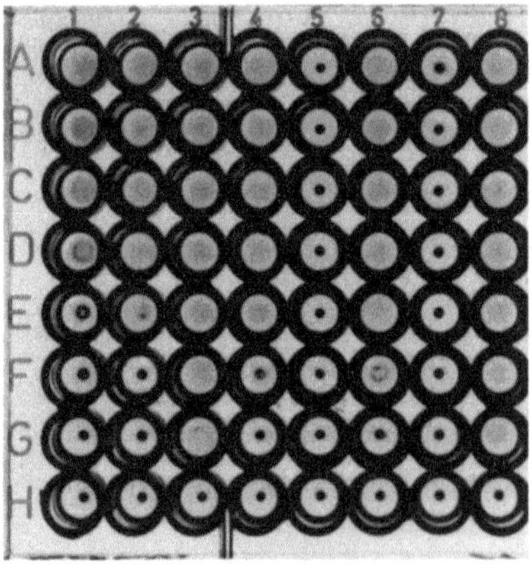

BLOOD GROUP A
CELLS

Table 1

Comparitive sensitivity of agglutination and haemadsorption

	Agglutination with 3 % cells	Haemadsorption with 0,25 % cells
Con A	1 µg/ml	0,5 µg/ml
RCA	1 µg/ml	1,0 µg/ml
WGA	10 µg/ml	0,1 µg/ml
Ulex lectin	1 µg/ml	1,0 µg/ml

Moreover, detection in HALT is always clear cut and is possible with the naked eye, whereas the interpretation of the agglutination at the limit of its sensitivity is somewhat subjective and is not possible without a microscope.

Efficiency of the lectin coating

The process of coating is extremely efficient and very fast. The process of coating needs less than 15 minutes. In fact, even after a coating period of 20 seconds, followed by extensive washing, a positive adsorption was detected. But maximum binding of lectin to the plastic was obtained after several hours. Therefore I recommend to coat the plates overnight.

Applications of HALT

In order to show the wide range of possible applications of the test, the following examples were chosen.

1. Effect of EDTA on the cell binding efficiency

Several lectins are known to need divalent ions such as Ca^{+2} or Mn^{+2} for optimal sugar recognition. This can easily be shown by the HALT. A dilution series of several lectins was allowed to be coated both in the presence and in the absence of 10 mg/ml EDTA. Therefore, two coating buffers were used :

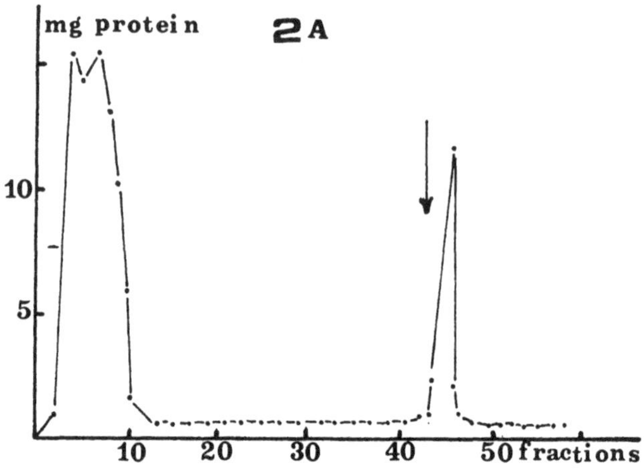

FIGURE 2
2 A = Elution profile. ARROW = specific elution with methyl α mannoside 20 mg/ml.
2 B = HALT on the undiluted fractions.
2 C = HALT on a dilution series of the fractions 40 - 51.

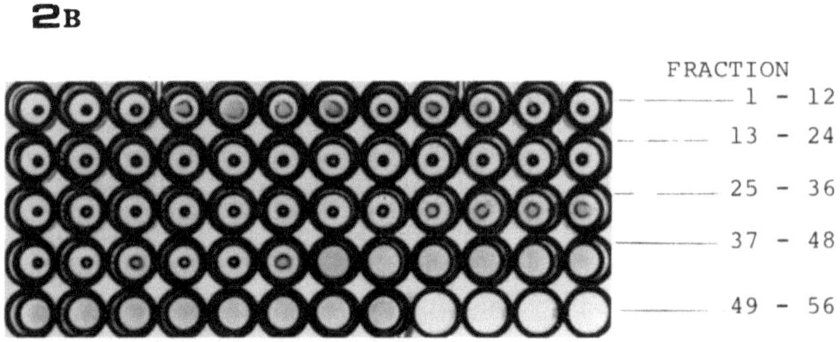

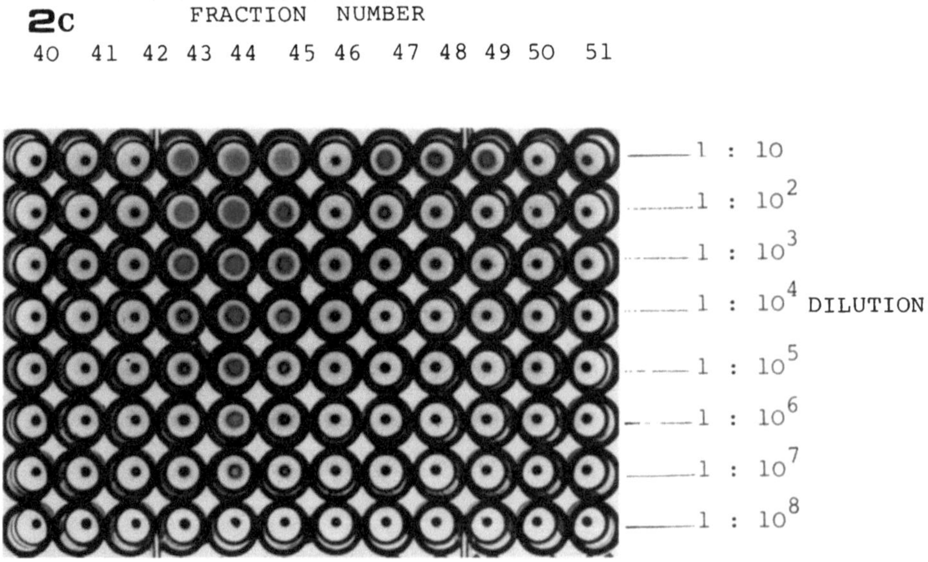

one as described in the Methods section and the other with the Ca^{+2} and Mn^{+2} replaced by EDTA 10 mg/ml.

Table 2

Haemadsorption in the presence of EDTA

Lowest detectable coating concentration	without EDTA	with EDTA
Con A	0,25 µg/ml	2,50 µg/ml
RCA	0,50 µg/ml	0,50 µg/ml
WGA	0,10 µg/ml	1,00 µg/ml
SBA	0,80 µg/ml	1,00 µg/ml
VeA	0,80 µg/ml	1,00 µg/ml

HALT was performed as described in the Methods section but with decreasing concentration of the 5 lectins. Two dilution series of each lectin were prepared in coating buffer, one with 10^{-3}M $CaCl_2$ and 10^{-3}M $MnCl_2$, the other with EDTA at 10 mg/ml.

Five lectins were tested : Con A, RCA, WGA, SBA and UeA. Two of them (Con A and WGA) showed a clearly reduced sensitivity in the presence of EDTA (see Table 2). With four other lectins, only a slight reduction in sensitive was seen when EDTA was added.

2. Detection of lectins in the eluate of affinity chromatography

Coating of the lectin proceeds independently of any sugar in the coating solution. HALT therefore is ideally adapted for detecting a lectin in the eluted fractions of affinity chromatography. Figure 2 shows the result of a chromatography of an extract of jack bean meal.
HALT could be performed immediately after the elution, without prior dialysis (Fig. 2B and 2C).
The photographs were taken two hours after the elution of the

last fraction. The whole experiment was carried out in one day. As can be seen from the figure, there is almost no haemadsorption in the main protein peak, while, after specific desorption with methyl-α-mannoside, all fractions are strongly positive for HALT (Figure 2B). By testing a dilution series of the positive fractions, one can even measure the lectin concentration (Figure 2C).

A similar experiment was carried out for RCA which was purified from castor beans, using Sepharose 4B as affinity medium and lactose as dissociating sugar. Again it was possible to monitor the fractions containing lectin immediately after elution (results not shown).

Discussion

Agglutination of red cells has some drawbacks which makes it often not very suitable for the study of lectins.

1. Its accuracy is not very high, especially at the limit of its sensitivity; a correct interpretation is sometimes difficult.
2. A microscope is needed to see the agglutinated cells.
3. Haemolytic proteins such as the complement system in serum disturb the agglutination.
4. Dissociating monosaccharides make it often impossible to use agglutination.

In the last case, prior dialysis is necessary in order to remove the inhibitory sugar. This may be time-consuming in affinity chromatography of lectins, when a large number of samples has to be analysed.

5. Monovalent lectins will not show any agglutination even if binding is optimal.

In this study I offer an alternative which does not show the shortcomings of agglutination. The present paper proves that, besides immunoglobulins, lectins also can be adsorbed in polystyrene and that after adsorption the selective sugar binding capacity is preserved.

Haemadsorption lectin test (HALT) seems to combine several features which makes it suitable as a general analytical tool for the study of lectins (sugar specificity, ion requirement, pH and temperature effects etc.) or for the study of glycoconjugates. The test combines the following advantages :

1. No instrument is needed. Only cheap microtest plates are necessary.
2. The test is extremely sensitive : 60 ng to 100 ng are sufficient, which is about two to ten times less than for the inhibition of agglutination.
3. The test is more reproducible and less subjective than the inhibition of agglutination test.
4. The presence of dissociating sugars does not interfere with the coating of the lectin. This property makes the test very suitable in affinity chromatography of lectins. No prior dialysis of the dissociating sugars is necessary.
5. Coated microtest plates can be kept at 4° C for at least three months without any change in sensitivity.
6. If needed, the incubation phase can be carried out in two phases. First, the glycoconjugates to be studied (e.g. human serum) are incubated in the precoated wells. After washing out the non-bound proteins, cells can then be added to detect the presence of lectin reactivity. By this two-way procedure, the influence of haemolytic proteins in the mixture is minimal and only the effect of molecules containing sugar is observed.

References

1. AGRAWEL B.B.L. and GOLDSTEIN I.J.
   Bioch. J. 96, 23-29, 1965.
2. CANTARERO L., BUTLER J.E. and OSBORNE J.W.
   Analytical Biochemistry 105, 375-382, 1980.
3. LIS H. and SHARON N.
   Methods Enzymol. 28, 360, 1972.

4. TOMITA M., KUROKAWA T., ONOZAKIK, ICHIKI N., OSAWA T. and UKITA T.
Experientia 28, 84-85, 1972.

5. VAN ELSEN A. and LEROY J.
Models for the study of inborn errors of metabolism, p. 329-332, 1979 - Ed. F.A. HOMMES - North Holland.

Part III

Animal Lectins

ENDOGENEOUS LECTINS ON HUMAN PERIPHERAL MONONUCLEAR LEUKOCYTES.

C. Kieda, M. Monsigny[+] and M.J. Waxdal

Laboratory of Immunology, National Institute of Allergy and Infectious Diseases, National Institutes of Health, Bethesda, Maryland 20205, USA and [+]Centre de Biophysique Moléculaire, C.N.R.S., 45045 Orléans cedex, France.

The presence of membrane lectins on murine lymphocytes has previously been described (1,2). These sugar receptors are specifically distributed on subpopulations of cells. We have also investigated their presence on human unfractionated peripheral blood mononuclear leukocytes (3). In order to further investigate the significance of the presence of such membrane lectins in the lymphoid system, we undertook to separate the cells according to the specificity of their sugar receptors. In this communication, we shall describe the use of a natural polyglucan to separate the monocytes from the lymphocytes. This polysaccharide (PWCH) was obtained from the roots of *Phytolacca americana*. The binding of this polysaccharide is characteristic of subpopulations of leukocytes, allows the agglutination of subsets of cells and their separation from the non agglutinated cells by differential sedimentation. Both populations can be recovered as free cells by reversal of the agglutination with α-methyl-D-mannopyranoside. Cytological and functional characterization of the subpopulations showed that the recognition of a polysaccharide by cells is a property of defined subpopulations with characteristic immunological behaviours.

Materials and Methods

Leukocytes were obtained by Ficoll-Hypaque density centrifugation (4) from heparinised venous blood given by normal healthy donors. T cells were identified by sheep erythrocyte rosetting (5). Monocytes and macrophages were detected by their morphology and non-specific esterase staining (6). B cells were identified with fluorescein labeled F(ab')2 fragments of anti human immunoglobulins (7).

Pokeweed Polysaccharide (PWCH). Polysaccharide preparation from the roots of *Phytolacca americana* (Pokeweed) were obtained from Potomac Biological Laboratories. This polysaccharide was characterised as an highly branched glucan containing 4 % xylose and 2 % arabinose. Pokeweed polysaccharide was shown to be free of any contamination by Pokeweed mitogens (Pa-1 to Pa-5) : Pokeweed polysaccharide did not agglutinate human red blood cells and did not stimulate human lymphocytes. Furthermore, pokeweed polysaccharide did not inhibit the agglutinating and mitogenic activities of Pa-2.

Separation of Human Peripheral Blood Leukocytes by Pokeweed Polysaccharide. This technique was previously described in detail (3). Briefly, 500 µl of a $5x10^8$ cells/ml suspension together with 50 µl of a $5x10^8$ cells/ml autologous red blood cells were allowed to agglutinate with 500 µl of a 1 mg/ml pokeweed polysaccharide solution in a culture medium (RPMI 1640). After 10 minutes at room temperature the cell suspension was layered on the top of a two step density gradient (5 % and 10 % bovine serum albumin in culture medium (RPMI 1640). The cells were sedimented at 1 g for 30 minutes, at room temperature. The upper non agglutinated ($PWCH^-$) and lower agglutinated ($PWCH^+$) fractions were collected, washed in culture medium (RPMI 1640) and the aggregates were dissociated by incubation in an isotonic solution of 0.05 M α-methyl-D-mannopyranoside. Both fractions were further fractionated by E-rosetting with sheep red blood cells (5), providing four distinct subpopulations :

$PWCH^-$ TCE = Pokeweed polysaccharide non agglutinable, T cell enriched ;
$PWCH^-$ TCD = Pokeweed polysaccharide non agglutinable, T cell depleted ;
$PWCH^+$ TCE = Pokeweed polysaccharide agglutinable, T cell enriched ;
$PWCH^+$ TCD = Pokeweed polysaccharide agglutinable, T cell depleted.

Flow Microfluorometry Analyses. For these analyses, monoclonal antibodies OKM-1 (binding to monocytes) ; OKT-3 (T cells) ; OKT-4 (helper T cells) ; OKT-8 (suppressor T cells) were obtained from Ortho Pharmaceutical Corp. ; Leu-1 (T cells) from Becton Dickinson ; 4F-2 (monocytes) a gift from Dr. A.S. Fauci. Affinity purified, fluorescein conjugated F(ab')2 goat anti-mouse immunoglobulins (goat 123) were a gift from Dr. R. Asofsky and B.J. Fowlkes and were used to evidence the mouse monoclonal antibodies specifically bound to the cell surface of cell subpopulations. Microfluorometry analyses of the cells were performed on a fluorescence activated cell sorter, FACS II (Becton Dickinson) as previously described (8).

Human Lymphocyte Stimulation. Leukocytes and each fraction obtained after pokeweed polysaccharide separation and E-rosetting were tested for their response to the pokeweed mitogen Pa-2 (9). The same assays were performed after mixing the various subpopulations to approximate their original proportion (reconstituted population). DNA synthesis in response to Pa-2 stimulation was assessed by incorporation of tritiated thymidine.

Immunoglobulin Production by B cells. The cells were stimulated by pokeweed mitogen Pa-2 (9) and immunoglobulin production was assessed after seven days of culture by a reverse plaque assay (10). In these assays, several concentrations of pokeweed mitogen Pa-2 (from 0.05 μg/ml to 1 μg/ml) were tested on leukocytes and on the various separated subpopulations, as well as on combinations of B cells with the other subpopulations added separately or together. Supernatants from $PWCH^+$ TCE ($2.5x10^6$ cells in 2.5 ml) were obtained at various times of culture. Supernatants were centrifuged, dialysed and stored at - 20°C. To check the influence of the supernatants, stimulation was achieved by adding 0.1 μg/ml Pa-2 and the supernatants were diluted four fold. In the sugar inhibition experiments, oligosaccharides and simple sugars were used at 50 mM and neoglycoproteins at 0.5 mg/ml (2).

Results

Pokeweed Polysaccharide Fractionation of Human Peripheral Blood Mononuclear Leukocytes. Agglutination and separation on the two step density gradient, two fractions were obtained (Table 1) : agglutinated cells ($PWCH^+$) and non agglutinated cells ($PWCH^-$). These data demonstrate that the pokeweed polysaccharide fractionation is not toxic to the cells because the viability and the final recovery are close to 95 %.
The $PWCH^-$ fraction is essentially devoid of esterase positive cells (monocytes). The $PWCH^+$ fraction contains approximately 70 % monocytes, the remaining 30 % are T cells which can be removed by E-rosetting ($PWCH^+$ TCE). The $PWCH^+$ non E-rosetting cells ($PWCH^+$ TCD) are predominantly esterase positive cells (monocytes). Furthermore, all these $PWCH^+$ TCD cells adhere to plastic culture plates.
The bulk of the E-rosette forming cells (T cells) are recovered from the $PWCH^-$ fraction. However, the $PWCH^+$ TCE are a subset specifically agglutinated by pokeweed polysaccharide and represent about 8 % of the human unfractionated peripheral blood mononuclear leukocytes.

TABLE 1

Pokeweed polysaccharide fractionation of human peripheral blood mononuclear leukocytes.

Cell fractions	Recovery	Viability	E-rosette	Esterase
Control	100 %	97 ± 2 %	53 ± 5 %	17 ± 2 %
PWCH$^-$	67 ± 3 %	94 ± 2 %	70 ± 4 %	< 0.1 %
PWCH$^-$ TCE	50 ± 4 %	90 ± 2 %	98 ± 3 %	< 0.1 %
PWCH$^-$ TCD	15 ± 3 %	95 ± 2 %	1 ± 1 %	< 0.1 %
PWCH$^+$	30 ± 3 %	96 ± 2 %	28 ± 2 %	70 ± 3 %
PWCH$^+$ TCE	8 ± 2 %	94 ± 2 %	98 ± 2 %	2 ± 1 %
PWCH$^+$ TCD	20 ± 2 %	96 ± 2 %	1 %	87 ± 3 %

Recovery is given as % of the original population. Viability is expressed as % of the isolated subpopulations. E-rosette expresses for a given subpopulation the % of cells forming rosettes with sheep red blood cells. Esterase expresses for an isolated subpopulation the % of esterase positive cells. Control was human unfractionated peripheral blood mononuclear leukocytes.

TABLE 2

Characterization of pokeweed polysaccharide fractionated cell populations with monoclonal anti differentiation antigens by flow cytofluorimetry.

Cell fractions	Antibodies					
	OKM-1	4F-2	OKT-3	Leu-1	OKT-4	OKT-8
Control	21	22	57	59	32	18
PWCH$^-$	1	1	67	65	43	20
PWCH$^+$	70	70	28	21	15	13
PWCH$^+$ TCE	2	2	78	ND	28	43
	Monocytes		T-cells		Helper cells	Suppressor cells

Cells were incubated in the presence of the various mouse monoclonal antibodies and then with fluorescein labeled F(ab')2 goat anti-mouse antibodies. Results are expressed as the % of positive cells related to the isolated cell subpopulations, (mean values of 23 separate experiments). Control was human unfractionated peripheral blood mononuclear leukocytes. ND : not determined.

Characterization of Pokeweed Polysaccharide Fractionated Cells with Monoclonal Antibodies by Flow Microfluorometry. Table 2 shows that pokeweed polysaccharide fractionation results in the separated monocytes (OKM-1$^+$, 4F-2$^+$) from the bulk of the lymphocytes. The PWCH$^-$ fraction is devoid of monocytes (< 0.1 %). The non monocytic cells contained in the PWCH$^+$ fraction are T cells. These T cells are enriched in suppressor cells (OKT-8$^+$).

Functional Properties of Pokeweed Polysaccharide Fractionated Cell Subpopulation. These assays were undertaken to determine if pokeweed polysaccharide fractionation activates, inhibits or destroys subpopulation functions. Measurements of the DNA synthesis induced by the pokeweed mitogen Pa-2 in the leukocytes or in the reconstituted population (see Materials and Methods) give similar dose response curves. Conversely, the separated subpopulations respond differently (data not presented). Immunoglobulin production was measured with PWCH$^-$ TCD population (B cells containing subsets) in the presence or in the absence of the other fractionated subpopulations. In all cases when PWCH$^+$ TCE cells were present, the immunoglobulin synthesis was totally suppressed. This result is illustrated in Figure 1.

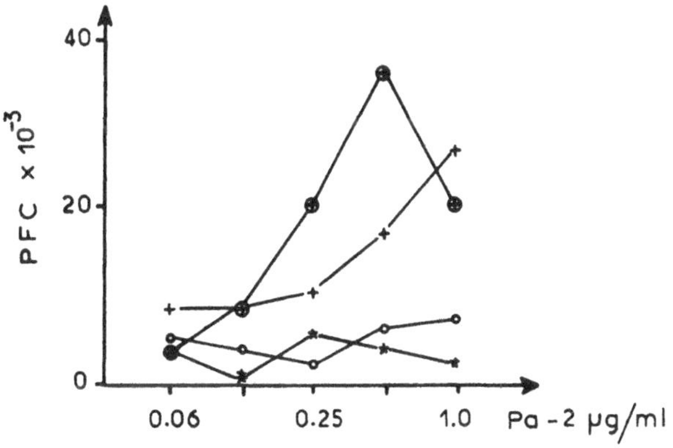

Fig. 1 - Immunoglobulin production by pokeweed polysaccharide fractionated cells expressed as the number of plaque forming cells (PFC) upon stimulation with the pokeweed mitogen Pa-2. ⊕ — ⊕ human unfractionated peripheral blood mononuclear leukocytes, + — + PWCH$^-$ TCD cells : pokeweed polysaccharide non agglutinable, T cell depleted, o — o PWCH$^-$ TCD cells + PWCH$^+$ TCE cells : pokeweed polysaccharide non agglutinable-T cell depleted cells mixed with pokeweed polysaccharide agglutinated-T cell enriched cells, * — * reconstituted population (see under Materials and Methods)

Property and Specificity of T cell enriched (PWCH$^+$ TCE) Supernatant. Culture supernatants from PWCH$^+$ TCE cells were prepared in order to determine whether these cells act *via* a cell-cell contact or *via* a soluble factor. $PWCH^+$ TCE supernatants contain a potent suppressor activity (Table 3). In control experiments, L-Rhamnose had no effect on the generation of PFC by Pa-2.

TABLE 3

Suppressor activity in supernatants of pokeweed polysaccharide agglutinated-T cell enriched cells ($PWCH^+$ TCE) expressed as the number of plaque forming cells (PFC) upon stimulation of various cell populations with pokeweed mitogen Pa-2.

Medium Adducts / Cell Population	None	$PWCH^+$ TCE Supernatant	$PWCH^+$ TCE Supernatant + L-Rhamnose
		ΔPFC	
Control	170	49	180
B cells	110	0	126
B cells + $PWCH^+$ TCD	90	9	109
B cells + $PWCH^-$ TCE	210	59	200
B cells + $PWCH^-$ TCE + $PWCH^+$ TCD	187	28	178

ΔPFC expresses the increase in the number of plaque forming cells in a cell population containing 10^5 cells due to the stimulation by pokeweed nitrogen Pa-2. Experiments are done in the absence of $PWCH^+$ TCE supernatants (first row) in the presence of $PWCH^+$ TCE supernatant (second row) and in the presence of $PWCH^+$ TCE supernatant and L-Rhamnose (third row). B cells were isolated on polystyrene tissue culture dishes coated with specifically purified rabbit anti-human immunoglobulin (11). Control was unfractionated peripheral blood mononuclear leukocytes. $PWCH^+$ TCD cells : pokeweed polysaccharide agglutinated, T cell depleted, were mainly monocytes. $PWCH^-$ TCE cells : pokeweed polysaccharide non agglutinable-T cell enriched were mainly T cells as shown by monoclonal antibodies characterization.

Conclusion

The finding of endogeneous lectins on human lymphoid cells extends the understanding of cell surface properties. Such characteristics may be used to identify and separate the cells on the basis of their sugar binding specificity. Cytological and functional characterizationsof the population described here suggest the following conclusions. Pokeweed polysaccharide non-agglutinated cells ($PWCH^-$) contains most of the T cells and all the B cells. Pokeweed polysaccharide agglutinated cells ($PWCH^+$) contains all the monocytes and about 8 % of the T cells. These T cells may be isolated by E-rosetting with sheep red blood cells. These $PWCH^+$ T cells are potent suppressors of the immunoglobulin production by B cells and act *via* a soluble suppressor factor, the activity of which is inhibited by L-Rhamnose.

References

1. Kieda, C.M., Bowles, D.J., Ravid, A. and Sharon, N., FEBS Letters 94, 391-396 (1978).
2. Kieda, C.M., Roche, A.C., Delmotte, F. and Monsigny, M., FEBS Letters 99, 329-332 (1979).
3. Kieda, C.M., Thomas, C.A., Monsigny, M. and Waxdal, M.J., submitted for publication (1982).
4. Boyum, A., Scand. J. Clin. Lab. Invest. 21, 1-9 (1968).
5. Bentwich, Z., Douglas, S.D., Siegal, F.P. and Kunkel, H.G., Clin. Immunol. Immunopathol. 1, 511-520 (1973).
6. Koski, I.R., Poplack, D.G. and Blaese, R.M., in In Vitro Methods in Cell Mediated and Tumor Immunity. (Bloom, B.R. and David, J.R., eds) Academic Press, N.Y., 359-362 (1976).
7. Johnson, G.D., Holborow, E.J. and Dorling, J., in Handbook of Experimental Immunology (Weir, D.M., ed.) Blackwell Scientific Oxford (1978).
8. Fowlkes, B.J., Waxdal, M.J., Sharrow, S.O., Thomas, C.A., Asofsky, R. and Mathieson, B.J., J. Immunol. 125, 623-628 (1980).
9. Janossy, G., Gomez de la Concha, E., Waxdal, M.J., Plasttsmill, T., Clin. Exp. Immunol. 26, 108-114 (1976).
10. Greene, W.C., Fleisher, T.A. and Waldmann, T.A., J. Immunol. 126, 1192-1197 (1981).
11. Mage, M.G., McHugh, L.L. and Rothstein, T.L., J. Immunol. Meth. 15, 47-55 (1977).

INVOLVEMENT OF A ß-GALACTOSE-SPECIFIC MACROPHAGE LECTIN IN BINDING AND PHAGOCYTOSIS OF ERYTHROCYTES

Ernst Müller, Cornelia Schröder, Marc W. Franco, Ashok K. Shukla and Roland Schauer
Biochemisches Institut, Christian-Albrechts-Universität, Olshausenstraße 40-60, D-2300 Kiel, West-Germany

Several lectins have been described with different sugar specificities, e.g. mammalian hepatic lectins specific for ß-Gal (1) and Fucα(1-3)GlcNAc groups (2), an ß-Gal-binding protein in human and rat lung (3), an avian hepatic lectin for GlcNAc (4) and an alveolar macrophage receptor for Man and GlcNAc residues (5). The biological role of lectins is not yet fully understood. However, a possible function is shown after sialidase treatment of erythrocytes of different species, which leads to the exposure of penultimate ß-Gal residues of the erythrocyte membrane and to interaction of these cells with a ß-Gal-specific lectin on liver and spleen macrophages (6), resulting - in vivo - in the rapid sequestration of the sialidase-treated erythrocytes from the blood stream (7, 8).
The aim of the present experiments was to investigate the prerequisites for binding and phagocytosis of sialidase-treated erythrocytes by homologous rat peritoneal macrophages using a new in vitro system (9).

Material and Methods

Peritoneal macrophages were isolated from adult CAP rats; for the in vivo experiments male White New Zealand rabbits were used. The experimental procedure for the in vitro and in vivo studies are described in ref. 8 and 9. Vibrio cholerae sialidase was obtained from Behringwerke (Marburg) and Arthrobacter ureafaciens sialidase from Nakarai Chemicals (Osaka). The immobilized V. cholerae sialidase was prepared by the method of Parker et al. (10). Rat IgG, rat IgM and anti-rat IgG was purchased from Paesel (Frankfurt) and guinea pig complement from Cordis (Zürich). Guinea pig complement (1 ml) was preabsorbed on 250 µl packed rat control erythrocytes at $0^{o}C$ for 1 h before testing in the macrophage assay. 2-Deoxy-2,3-dehydro-N-acetylneuraminic acid (Neu5Ac2en) was obtained from Boehringer (Mannheim), heat-aggregated rat IgG was prepared by heating at $65^{o}C$ for 1 h and Fc-fragments of human IgG were a gift of Dr. Hofstätter, Behringwerke (Marburg). The lectins were donated by Dr. Sharon, Rehovot. In the macrophage assay, they were added instead of serum if not otherwise indicated. Sialic acids were determined with the orcinol method after hydrolysis and purification (11) or by a new fluorimetric assay directly in the erythrocyte membranes (12).

Results

Incubation of rat erythrocytes with various concentrations of V. cholerae sialidase resulted in the release of an increasing amount of sialic acids (Neu5Ac, Neu5,9Ac_2, Neu5,7,9Ac_3) from the erythrocyte membrane with a maximum value of about 150 µg/ml packed cells. This corresponds to about 70 % of total, colorimetrically measured sialic acids (220 µg/ml packed cells). After cleavage of 10-20 µg (5-10 %) sialic acids/ml packed cells, binding of sialidase-treated erythrocytes to homologous rat peritoneal macrophages was significantly increased. The number of attached cells further increased according to the amount of enzymically released sialic acids. The extent of phagocytosis depends on the amount of hydrolyzed sialic acids, too, but for a measureable induction of this step more sialic acids must be removed from the erythrocyte membrane [50-65 µg (25-30 %)/ml packed cells] (Fig. 1).

Incubation of erythrocytes with A. ureafaciens sialidase resulted in the cleavage of a higher amount of sialic acids,

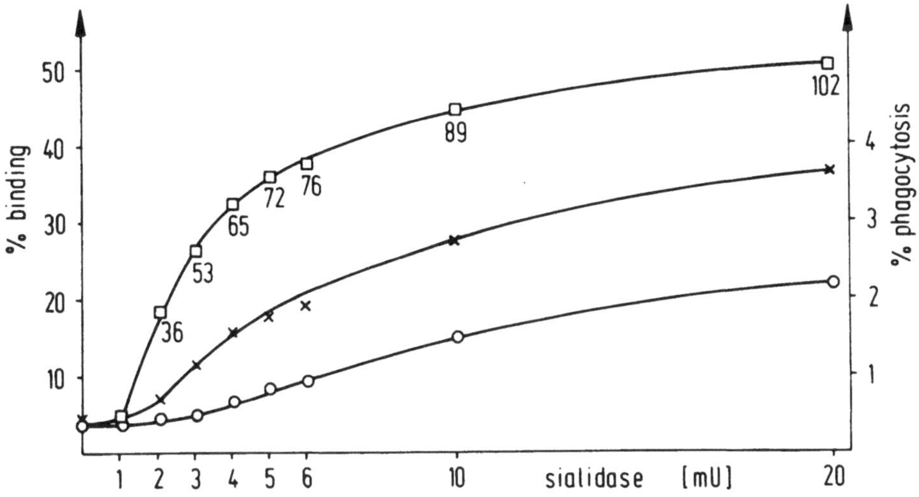

Fig. 1. Relationship between enzymically released sialic acids (□-□-□ ; the numbers are the amount of sialic acids released by V. cholerae sialidase [µg/1 ml packed cells]), binding (x-x-x) and phagocytosis (O-O-O) of erythrocytes with homologous peritoneal rat macrophages.

followed by stronger binding and phagocytosis of these erythrocytes in the macrophage assay, when compared with cells incubated with V. cholerae sialidase.

Although it seems to be established that the enzymic release of sialic acids from erythrocyte membranes is the responsible factor for binding and phagocytosis, as will be discussed below, a role of sialidase molecules attached to the erythrocyte surface for the sequestration step should also be taken into consideration. The following experiments, however, do not support this possibility and show that only the release of sialic acids is the responsible factor for the sequestration process.

1. If sialidase concentrations up to 0.5 mU/ml (not releasing sialic acids) were present during the whole incubation time of rat peritoneal macrophages with enzymically untreated erythrocytes, binding and phagocytosis of the cells was not increased.

2. Simultaneous incubation of erythrocytes with V. cholerae

sialidase and Neu5Ac2en (30 mM) prevented release of sialic acids from the erythrocyte membrane by inhibition of enzyme action. These cells are not stronger bound and phagocytosed than control cells.

3. Treatment of erythrocytes with immobilized V. cholerae sialidase, which enabled complete removal of the enzyme from erythrocytes after incubation, also induced binding and phagocytosis, but to a lower level as when using the soluble enzyme, according to the lower cleavage rate of sialic acids.

It has been shown in several experiments that hydrolysis of sialic acids from the erythrocyte membrane leads to the exposure of penultimate ß-Gal residues (13), which are recognized by a specific macrophage lectin (14) resulting in binding of the sialidase-treated erythrocytes to macrophages. Correspondingly, preincubation of macrophages with oligosaccharides, natural and synthetic glycoproteins and glycolipids with terminal ß-Gal residues results in a blockade of the macrophage lectin and the inhibition of binding and also phagocytosis of sialidase-treated erythrocytes. (Human asialo-glycophorin was the best inhibitor: 1 µM inhibited binding by 38 % and phagocytosis by 51 %). In vivo injections of asialo-fetuin and lactose into the blood stream of rabbits also prevented sequestration of sialidase-treated erythrocytes, whereas native fetuin and sialyllactose showed no significant effect (8).

Küster and Schauer (9) reported that binding of sialidase-treated erythrocytes already occurs in buffer, whereas for phagocytosis the presence of serum is required. Therefore, the involvement of immunoglobulins and complement factors for the phagocytosis step was assumed. However, on the basis of various experiments, the immunoglobulins IgG and IgM can be excluded to be the responsible substances for phagocytosis, as shown in the following scheme:

no influence on phagocytosis after

- the addition of
  - rat IgG
  - anti rat IgG
  - rat IgM
- the blockade of the Fc-receptor by
  - Fc-fragments of human IgG
  - heat-aggregated rat IgG

The role of complement factors has not yet completely been elucidated: heat inactivation of serum (56°C, 30 min) led to a reduction of the phagocytosis rate of sialidase-treated erythrocytes by about 50 %, but guinea pig complement preabsorbed with control erythrocytes did not restore the phagocytosis rate to the original level. Inhibition studies of the macrophage complement receptor (2-deoxy-D-glucose, trypan blue, α_1-anti-trypsin) also showed no convincing effects.

Further investigations of these phenomena have shown that phagocytosis can increase already without serum: incubation of rat peritoneal macrophages with different amounts of sialidase-treated erythrocytes in mere buffer led, within a wide range, to a linear increase in the absolute number of erythrocytes bound and phagocytosed by macrophages (Fig. 2). Both binding and phagocytosis of erythrocytes can increasingly be stimulated by the addition of increasing amounts of native serum. Thus, the same number of cells are bound and phagocytosed either by the addition of more erythrocytes or of serum.

From these experiments can be concluded: 1. Serum contains (a) component(s) of not yet known nature which has (have) a stimulating influence first of all on binding. 2. The higher the binding rate the higher the phagocytosis activity.

Lectins specific for ß-Gal residues, i.e. peanut agglutinin (PNA: Galß(1-3)GalNAc-specific) and Erythrina crystagalli agglutinin (ECA: Galß(1-4)GlcNAc-specific) were added to the incuba-

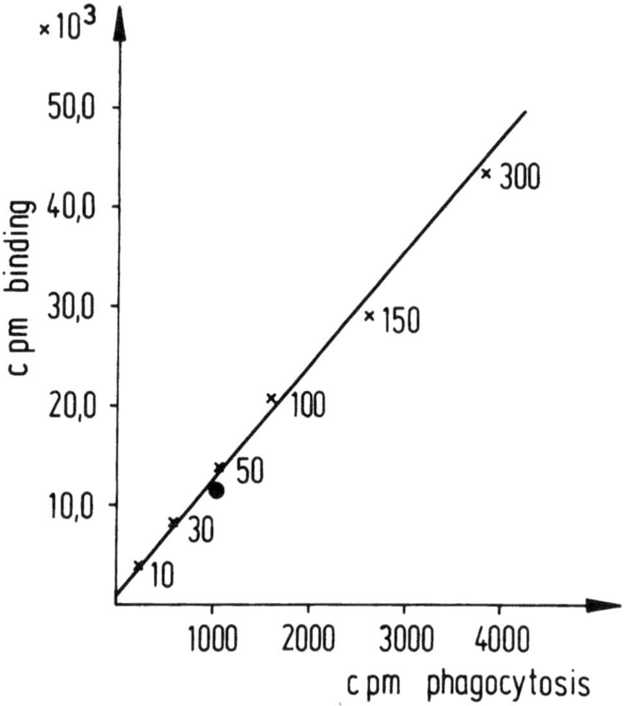

Fig. 2. Relationship between the number of bound and phagocytosed sialidase-treated rat erythrocytes by homologous peritoneal macrophages in dependence on the amount of added erythrocytes after incubation in buffer only. The numbers in the figure are µl of a 1:50 dilution of packed erythrocytes. The filled circle represents the value of the standard 30 µl erythrocyte cell suspension, incubated in the presence of 30 % native rat serum.

tion buffer in subagglutinating concentrations. They slightly stimulated binding and strongly induced phagocytosis of sialidase-treated erythrocytes. Soybean agglutinin (SBA: GalNAc-specific), however, had only little influence. The effect of the tested lectins on control erythrocytes was negligible (Table).

It is noteworthy that the lectins must be present during the whole incubation time of erythrocytes with macrophages. Preincubation of macrophages with the lectins for 15 min, followed by thorough washing before the addition of sialidase-treated erythrocytes nearly abolished the capacity of the lectins to

	Buffer		PNA		ECA		SBA	
	C	SA	C	SA	C	SA	C	SA
% Binding	7,7 ± 2,0	26,2 ± 2,4	12,5 ± 4,0	65,2 ± 2,9	15,3 ± 2,8	57,3 ± 2,3	12,5 ± 2,3	34,3 ± 0,8
% Phagocytosis	1,0 ± 0,2	2,9 ± 0,4	0,8 ± 0,2	20,3 ± 2,4	1,6 ± 0,1	18,4 ± 1,1	1,4 ± 0,2	4,2 ± 0,2

Table. Influence of different lectins, added to the incubation buffer, on binding and phagocytosis of control (C) and sialidase-treated (SA), ^{51}Cr-labelled, rat erythrocytes to homologous peritoneal macrophages. The lectins were used in subagglutinating concentrations (PNA 12.5 µg/ml, ECA 6.25 µg/ml, SBA 3.1 µg/ml).

stimulate phagocytosis. (In the case of ECA: binding decreased from 57.3 % to 44.8 % and phagocytosis from 18.4 % to 4.1 %.)

Concluding remarks

The data presented here indicate that only the amount of released sialic acids is the decisive factor for binding and phagocytosis of sialidase-treated erythrocytes. After exposure of ß-Gal residues the sequestration mechanism is initiated by the interaction of sialidase-treated erythrocytes with a ß-Gal-specific lectin existing on macrophages. The presence of serum increases both binding and phagocytosis. However, the nature and mechanism of action of the serum factor(s) are still unknown. The phagocytosis-stimulating effect of ß-Gal-specific plant lectins may provide a model for the action of such serum components. It is imaginable that lectins induce conformational changes of the glycoconjugate coat on the erythrocyte membrane. This may lead to the exposure of membrane sites which increase the binding strength of the erythrocytes to the macrophages, perhaps by closer cell contacts, or directly stimulate the phagocytosis activity of these cells.

Acknowledgements

The authors thank S. Wiese for excellent technical assistance. We are grateful to Dr. Sharon for the lectins and to Dr. Hofstätter for the Fc-fragments. The financial support of the project by the Deutsche Forschungsgemeinschaft (grant Scha 202/11-1) and the Fonds der Chemischen Industrie is acknowledged.

References

1. Hudgin, R.L., Pricer, W.E., and Ashwell, G.: J. Biol. Chem. 249, 5536-5543 (1974).
2. Prieels, J.-P., Pizzo, S.V., Glasgow, L.R., Paulson, J.C., and Hill, R.L.: Proc. Natl. Acad. Sci. U.S.A. 75, 2215-2219 (1978).
3. Powell, J.T.: Biochem. J. 187, 123-129 (1980).
4. Kawasaki, T., and Ashwell, G.: J. Biol. Chem. 252, 6536-6543 (1977).
5. Schlesinger, P.H., Rodman, J.S., Doebber, T.W., Stahl, P.D., Lee, Y.C., Stowell, C.P., and Kuhlenschmidt, T.B.: Biochem. J. 192, 597-606 (1980).
6. Jancik, J.M., Andres, K.H., von Düring, M., and Schauer, R.: Cell Tiss. Res. 186, 209-226 (1978).
7. Kolb, H., Friedrich, E., and Süss, R.: Hoppe-Seyler's Z. Physiol. Chem. 362, 1609-1614 (1981).
8. Müller, E., Franco, M.W., and Schauer, R.: Hoppe-Seyler's Z. Physiol. Chem. 362, 1615-1620 (1981).
9. Küster, J., and Schauer, R.: Hoppe-Seyler's Z. Physiol. Chem. 362, 1507-1514 (1981).
10. Parker, T.L., Corfield, A.P., Veh, R.W., and Schauer, R.: Hoppe-Seyler's Z. Physiol. Chem. 358, 789-795 (1977).
11. Schauer, R.: Methods Enzymol. 50C, 64-89 (1978).
12. Shukla, A.K., and Schauer, R.: Hoppe-Seyler's Z. Physiol. Chem. 363, 255-262 (1982).
13. Pfannschmidt, G., and Schauer, R.: Hoppe-Seyler's Z. Physiol. Chem. 361, 1683-1695 (1980).
14. Nagamura, Y., and Kolb, H.: Lectins - Biology, Biochemistry, Clinical Biochemistry (Bøg-Hansen, T.C.,ed.) Vol. 1, pp. 33-42, Walter de Gruyter, Berlin · New York 1981.

ISOLATION OF A MEMBRANE LECTIN FROM THE MARMOSET LYMPHOID CELL LINE 70N2

Janet M. Decker
Department of Biochemistry, Medical University of South Carolina, Charleston, South Carolina, 29425, U.S.A.

Membrane lectins have been detected on many different vertebrate cell types, and are under intense study as possible mediators of cell-cell interactions (c.f. 1-11). Because lymphocyte function is regulated by interactions between different cell types, the lectins which have been found on murine T and B lymphocytes and on macrophages are of interest as candidates for the recognition molecules on these cells (12-15). Previously, a group of lectin-like molecules was identified on murine lymphocytes by their binding to the oligosaccharide portion of the fetal serum α-globulin fetuin (13) and isolated by fetuin-sepharose affinity chromatography. Fetuin was chosen as a glycoprotein adsorbant for lymphocyte membrane lectin because it contains relatively high amounts of two oligosaccharides whose structures are known (16-17) and which are fairly typical of the oligosaccharides found on mammalian cell glycoproteins (18). One of the oligosaccharides is a serine-threonine linked structure composed of NAc-galactosamine, galactose and sialic acid; the other is an asparagine linked complex oligosaccharide with the (NAc-glucosamine)$_2$-mannose core and three branches of mannose, NAc-glucosamine, galactose and sialic acid.

I have chosen to investigate the molecular structure of fetuin-binding molecules from a T lymphocyte line (70N2) of the cotton-topped marmoset *Saguinus oedipus*, a New World squirrel monkey. 70N2 cells display membrane markers which are immunologically cross-reactive with those of human helper T lymphocytes: ability to form spontaneous rosettes with erythrocytes, T cell (OKT-11a) and helper T cell (Leu 3A) markers, the HLA-Dr antigen, and the absence of surface immunoglobulin constant regions and of the EAC receptor (19-20). 70N2 cells are also able to provide an accessory cell signal necessary for human B lymphocytes to secrete immunoglobulin *in vitro* (19). Finally, the cells can be grown in large quantities, facilitating biochemical studies. This paper describes a lectin isolated from the 70N2 cell line with an oligosaccharide binding specificity that includes sialic acid residues.

Materials and Methods

Materials. The cell line 70N2 was the generous gift of Dr. H. Rabin, Molecular Oncology Program, Frederick Cancer Research Center, Frederick, Maryland. The cell line was derived from a *Herpesvirus saimiri*-induced tumor in a cotton-topped marmoset (21). Fetuin was purchased from GIBCO (purified by the Spiro-technique); CNBr-activated sepharose came from Pharmacia.

Fetuin was coupled to the activated sepharose according to the procedure recommended by Pharmacia at a ratio of 1 mg fetuin/ml sepharose. This ratio was chosen because adsorbants prepared at a higher fetuin: sepharose ratio tended to leach fetuin along with the eluted lectin. Other glycoproteins and saccharides were purchased from Sigma Chemical Company and Calbiochem.

Affinity purification of 70N2 lectin. Lectin was isolated from exponentially growing cells washed several times to remove passively adsorbed fetal calf serum components. Membrane proteins were extracted with 0.1% Triton X-100 in 0.05*M* tris-buffered saline containing 10m*M* $CaCl_2$ (TBS-Ca) for 30 min at 0°C. Nuclei and insoluble material were removed by centrifugation (13). Extracted material was passed over a column of fetuin-sepharose and the bound material eluted with either TBS containing 10 m*M* EDTA or with 3.5 *M* NaSCN. Eluted material was dialyzed overnight into TBS-Ca and an aliquot was iodinated by the chloramine T method (22) and analyzed by polyacrylamide gel electrophoresis in SDS-containing buffer (13). Gels were cut into 2mm slices which were counted in a Beckman Gamma 7000 gamma counter.

Determination of binding specificity. Binding specificity was determined by blocking experiments with fetuin, other glycoproteins, and monosaccharides. Since affinity-purified lectin showed only very low binding activity, most of these experiments were performed with a crude Triton X-100 extract from 70N2 cells which had previously been iodinated by the lactoperoxidase technique, which labels only external membrane proteins of viable cells (23). Aliquots of extract were incubated with inhibitor for 30 min at room temperature, followed by incubation of the extract-inhibitor mixture with a small aliquot of fetuin-sepharose for a further 30 min. Samples were washed 3 times with TBS-Ca and the counts/min bound to the adsorbant were determined. Background values of cpm bound to sepharose 4B were subtracted from all values. Percent inhibition of binding was calculated as

$$100 \text{ minus } \frac{\text{cpm bound to fetuin sepharose in the presence of inhibitor}}{\text{cpm bound to fetuin sepharose in the presence of TBS-Ca}} .$$

Results

The electrophoretic mobility of the lectin isolated by fetuin affinity chromatography is shown in Fig. 1. Only one major component is seen when fetuin-binding material is eluted, iodinated and subjected to polyacrylamide gel electrophoresis under reducing conditions as described in the Methods section. Fig. 1B represents material eluted from fetuin-sepharose with 3.5 M NaSCN; however, similar patterns are seen with EDTA-eluted material. When viable 70N2 cells are iodinated, extracted with Triton X-100 and the unpurified extract incubated with fetuin sepharose, a molecule with the same relative electrophoretic mobility is recovered from the affinity adsorbant (Fig. 1A). The accessibility of the lectin to

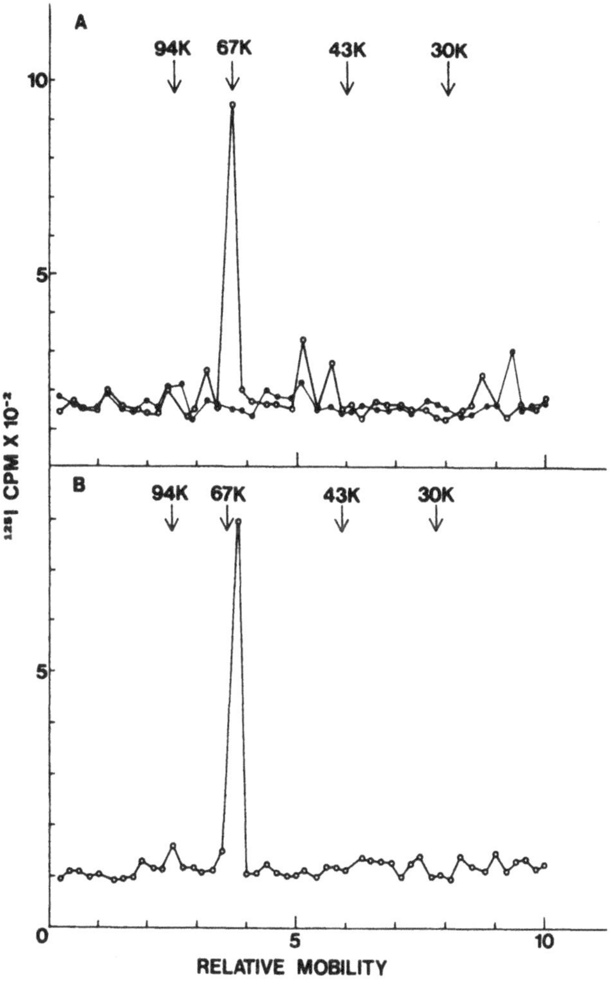

Figure 1. Polyacrylamide gel electrophoretic analysis of fetuin-sepharose-binding material from marmoset T-cell line 70N2. A. Cells were radioiodinated and extracted with Triton X-100, and the extracted material was incubated with either fetuin-sepharose (o——o) or sepharose 4B (●——●). B. Material bound to fetuin-sepharose was eluted with NaSCN and radioiodinated. Gels in both cases contained 10% acrylamide; samples were loaded in buffer containing 2-mercaptoethanol. External standards were phosphorylase A (94,000 daltons), bovine serum albumin (67K), ovalbumin (43K) and carbonic anhydrase (30K).

iodination before detergent extraction supports the hypothesis that it is an external membrane molecule.

Unpurified Triton X-100 extract from 70N2 cells was passed over a column of Concanavalin A sepharose and the bound material eluted, dialyzed, and passed over fetuin sepharose. The molecules which bound to fetuin sepharose had the same relative electrophoretic mobility as lectin not exposed to Concanavalin A sepharose. Since Concanavalin A is not glycosylated and would not be bound by the sugar-binding site of the 70N2 lectin these findings suggest that the 70N2 lectin is a glycoprotein which is bound to the lectin site of Concanavalin A.

The 70N2 fetuin-binding material appears to be unstable in its binding characteristics. Material bound to fetuin-sepharose immediately after extraction from 70N2 cells can generally be completely eluted with TBS-10m_M_ EDTA. The longer extracted material is kept before fetuin-sepharose affinity chromatography, the higher the percentage of material which does not elute with TBS-10m_M_ EDTA but requires 3.5_M_ NaSCN to pull it off the column. Binding of 70N2 lectin (either purified material or crude extract from radioiodinated cells) also decreases if the material is stored for as short a time as one day. Material eluted from fetuin sepharose only rarely retained its ability to bind fetuin. The loss of activity did not appear to be due to fetuin co-eluting from the column, since no material with an electrophoretic mobility comparable to that of an external fetuin sample was observed.

Specificity studies have been hampered by the lability of the 70N2 lectin binding activity. Such studies have however been carried out with the Triton X-100 extract from radioiodinated cells. Results of a representative experiment are shown in Table 1 below.

TABLE 1

SPECIFICITY OF 70N2 LECTIN BINDING TO FETUIN-SEPHAROSE

INHIBITOR	% INHIBITION
Bovine submaxillary mucin	100
Thyroglobulin	77
Fetuin	55
Ovalbumin	34
Ovomucoid	0
Galactose	2
Glucose	-6
Mannose	-7
Raffinose (Gal-Glc-Fructose)	4
NAc-Neuramine lactose (NAN-Gal-Clc)	95

Fetuin blocking was low in this particular experiment; it is more usually 80%. A mixture of gangliosdies was also a good blocker of fetuin binding, as were the glycopeptides from pronase-treated fetuin. Not surprisingly, glycoproteins with oligosaccharides similar to those of fetuin are good inhibitors while glycoproteins with dissimilar oligosaccharides are poor inhibitors of fetuin binding.

Discussion

These results demonstrate that there is a molecule with an electrophoretic mobility corresponding to 67,000 daltons which can be isolated from the marmoset T cell line 70N2. The molecule is located in the plasma membrane (accessibility to lactoperoxidase-catalyzed iodination of viable cells) and can also be biosynthetically labeled by culturing the cell line in ^{75}Selenomethionine (data not shown). The lectin is probably glycosylated, since it binds to Concanavalin A. Binding specificity of the 70N2 lectin has not yet been satisfactorily determined because of the lability of binding activity; however, sialic acid would appear to be part of the recognition site since all the glycoconjugates which are good inhibitors of fetuin binding contain sialic acid, and it is clear the molecule does recognize saccharide rather than protein determinants.

These results agree with other data on lymphocyte lectins presented at this meeting. The lectin of Kieda _et_ _al_. (24) from human peripheral monocytes is also specific for a complex oligosaccharide structure, and binding to this structure can not be efficiently blocked with monosaccharides. Kolb (25) has recently reported that a lectin found on murine T and B lymphocytes and responsible for those cells' ability to rosette autologous erythrocytes has binding specificity for gangliosides.

Recent reports demonstrating that spontaneous cytotoxicity by human NK lymphocytes can be blocked by saccharides (26-27) supports the hypothesis that lectin: saccharide recognition is important to immune function. Further characterization of lymphocyte membrane lectins and the molecules they recognize will further our understanding of the mechanism of these interactions.

This work was supported by PHS Grant Number CA28992, awarded by the National Cancer Institute, DHHS, and by the Medical University of South Carolina State Appropriations for Research 1981-1982.

References

1. Stockert, R.J., Morell, A.G. and Scheinberg, I.H. Science 186: 365-366 (1974).

2. Kawasaki, T. and Ashwell, G. J. Biol. Chem. 252: 6536-6543 (1977).

3. Kolb, H., Kolb-Bachofen, V. and Schlepper-Schäfer, J. Biol. Cellulaire 36: 301-308 (1979).

4. Beyer, E.C., Tokuyasu, K.T. and Barondes, S.H. J. Cell Biol. 82: 565-571 (1979).

5. Baenziger, J.U. and Maynard, Y. J. Biol. Chem. 255: 4607-4613 (1980).

6. Bertolotti, R., Rutishauser, U. and Edelman, G.M. Proc. Natl. Acad. Sci., USA 77: 4831-4835 (1980).

7. Nowak, T.P., Kobiler, D., Roel, L.E. and Barondes, S.H. J. Biol. Chem. 252: 6026-6030 (1977).

8. Schwartz, A.L. Fridovich, S.E., Knowles, B.B. and Lodish, H.F. J. Biol. Chem. 256: 8878-8881 (1981).

9. Wieser, R., Golecki, J.R. and Brunner, G. Biochem. Biophys. Acta 648: 275-283 (1981).

10. Harrison, F.L. and Chesterton, C.J. Nature 286: 502-504 (1980).

11. Grabel, L.B., Kosen, S.D. and Martin, G.R. Cell 17: 477-484 (1979).

12. Kieda, C.M.T., Bowles, D.J., Ravid, A. and Sharon, N. FEBS Lett. 94: 391-396 (1978).

13. Decker, J.M. Mol. Immunol. 17: 803-808 (1980).

14. Kolb, H. and Kolb-Bachofen, V. Biochem. Biophys. Res. Comm. 85: 678-683 (1978).

15. Nagamura, Y. and Kolb, H. FEBS Lett. 115: 59-62 (1980).

16. Baenziger, J.U. and Fiete, D. J. Biol. Chem. 254: 789-795 (1979).

17. Nilsson, B., Norden, N.E. and Svenson, S.J. Biol. Chem. 254: 4545-4553 (1979).

18. Kornfeld, R. and Kornfeld, S. Ann. Rev. Biochem. 45: 217-237 (1976).

19. Marchalonis, J.J. and Strelkauskas, A.J. J. Immunogenetics 8: 477-483 (1981).

20. Robinson, J.E., Andiman, W.A., Henderson, E. and Miller, G. Proc. Natl. Acad. Sci., USA 74: 749-753 (1977).

21. Strand, B.C., Newbauer, R.H. and Rabin, H. Int. J. Cancer 23: 76-81 (1979).

22. Greenwood, F.C., Hunter, W.M. and Glover, J.S. Biochem. J. 89: 114 (1963).

23. Marchalonis, J.J., Cone, R.E. and Santer, V. Biochem. J. 124: 921-927 (1971).

24. Kieda, C.M.T., Monsigny, M., and Waxdal, M., this volume

25. Kolb, H. Biochem. Biophys. Res. Commun. 105: 1488-1495 (1982).

26. Ades, E.W., Hinson, A. and Decker, J.M. Immunobiol. 160: 248-258 (1981).

27. Muchmore, A.V., Decker, J.M. and Blaese, R.M. Immunobiol. 158: 191-206 (1981).

STUDIES ON LECTINS

LV. SUBCELLULAR LOCALIZATION OF AN ENDOGENOUS LECTIN IN FISH OOCYTE

Jindřich Nosek

Department of Developmental Biology, Charles University,
Vinicná 7, 128 44 Praha 2, Czechoslovakia

Alexander Krajhanzl and Jan Kocourek

Department of Biochemistry, Charles University,
Albertov 2030, 128 40 Praha 2, Czechoslovakia

Introduction

Lectins and lectin-like molecules are widespread in ova of lower vertebrates as well as in gametes of many invertebrates, plants and fungi, and it is generally assumed that they play some role in gamete mating and at the postfertilization events (for reviews see 1.2). As revealed by erythroagglutination screening tests of saline extracts, a particularly rich source of lectins are ova of amphibians (3,4,5), the ripe roe of several species of freshwater fish (6,7,8) and the roe of salmon (9,10). Several fish lectins have been recently purified by affinity chromatography and partially characterized (11,12,13,14), but the subcellular distribution of lectins, their biological role and their fate during the oocyte maturation have not been investigated and still have to be clarified. These questions became particularly important in view of recent findings that the erythroagglutinating activity (15) and physicochemical properties (Krajhanzl and Kocourek, in preparation) of certain fish lectins show seasonal variations in dependence on the phase of oocyte growth and maturation.

The present paper brings a brief summary of the properties of the lectin isolated from the ovary of perch (Perca fluviatilis L.) at different stages of maturity and an immunofluorescence study on the localization of the perch lectin in previtellogenic oocytes.

Lectins, Vol. III

Results

Isolation and characterization of lectin.

The lectin was isolated from frozen immature ovaries obtained in October. Our original procedure for the isolation of lectin (12,13) has been modified as follows: Lectin was purified by affinity chromatography on O-α-D-glucopyranosyl-Separon H-1000 equilibrated in 0.05 M acetate buffer pH 5.5, 0.25 M NaCl and eluted with the same buffer containing 0.1 M D-glucose. The lectin was extensively dialyzed against distilled water and lyophilized.

The preparation obtained by this procedure contained lectin composed of 10 or 15 globular subunits of M_r 13 000 (designated "A" subunits). The high molecular weight associates of M_r values 130 000 and 190 000 have sedimentation coefficients $s_{20,w}$ of 7.02 and 9.83, respectively.

In addition to the main subunit component A, we also found a minor amount of another protein of lectin character with a M_r of 16 800 (designated "B") which was observed previously by Entlicher and Bažil (unpublished results). Under the condition of discontinuous electrophoresis in the presence of SDS (16) this minor component forms dimers and mixed aggregates with subunit A.

The amino acid composition of the preparation is in good agreement with the results published previously (13).

In contrast to the preparation isolated from O-α-L-fucosyl polyacrylamide gel by elution with 0.3 M L-fucose solution (13) which, in addition to subunit A, contained also a significantly higher proportion of protein B (Entlicher, unpublished results), the isolation procedure using O-α-D-glucopyranosyl-Separon H 1000 yielded an almost pure lectin formed by subunits A with a negligible amount of protein B. This was also reflected in a lower content of neutral sugar (0.8% by weight), a higher extinction coefficient ($A^{1\%}_{284}$ = 30.2) and altered agglutinating properties (preferable agglutination of human erythrocytes O and A_2). The agglutination is inhibited by following sugars (in order of decreasing activity): D-glucose, L-fucose, methyl α-D-mannopyranoside, D-xylose and D-arabinose.

Lectins present in a preparation, obtained from mature oocytes and spawned roe by the same isolation procedure, have somewhat different properties when compared with lectins isolated from immature oocytes (altered amino acid composition, M_r of the individual components, higher content of neutral sugar and a significantly lower solubility (Krajhanzl and Kocourek, in preparation).

Localization of lectin in the ovary.

For comparison, paraffin sections of immature perch ovaries obtained during autumn were first stained with hematoxylin-eosin, PAS (periodic acid - Schiff), and azofuchsin-anilin blue-orange G. Ovaries contain only previtellogenic oocytes, classified by morphological creteria as stage I and II (after 17). Stage-I oocytes have a relatively large nucleus and a small amount of undifferentiated cytoplasm. Stage-II oocytes are larger, with well differentiated and stratified cytoplasm and a germinal vesicle with peripheral nucleoli. The cytoplasm still does not contain any appreciable amount of yolk granules. The most prominent feature of Stage-II oocytes are numerous cortical granules, which form several layers at the cortical cytoplasm (Fig. 1a). In the vitellogenic oocyte cortical granules form only a thin discontinyous layer at the periphery of central yolk mass (18) and are much less conspicuous. The content of cortical granules stains for saccharides by the PAS reaction.

For the detection of lectin in the paraffin sections of the immature ovary, an indirect immunoflourescence method with rabbit antibody against lectin and fluorescein isothiocynate (FITC)-conjugated porcine antibody against rabbit IgG has been employed. The most intense fluorescent staining was observed inside the cortical granules of Stage-II oocytes (Fig. 1b,c). The cytoplasm of Stage-II oocytes and the follicular epithelium were less fluorescent, while the Stage-I oocytes, the outer envelope of oocytes (zona radiata) and the surrounding muscle tissue and blood capillaries remained dark. Some sera stained also the germinal vesicle.

Control sections of the same tissue treated either with the preimmune rabbit serum instead of antibody, or with the antibody absorbed by acetone powder of perch ovary or with the FITC-conjugated porcine antibody alone, confirmed that the reaction of the antibody with the contents of the cortical granules and the cytoplasm of Stage-II oocytes was specific. Only a light fluorescence of the follicular epithelium and the germinal vesicle was seen in control sections, apparently resulting from a nonspecific reaction with porcine antibody or from autofluorescence (Fig. 1d).

Recently, we have confirmed that the distribution of lectin follows the distribution of cortical granules also in cryostat sections of the ripe vitellogenic perch oocytes. Other studies from our laboratory (Nosek et al., in preparation) indicate, that lectins are closely associated with cortical granules also in several species of Cyprinidae.

Our results show that the first appearance of lectin is related to the formation of cortical granules by previtellogenic oocytes. As in the sea urchin (19), cortical granules are the first differentiated structures in the cytoplasm of a growing perch oocyte. In some species of fish, cortical granules fill the whole cytoplasm of the Stage-II oocyte, but only those located near the cell periphery are fully mature (17). Thus the fluorescent staining of the perinuclear cytoplasm of Stage-II oocytes indicates either the site of lectin synthesis and formation of cortical granules, or the involvement of lectin in the transport of glycosylated molecules (20) from the cytoplasm to the cortical granules. Since the content of the cortical granules is realesed to the extracellular space upon fertilization, we must consider also an alternative possibility, namely that the sugar binding properties will play some role in postfertilization events. For the lectins present in female gonads of various ani-

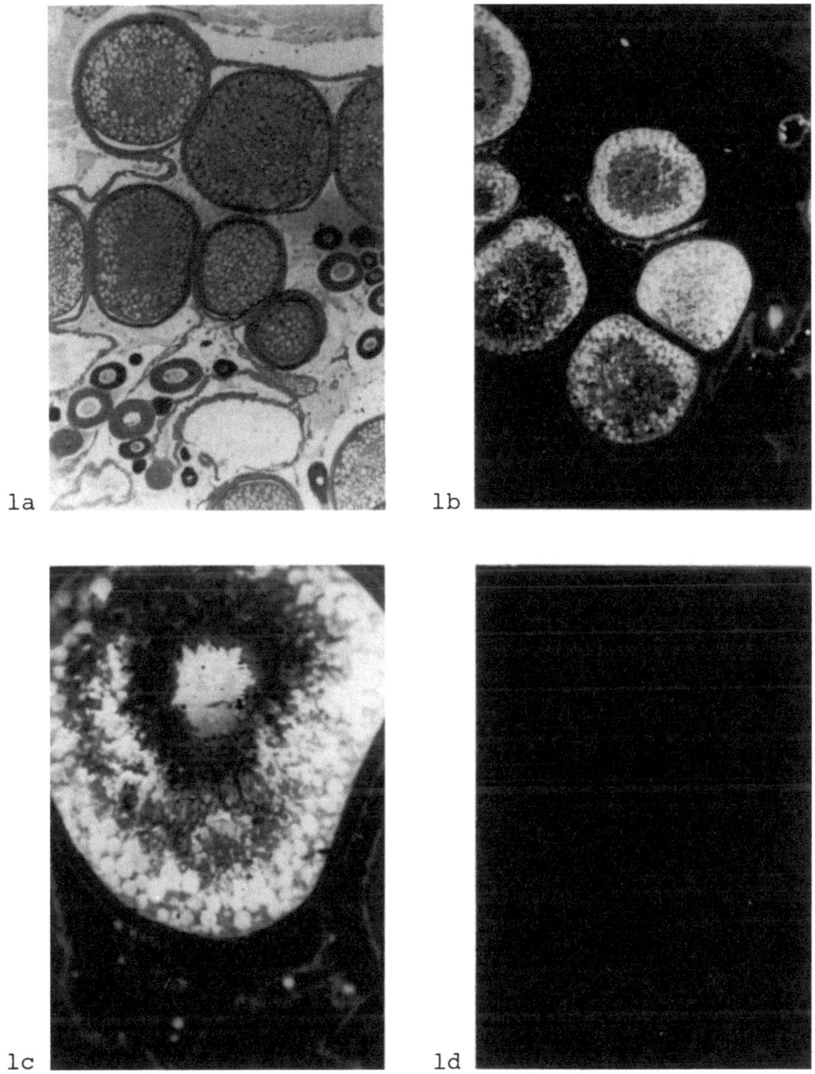

Fig. 1. Ovary of the perch (Perca fluviatilis L.)
a. Section stained with hematoxylin-eosin, showing several Stage-II oocytes with numerous cortical granules, and the surrounding tissue (magnification 80x).
b. Indirect fluorescein staining of the section of perch ovary, showing the distribution of lectin (80x).
c. Same as in b at larger magnification (320x) The most intensive fluorescent staining is in the cortical granules.
d. Control section treated with the rabbit preimmune serum and FITC-conjugate, showing the nonspecific reaction of follicular epithelium with the FITC-conjugate (80x).

mal species, the name "protectins" has been proposed by Prokop and his co-workers (21), to indicate their protective role. This assumption, however, was based only on the ability of lectins to agglutinate several bacteria and will need some stronger experimental support to prove its validity.

The most probable explanation for the presence of lectin in the cortical granules has been suggested by Hedrick and his collaborators (4,22,23), who studied the fate of the cortical granule lectin of an amphibian, Xenopus laevis, during fertilization. These authors reported that after the release of lectin from the cortical granules, the lectin penetrates the vitelline membrane and cross-links the proximal layer of the jelly coat, forming a permanent barrier to polyspermy. Other consequences of this reaction are the elevation of the fertilization membrane and formation of the perivitelline space. The perch lectin is bound by the gel layer when ripe roe is used for its isolation. In unfixed cryostat sections of mature oocytes, lectin released from the cortical granules is bound to the outer gel layer and can be detected by fluorescent glycosulated markers (Krajhanzl, unpublished results) throughout this layer. This indicates that the biological role of the perch lectin may be similar to the cortical lectin of Xenopus laevis.

References

1. Gold, E.R., Balding, P.: Receptor-Specific Proteins. Plant and Animal Lectins, Excerpta Medica, Amsterdam (1975).
2. Nosek, J.: Biol.Listy 45, 180-195 (1980) in Czech.
3. Schenkel-Brunner, H., Kothbauer, H., Luttenberger, F., personal communication quoted in ref. 1.
4. Wyrick, R.E., Nishihara, T., Hedrick, J.L.: Proc.Natl.Acad. Sci. U.S.A. 71, 2067-2071 (1974).
5. Sakakibara, F., Takayanagi, G., Kawauchi, H., Watanabe, K., Hakomori, S.: Biochim.Biophys.Acta 444, 386-395 (1976).
6. Prokop, O., Schnitzler, S., Uhlenbruck, G.: Acta.Biol.Med. Ger. 18, K7 (1967)
7. Todd, G.M.: Vox Sang. 21, 451-454 (1971).

8. Anstee, D.J., Holt, P.D.J., Pardoe, G.I.: Vox Sang. 25,347-360 (1973).

9. Uhlenbruck, G., Prokop, O.: Vox Sang. 12, 465-466 (1967).

10. Prokop, O., Schlesinger, D., Geserick, G.: Z. Immunitätsforsch. 132, 491-494 (1967).

11. Voak, D., Todd, G.M., Pardoe, G.I.: Vox Sang. 26, 176-188 (1974).

12. Krajhanzl, A., Hořejší, V., Kocourek, J.: Biochim.Biophys. Acta 532, 209-214 (1978).

13. Krajhanzl, A., Hořejší, V., Kocourek, J.: Biochim.Biophys. Acta 532, 215-224 (1978).

14. Voss, E.W., Fryer, J.L., Banowetz, G.M.: Arch. Biochem.Biophys. 186, 25-34 (1978).

15. Schenkel-Brunner, H., Kothbauer, H.: Z. Immunitätsforsch. 154, 155-161 (1978).

16. Laemmli, U.K.: Nature (London) 227, 680-685 (1970).

17. Arndt, E.A.: Zellforsch. 51, 209-224 (1960).

18. Devillers, C.: Adv.Morphogen. 1, 379-428 (1961).

19. Anderson, E.: J.Cell Biol. 37, 514-539 (1968).

20. Krüpe, M.: Blutgruppenspezifische Pflanzliche Eiweisskörper (Phytoagglutinine), F. Enke, Stuttgart (1956).

21. Prokop, O., Uhlenbruck, G., Rothe, A., Cohen, E.: Ann.N.Y. Acad.Sci. 234, 228-231 (1974).

22. Greve, L.C., Hedrick, J.L.: Gamete Res. 1, 13-18 (1978).

23. Wolf, D.P., Nishihara, T., West, D.M., Wyrick, R.E., Hedrick, J.L.: Biochemistry 15, 3671-3677 (1976).

LECTINS FROM TUNICATES AND CYCLOSTOMES
A BIOCHEMICAL CHARACTERIZATION

Gerardo R. Vasta and John J. Marchalonis

Department of Biochemistry, Medical University of South Carolina, Charleston, South Carolina 29425, USA.

Introduction

Although the possibility has been considered that lectins of invertebrate species can act as part of a non-self recognition system and might be evolutionary precursors of the vertebrate immune system, little direct evidence for the hypothesis has been obtained. In the first place, it has been difficult to ascribe definite physiological functions to invertebrate lectins. In the second place, the lectins which have been studied in some detail bear little resemblance, if any, to the structures exhibited by classical vertebrate immunoglobulins (1,2). Because of the presumed ancestral relationship of tunicates to vertebrate species (3), we chose to investigate lectins of tunicates with the goal of establishing their specificities, determining whether or not there is a species distribution of certain specificities, and elucidating the structures of some of the more promising lectins (4). In the course of these studies we identified and developed procedures for the isolation of a lectin from Didemnum candidum which was specific for D-galactose and a Ca^{++}-dependent lectin from Halocynthia pyriformis which was specific for sialoconjugates (4). In this communication, we compare the isolated tunicate lectins on the basis of apparent mass as judged by polyacrylamide gel electrophoresis and amino acid composition analysis with the sialic acid-specific lectins isolated from the eggs and serum of the sea lamprey, Petromyzon marinus. The lectins of the sea lamprey were chosen for this study because the lamprey is a primitive chordate which would be expected to show more relationship to the ancestral tunicates than would the more highly evolved higher vertebrates. In addition, we have previously shown that the serum of the sea lamprey contains a powerful agglutinin for horse erythrocytes (5). Further work documented here has shown that this molecule has specificity for sialoconjugates and requires Ca^{++}.

Materials and Methods

Source of specimens: Species and number of specimens studied are as follows: Didemnum candidum (10 colonies): Gulf Specimen Company, Panacea, Florida; Halocynthia pyriformis (200 specimens) Marine Research Associates, New Brunswick, Canada, and Petromyzon marinus (10 males and 10 females) Hammond Bay Fisheries Laboratory, U.S. Department of the Interior, Hammond Bay, Michigan.

Collection of blood: Immediately upon receipt, tunicates were cleaned of adherent materials, gently squeezed to eliminate sea water through the syphons, and kept on absorbent paper at 6°C for 2 hs to allow release of remaining sea water. Halocynthia was bled by cutting the tunic in the area of attachment to the substrate with a scalpel, without injuring internal organs. In Didemnum specimens, several incisions were made in the colony wall. Halocynthia and Didemnum specimens were placed separately in plastic funnels at 6°C and the body fluids, mainly blood, allowed to flow for two hours and collected on ice. Blood pools were cleared by centrifugation at 10,000 x g at 4°C, aliquoted and stored at -25°C until use. Lampreys were bled by cardiac puncture with a 10 ml hypodermic syringe provided with a 19 gauge needle. Blood samples were stored at 4°C for 8 hs, centrifuged at 800 x g for 5 min and the sera separated, pooled and stored at -25°C.

Preparation of egg extract: 50 gr of lamprey eggs were thoroughly washed with 0.85% NaCl and homogenized in 50 ml of tris-buffered saline (TBS) (50 mM tris-HCl, 100 mM NaCl, 20 mM Ca^{++}, pH 7.6), in a Waring blender at 4°C for 5 min. The extract was centrifuged at 10,000 x g and the aqueous supernatant separated, cleared by centrifugation (20 min at 10,000 x g) and stored at -25°C.

Preparation of immunoadsorbants: Horse erythrocytes (RBC), collected in Alsever's, from Chisolm Biological Laboratories, Seabrook, South Carolina, were formaldehyde-fixed according to the method of Nowak and Barondes (6). Fetuin (Gibco, Grand Island, New York) was hydrolyzed for 1 hr at 80°C in 0.1 N H_2SO_4 and 78.1% of the total sialic acid was released as measured by the resorcinol method (7). Asialo-fetuin was bound to CNBr-activated Sepharose 4B (Pharmacia, Uppsala, Sweden) following instructions of the manufacturer, at a ratio of 10 mg/ml of gel.

Purification of lectins: Halocynthia and Petromyzon (serum and eggs) lectins were purified as follows: 45 ml of tunicate plasma or lamprey serum (or egg extract) were incubated with 15 ml of thoroughly washed and packed formaldehyde fixed horse RBC for 1 hr at 4°C with gentle agitation every 15 min. The complexes lectin-RBC were washed with TBS until OD at 280 nm was <0.020 and eluted with 30 ml of calcium-free TBS + 10 mM EDTA. Eluates were dialyzed exhaustively against TBS, distilled water and lyophilized. Didemnum lectin was purified as follows: 45 ml of plasma was passed through a column (25 mm x 70 mm) of asialo fetuin-Sepharose 4B previously equilibrated with TBS, monitored at 280 nm. The column was washed to the buffer baseline with TBS and eluted with TBS + 50 mM D-galactose. Fractions correspon-ding to the major peak were pooled, dialyzed exhaustively against TBS, distilled water and lyophilized. Didemnum, Halocynthia and Petromyzon purified lectins exhibited specificities and specific activities as expected from our previous work.

Sodium dodecyl sulfate polyacrylamide gel electrophoresis (SDS-PAGE): SDS-PAGE was performed under reducing conditions (1% 2-mercaptoethanol) in 10% acrylamide-0.2% bisacrylamide gels running in parallel well characterized molecular weight standards.

Molecular weight determinations of lectin subunits: Approximate molecular weight of lectin subunits were estimated in a Log MW vs. mobility graph from SDS-PAGE data. Values were consistent with MW estimates determined by computation of the amino acid composition as performed by the method of Alt et al. (8).

Amino acid analysis: Samples were hydrolyzed with 6N HCl in sealed tubes under vacuum at 110°C for 24 hr. Amino acid composition of the hydrolysate was determined on a Durrum D-500 amino acid analyzer.

S ΔQ analysis of relatedness among the purified lectins: degrees of difference between lectins on the basis of amino acid composition were assessed by the parameter S ΔQ of Marchalonis and Weltman (9), according to the formula:

$$S\Delta Q = (Xi,j - Xk,j)^2$$

where i and k identify the particular proteins compared and Xj is the content (mole %) of a given amino acid of type j.

Table 1: Specificities and approximate subunit molecular weights of purified lectins

Subphylum	Species	Specificity of major lectin	Approximate mol. weight
Tunicata (Urochordata)	Didemnum candidum	D-galactose*	16000
	Halocynthia pyriformis	Sialoconjugates*	40000
Agnatha	Petromyzon marinus (eggs)	Sialoconjugates	43000
	Petromyzon marinus (serum)	Sialoconjugates	40000 + 43000

* Data from Vasta et al. (reference 4)

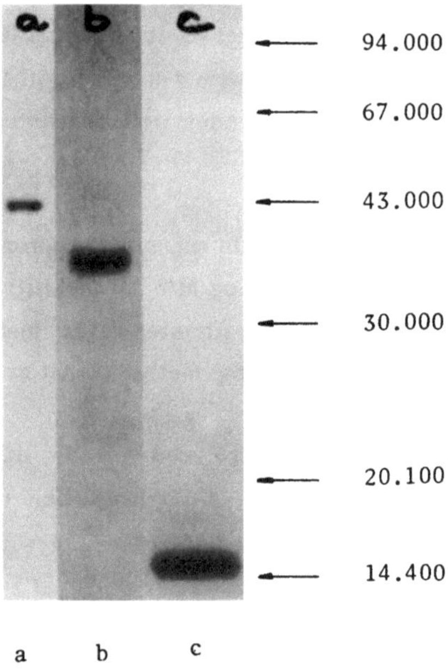

Figure 1: Sodium dodecyl sulphate polyacrylamide gel electrophoresis of reduced (1% 2-mercaptoethanol) lectins: a: Petromyzon egg lectin (affinity purified on formaldehyde fixed horse RBC); b: Halocynthia lectin (affinity purified on formaldehyde fixed horse RBC); c: Didemnum lectin (affinity purified on asialo-fetuin-Sepharose 4B). Arrows indicate molecular weight standards.

Results:

Our previous work about occurrence and specificity of lectins in tunicate species has shown a range of specificities in the serum lectins of ten species studied (4). The two species selected for the present report, Didemnum candidum and Halocynthia pyriformis exhibit major specificities for D-galactose and sialoconjugates respectively (4). On the other hand the cyclostome Petromyzon marinus exhibits serum and egg lectins which bind sialic acid-containing receptors . As suggested by Table 1, tunicate and cyclostome lectin subunits are different in their molecular weights and except for lamprey serum lectin which shows two bands of approximately 43,000 and 40,000 daltons (data not shown) each purified lectin appears to be constituted by a single type of subunit on the basis of apparent mass as shown by Figure 1. Data on the amino acid compositions of these subunits are given in Table 2, expressed as mole %. These amino acid compositions are compared by the $S\Delta Q$ parameter (Table 3): The two tunicate lectins differ only in 23 $S\Delta Q$ units and both

Table 2: Amino acid composition of purified lectins

Amino acid	Didemnum candidum	Halocynthia pyriformis	Petromyzon marinus Serum**	Petromyzon marinus Egg
Asp	11.9	13.5	14.3	11.1
Thr	5.9	5.6	6.7	4.4
Ser	7.9	8.0	6.3	13.3
Glu	7.9	8.7	12.0	12.9
Pro	3.9	4.3	4.3	7.1
Gly	11.2	10.2	5.7	8.6
Ala	9.9	9.1	5.5	5.9
Val	8.6	6.8	4.4	6.1
Met	0.6	1.9	0.5	0.7
Ile	3.9	4.5	4.6	3.2
Leu	6.6	6.8	8.0	6.9
Tyr	4.6	4.9	6.0	3.2
Phe	5.9	3.8	7.8	3.4
His	1.3	3.5	2.6	2.9
Lys	4.6	4.9	6.5	6.1
Arg	3.3	4.6	4.8	4.4

* Values expressed as residues % residues of carbohydrate-free protein. Cystein and tryptophan are not included.

** Data from Marchalonis and Edelman (reference 5)

Table 3: S ΔQ comparisons of purified lectins with potentially related proteins

	Didemnum lectin	Halocynthia lectin	Petromyzon serum lectin	Petromyzon egg lectin
Didemnum lectin	0			
Halocynthia lectin	23	0		
Petromyzon serum lectin	37	24	0	
Petromyzon egg lectin	103	78	74	0
Petromyzon μ chain	46	42	40	68
Stingray μ chain	105	93	40	40
Rabbit C-reactive protein	52	57	45	54
Mouse C-reactive protein	58	56	43	38

are closely related to the cyclostome serum lectin. Lamprey egg lectin appears to be distantly related to the serum lectin, which appears to be marginally related to the lamprey μ chain from serum immunoglobulin. This statistical difference stems largely from the relative abundance of serine in the egg lectin, and a homologous relationship between the lamprey lectins is not conclusively excluded. Tunicate lectins exhibit a large S ΔQ value when compared to the stingray μ chain, but appears to have some relatedness to lamprey lectins. Finally, the tunicate and cyclostome lectins show a marginal degree of relatedness to two vertebrate C-reactive proteins, humoral proteins specific for phosphorylcholine which can also precipitate glycosubstances (10).

Discussion:

Except for conservative taxa such as the class Merostomata (horseshoe crabs) and order Scorpiones which exhibit a certain homogeneity in the major specificity of the serum lectins of different species (11), no general correlation has been found between lectin specificity and taxonomic classifications. This statement holds for the tunicates as shown by our previous studies involving a screening of lectins of ten distinct tunicate species (4). In addition, the subunit molecular weights are quite diverse when comparing lectins from different tunicate and cyclostome species. Despite these size differences, it is still possible that evidence of a homologous relationship among lectins of tunicates and even lower vertebrates might emerge from a comparison of their primary amino acid sequence. The diversity in apparent mass among lectins of tunicate species and cyclostomes might be misleading because these molecules could represent a family of informational macromolecules which resembles immunoglobulins in evolving by duplication of a relatively small precursor gene. Since information on the primary structure of tunicate lectins and lower vertebrate immunoglobulins is not available at this time, we estimated the likelihood of relatedness of the basis of comparisons of their amino acid compositions as assessed by the $S\Delta Q$ parameter derived by Marchalonis and Weltman (9). This approach has given a reasonable correlation with amino acid sequence differences in the case of families of known proteins and has proven a useful method for predicting in a semi-quantititative sense whether or not two proteins might be related. The $S\Delta Q$ parameter was suitable for the present analysis because it is independent of the size of the molecules; another similar parameter $S\Delta n$ (12) allows a fairly precise estimate of relatedness but requires that proteins be the same size for the analysis to be valid. Our present results indicate that the two tunicate lectins are closely related in their amino acid composition, and presumably their sequences, although they are very different in combining site specificity and in subunit molecular weight. In addition, both of these are similar to the lamprey serum lectin in amino acid composition, but differ from the lamprey egg lectin. This is interesting because the specificities of Halocynthia and Petromyzon serum and egg lectins are very similar. However, the major amino acid difference is in serine which accounts for the relatively large observed differences. It is still possible that the sialoconjugate-specific lectins show a high degree of amino acid sequence homology. The lamprey serum lectin, in addition, shows a certain relatedness to lamprey serum immunoglobulin μ-chain and to the μ-chain of the elasmobranch stingray. It is interesting that the lamprey egg lectin also shows a fair degree of similarity to the stingray μ-chain. It is also noteworthy that the lamprey lectins show marginal relatedness to vertebrate C-reactive proteins, which might be considered a type of lectin, found in serum of higher vertebrates (10).

These results support our previous hypothesis that a super family of related recognition molecules exists which has evolved from a common precursor ancestral to such diverse classes of molecules as immunoglobulins, C-reactive proteins, complement components, products of the major histocompatibility complex, mammalian T-cell receptors, and humoral lectins of chordate species (13). The possibility of clear-cut homologies in recognition molecules of tunicates and lower vertebrates warrants further attention.

Acknowledgement: G.R.V., from the Consejo Nacional de Investigaciones Cientificas y Tecnicas de la Republica Argentina, is a recipient of a Fogarty International Fellowship (NIH) 1F05 T W 03155-01. This research was supported in part by the State of South Carolina Appropriations for Biomedical Research. We thank Mrs. Joan Eynon for typing the manuscript.

References:

1. Marchalonis, J.J.: Immunity in Evolution, Harvard University Press, Cambridge, 1977.

2. Barondes, S.H.: Ann. Rev. Biochem. 50, 207-231 (1981).

3. Berrill, N.J.: The Origin of Vertebrates. Oxford University Press, New York, 1955.

4. Vasta, G.R., Warr, G.W., Marchalonis, J.J.: Comp. Biochem. Physiol. In press.

5. Marchalonis, J.J., Edelman, G.M.: J. Mol. Biol. 32, 453-465 (1968).

6. Nowak, T.P., Barondes, S.H.: Biochim. Biophys. Acta 393, 115-123 (1975).

7. Svennerholm, L.: Biochim. Biophys. Acta 24, 604- (1957).

8. Alt, J., Heymann, E., Krish, K.: Eur. J. Biochem. 53, 357-369 (1975).

9. Marchalonis, J.J., Weltman, J.K.: Comp. Biochem. Physiol. 38B, 609-625 (1971).

10. Uhlenbruck, G., Karduck, D., Haupt, D., Schwick, H.G.: Z. Immun. Forsch. 155, 262-266 (1979).

11. Vasta, G.R., Marchalonis, J.J.: Am. Zool. In press.

12. Cornish-Bowden, A.: Biochem. J. 191, 349-354 (1980).

13. Marchalonis, J.J., Warr, G.W., Vasta, G.R., Ledford, B.E.: In "Receptors in Biological Systems" Ed. by R. Gorczynski, Marcel Dekker, New York. In press.

IS THE SELECTIVITY OF PROTEID YOLK PRECURSOR UPTAKE IN INSECTS BASED ON A LECTIN-TYPE BINDING REACTION ?

Dirk Stynen and Arnold De Loof

Zoological Institute, Catholic University of Leuven, Naamsestraat 59,
B-3000 Leuven, Belgium

Introduction

The process of proteid yolk formation during vitellogenesis in insects very much resembles that in egg laying vertebrates. Under appropriate conditions, endocrine events trigger the synthesis of proteid yolk precursors (vitellogenins) by the fat body, which is the main protein synthesizing tissue in insects. The haemolymph transports the vitellogenins to the ovary where they are selectively accumulated by the developing oocytes through micropinocytosis (for review see 1). The selectivity of vitellogenin endocytosis has already been demonstrated several times (2,3), however, the underlying mechanism is not yet understood. Obviously, vitellogenic proteins possess some properties which enable their selective uptake and which distinguish them from other haemolymph proteins.

In previous papers we dealt with the haemagglutinin activity in the haemolymph of Colorado potato beetle larvae (4) and reproducing females (5). The latter study demonstrated, by means of haemagglutination and precipitation experiments, that the major vitellogenin, vitellogenin 1, has an affinity for hexosamines and sulphated polysaccharides. Another vitellogenic protein, chromoprotein 2, was also shown to react with sulphated polysaccharides.
In the present report we give some additional data and discuss how our findings may contribute to a better understanding of the mechanism of yolk formation.

Lectins, Vol. III

Materials and Methods

Animals. Reproducing Colorado potato beetles (Leptinotarsa decemlineata Say) were reared under long-day conditions (LD) (16 h photoperiod). Beetles exposed to a short photoperiod (SD) (10 h photoperiod) do not reproduce.

Haemagglutination and haemagglutination inhibition experiments. Eighty µl of a 1 % human erythrocyte suspension in phosphate buffered saline (PBS) was added to 80 µl (0.15 M, pH = 7.1) of twofold serial dilutions of centrifuged haemolymph in microtitrator plates. Results were read after the mixtures had been incubated for 1 h at 37°C and overnight at 4°C. The haemagglutinin titre is defined as the reciprocal of the highest reactive haemolymph dilution. The complete list and the origin of the 15 monosaccharides, 3 disaccharides, 8 polysaccharides and 2 glycoproteins we used in haemagglutination inhibition tests have already been published (5). This paper also fully describes the preparation method of the ovary extract. This extract only contains the non dialyzable, heat-stable substances of the active ovary. Glycosubstances were dissolved in PBS. Eighty µl of the solution under test were added to 40 µl of twofold serial dilutions of haemolymph. After incubation during 1 h at 37°C, 40 µl of a 2 % erythrocyte (group 0) suspension were added, and incubation continued for a further hour. Inhibitory effect is expressed as the final sugar concentration that reduces the haemagglutinating activity of the haemolymph by one dilution. Glycosubstances not affecting haemagglutination at concentrations of 0.2 M (mono- and disaccharides) or 5 mg/ml (polysaccharides and glycoproteins) were considered as non inhibitory.

Immunological techniques and precipitation tests. An antiserum against the haemagglutinins was prepared by injecting agglutinated red blood cells into a rabbit (for details see 4). The preparation and specificity of the antiserum against chromoprotein 2 have been described by Peferoen et al. (6). Radial immunodiffusion was carried out according to Mancini et al. (7). Immunoelectrophoresis was performed in a Corning ACI apparatus on Corning ACI Universal ready to use agarose gels using sodium barbital-EDTA buffer (0.05 M, pH = 8.6) as the electrolyte solution. Gels were stained in Amido Black 10 B.

TABLE 1

Agglutination of human erythrocytes by the haemolymph of reproducing female Colorado potato beetles.

Erythrocytes	Haemagglutinin titre
Blood-group A	256
B	256
AB	256
O	256

Results

The haemagglutinin activity in the haemolymph of reproducing female Colorado beetles does not show blood-group specificity. Erythrocytes of all blood-groups of the ABO system are agglutinated to the same extent. The haemagglutinin titre is 256 in all cases (Table 1). Haemagglutination inhibition tests show that the haemagglutinins have a low affinity for hexosamines and a higher one for sulphated polysaccharides (Table 2). An ovary extract, containing the heat-stable, non dialyzable substances, also inhibits the haemagglutination. This indicates that in the ovary, substances occur which react with the haemagglutinins.

TABLE 2

Inhibitory effect of saccharides on haemagglutination.

D(+)-Glucosamine	30 mM *
D-Galactosamine	30 mM
D-Mannosamine	30 mM
Heparin	0.25 %
Mucin	0.17 %
Dextran sulphate	0.0038 %

* Results expressed as the concentration that decreases haemagglutinin activity of the haemolymph by one dilution.

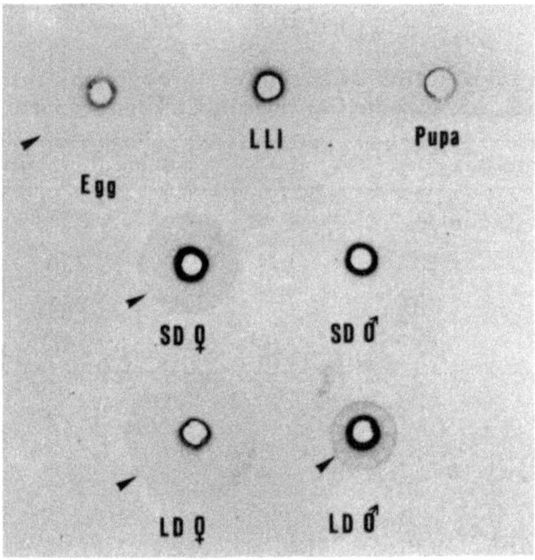

Fig. 1. Radial immunodiffusion of egg homogenate and haemolymph of different developmental stages. In the agarose gel : antiserum against haemagglutinins of reproducing female Colorado beetles. LLI : last larval instar. LD : reproducing animals. SD : non reproducing animals. Immunoprecipitates are indicated by arrows. Other precipitates are not specific.

By using an antiserum against the haemagglutinin in a radial immunodiffusion test, we obtained a semiquantitative notion as to the presence of the haemagglutinin in the different developmental stages (Fig. 1). The results are typical for a vitellogenic protein : high amounts in egg homogenate and reproducing females, low amounts or absence in males, non reproducing adults and juvenile animals.

Fig. 2 supports our earlier observation (5) that chromoprotein 2, another vitellogenic protein, reacts with dextran sulphate. The haemolymph protein precipitated by dextran sulphate has, indeed, the same electrophoretic mobility as chromoprotein 2.

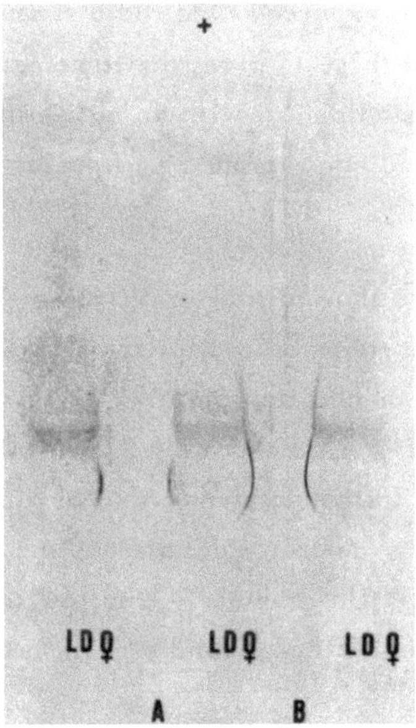

Fig. 2. Electrophoresis of a haemolymph protein having the same electrophoretic mobility as chromoprotein 2. Sample : haemolymph of reproducing female Colorado beetles (LD ♀). A : Affinity precipitation with dextran sulphate solution (0.25 %). B : Immunoprecipitation with antiserum against chromoprotein 2.

Discussion

Unlike the lectins of the Gypsy moth, for example, (8) and the Californian fleshfly (unpublished observation), the haemagglutinins of the Colorado potato beetle do not distinguish between the human blood-groups of the ABO system. In this respect they more resemble the haemagglutinins of the cricket Teleogryllus commodus (9).

Immunological experiments with the antiserum against haemagglutinins already revealed that the haemagglutinin in female haemolymph is immunologically identical to the vitellogenin 1 (5), which is the haemolymph precursor of the major yolk protein, constituting more than 60 % of the total amount of proteid yolk (6). So, as could be expected, the radial immunodiffusion with the anti-haemagglutinin antiserum (Fig. 1) gives the same result as was described by Peferoen et al. (6) in a similar test with an antiserum against vitellogenin 1. Juvenile stages, males and non reproducing females have fewer or no vitellogenins in the haemolymph.

In a previous paper (5) we showed that in diffusion tests in agarose gels 2 vitellogenic haemolymph proteins precipitated with dextran sulphate. One of them could be identified as chromoprotein 2, since the precipitation lines had the typical orange colour of chromoprotein 2, which is due to its carotenoid moiety. Fig. 2 provides further evidence that our identification was correct: the protein precipitated by dextran sulphate has the same electrophoretic mobility as chromoprotein 2. Unfortunately, the low ionic strength of the 'ready to use' agarose gels prevented the second protein precipitating after electrophoretic separation.

Summarizing, our study shows that vitellogenin 1 and chromoprotein 2, two important yolk precursors, have an affinity for hexosamines and sulphated polysaccharides. This property may be important in ensuring the selectivity of the accumulation of vitellogenic proteins by the oocyte. A lectin-type binding reaction would be more efficient than other mechanisms which were earlier proposed. Kunkel and Pan (3) studied the uptake of the vitellogenins of Blatella germanica and Hyalophora cecropia by each other's oocytes. Although both vitellogenins have a pronounced negative charge at physiological pH and other common chemical properties (sedimentation coefficient, Stokes' radius, subunit structure ...), no mutual accumulation occurs. This observation and the fact that the vitellogenins are not the only negatively charged haemolymph proteins suggest that the charge and other physical characteristics of the vitellogenin molecule are not the main determinants governing their selective endocytosis. Analysis of the carbohydrate structure of the Blatella germanica vitellin did not yield any direct evidence that this would be the crucial factor (10). Since we show that vitellogenins are themselves lectins, it is the carbohydrate structure of the vitellogenin receptor, rather than that of the vitellogenin, which is important.

The exact location of the specific reaction in the ovary between vitellogenins and saccharides remains unknown. A glycoprotein or glycolipid in the coated pits of the oolemma could function as a receptor molecule and mediate the internalisation of the yolk precursor. This mechanism is perhaps the most obvious. However, evidence can be found for yet another possibility.

Relying on electronmicroscopic observations, De Loof et al. (11) suggested that, prior to pinocytotic uptake, the vitellogenic proteins are precipitated by mucopolysaccharides, which are secreted by the follicle cells into the intercellular space between follicle epithelium and the oocyte. Our results clearly show that Colorado beetle vitellogenins precipitate with polysaccharides. Anderson and Telfer (12,13) also attribute an important role to a follicle cell product which would function as a blood protein binding agent. It is noteworthy that this product can be labeled with tritiated glucosamine, one of the inhibitors of the haemagglutination reaction (Table 2).

Further experiments are required to elucidate the whole mechanism of the selective yolk protein endocytosis, on the nature and function of the heat-stable macromolecular ovarian substance which inhibits the haemagglutinin reaction, on the localization and nature of the vitellogenin receptor and on the role of the follicle cells.

Acknowledgements

Dirk Stynen acknowledges a research scholarship of the Institute for Research in Industry and Agriculture (I.W.O.N.L.) of Belgium.

References

1. Hagedorn, H.H., Kunkel, J.G.: Ann. Rev. Entomol. 24, 475-505 (1979).
2. Telfer, W.H.: Biol. Bull. 118, 338-351 (1960).
3. Kunkel, J.G., Pan, M.L.: J. Insect Physiol. 22, 809-818 (1976).

4. Stynen, D., Peferoen, M., De Loof, A.: J. Insect Physiol. 28, 465-470 (1982).

5. Stynen, D., De Loof, A.: Wilhelm Roux's Archives (1982) (in press).

6. Peferoen, M., Stynen, D., De Loof, A.: Comp. Bioch. Physiol. (1982) (in press).

7. Mancini, G., Carbonara, A.O., Heremans, J.F.: Immunocytochemistry 2, 235-254 (1965).

8. Umetsu, K., Kosaka, S., Kashimura, S., Suzuki, T.: J. exp. Med. 129, 161-167 (1979).

9. Hapner, K.D., Jermyn, M.A.: Insect Biochem. 11, 287-295 (1981).

10. Kunkel, J.G., Shepard, G.L., McCarthy, R.A., Ethier, D.B., Nordin, J.H.: Insect Biochem. 10, 703-714 (1980).

11. De Loof, A., Lagasse, A., Bohyn, W.: Proc. K. Ned. Akad. Wet. Ser. C 75, 125-143 (1972).

12. Anderson, L.M., Telfer, W.H.: Tissue and Cell 1, 633-644 (1969).

13. Anderson, L.M., Telfer, W.H.: J. Cell Physiol. 76, 37-54 (1970).

Part IV

Microbial Lectins

BACTERIAL SURFACE LECTINS

T. Wadström

Department of Bacteriology and Epizootology, Biomedicum, Box 583, S-751 23 Uppsala, Sweden

T. J. Trust

Department of Biochemistry and Microbiology, University of Victoria, Victoria, British Columbia, Canada

D. E. Brooks

Departments of Pathology and Chemistry, University of British Columbia, Vancouver, British Columbia, Canada

Despite the many studies on plant lectins and their wide use in many fields of biological research (1, 2), it has not been widely realized that bacteria also produce lectin-like proteins. Some of these proteins have important roles in microbial pathogenicity and include exotoxins and surface proteins that are involved in enhanced specific adhesion of the bacterial cell to the eucaryotic cell. This minireview illustrates the nature and importance of these adhesive proteins, and defines these carbohydrate-recognizing bacterial proteins as lectins.

Bacterial surface lectins are defined as carbohydrate-binding proteins involved in cellular adhesion, which may or may not exhibit cell-aggregating activity, and which may or may not be inhibited by monosaccharides. The majority of these surface lectins (adhesins or colonization factors) appears to recognize complex oligomers on the eucaryotic cell surface. Indeed, the identity of most of these carbohydrates has yet to be identified. These bacterial surface lectins include fimbriae (pili)

Lectins, Vol. III

and certain outer membrane proteins of bacteria, as well as the fimbriae-like appendages (fibrillae, or fuzzy layers) of Gram positives.

Methods

Assays for bacterial surface lectins. The haemagglutination assay has most commonly been used to demonstrate bacterial adherence, i.e., the simultaneous adhesion of bacteria to the surface of two or more erythrocytes causing red cell aggregation via a bridging mechanism. This aggregation is detected visually on a slide, in a tube, or in a microtitre plate. Demonstration of lectin participation in this adherence has required that the haemagglutination is inhibited or reversed by monosaccharides or by more complex carbohydrates (3, 4).

Recently, an assay of haemagglutination which provides quantitative information on the kinetics, reversibility, and adhesive strength of the attachment process has been developed. The technique is an adaptation of a viscometric assay which has also been applied to the study of lectin-induced agglutination (5, 6). It is based on the fact that aggregated suspensions exhibit a higher apparent viscosity than otherwise equivalent but non-aggregated samples. The technique has been used to demonstrate that the agglutination reaction can proceed in two phases, the second of which is induced by shear forces (D.E. Brooks and T.J. Trust, to be published). This shear-induced bacterial haemagglutination produced by a mannose-sensitive haemagglutination (MSHA) strain of Aeromonas salmonicida and its incomplete reversal by α-methyl-mannoside closely parallells the behaviour of the tetravalent lectin concanavalin A in the viscometric assay (5). These findings have provided strong evidence that a major mechanism by which bacteria can adhere to cell surfaces is via lectin interaction. In this haemagglutination assay, the bacterial cell clearly acts as a multi-valent lectin, with several bridging sites.

A variety of other assay techniques has been employed to examine the bacterial attachment to many different cell types (7). The most commonly used eucaryotic cells have been obtained by mucosal scrapings in the oral cavity (bucca and pharynx) and desquamated urine sedimented cells. Preparations of gastrointestinal single cells or mucosal brush borders (8). For example, rabbits employed in studies on the attachment of human enterotoxigenic E. coli, while porcine brush borders have been used with porcine enterotoxigenic E. coli with the K88 surface fimbrial lectin-like antigen porcine (9). Several non-biological systems have been especially useful for studies on receptor ligand interactions. For example, the demonstration that haemagglutinating E. coli display hydrophobic interactions allowed the development of a rapid simple test for the detection of non-haemagglutinating adhesins, such as the so-called 987p lectin of porcine enterotoxigenic E. coli (10-13). Most recently, the development of an autoagglutination test in salt solutions of different molarities (the "salting out principle") (14) has produced a simple test to rapidly screen enterobacteria for the presence of lectin-like surface protein adhesins and to quantitate their relative surface hydrophobicities.

Results and Discussion

The bacterial surface lectin. The appendages found on the surface of most adhesive bacteria (fimbriae or pili) have been implicated as the bacterial ligands binding to host receptors. The best characterized case is type I fimbriae binding to mannose residues. These fimbriae are produced by members of the Enterobacteriaceae, and the type I fimbriae of E. coli have received the most attention. These fimbriae are arranged continously around the whole cell, and there are between 100 and 250 per cell. The type I fimbriae are about 7.0 nm in width and up to 100 nm in length (Table 1).

TABLE 1

Bacterial Surface Fimbrial Lectins[1]

Fimbriae type	Diameter	Haemagglutination	Cell adhesin
Type 1	7.0 nm	MSHA[2]	?[3]
2	7.0 nm	HA negative	
3	4.8 nm	HA negative	fungal and plant cells
4	4.0 nm	MRHA	*Proteus mirabilis* adhesin
E. coli F7[4]	7.0 nm	MRHA	p^k cell receptors (erythrocytes and urinary tract cells
E. coli CFAI	7.0 nm	MRHA	GM_2-like glycoconjugate
Pr. mirabilis (second type)	7.0 nm	MSHA?	

1) Modified from Wadström and Trust (15)

2) MRHA: mannose-resistant haemagglutination
MSHA: mannose-sensitive haemagglutination

3) Not proven to be an adhesin in *E. coli* mucosal infections promoting adhesion to epithelial cells (20).

4) F7 according to the recently proposed nomenclature for bacterial fimbriae (20) after a modified scheme originally proposed by Duguid (3, 4). See also references 16-20.

Sweeney and Freer (21) have obtained evidence for the presence of multiple binding sites on these haemagglutinating fimbriae which were progressively shortened by treatment with ultrasound. As the fimbriae were shortened, so their haemagglutinating capacity was decreased, but their ability to bind to erythrocytes did not decrease to the same extent. The results provide strong evidence that binding sites are located along the length of the pili, rather than on a single terminal binding site. Furthermore, isolated pili did not agglutinate inside-out vesicles prepared from horse erythrocytes, or liposomes, suggesting that the binding mechanism was not based on non-specific hydrophobic interactions.

While many bacterial species are now known to produce mannose-sensitive haemagglutinating fimbriae (Table 2), the structural relationships between these fimbriae are still not clear. Although the adhesive activities of the various type 1 fimbriae are similar, antigenic differences between these type 1 fimbriae suggest chemical and structural differences (3, 4, 7, 16).

TABLE 2

Monosaccharide specificities for some bacterial lectins[1]

Source	
D-Mannose	Escherichia coli Vibrio cholerae Klebsiella pneumoniae Salmonella typhimurium Aeromonas hydrophila
L-Fucose	Vibrio cholerae Aeromonas hydrophila
D-galactose	Pseudomonas aeruginosa Aeromonas hydrophila Fusobacterium nucleatum Actinomyces viscosus Streptococcus pyogenes
N-acetylgalactoseamine	Pseudomonas aeruginosa Actinomyces viscosus Outer bacterial fimbriae[2]
Sialic acid[3]	Mycoplasma pneumoniae Neiserria gonorrhoeae Bordetella pertussis

1) Modified from Wadström and Trust (15).

2) Terminal N-acetylgalactoseamine residues of gangliosides were proposed as receptor structure for gonococcal fimbriae (Buchanan et al., 1978), and fimbriae of human and animal enterotoxigenic E. coli.

3) CFAI and other fimbriae of enterotoxigenic E. coli recognize sialic acid residues on complex glycoconjugates (Wadström et al., in press).

The sugar inhibition pattern of the various type 1 fimbriae is similar. Old (22) studied the carbohydrate inhibition of haemagglutination by fimbriated *S. typhimurium* and *Shigella flexneri* cells and showed that the α-configuration at the C-1 position in the mannopyranoside molecule and unmodified hydroxyl groups at C-2, C-3, C-4, and C-6 of the D-mannose molecule were required for maximum binding to the bacterial fimbrial lectin sites. Similar results have since been reported with purified type 1 *E. coli* fimbriae (23), and purified *S. typhimurium* type 1 fimbriae (T. Korhonen, personal communication). Old (22) was in fact the first worker to notice and comment on the remarkable similarity in the sugar inhibition pattern of the type 1 fimbriae and the plant lectin concanavalin A. The monosaccharide specificities for some bacterial fimbrial lectins are summarized in Table 2.

While some species and strains probably produce only single fimbriae types, there is increasing evidence that a single bacterial strain can produce a variety of fimbriae types. In the case of *E. coli*, fimbriae production can be plasmid-mediated. This is true for the K88, K99, CFAI and CFAII fimbriae types.

Perhaps the best illustration of the ability of a single strain to produce different fimbriae types is *Neisseria gonorrhoeae*, strain P9, although there is no evidence for plasmid involvement in fimbriation: This organism can exist in at least one of four states (24-26). Isogenic variants of P9 have been shown to produce fimbriae with subunit molecular weights 19,500, 20,500, 21,000, and 18,500. These fimbriae are designated *a*, *B*, *D*, *G*, respectively. *D* and *G* pili were isolated from *a* and *B* fimbriated gonococci surviving after 7 days in guinea pig subcutaneous chambers. Despite the differences in molecular weight, *G* and *D* fimbriae show antigenic cross-reactivity with *a* and *B* fimbriae. Studies utilizing peptide mapping have shown that *a* and *B* subunits have regions of considerable structural homology and differ only in a small region of the molecule (26). It is likely that G and D fimbriae represent further modifications to a basic common structural component. The binding of the α and β fimbriae to cells has been shown to differ. Removal of sialic acid residues from buccal cell surface carbohydrates by treatment with neuraminidase markedly affected the binding of a fimbriae but had minimal effect on B fimbrial adhesion. Subsequently, treatment of neuraminidase-modified buccal cells with mixed exoglycosidases further drastically reduced the binding of a fimbriae, but not the binding of B fimbriae. These findings suggest that a fimbriae bind specifically to a receptor on buccal cells and that this receoptor contains both sialic acid and other sugars. The binding of B fimbriae to buccal and blood cells and of a fimbriae to blood cells appears not to involve carbohydrates and may result from hydrophobic interactions.

TABLE 3

Physical and chemical properties of some bacterial fimbrial lectins

Fimbrial type	Subunit molecular weight	Isoelectric point	% Carbohydrate
E. coli(type 1)	16,600-17,500	4.5-5.1	None
E. coli (K99)	2 subunits (Major-22,500) (Minor-29,500)	10.1	0.6
E. coli (K88)	25,000	-	1.0
N. gonorrhoeae	19,000±2,500	5.3 (minor band at 4.9)	1.3-2.0 (1-2 hexoses/ subunit)
P. aeruginosa K (PAK)	17,800±300	3.9	None
M. nonliquefaciens	17,000	-	-
C. renale (Strain type II)	19,000-19,400	4.35	None

Modified and abbreviated
After Pearce and Buchanan (27), and Wadström and Trust (15).

Fimbriae are not the only surface components mediating attachment. The outer membrane of N. gonorrhoeae also contains a protein, Protein II, which appears to mediate attachment. Protein II comprises a family of proteins with molecular weights ranging from 27,500 to 29,000 (24). The presence in the outer membrane of any one of the Protein II family increases attachment to buccal cells, Protein II giving the best increase in specific attachment. As was the case with purified fimbriae, a variety of simple sugars do not inhibit the attachment mediated by Protein II (28), but further studies are required to show whether a more complex carbohydrate may be involved in membrane receptor recognition.

Another pathogen where a membrane protein lectin likely mediates attachment is Mycoplasma. These organisms lack the cell wall of Gram-positive bacteria, but do have a sterol-containing membrane.

Mycoplasma are extracellular parasites when they grow in vivo. In order to obtain nutrients essential for growth they attach to the surfaces of animal cells. They are excellent model organisms for studies on bacterial lectins, because they must adhere to cells and because they recognize carbohydrate moieties in this attachment process. The close association of mycoplasma organisms with eucaryotic host cells has been well described using electron microscopy. Specialized electron dense structures at the tips of many mycoplasma cells appear to function as attachment organelles giving rise to polar or "end-on" attachment (29, 30). Scanning electron microscopy studies have shown that some forms of M. pneumoniae can also adhere to erythrocytes via membrane sites other than the tip structure. These adherent cells are rounded, pear-shaped, or irregular organisms which lack tip structure (31).

Glycoconjugates as lectin receptors. It has been known for many years that certain pathogens display a predilection for specific tissues. It seems likely that surface lectins of pathogenic bacteria will contribute to the tissue specificity by recognizing and binding to specific carbohydrates on the surface of eucaryotic cells. The surface of the cell which the bacteria encounters is composed of many different types of compounds, e.g., phospholipids, cholesterol, glycolipids and glycoproteins. The presence of the sugar sequences of the glycoconjugates at the external cell surface places them in a key position to function in receptor phenomena(32-35). In fact, carbohydrates are especially good candidates for participation in such recognition functions as receptors. A simple glucose disaccharide can exist in 11 distinct forms, and a glucose trisaccharide yield 176 anomeric configurations. A mixed trisaccharide obviously allows many more structural variants. Typical oligosaccharides of mammalian cell membranes are composed of between 4 and 20 monosaccharide residues, most commonly of glucose, galactose, mannose, fucose, N-acetylglucosamine, N-acetylgalactosamine, and N-acetylneuraminic acid (35). This allows for a multitude of different possible receptors.

Most recently, considerable attention has been focused on the role of glycolipids as receptors for bacterial lectins (18, 19). Certainly, the glycosphingolipid composition varies extensively for one tissue in different species, for different tissues of one species, and between individuals of the same species. The glycolipid diversity is clearly sufficient to make such compounds candidates as specific receptors in many systems.

In the urinary tract epithelium, certain glycolipids do seem to act as receptors for the attaching E. coli bacteria. The glycopipids involved are globotetraosylceramides, and the structural site recognized seems to include as part the sequence Ga1α1 - 4Ga1, but the detailed specificity of the recognition has not yet been characterized (18, 19).

The evidence that the receptor specificity of the bacterial

surface lectin and the presence of receptors on the host tissue have clinical consequences has best been established for E.coli carrying the adhesin K88, which cause small intestine infection in neonatal pigs. The K88 antigen confers the ability to the bacteria to bind to the intestinal epithelium. Different herds of pigs vary in susceptibility to neonatal diarrhoea, and susceptibility to K88 colonization is inherited as a mendelian dominant. Susceptibility to infection clearly correlates with susceptibility to bacterial attachment and glycolipid analyses of brush border preparations from susceptible and resistant piglets have suggested a specific minor glycolipid as the receptor for the K88 adhesin. This glycolipid is missing in resistant animals (H. Leffler, personal communication).

There is now evidence for involvement of glycoconjugates in the receptors for K99 and CFAI adhesins (11, 36). We have systematically investigated inhibition of haemagglutination by K99 and CFA1 enterotoxigenic E. coli strains with crude and purified ganglioside preparations. The observation that only monosialogangliosides gave inhibition of K99 and CFAI enterotoxigenic E. coli with no inhibition at concentrations much above the critical micellar concentration prompted inhibition studies with highly purified specific gangliosides. These results suggest that both K99 and CFAI recognize terminal galactosylresidues on glycoconjugates in the erythrocyte membrane. This binding may well involve both glycoproteins and glycolipids. More recent studies by others have confirmed these observations with CFAI enterotoxigenic E. coli. Glycoproteins have also been implicated in recognition; for example, there is strong evidence that sialic acid residues of cell membrane glycoconjugates act as specific receptor sites for Mycoplasma pneumoniae and Mycoplasma gallisepticum (37). Indeed, the predominant sialoglycoprotein of human erythrocytes, glycophorin, has been shown to be a major receptor for M. gallinarum and for M. pneumoniae. In the case of M. pneumoniae, there also appears to be additional receptors on human erythrocytes. These as yet unidentified receptors are destroyed by trypsin. In the case of M. gallisepticum, it has been demonstrated that neuraminidase-treated glycophorin did not bind. The best inhibitors had a mucin-like structure, with high molecular weights and high sialic acid contents. N-acetylneuraminic acid appeared to be the favored sialic acid structure for binding, but there was no strict specificity for its anomeric linkage.

The glycoconjugate recognized by the bacterial surface lectin does not always have to be of host origin. Some bacteria may in fact only bind to virus-infected cells. This has been demonstrated in the case of Group B streptococci. These organisms bind to kidney epithelial cells but only after these kidney cells have been infected with influenza virus (38). Indeed, the adherence of the streptococci appears to involve recognition of viral haemagglutinin, a glycoprotein which is inserted into the mammalian cell membrane before exit of mature virus. It is likely that more examples of this phenomenon will be found.

Prospects

It is now quite clear that there is a very large number of different bacterial surface lectins. Quite a deal of information has been accumulated on a small number of these, however, in no case has the actual lectin portion of the molecule been identified. This gap in our knowledge concerning the bacterial surface lectin (adhesin) needs to be filled. Studies are urgently needed to elucidate the composition, structure, and arrangement of those portions of the adhesive molecule which participate in the lectin interaction. We need to know how many lectin sites are present per molecule, and per bacterial cell.

The question of disease prophylaxis and therapy will obviously command a great deal of attention. Immunization with critical virulence lectins has met with success. Indeed, vaccines which protect against K88 and K99 antigens are commercially available. In these cases, the adult female animal is immunized and transfers protective anti-lectin (K88 or K99) antibody to the suckling neonates via colostrum (13).

The possibility that purified bacterial surface lectins can be used for blocking receptors on target surfaces has also to be explored. One of the few examples of this to date is the protection afforded to mice by oral administration of "cholera lectin", when the animals were subjected to oral challenge with V. cholerae (39). However, given the diversity of bacterial surface lectins and the probable complexity of their carbohydrate receptor structures (40), it may be more profitable to attempt to prevent mucosal surface colonization by nonspecific interactions based on lectin carbohydrate binding characteristics. There is already one good example of this line of thought. Based on the concept that bacterial surface lectins of fimbrial nature show pronounced surface hydrophobic properties, hydrophobic gels were used quite sucessfully to prevent intestinal infections in an infant rabbit model with human enterotoxigenic E. coli (15, 41).

Another approach might be to use receptor megatherapy (32, 33). This may be most appropriate for gastrointestinal infections. Here it may be possible for oral administration of compounds with sufficiently great affinity for a bacterial adherence factor to compete with the natural epithelial cell receptor for binding to the organism. Receptor megatherapy has been used experimentally in the prevention of colonization of the urinary tract of mice with E. coli. In the case of simple sugars, it may even be reasonable to attempt to displace an organism from its binding site by using the appropriate sugar for elution. Aronson et al. (42) have reported that the injection of mannose-sensitive E. coli in the presence of methyl-α-D-mannopyranoside resulted in a considerable reduction in the number of bacteriuric mice. A simple variation of receptor megatherapy might also be useful as a prophylactic measure in some cases (32), e.g., an excess of some non-toxic molecule capable of binding to the eucaryotic receptor. By occupying the site it would block attachment.

TABLE 4

Some approaches to identify cell surface receptors for bacterial lectins

Inhibition studies with	Example
Monosaccharides	Type 1 fimbriae
Oligosaccharides	Salivary glycoproteins and peptides (Streptococcus mutans, Actinomyces viscosus)
Monosaccharide coupled to a solid phase	Adhesin(s) of V. cholerae (L-fucose to Sephadex beads)
Glycoconjugates	CFAI and K99 adhesin of enterotoxigenic E. coli
Erythrocytes with different blood group antigens	F7 fimbriae E. coli isolated from urinary tract infections
Blocking experiments with plant lectins	Type 1 fimbriae of E. coli K88 adhesin of porcine enterotoxigenic E. coli Gonococcal fimbriae Schistosoma haemopharasites

Modified from Wadström and Trust (15). See also references 4, 7, 9, 23, 40-43.

References

1. Nicolson, G.L.: Intern. Rev. Cytol, 39, 89-190 (1974).

2. Lis, H., Sharon, N.: In "The Biochemistry of Plants: A Comprenensive Treatise" (P.K. Stumpf and E.E. Conn, eds.) Vol. VI, Proteins and Nucleic Acids (A. Marcus, ed.), Academic Press, New York, in press.

3. Duguid, J.P., Campbell, D.: J. med. Microbiol. 2, 535-553 (1969).

4. Duguid, J.P., Old, D.C.: In "Receptors and Recognition", Series B, Vol. 6, Chapman and Hall, London, pp. 185-218 (1980).

5. Greig, R.G., Brooks, D.E.: Nature 282, 738-739 (1979).

6. Greig, R.G., Brooks, D.E.: Biochim. Biophys. Acta 641, 410-415 (1981).

7. Jones, G.W.: In "Microbial Interactions" (J.L. Reissig, ed.). In "Receptors and Recognition", Series B., Vol. 3, Chapman and Hall, London, pp. 139-176 (1977).

8. Wadström, T., Faris, A., Freer, J.H., Habte, D., Hallberg, D, Lungh, Å.: Scand. J. Infect. Dis. Suppl. 24, pp.148-153 (1980).

9. Sellwood, R.:Biochim, Biophys, Acta 632, 326-335 (1980).

10. Smyth, C.J., Jonsson, P., Olsson, E., Söderlind, O., Hjertén, S., Wadström, T.: Infect. Immun. 22, 462-472 (1978).

11. Faris, A., Lindahl, M., Wadström, T.: FEMS Microbiology letters 7, 265-269 (1980).

12. Faris, A., Wadström, T., Freer, J.H.: Current Microbiol. 5, 67-72 (1981).

13. Moon, H.W., Kohler, E.M., Schneider, R.A., Whipp, S.C.: Infect. Immun. 27, 222-230 (1980).

14. Lindahl, M., Faris, A., Wadström, T., Hjertén, S.: Biochim. Biophys. Acta 677, 471-476 (1981).

15. Wadström, T., Trust, T.J.: In "Medical Microbiology" (J. Jeljaszewicz and C.F.S. Easmon, eds.), Academic Press, London, Vol. 1, in press.

16. Eshdat, Y., Sharon, N.: Methods in Enzym. 83, 386-391 (1982).

17. Gilboa-Garber, N.M.: Methods in Enzym. 83, 378-385 (1982).

18. Källenius, G., Möllby, R., Svensson, S.B., Winberg, J., Lundblad, A., Svensson, S., Cedergren, B.: FEMS Microbiol. letters 7, 297-302 (1980).

19. Leffler, H., Svanborg-Edén, C.: FEMS Microbiol. letters 8, 127-143 (1980).

20. Ørskov, L., Ørskov, F., Birch-Andersen, A.: Infect. Immun. 27, 657-666 (1980).

21. Sweeney, G., Freer, J.H.: J. gen. Microbiol. 112, 321-329 (1979).

22. Salit, I.E., Gotschlich, E.C.: J. Exp. Med. 146, 1182-1193 (1977).

23. Old, D.C.: J. gen. Microbiol. 71, 149-157 (1972).

24. Lambden, P.R. Heckels, J.E., James. L.T., Watt, P.J.: J. gen. Microbiol. 114, 305-310 (1979).

25. Lambden, P.R., Robertson, J.N., Watt. P.J.: J. Bacteriol. 141, 393-396 (1980).

26. Lambden, P.R., Heckels, J.E., McBride, H., Watt, P.J.: FEMS Microbiol. letters 10, 339-341 (1981).

27. Pearce, W.A., Buchanan, T.M.: In "Bacterial Adherence" (E.H. Beachey, ed.) In "Receptors and Recognition", Series B., Vol. 6, Chapman and Hall, London, pp. 289-344 (1980).

28. Watt, P.J.: In "Cell Adhesion and Motility". The Third Symposium of the British Society for Cell Biology (A.S.G. Curtis and J.D. Pitts, eds.), Cambridge University Press, Cambridge, pp. 137-170 (1980).

29. Bredt, W., Feldner, J., Hanne, I.: In "Adhesion and Microorganism Pathogenicity", CIBA Foundation Symposium 80, Pitman Medical, Bath, England, pp. 3-16 (1980).

30. Collier, A.M.: In "Bacterial Adherence". In "Receptors and Recognition", Series B, Vol. 6, Chapman and Hall, London, pp. 159-184 (1980).

31. Razin, S., Kahane, I., Banai, M., Bredt, W. In "Adhesion and Microorganism Pathogenicity", CIBA Foundations Symposium 80, Pitman Medical, Bath, England, pp. 98-118 (1981).

32. Keusch, G.T.: Am. J. Clin. Nutrition 31, 2208-2218 (1978).

33. Keusch, G.T.: Rev. Infect. Dis. 1, 517-529 (1979).

34. Hart, D.A.: Am. J. Clin. Nutrition 33, 2416-2439 (1980).

35. Barondes, S.H.: Ann. Rev. Biochem. 50, 207-231 (1981).

36. Lindahl, M., Faris, A., Wadström, T.: Attachment of microorganisms to the gastrointestinal mucosal surface (E.C. Boedeker, ed.), CRC Press Inc., Boca Raton, U.S.A., in press.

37. Banai, M., Razin, S., Bredt, W., Kahane, I.: Infect. Immun. 32, 628-634 (1980).

38. Pan, Y.T., Schmitt, J.W., Sanford, B.A., Elbein, A.D.: J. Bacter. 139, 507-514 (1979).

39. Finkelstein, R.A., Arita, M., Clements, J.D., Nelson, E.T.: Proc. 13th Joint Cholera Res. Conf., Atlanta, Georgia, U.S.A. Sept. 19-21, U.S. Governm. Printing Office, Washington, D.C. pp, 137-151 (1977).

40. Atkinson, H.M., Trust, T.J.: Infect. Immun. 27, 928-946 (1980).

41. Wadström, T., Faris, A., Lindahl, M., Hjertén, S., Ågerup, B.: Scand. J. Infect. Dis. 13, 129-132 (1981).

42. Aronson, M., Medalea, O., Schori, L., Mirelman, D., Sharon, N., Ofek, I.: J. Infect. Dis. 139, 329-332 (1979).

43. Buchanan, T.M., Pearce, W.A., Chen, K.C.S.: In "Immunobiology of Neisseria gonorrhoea", ASM, Washington, D.C., pp. 242-249 (1978).

THE BIOLOGICAL FUNCTIONS OF PSEUDOMONAS AERUGINOSA LECTINS

Nechama Gilboa-Garber
Department of Life Sciences, Bar-Ilan University
Ramat-Gan, 52 100, Israel

Introduction

Pseudomonas aeruginosa is one of the most important bacterial pathogens involved in human infections. Patients with malignancies and individuals whose immune status has been compromised are very susceptible to infection by it (1). The spectrum of clinical diseases produced by *P. aeruginosa* ranges from urinary tract infection to septicemia, pneumonia, meningitis and infections of drainage sites resulting from surgery or trauma. Cystic fibrosis patients are prone to this bacterium which is associated with progressive pulmonary infections in them (1). Most of the clinical isolates of *P. aeruginosa* produce several extracellular factors which may contribute to their high pathogenicity. These factors have been shown to exhibit cytotoxicity *in vivo* or *in vitro*, or both. They include: protein exotoxins, which inhibit protein synthesis due to catalysis of NAD - adenosine diphosphate ribosyltransfer to elongation factor 2 (2), proteases (3), elastase (3) lipase, phospholipase, hemolysin, slime and pyocyanin. These substances cause severe damage in the host (4) by decomposition of proteins, inhibition of new protein synthesis (5), lysis of blood cells and vessels (4) and necrosis of the tissues around them . Most of the clinical isolates of *P. aeruginosa* also contain considerable activities of special cellular substances such as the enzyme acetylcholinesterase (6), which is rare in bacteria and a high hemagglutinat-

Lectins, Vol. III

ing activity (7). The hemagglutinating activity, found in extracts of the bacterium is due to the presence of two lectins: Lectin PA-I is specific for D-galactose and its derivatives (α-D-Gal > D-Gal; ßD-Gal > D-Gal NAc) and lectin PA-II binds L-Fuc > D-man (and mannan); L-Gal > D-Fru (8). Both lectins were purified by affinity chromatography (8).

The Pseudomonas lectins were shown to be located mainly inside the bacterial cells and in low amounts on the cell surface, in the periplasmic space and on the cytoplasmic membrane (9). The levels on the cell surface and in the cell extracts vary during the course of the Pseudomonas cells growth in the culture and are dependent on the composition of the medium (8). There is no correlation between the lectin activity in the cell extracts and the hemagglutinating activity of the intact cells from which the extracts were prepared. Hemagglutinating activity of the intact cells is highest during the first two days of the culture growth, whereas the extract lectin activity is highest in 3-4-day-old cells. The two lectins interact with many cell types. Their activity is usually measured by using papain-treated human erythrocytes (8). They were shown to exhibit diverse biological activities such as mitogenic stimulation of human peripheral lymphocytes (10) with their blastogenic transformation; enhancement of growth rate and phagocytosis in unicellular organisms (11); inhibition or augmentation (depending on the system conditions) of bacteria phagocytosis by PMN leukocytes, (12) interaction with serum glycoproteins, etc. They have also been used for specific agglutination of certain enteropathogenic E. coli strains and for the preparation of a vaccine against Pseudomonas. Injection of the purified Pseudomonas lectin preparations to mice fully protected them from lethal infection with the autologous live bacteria (13). The lectin interaction with various cells may serve the bacterium, which bears them on its surface, in binding to the host cells. The adherence of bacteria to the host cells may be important for its invasion into it as was also shown in the case of E. coli surface lectins (14,15). Subsequently the Pseudomonas lectin interaction with the host plasma proteins and PMN leukocytes may protect it from the immune mechanisms of the host. The present communication describes a new aspect of the lectin contribution to the Pseudomonas pathogenicity. We have found a close linkage, or correlation, between the lectin level inside the bacterium cells and the ability of the cells to secrete the three exocellular toxic factors: protease, hemolysin and pyocyanin.

Materials and Methods

Twelve P. aeruginosa strains which differ serologically and two mutants of one of them (strain No. 1 in Tables I and II) have been examined. They were grown with shaking at 28°C for 3 days (8) in choline-containing nutrient broth and then the cells were separated from the culture medium by centrifugation. The cells were disintegrated by ultrasonic vibrations and their cell-free extracts were examined for lectin activity. The medium supernatant was used for determinations of protease, hemolysin and pyocyanin. The lectin activity was determined by hemagglutination assay using papain-treated human erythrocytes as previously described (8).

The protease activity in the supernatants of the Pseudomonas culture medium was determined spectrophotometrically, with azo-casein as a substrate, and by measuring the clearing zones around wells filled with the examined samples in 30% w/v skim milk (Difco) - containing agar-plates. The reaction mixture for the spectrophotometric method contained 1.0mL of azo-casein reagent (0.5% solution in distilled water adjusted to pH 7.5, 1.0mL of 0.02M Tris-HCl buffer (pH 7.5) containing 0.002M $CaCl_2$, and 10 µL of the enzyme solution. The reaction mixture was stopped after 15 min of incubation at 37°C by the addition of 2.0ml of 10% trichloroacetic acid. After standing for 30 min at room temperature, the precipitate was removed by centrifugation and the absorbance at 370nm of the supernatant fluid was measured.

Hemolysin activity was determined in tubes containing 1.5ml of 5% human erythrocyte suspension in physiological NaCl solution mixed with equal vol of double dilutions of the Pseudomonas culture supernatant. After incubation at room temperature for 1h and centrifugation, the concentration of hemoglobin in the supernatant fluid was determined spectrophotometrically at 540 nm. The pyocyanin secretion to the medium was determined by measuring its green colour spectrophotometrically at 660 nm.

Results and Discussion

The strains were found to vary considerably in lectin, protease, hemolysin and pyocyanin production from high to no production. The strong indication for the linkage between the four substances was based on the following observations:

a) a significant correlation between lectin levels in the cell extracts of various strains and the levels of protease, hemolysin and pyocyanin in their culture supernatant medium (Table I).

The activities of the cellular enzymes acetylcholinesterase, 5' nucleotidase and alkaline phosphatase in these strains did not correlate with their lectin levels.

b) Concomitant reduction of the lectin and the three exocellular factor levels in two mutants obtained from P. aeruginosa wild type which produces them at a high level (Table II, Fig. 1).

TABLE I.

Levels of lectins and the three exocellular factors (protease, hemolysin and pyocyanin) in various *P. aeruginosa* strains

	P.aeruginosa strain:											
	1	2	3	4	5	6	7	8	9	10	11	12
Lect.	256	64	64	4	0	16	128	512	64	512	0	8
Prot.	20	14	20	4	0	12	20	22	4	22	0	10
Hem.	227	200	275	22	0	166	260	210	142	270	0	81
Pyo.	192	41	200	12	0	72	148	82	18	72	0	192

Lect. = Lectin titer reciprocal; Prot. = Protease activity (units); Hem. = Hemolysin (units); Pyo. = Pyocyanin (units)

c) Maximal activities of both the lectins and the three exocellular factors at the late stationary phase of the cell growth.

d) similar repression of the production of both the lectins and the three exocellular factors without inhibition of the bacteria growth was obtained by the addition of the sugars D-glucose, D-galactose or D-mannose (at 2% concentration) to the culture medium. The repression was highest with D-mannose (D-Man > D-Gal > D-glc).

TABLE II

Levels of lectins and the three exocellular factors in *P. aeruginosa* wild type and its two deficient mutants

P. aeruginosa strain:	Wild Type	Mutant 1	Mutant 2
Lectin titer reciprocal	256	8	16
Protease (u)	20	10	12
Hemolysin (u)	227	62	180
Pyocyanin (u)	192	30	63

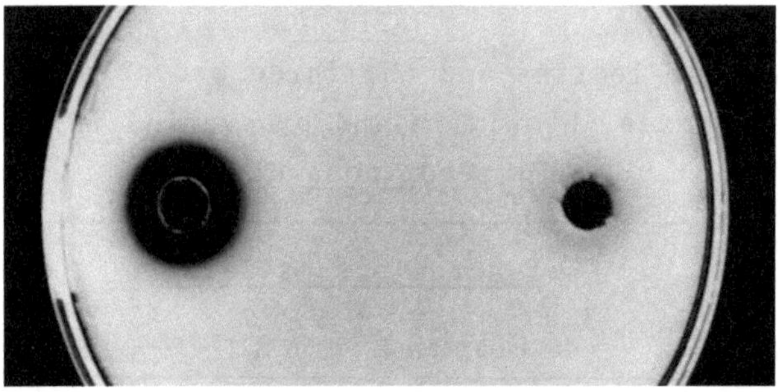

Fig. 1. Protease activity clearing zones, in 3% skim milk-containing agar plates, around wells filled with the examined supernatants of culture medium of two P. aeruginosa strains (see Materials and Methods). Left well contains supernatant fluid of strain (No. 2 in Table I) which exhibits a considerable lectin activity together with high hemolysin, protease and pyocyanin level, whereas the right well contains the supernatant of the culture of a strain (No. 5 in Table I) which lacks all these activities.

The linkage between the lectin level and the secretion of the three exocellular factors may be due to one of the following:

a. A common genetic entity (plasmid, etc.) contains the information for all of them. Elimination or repression of this entity may stop their production.

b. Alteration in the cell envelope structure or function affects the production of all the factors.

c. A common regulatory mechanism controls their production.

d. The lectin is involved in either the envelope organization or the production, transport or secretion of the three exocellular factors.

No correlation between lectin level and exotoxin A production was observed (a strain which lacks lectins, protease and pyocyanin may still produce the exotoxin). If the presence of lectin is crucial for the secretion of the proteases, hemolysin and pyocyanin, then its role is not limited to adhesion phenomena or to overcoming the host immunological defence mechanisms, but it is also important for the cytotoxicity and tissue damaging activity of the invading bacterium by enabling it to secrete these exocellular toxic factors.

Acknowledgements

The author wishes to thank Miss L. Mizrahi, Mrs. D. Sudakevitz, Miss D. Avichezer and Mrs. R. Buxenbaum for helpful assistance in the study on the Pseudomonas lectins, and Mrs. B. Lederhendler for skillful typing of the manuscript.

References

1. Janda, J.M., Bottone, E.J.: J. Clin. Microbiol. 14, 55-60 (1981).
2. Iglewski, B.H., Kabat, D.: Proc. Nat. Acad. Sci. U.S.A. 72, 2284-2288 (1975).
3. Morihara, K., Tsuzuki, H.: Infec. Immun. 15, 679-685 (1977).
4. Liu, P.V.: J. Infect. Dis. 116, 481-489 (1966).
5. Snell, K., Holder, I.A., Leppla, S.A., Saelinger, C.B.: Infect. Immun. 19, 839-845 (1978).
6. Gilboa-Garber, N., Zakut, V., Mizrahi, L.: Biochim. Biophys. Acta 297, 120-124 (1973).
7. Gilboa-Garber, N.: Biochim. Biophys. Acta 273, 165-173 (1972).

8. Gilboa-Garber, N. in Colowick and Kaplan (Ginsburg, ed). Methods in Enzymology 83, 378-385, Academic Press, Inc. New York (1982).

9. Glick, J., Malik, Z., Garber,N.: Microbios 32, 181-188 (1981).

10. Sharabi, Y., Gilboa-Garber, N.: FEMS Lett. 5, 273-276 (1979).

11. Gilboa-Garber, N., Sharabi, Y.: J. Protozool. 27, 209-212 (1980).

12. Sudakevitz, D., Gilboa-Garber, N.: Microbios, in press.

13. Gilboa-Garber, N., Sudakevitz, D.: XI Trienniel World Congress of Pathology Proc., p. 97 (1981).

14. Wadström, T., Faris, A., Lindahl, M., Lönnerdahl, B., Ersson, B.: This volume, p. (1982).

15. Eshdat, Y., Sharon, N.: This volume, p. (1982).

NONMANNOSE-SPECIFIC LECTINS ON ENTEROTOXIGENIC E. COLI RECOGNIZING DIFFERENT GLYCOCONJUGATES

Torkel Wadström, Ahmed Faris, Mats Lindahl

Department of Bacteriology and Epizootology, Biomedicum, Box 583, S-751 23 Uppsala, Sweden

Bo Lönnerdahl

Department of Human Nutrition, University of California at Davis, California

Bo Ersson

Biochemical Separation Center, Biomedicum, Box 577, S-751 23 Uppsala, Sweden

Introduction

In 1955, Coller and de Miranda (1) defined the first receptor of E. coli fimbriae (type 1). Type 1 fimbriae were shown to recognize D-mannose-specific lectins. Other fimbriae on the E. coli surface have been characterized more recently, e.g., colonization factor antigens I and II (CFAI, CFAII) and K88, K99 (3-7). E. coli possessing such fimbriae haemagglutinate erythrocytes of a number of animal species and cause a species-specific colonization of the small intestine of young animals as well as humans (2,6,7). However, strains with these non-mannose-specific fimbriae can simultaneously also possess type 1 mannose-specific fimbriae. In other words, more than one type of fimbriae can be found on the surface of single E. coli strains.

During the last decade, much work has been done to define the receptors for other E. coli fimbriae causing nonmannose-specific haemagglutination. Gibbons et al. (8) tried different glycoconjugates to inhibit haemagglutination with K88 fimbriae, which are present on the surface of enterotoxigenic E. coli. Such strains cause an increase in fluid secretion in the small intestine and a profuse water diarrhoea. They could not define the receptor but concluded that D-galactose is involved in the receptor structure, since the strongest inhibitors all possessed D-galactose, as recently confirmed (2). The only receptor defined so far, for a nonmannose-specific fimbrial haemagglutinin, is that of an E. coli isolated from urinary tract infection. Källenius et al. (3) and Leffler et al. (4) showed that the E. coli haemagglutinins recognized a di-galactose moiety of a glycolipid present on human erythrocytes (p^k antigen) and epithelial cells from the urinary tract. We recently found that enterotoxigenic E. coli with CFAI fimbriae cause haemagglutination which is insensitive to a great number of monosaccharides at concentrations of 2 mM (5). Moreover, neuraminidase-treated erythrocytes of human origin were not haemagglutinated by E. coli with CFAI fimbriae, while untreated erythrocytes were haemagglutinated (6). This finding suggested that N-acetylneuraminic acid probably is a constituent of CFAI erythrocyte receptor. In this study we have used different glycoconjugates to inhibit haemagglutination of CFAI fimbriated E. coli cells, in search for a possible receptor structure.

Materials and Methods

Bacterial culture conditions. Strains were grown on improved CFA agar medium (7).

Haemagglutination tests. Haemagglutination was performed in an aggregometer (Payton 300 A). The increase in transmittance of light (T_{609nm}) through the bacterial and erythrocyte cell mixture, due to a decrease in light scattering as erythrocyte aggregate into larger clumps, was measured. In the standard procedure bacterial cells (10^6-10^8) were mixed with 10^7 erythrocytes in a 0.5 ml α-methyl-mannoside (5 mg/ml) containing PBS, pH 6.8, and 0.30 M sucrose (1:1) (6). Erythrocytes were added to the solution with or without potential inhibitors, and the mixture was stirred (800 rpm). As soon as the transmittance stabilized, bacterial cells were added and haemagglutination reaction was recorded. The reaction was followed to completion, i.e., maximum transmittance, after 15-30 minutes.

Fractionation of milk glycoconjugates. Glycoconjugates were fractioned from human milk. Columns were prepared with lectins covalently bound to the matrix by the use of the cyanogen bromide method (9). Two affinity-columns were prepared, one with concanavalin A immobilized on the gel, reacting with mannose-containing carbohydrates, and one with *Crotalaria juncea* lectin which reacts with galactose-containing carbohydrates. A volume of 25 ml skimmed and decaseinized milk, dialysed aginst PBS, pH 7.4, was applied to a 25 ml Con A- or *Crotalaria juncea*-Sepharose column, at a flow rate of 5 ml/cm^2/h. The adsorbed material was eluted with 0.1 Mα-methyl-mannoside (Con A-column) or 0.1 M lactose (*C. juncea*-column) in PBS, pH 7.4. Eluted fractions were dialysed against PBS, pH 7.4, and their effect on CFAI haemagglutination was studied in the aggregometer. The material eluted from the Con A- and the *C. juncea*-column was used in the inhibition test at a protein concentration of 0.1 mg/ml and 0.01 mg/ml, resepctively.

Chemicals. Other glycoconjugates, fetuin (type III9) ovalbumin, and N-acetylneuraminic acid (Sigma Chem. Co., St. Louis, Mo., U.S.A.), bovine submaxillary mucin (Boehringer, Mannheim, W. Germany), orosomucoid (a gift from KABI, Stockholm, Sweden), and purified *E. coli* K1 capsule polymer (poly-α-2-8 N-acetylneuraminic acid, a kind gift from Dr. Klaus Jann, Max-Planck Institute, Freiburg, W. Germany) were also used for inhibition experiments.

TABLE 1

Haemagglutination inhibition of CFAI with glycoconjugates

Substance	Lowest concentration (μg/ml) giving inhibition
Poly-α-2-8 N-acetyl-neuraminic acid	400
Bovine submaxillary mucin	>8,000
Orosomucoid	>8,000
Fetuin	>8,000
Ovalbumin	>8,000

Results and Discussions

The haemagglutination of human erythrocytes with E. coli possessing CFAI fimbriae was followed in the aggregometer in the presence or absence of glycoconjugates. Poly-α-2-8 N-acetylneuraminic acid and milk glycoconjugate fractions were found to inhibit haemagglutination with CFAI fimbriae (Tables 1 and 2). Bovine submaxillary mucin, orosomucoid, fetuin, and ovalbumin did not inhibit haemagglutination (Table 1). N-acetylneuraminic acid monosaccharides did not inhibit haemagglutination even at high concentrations (3 mg/ml). This suggests that the receptor for this CFAI fimbrial haemagglutination is not a monosaccharide. However, the receptor probably contains N-acetylneuraminic acid, since poly-α-2-8 N-acetylneuraminic acid inhibits haemagglutination.

Surprisingly, other glycoconjugates containing N-acetylneuraminic acid of different relative percentage did not inhibit the haemagglutination, even though they contain common sialodisaccharides (17). Hence, we propose that the inhibition by poly-α-2-8 N-acetylneuraminic acid is due to a specific location of the monomer in a glycoconjugate, and the finding that the glycoconjugates from human milk inhibits the haemagglutination with CFAI fimbriae suggests to us that body secretions, such as milk, contain substances which behave as receptor analogues for the CFAI fimbrial haemagglutination.

Glöckner et al. (13) and Newman and Uhlenbruck (12) reported that crude glycoprotein fractions of human milk fat globule membranes contain blood group substances which react with different lectins of plant origin. Our finding that CFAI fimbriae, which gives haemagglutination of human erythrocytes, is inhibitable by glycoconjugate fractions from human milk, is an interesting finding. The complexity of glycoconjugate mixtures in milk and other body secretions can now probably be used to isolate and define subtypes of CFAI fimbriae and other bacterial lectins (5). Thus, a difference in haemagglutination inhibition with different purified glycoproteins or other glycoconjugates from human milk indicates that CFAI and probably also other bacterial haemagglutinins can be defined by showing different glycoconjugate specificity (Tables 1 and 2). However, it cannot be ruled out that these E. coli strains possess more than one nonmannose-specific surface lectin, as recently described in one strain of Neisseria gonorrhoeae (10).

In conclusion, we propose that

1) The receptor structure for CFAI fimbriae probably contains N-acetylneuraminic acid which is present either in a certain position, as in poly-α-2-8 N-acetylneuraminic acid, or as a moiety of more complex carbohydrate receptor structure.

2) Glycoconjugate fractions from human milk contain oligosaccharide structures which mimic the CFAI receptors on erythrocytes and probably also on epithelial surfaces in the small intestine which enables intestinal colonization and development of an acute cholera-like diarrhoeal disease.

3) Only N-acetylneuraminic acid as an α-2-8 polymer (colominic acid) inhibits haemagglutination, while neuraminic acid in an α-2-6-linkage, such as in bovine submaxillary mucin, or in an α-2-3-linkage, such as in fetuin (11), does not (Table 1 and Figure 1). This observation suggests that the exact position of N-acetylneuraminic acid in a glycoconjugate structure is critical in order to be a part of a receptor for the CFAI fimbrial haemagglutinin and for other bacterial lectins, such as CFAII (unpublished).

FIGURE 1

Oligosaccharide structures

polyα2-8NANA (100%)

2-8NANA(α2-8 NANA)9α2-8 NANA

Bovine submaxillary mucin (10%)

NANAα2-6GalNAc-Ser (or Thr)

Orosomucoid (10%)

NANA-Galβ1-4GlcNAcβ1-4
NANA-Galβ1-4GlcNAcβ1-2Manα1-3
Manβ(1-4)GlcNAcβ(1-4)GlcNAcβ1-Asn
NANA-Galβ1-4GlcNAcβ1-2Manα1-6
NANA-Galβ1-4GlcNAcα1-6

Fetuin (6,2%)

NANAα2-3Galβ1-4GlcNAcβ1-4
Manα1-6
NANAα2-6Galβ1-4GlcNAcβ1-2
Manβ1-4GlcNAcβ1-4GlcNAc-Asn
NANAα2-3Galβ1-4GlcNAcβ1-2Manα1-3

Ovalbumin (0%)

Manα1-3
Manα1-6
Manα1-6
1βGlcNAc
4
Manβ1-4GlcNAcβ1-4 GlcNAc-Asn
Galβ1-4 GlcNAcβ1-4 Manα1-3
GlcNAcβ1-2

TABLE 2

Inhibition of haemagglutination of E. coli possessing CFAI fimbriae with glycoconjugates from human milk.

Strain No.	Inhibition Con-A fraction	C. juncea fraction
E 1049a-13	weak	moderate
E 1058a-2	weak	moderate
1628-15	weak	strong

Acknowledgements

This study was supported by a grant from the Swedish Medical Research Council (16X-04723).

References

1. Collier, W.A., deMiranda, J.C.: Antoine van Leeuwenhoek, J. Microbiol. serol. 21, 135-140 (1955).
2. Sellwood, R.: Biochim. Biophys. Acta 632, 326-335 (1980).
3. Källenius, G., Möllby, R., Svenson, S.B., Winberg, J., Lundblad, A., Svensson S., Cedergren, B.: FEMS Microbiol. Lett. 7, 297-302 (1980).
4. Leffler, H., Svanborg-Edén, C.: FEMS Microbiol. Lett. 8, 127-134 (1980).
5. Lindahl, M., Faris, A., Wadström, T.: Attachment of micro-organisms to the gastrointestinal mucosal surface. Proceedings of the October 1981 workshop. Ed. Boedeker, E.C., CRC Press Inc., Boca Raton, U.S.A. In press.
6. Faris, A., Lindahl, M., Wadström, T.: FEMS Microbiol. Lett. 7, 265-269 (1980).
7. Evans, D.G., Evans Jr., D.J., Clegg, S., Pauley, J.A.: Inf. Immun. 25, 738-748 (1979).
8. Gibbons, R.A., Jones, G.W., Sellwood R.: J. Gen. Microbiol. 86, 228-240 (1975).
9. Axén, R., Porath, J., Ernback, S.: Nature 214, 1302-1304 (1967).

10. Lambden, P.R., Robertson, J.N., Watt, P.J.: J. Gen. Microbiol. 124, 109-117 (1981).

11. Schneider, R.A., To, S.C.M.: Inf. Immun. 36, 417-418 (1982).

12. Newman, R.A., Uhlenbruck, G.C.: Europ. J. Biochem. 76, 149-155 (1977).

13. Glöckner, W.M., Newman, R.A. Dahr, W. Uhlenbruck, G.C.: Biochim, Biophys, Acta 443, 402-413 (1976).

14. Montreuil, J.: Adv. Carbohydr. Chem. Biochem. 37, 157-222 (1980).

Part V

Plant Lectins

DISTRIBUTION AND FUNCTION OF THE 120,000 DALTON SOYBEAN LECTIN IN THE GENUS GLYCINE

Steven G. Pueppke
Department of Plant Pathology, University of Florida, Gainesville, Florida 32611 USA

Introduction

Although the 120,000 dalton soybean lectin (SBL) is used widely as a biochemical tool, knowledge of the distribution and natural function of SBL is incomplete (6). Most genetic lines of the cultivated soybean, *Glycine max* (L.)Merr., contain the lectin, but seeds of several are devoid both of hemagglutinating activity and of materials that react with anti-SBL antibodies (15, 17). Five of the lectin-negative lines (Columbia, Norredo, Sooty, T-102, Wilson 5) have been characterized (14, 15, 20), and SBL is found neither in seeds nor in vegetative tissues of such plants (14, 20, 21). Each line is homozygous recessive at the locus designated *le*, which is responsible for the lack of SBL (10, 18). The status of lectins in the botanical relatives of soybean is more tentative. *Glycine soja* Sieb. & Zucc., the wild soybean, is closely related to *G. max* (7). Fewer than half of the tested *G. soja* lines contain materials that react with anti-SBL antibodies (19), and although a lectin has been isolated from one *G. soja* line, its biochemical relationship to SBL was not examined (16). The remaining species of the genus *Glycine* (all classified in the subgenus *Glycine*) have received even less attention, probably because they are not cultivated and are difficult to obtain (7).

Lectin-negative soybean lines are ideal materials for examining the putative role of SBL in the nitrogen-fixing, soybean-*Rhizobium* symbiosis. In 1974, Bohlool and Schmidt (3) proposed

Lectins, Vol. III

that SBL has an enabling function in this association. Fluorescent-labeled SBL bound to 22 of 25 strains of *R. japonicum* that nodulated soybean, but not to any of 23 non-nodulating strains. It thus seemed reasonable that SBL might allow soybean to distinguish nodulating from non-nodulating *Rhizobium* strains by binding specifically to cells of the former. All of the examined soybean lines known to lack SBL, however, are nodulated by the typical soybean rhizobia (15, 20). Although such data are not in accord with the hypothesis that SBL has an essential role in the recognition of symbiotic rhizobia, it is not known if *le* soybeans actually are infected in the usual way, i.e. by means of infection threads in root hairs. Indeed, Dazzo (4) has argued that rhizobia may enter *le* soybean roots at alternate sites, such as void spaces in the epidermis or at zones of lateral root emergence.

The experiments reported here had two objectives dealing with the distribution and function of SBL. The first was to complete the inventory of the genus *Glycine* for SBL and to establish the relationship of the lectin from *G. soja* to SBL. The second was to determine if the capacity of root hairs to recognize and to be infected by rhizobia is maintained in soybean lines devoid of SBL.

Materials and methods

Seed lectin from *G. soja* P.I. 342619 was purified by affinity chromatography on a column of lactose coupled to Sepharose 4B by the divinylsulfone method (1). SBL prepared from the soybean cultivar Pickett served as the control. Affinity-purified lectin preparations were dialyzed against phosphate-buffered saline (PBS) and stored frozen. Protein (9), neutral sugar (8), hemagglutinating activity (11), and the inhibition of hemagglutinating activity (11) were determined according to standard methods. Lectin samples were isoelectric focused in the 4 to 9 pH range in polyacrylamide containing 9.2 M urea and 2% nonidet P-40 according to the protocol of Horst *et al.* (5). Samples of the *G. soja* lectin and SBL were chromatographed on a 1.5 X 70 cm column of Bio-Gel P-150 in PBS. The flow rate was 6.4 ml per hour, the fraction size was 2.5 ml, and the column was run at room temperature.

Seeds of *G. canescens*, *G. clandestina*, *G. falcata*, *G. latifolia*, *G. tabacina*, and *G. tomentella* (from Dr. T. Hymowitz, University of Illinois) were crushed and extracted with PBS (4 to 10 seeds per ml) using a Polytron Homogenizer. The extracts were centrifuged at low speed, and the supernatant solutions dialyzed prior to analysis by hemagglutination (11) and by diffusion in agarose against anti-SBL antibodies (20).

In studies of infection threads, soybean and *G. soja* seeds were surface-sterilized and germinated on water agar. Roots of three-day-old seedlings were dipped into suspensions of rhizobia (10^8 cells per ml) and transferred to plastic growth pouches, which were prepared essentially as described by Bhuvaneswari *et al.* (2). Four days after inoculation, seedlings were harvested, and roots were plunged for about 15 seconds into 0.05% toluidine blue in PBS. The roots were rinsed in PBS, longitudinal sections of surface tissues were excised, and the sections were examined at 100X using bright field optics and at 400X using interference contrast optics.

Results and discussion

Lectins were purified from defatted seed meals of *G. soja* and of soybean (as a control) by affinity chromatography on columns of immobilized lactose (elution was with 0.1 M D-galactose). Isoelectric focused lectins were resolved into a major (pI 5.9 $\pm$ 0.1) and a minor (pI 6.1 $\pm$ 0.1) isolectin. For the *G. soja* lectin, a 91-fold purification of specific hemagglutinating

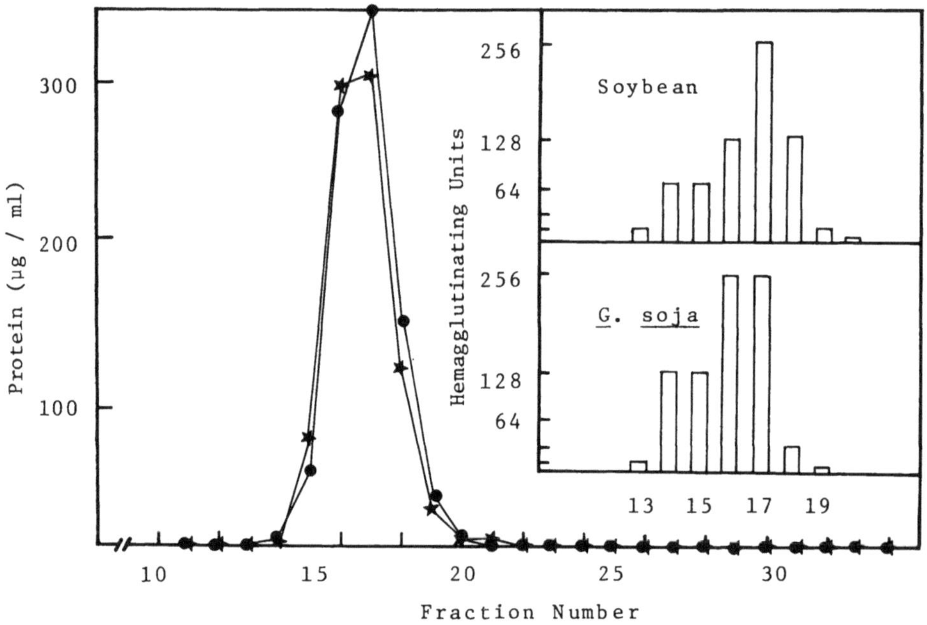

Figure 1. Chromatography of G. soja seed lectin (★) and soybean lectin (●) on a column of Bio-Gel P-150. Hemagglutinating activity was determined with trypsinized human Type A erythrocytes.

activity was achieved by the one-step affinity procedure (yield: 211% of initial hemagglutinating activity), and the purified lectin was a glycoprotein containing 5.7% neutral carbohydrate. Hemagglutinating activity of the G. soja lectin and of Pickett SBL was inhibited to an equal extent by a series of haptens, N-acetyl-D-galactosamine (the typical SBL hapten) having the most potency of the 11 tested sugars. Both lectins also eluted as coincident symmetrical peaks of protein and hemagglutinating activity from a column of Bio-Gel P-150 (Fig. 1). The wild soybean thus produces a seed lectin that is indistinguishable from the SBL of cultivated soybean.

The current inventory of SBL in the genus Glycine is in Table 1. Seeds of the 55 tested lines contained in the subgenus Glycine all lacked materials that reacted with anti-SBL antibodies or that agglutinated native, trypsinized, or desialyzed

Table 1. Distribution of SBL in the genus Glycine

Species	No. of Lines	Soybean Lectin: Present	Absent
Subgenus Soja			
Glycine max†	2,284	2,270	14
Glycine soja†	599	285	314
SUBTOTAL	2,883	2,555	328
Subgenus Glycine			
Glycine canescens	11	0	11
Glycine clandestina	11	0	11
Glycine falcata	3	0	3
Glycine latifolia	10	0	10
Glycine tabacina	10	0	10
Glycine tomentella	10	0	10
SUBTOTAL	55	0	55

† Data compiled from Ref. 12, 17, and 19.

human erythrocytes. With the exception of Glycine latrobeana, a rare Australian species (7), all species of the genus Glycine now have been examined for the lectin. SBL is restricted to G. max and G. soja, two species of the subgenus Soja, and even in these species the lectin is not essential.

The data on infection threads in le soybean are in Table 2. Although threads formed in roots of both le and normal Le soybean, there was no differential infection by the 12 tested Rhizobium strains, i.e. a given strain infected either none or all of the plants. Infected root hairs typically were hook-shaped and contained branched threads apparently originating from the concavity of the hook (Fig. 2). There existed an absolute correspondence between the presence of infection threads and the appearance of nodules (see Ref. 13 and 20 for nodulation data). Plants of the le genotype thus are infected and nodulated by the usual Rhizobium strains in the usual way,

Table 2. Infection of *Le* and *le* soybean by rhizobia

Strain	Infection threads in† Le soybean	le soybean
R. japonicum 61	+	+
R. japonicum 83	+	+
R. japonicum 138	+	+
R. japonicum 143	+	+
R. japonicum 110I	+	+
R. japonicum 110Y	+	+
R. lupini 96B9	-	-
R. sp. 3G4b5	-	-
R. sp. 3G4b16	+	+
R. sp. 3G4b20	-	-
R. sp. 229	-	-
R. sp. 132B13	-	-

† The *Le* lines were Hardee, Harosoy 63, and Viçoja. The *le* lines were Columbia, Norredo, Sooty, T-102, and Wilson 5.

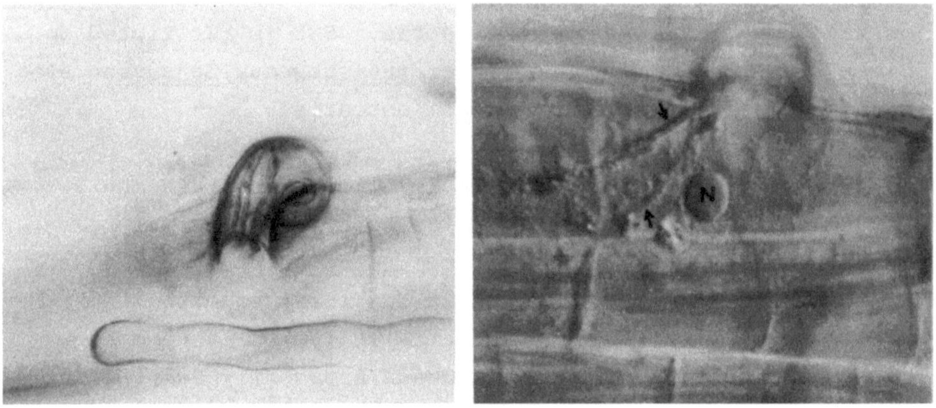

Figure 2. Infection threads of *Rhizobium* sp. 3G4b16 in roots of Wilson 5 soybean, a *le* line. Left: short, curled root hair containing four infection thread branches. Right: infection threads (arrows) elongating in the base of a root hair cell. The host cell nucleus (**N**) is in attendance. Both photographs were taken four days after inoculation using a Zeiss interference contrast microscope. The tissues were unstained.

even though they genetically lack SBL.

This research was supported by funds from the Florida Agricultural Experiment Station.

References

1. Allen, H. J., Johnson, E. A. Z.: Carbohydr. Res. 58, 253-265 (1977).

2. Bhuvaneswari, T. V., Turgeon, B. G., Bauer, W. D.: Plant Physiol. 66, 1027-1031 (1980).

3. Bohlool, B. B., Schmidt, E. L.: Science 185, 269-271 (1974).

4. Dazzo, F. B.: Lectins and their Saccharide Receptors as Determinants of Specificity in the Rhizobium-legume Symbiosis, (Subtelny, S., Wessels, N. K. eds.) pp. 277-304. Academic Press, New York (1980).

5. Horst, M. N., Basha, S. M. M., Baumbach, G. A., Mansfield, E. H., Roberts, R. M.: Analyt. Biochem. 102, 399-408 (1980).

6. Hymowitz, T.: Rec. Adv. Phytochem. 14, 157-179 (1980).

7. Hymowitz, T., Newell, C. A.: Taxonomy, Speciation, Domestication, Dissemination, Germplasm Resources and Variation in the Genus Glycine, (Summerfield, R. J., Bunting, A. H. eds.) pp. 251-264. Royal Botanic Gardens, Kew (1980).

8. Kabat, E. A.: Kabat and Mayer's Experimental Immunochemistry, Charles Thomas Co., Springfield, Illinois, USA (1961).

9. Lowry, O. H., Rosebrough, N. J., Farr, A. L., Randall, R. J.: J. Biol. Chem. 193, 265-275 (1951).

10. Orf, J. H., Hymowitz, T., Pull, S. P., Pueppke, S. G.: Crop Sci. 18, 899-900 (1978).

11. Pueppke, S. G.: Plant Physiol. 64, 575-580 (1979).

12. Pueppke, S. G., Benny, U. K., Hymowitz, T.: Plant Sci. Lett., In press.

13. Pueppke, S. G., Freund, T. G., Schulz, B. C., Friedman, H. P.: Can. J. Microbiol. 26, 1489-1497 (1980).

14. Pueppke, S. G., Friedman, H. P., Su, L.-C.: Plant Physiol. 68, 905-909 (1981).

15. Pull, S. P., Pueppke, S. G., Hymowitz, T., Orf, J. H.: Science 200, 1277-1279 (1978).

16. Stacey, G., Paau, A. S., Brill, W. J.: Plant Physiol. 66, 609-614 (1980).

17. Stahlhut, R. W., Hymowitz, T.: Soybean Genet. Newsl. 7, 41-43 (1980).

18. Stahlhut, R. W., Hymowitz, T.: Soybean Genet. Newsl. 8, 33-34 (1981).

19. Stahlhut, R. W., Hymowitz, T., Orf, J. H.: Crop Sci. 21, 110-112 (1981).

20. Su, L.-C., Pueppke, S. G., Friedman, H. P.: Biochim. Biophys. Acta 629, 292-304 (1980).

21. Vodkin, L.: Plant Physiol. 68, 766-771 (1981).

MANNOSE-SPECIFIC LECTINS AND THE RECOGNITION OF PEA ROOTS BY RHIZOBIUM LEGUMINOSARUM.

Jan W. Kijne, Ineke A.M. van der Schaal, Clara L. Díaz, Frank van Iren.

Botanisch Laboratorium, Rijksuniversiteit Leiden,
Nonnensteeg 3, 2311 VJ LEIDEN, The Netherlands.

INTRODUCTION

Rhizobium leguminosarum infects root hairs of its host plants (Pisum, Vicia, Lens, Lathyrus) as an early step in the development of a nitrogen-fixing root nodule symbiosis. Hamblin & Kent (15) and Bohlool & Schmidt (2) suggested that attachment of Rhizobium to the legume roots could be mediated by specific lectin-polysaccharide interactions as a basis of the observed host specificity in this symbiosis. Specific attachment of Rh. trifolii to clover root hairs has been demonstrated by Dazzo and co-workers (7), and following research clearly revealed a specific role for a clover lectin (trifoliin A) and rhizobial surface trifoliin A-receptors (6,8,17) in the first root attachment stages (Phase I- attachment, 6). Paau et al. (23) found evidence for a specificity-determining role of a non-hemagglutinating Medicago (alfalfa)- lectin in the attachment phase of the Medicago - Rh. meliloti - symbiosis.

A specific role for lectins during infection of Pisum-roots by Rh. leguminosarum has been suggested (19,22) but has never been demonstrated conclusively. Quantification of bacterial attachment to pea root hairs with use of the light microscope is severely hampered by the relatively large dimensions of the root. Moreover, in our experience Rh. leguminosarum does not induce nodules on pea roots in agar or water culture. Use of a coarse gravel culture, which gives optimal conditions for infection, results in root hair damage during root harvest for attachment experiments. Indirect measurements of bacterial attachment to pea roots with use of ^{14}C-labeled rhizobia (20) suggest that host lectin - bacterial receptor interactions might be involved only partially in root hair adsorption. Other (non-specific?) binding mechanisms might be involved as well. Non-specific attachment of various rhizobia to soybean roots has been suggested in a recent communication of Bauer (1).

However, the root infection mechanism of the fast-growing rhizobia (a.o. Rh. leguminosarum, Rh. trifolii, Rh. meliloti) seems to be very similar (4). When host lectin binding determines the specificity of Rh. trifolii (6) and Rh. meliloti (23) a completely other attachment mechanism for Rh. leguminosarum seems to be improbable. Present research into the possible involvement of lectins in the first developmental stages of the Rh. leguminosarum - pea - symbiosis is characterized by the analysis of essential elements: characterization and localization of lectins, characterization and localization of lectin receptors, and comparative work with different host plants and rhizobial species and strains. So far only host plant lectins and rhizobial lectin receptors have been taken into consideration. We present preliminary evidence for a mannose-specific agglutinin from Rh. leguminosarum, in addition to data on pea root lectin characterization and localization.

MATERIALS AND METHODS

Pea seed (var. Finale, Cebeco, Rotterdam) lectins have been purified following methods of Entlicher et al.(10) and Van Driessche et al.(25). Sephadex G-75 affinity chromatography was followed by separation of the three main lectin fractions by chromatofocusing (Pharmacia, manufacturer's instructions). Pea root surface lectin was isolated from 7-days old pea seedling roots (gravel-grown, N-free medium) as described before (22) with use of an acid glycine/HCl-buffered washing solution with the following modifications: 0.5 M NaCl instead of 0.9% w/v, and addition of 1% w/v $CaCl_2$ and 1% w/v $MnCl_2$.

Rabbits were injected subcutaneously with purified seed lectin 2 (1 mg/ml, + 1 ml complete Freund's adjuvant), and lectin antibodies were partially purified from crude antiserum by ammonium sulphate precipitation. Lectin-antibody-precipitation was demonstrated with the Ouchterlony- double diffusion technique.

Adsorption of lectin 2 to leguminous roots was visualized after incubating appr. 30 seedling roots for 15-20 min in 1 ml of a 0.9% w/v NaCl-solution of lectin 2 (62.5 µg/ml, titer 4) by measuring the hemagglutination titer (human A^+ erythrocytes) after removal of the roots.

Pea root cell walls were isolated following Buchala & Franz (3) from 7-days old pea seedling roots (agar-grown). The cell walls were extracted with 0.9% NaCl (80°C, 2 hrs)(9). Polysaccharides were precipitated with 3 vol. cold ethanol, and the resulting supernatant (s_1) was assayed for pea lectin activity.

Non-SDS disc gel electrophoresis (PAGE) was done according to Davis & Ornstein (5). Analytical electrofocusing was carried out with prefabricated Ampholine PAG plates (5%, pH 5-8) (LKB-Multiphor) following the instructions of the LKB-manual.

Rh.leguminosarum RBL1 was grown in A^+-mannitol medium (mannitol 3.5 g/l, NaCl 0.2 g/l, $CaCl_2 \cdot 2H_2O$ 1.0 g/l, $MgSO_4 \cdot 7H_2O$ 0.2 g/l, yeast extract (Difco) 0.8 g/l, KH_2PO_4 0.99 g/l, K_2HPO_4 0.32 g/l, demineralized water, pH 6.2) at 28°C and 180 rpm. Prior to autagglutination UV-sterilized sugars were added in a final concentration of 0.2 M. Crude rhizobial agglutinin was isolated according to Eshdat et al.(11), and found to be present in the 50,000 g - pellet.

RESULTS AND DISCUSSION

Both Entlicher's dry meal - buffer extraction and Van Driesche's soaked seeds - salt extraction result in the isolation of three main hemagglutinating isolectin fractions from pea seeds after affinity chromatography and chromatofocusing. Sugar specificity of the three fractions is the same (22,26), haptens being "mannose/glucose type" hexoses. Rabbit antibodies raised against lectin 2 gave an identical precipitation pattern in an Ouchterlony double diffusion assay with each of the three isolectin fractions: 2 precipitin lines. The isolectin fractions were characterized by the different isoelectric point of the main protein components: lectin 1: 7.5, lectin 2: 6.5, lectin 3: several bands in the pI range of 5.5.

Pea root lectin can be washed from intact seedling roots with an acid glycine/HCl buffered solution (20,22), and purified with Sephadex G-75 affinity chromatography. Sugar specificity of the root surface lectin is the same as reported for pea seed lectins (20, own observations). A different sugar specificity for a lectin fraction from pea root homogenate has been described (14). Non-SDS PAGE (22) as well as analytical electrofocusing indicate that pea root surface lectin might be related to seed lectin 2 (table 1). It remains to be established where exactly this root surface lectin is located, and if presence of a mannose/glucose - lectin in/on the root surface is a characteristic of all host plants of Rh. leguminosarum.

table 1: pI of seed lectin fractions and pea root surface lectin (see Materials & Methods). Major protein bands are encircled.

lectin 1			6.9	(7.5)
lectin 2		(6.5)	6.9	
lectin 3	(5.5)	6.5	6.9	
root surface lectin	5.5	(6.5)		
(Van Driessche et al.,25:				
pea lectin (Seph. G-75)	5.5	6.4	6.8	7.5)

Association of lectins with plant cell walls has been suggested frequently (e.g. 21,22). A salt extract of isolated cell walls from 7-days old pea seedling roots exhibited mannose/glucose specific lectin activity, when the seedlings were grown in a N-free medium. When grown in presence of 20 mM NO_3^- (nodulation of pea roots is almost completely inhibited under these conditions) pea root cell walls show no lectin activity in the s_1-extract. It is unclear if the lectin activity is structurally cell wall associated or if the extraction procedure simply dissociates a lectin-cell wall complex formed during tissue homogenization.

Vicia-, *Lathyrus*- and, to a lesser extent, *Lens*- seedling roots bind pea seed lectin 2 when incubated in a diluted lectin 2 - solution, while *Trifolium*-, *Medicago*- and *Cicer*-roots do not (table 2). These results give a first indication of the specific presence of "mannose/glucose - lectin"- receptors on the surface of host plants of *Rh. leguminosarum*. The "mannose/glucose-lectins" of *Vicia*, *Pisum*, *Lathyrus* and *Lens* are very similar (e.g. 16,24), and it is attractive to speculate on the similarity of the corresponding lectin receptors.

table 2: Pea lectin adsorption to leguminous roots. (see Materials & Methods)

	titer	
	before incubation	after removal
Vicia hirsuta [a]	4	1
Vicia sativa [a]	4	1
Lathyrus [a]	4	1
Lens [a]	4	2
Cicer	4	4
Trifolium	4	4
Melilotus	4	4
filter paper (1.5 cm^2)	4	4

[a] host plant of *Rh. leguminosarum*.

Rh. leguminosarum RBL1 shows massive autagglutination on a yeast extract/mannitol/salts medium at an OD_{620} of 0.75 (mid-late log-phase). Addition of mannose (final concentration 0.2M) prior to visible autagglutination inhibits flocculation of the bacterial cells almost completely. Glucose, galactose and mannitol were inactive (fig. 1). Addition of sugar or acid does not dissolve the aggregates once formed, which could indicate that the firm association of the rhizobial cells is started by a mannose-inhibitable step. Following the method of Eshdat et al.(11) we could isolate a crude mannose- (and not glucose-) specific agglutinin from just-agglutinating rhizobia. The agglutinin could be pelleted at 50,000 g, which points to a large molecular weight. The possibility exists that, comparable with

fig. 1: Autagglutination of Rh. leguminosarum RBL1: inhibition by 0.2 M mannose (see Materials & Methods). Autagglutination is visualized by stagnation of the increase in optical density.

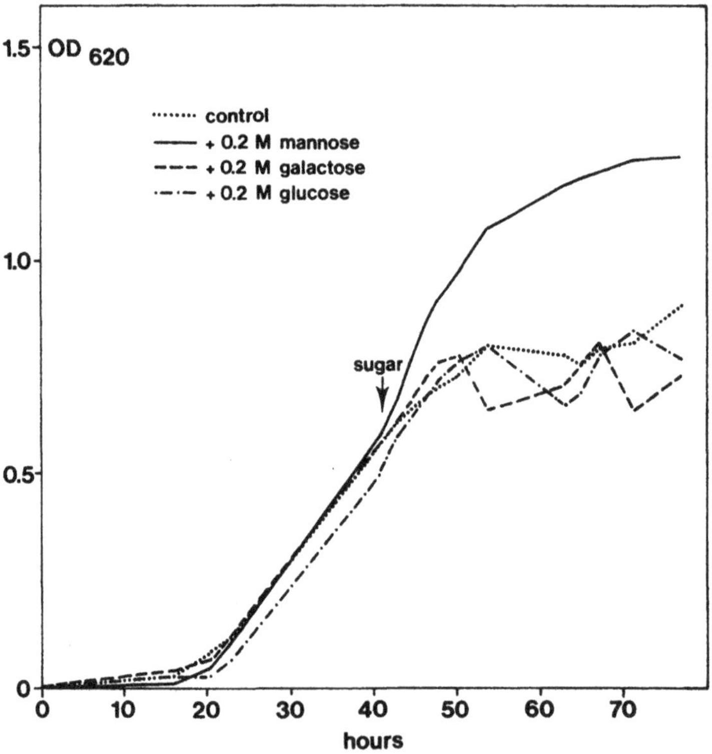

E. coli (11,12,13), Rh. leguminosarum produces "mannose-type"-flagellae/fimbriae/pili under certain cultural conditions. Presence of these appendages together with mannose-containing pea lectin receptors in the rhizobial capsule (see Van der Schaal et al.,this issue) could lead to autagglutination. This suggestion is accentuated by the observation that the rhizobial aggregates only consist of capsulated cells (light microscope, Indian ink negative staining). In a comparable way Rh. leguminosarum can bind to a plant root surface exhibiting pea lectin receptors (see table 2).

CONCLUDING REMARKS

The data given in this communication should be considered as introductory results setting a framework for more detailed and directed studies. The demonstration of a rhizobial agglutinin enlarges the scope of the original lectin hypothesis in Rhizobium - legume recognition research. The need of information on lectin receptors of plant origin is obvious. This information will add to a better understanding of the role of plant lectins in vivo. Change of the lectin hypothesis into a fundamental theory depends furthermore on the answering of several questions, e.g.:

- is transfer of root-infection ability from one Rhizobium-strain to another (18) always coupled to transfer of the ability to produce host plant lectin receptors and/or homologous bacterial agglutinin?
- are plant lectins and/or lectin receptors always exhibited where rhizobial infection takes place: the growing root hair (branch) tip?
- do the legume rhizosphere and the root mucigel offer the right culture conditions for the production of rhizobial agglutinins and/or host lectin receptors?

REFERENCES

1. Bauer, W.D.: Plant Physiol. 69, suppl., 143 (abstr.)(1982).
2. Bohlool, B.B., Schmidt, E.L.: Science 185, 269 (1974).
3. Buchala, A.J., Franz, G.: Phytochemistry 13, 1887 (1974).
4. Dart, P.J.: in Quispel, A. (ed.), The biology of nitrogen fixation, North Holland Publ. Cy., Amsterdam, 381 (1974).
5. Davis, B.J., Ornstein, L.: in Williams, C.A., Chase, M.W. (eds), Methods in immunology and immunochemistry, vol.2, Academic Press, New York-London, 38 (1968).
6. Dazzo, F.B.: J. Supramol. Struct. Cell Biochem. 16, 29 (1981).
7. Dazzo, F.B., Napoli, C.A., Hubbell, D.H.: Appl. Environ. Microbiol. 32, 166 (1976).
8. Dazzo, F.B., Hrabak, E.M., Urbano, M.R., Sherwood, J.E., Truchet, G.: in Gibson, A.H., Newton, W.E.(eds), Current perspectives in nitrogen fixation, Australian Academy of Sciences, 292 (1981).
9. Díaz, C., Kijne, J.W., Quispel, A.: in Gibson, A.H.,Newton, W.E.(eds), Current perspectives in nitrogen fixation, Australian Academy of Sciences, 426 (1981).
10. Entlicher, G., Kostir, J.V., Kocourek, J.: Biochim.Biophys. Acta 221, 272 (1970).
11. Eshdat, Y., Ofek, I., Yashouv-Gan, Y., Sharon, N., Mirelman, D.: Biochem. Biophys. Res. Comm. 85, 1551 (1978).
12. Eshdat, Y., Sharon, N., Ofek, I., Mirelman, D.: Isr. J. Med. Sci. 16, 479 (1980).
13. Eshdat, Y., Silverblatt, F.J., Sharon, N.: J. Bacteriol. 148, 308 (1981).
14. Gatehouse, J.A., Boulter, D.: Physiol. Plant. 49, 437 (1980).
15. Hamblin, J., Kent, S.P.: Nature New Biol. 245, 28 (1973).
16. Hemperly, J.J., Hopp, T.P., Becker, J.W., Cunningham, B.A., J. Biol. Chem. 254, 6803 (1979).
17. Hrabak, E.M., Urbano, M.R., Dazzo, F.B.: J. Bacteriol. 148, 697 (1981).

18. Johnston, A.W.B., Beynon, J.L., Buchanan-Wollaston, A.V., Setchell, S.M., Hirsch, P.R., Beringer, J.E.: Nature 276, 634 (1978).

19. Kato, G., Maruyama, Y., Nakamura, M.: Agric. Biol. Chem. 43, 1085 (1979).

20. Kato, G., Maruyama, Y., Nakamura, M.: Plant Cell Physiol. 22, 759 (1981).

21. Kauss, H., Glaser, C.: FEBS Lett. 45, 304 (1974).

22. Kijne, J.W., Van der Schaal, C.A.M., De Vries, G.E.: Plant Sci. Lett. 18, 65 (1980).

23. Paau, A.S., Leps, W.T., Brill, W.J.: Science 213, 1513 (1981).

24. Sakakibara, M., Noguchi, H., Makino, S.: Agric. Biol. Chem. 43, 1647 (1979).

25. Van Driessche, E., Vandenbranden, S., Kanarek, L.: Arch. Int. Physiol. Biochem. 86, 963 (1978).

26. Van Wauwe, J.P., Loontiens, F.G., De Bruyne, C.K.: Biochim. Biophys. Acta 379, 456 (1975).

PEA LECTIN BINDING BY RHIZOBIUM

Ineke van der Schaal, Jan Kijne, Clara Díaz and Frank van Iren
Botanical Laboratory, State University of Leiden, Nonnensteeg 3
2311 VJ Leiden, The Netherlands

Introduction

The bacterial genus Rhizobium is capable of infecting the roots of leguminous plants. After adhering to the roothairs the bacteria penetrate the roots via a so-called infection thread, which grows centripetally to the cortex of the root. The root cortical cells start dividing and a root nodule is formed. In this nodule atmospheric nitrogen is fixed by the rhizobia into ammonia (1,2). The infection process is restricted to definite combinations of rhizobia and legumes. E.g., Rh. leguminosarum infects pea, lentil, vetch and lathyrus, Rh. trifolii infects clover, Rh. phaseoli infects bean and Rh. japonicum infects soybean (3).

Much attention has been paid to the role of plant lectins in this host range specificity. Hamblin and Kent (4) first described binding of seed lectin from beans (Phaseolus vulgaris) to its symbiont Rh. phaseoli. By demonstrating the binding of red blood cells to the roothairs of the plant they suggested the presence of a phytohemagglutinin on the roothairs, although sugar specificity of the binding was not determined. Using fluorescein labeled seed lectins Bohlool and Schmidt (5) first demonstrated a specific interaction of lectins with rhizobia: soybean (Glycine max) lectin bound to 22 out of 25 strains of the soybean nodulating bacterium Rh. japonicum, but not to any other rhizobial species. These results in combination with the finding that only infective cells of Rh. trifolii were bound by the lectin of their host Trifölium repens led Dazzo and Hubbell (6) to propose their "common-antigen" theory. They suggested a specific adsorption of rhizobia to the surface of their host roothairs by cross-bridging of common surface antigens by a multivalent host lectin. The specific role of lectins in the adsortion of bacteria to their host roothairs has been indicated by many others (7-14). However, also contradictory results have been obtained; both non-binding of host lectin and binding of non-host lectin to rhizobia have been found (15-19). Some of these discrepancies can be explained by the fact that the presence and attainability of bacterial lectin receptors are dependent on the growth-phase and nutrient supply of the culture (7,8,20,21). Bhuvaneswari and Bauer (7) and Dazzo et al. (21) did show this growth-phase dependency of the host lectin binding of respectively Rh. japonicum and Rh. trifolii. In this communication we report this phenomenon for Rh. leguminosarum. For this purpose we developped an agglutination assay based on the decrease of the optical density of a bacterial suspension after lectin-agglutinated bacterial complexes have been filtered from the suspension. Besides the lectin mediated agglutination of rhizobia we also looked at the binding of pea seed lectin employing the more widely used fluorescein labeled lectin method (5,7,8,15,16,19). Discrepancies between the results obtained with these two methods with strains from different cross-inoculation groups, and the meaning of these results for the specific role of lectins in the Rhizobium-legume symbiosis will be discussed.

Materials and methods

Rhizobial strains. Rhizobium leguminosarum strain A171 (original code) = LPR 17 (code Biochemical Laboratory, Leiden), $inf^{+}eff^{+}$ on peas; Rhizobium phaseoli strain 1233, $inf^{+}eff^{+}$ on beans; Rhizobium phaseoli strain 148d. This strain is a Rh.phaseoli ex-conjugant carrying the pea-nodulation plasmid pJB5JI, but missing the bean-nodulation plasmid pRP1JI, $inf^{+}eff^{+}$ on peas, inf^{-} on beans (22).

Culture conditions. Media used were TY medium (trypton (Difco) 5.0; yeast extract (Difco) 3.0; $CaCl.2H_2O$ 1.0g/l), A^+medium (yeast extract 0.8; glucose 10.0; mannitol 3.5; $MgSO_4.7H_2O$ 0.2; NaCl 0.2; $CaCl_2.2H_2O$ 0.1; KH_2PO_4 0.993; K_2HPO_4 0.318g/l) and B^- medium (mannitol 3.5; KNO_3 1.52; $Ca(NO_3)_2.4H_2O$ 0.13; $MgSO_4.7H_2O$ 0.5; KH_2PO_4 0.993; K_2HPO_4 0.318g/l, trace elements: $MnSO_4.4H_2O$ 1.5; $ZnSO_4$ 0.25; H_3BO_4 3.2; Na_2MoO_4 1.0; NaFe-sequestrene 33.0mg/l, biotine 100; thiamine 100; $CuSO_4.5H_2O$ 88μg/l). For experiments a 1% inoculum was made from a stationary phase culture grown in TY medium. The cultures were shaken at 28°C at 180rpm.

Lectins. Pea seeds (Pisum sativum cv. Rondo) were purchased from Cebeco, Rotterdam, Holland). Pea seed lectins (PL) were isolated according to Kijne et al.(23).

Preparation of fluorescein isothiocyanate - pea lectin (FITC-PL) conjugate. 100μl FITC solution (5mg/ml in 0.5M sodium carbonate buffer pH 9.5) was added to 1ml of a 1% lectin solution in 0.15M NaCl. After 16-24hr incubation in the dark at room temperature the conjugate was extensively dialysed against phosphate buffered saline (PBS: 0.05M sodium phosphate pH 7.2, 0.14M NaCl) The conjugate was centrifuged (2min x 14,000g) and stored at -20°C. This procedure gave an optical density ratio OD_{495}/OD_{280} = 0.8. No loss of hemagglutinating activity owing to the labeling procedure was observed.

Agglutination assay. At distinct times samples were taken from the culture, centrifuged (15min x 40,000g, 4°C) and washed three times with cold PBS. The pellet was resuspended in the same buffer, filtered through filterpaper (Schut Papier bv., Heelsum, Holland) to remove nondispersible flocs and adjusted to an optical density at 620 of 1.0 (1cm pathway, about 10^9 cells/ml). Equal amounts of the suspension and a lectin solution in PBS were mixed. After incubation (15min, 0°C) the OD_{620} of the suspension was determined (A_1). The suspension was passed through filterpaper and the OD_{620} of the filtered suspension was measured (A_2). The percentage of agglutination can be calculated from the equation:

$$\% \text{ agglutination} = \frac{1 - \dfrac{A_2}{A_1} - \left(1 - \dfrac{A_2^0}{A_1^0}\right)}{1 - \left(1 - \dfrac{A_2^0}{A_1^0}\right)} \times 100\%$$

in which A_1^0 and A_2^0 are the OD_{620} values of the zero control when PBS without lectin was used. These values can change with different lots of filterpaper. The sugar specificity of the agglutination was checked by adding 0.1M glucose (a PL hapten) or 0.1M galactose (no PL hapten) to the incubation-mixture.
Fluorescence microscopy. Samples were taken from the culture at distinct time intervals, washed in PBS or directly diluted in PBS to $OD_{620} = \pm 0.05$. Drops of this suspension were placed on 8 spot-coated microscope slides (Cooke, no CL 800-21424). After drying under a cold fan the slides were heat-fixed and washed gently in distilled water. For incubation the FITC-PL conjugate was diluted 10 times in PBS, 15µl drops were applied to each spot and incubated for 2hr at 28°C in a wet chamber. After incubation the slides were shaken dry and washed 3 times for 20min in PBS. To check the sugar specificity of the binding 0.1M mannose (a PL hapten) or 0.1M galactose was added both to the incubation mixture and to the washing buffer. As a mounting solution we used a solution made of 10ml PBS, 90ml glycerol, 100mg p-phenylenediamine, 50mg propidium iodide, adjusted to pH 8.0 with NaOH. The p-phenylenediamine was added to prevent quick fading of the fluorescence upon illumination (24), while the propidium iodide acts as a counterstain that intercalates into the DNA of the bacteria, staining them red. This counterstain enables us to see also non-FITC bound bacteria during dark field illumination and can easily be filtered out with a KP 560 barrier filter (Leitz). Reading was done at 900x with a standard fluorescence microscope equipped with incident illumination using a xenon arc lamp for excitation and a 90/1.20 water immersion objective.

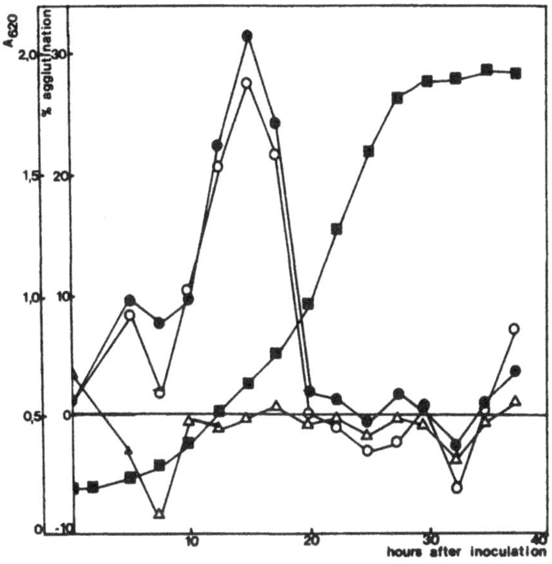

Quantitative agglutination of Rh. leguminosarum A171 in relation with growth (■-■) in presence of pea lectin 2.5µg/ml (●-●), pea lectin + galactose 0.1 M (○-○), pea lectin + glucose 0.1 M (△-△).

Results

As can be seen from the figure Rh. leguminosarum strain A171 growing on A^+ medium agglutinates with PL (2.5ug/ml; both lectin 1 and lectin 2 (23,25)) only in the first part of the log-phase with a peak of 32% agglutination. The sugar specificity of this agglutination was proven by the inhibition of the reaction by glucose but not by galactose. The Rh. phaseoli ex-conjugant 148d also agglutinates with PL although a 10 times higher lectin concentration had to be used. The highest percentage of agglutination obtained with this strain was 14%. Its mother strain Rh. phaseoli 1233 never showed any agglutination.

All three strains however bound PL with use of FITC-PL. This binding was not uniformly around the cell. Most positive cells showed fluorescence at one or both ends of the bacterium often accompanied by one ore more bands in the middle of the cell.

The number of positive cells could reach over 50%, however there was no clear correlation between lectin binding and culture age. The binding could be inhibited by mannose and not by galactose.
When the washing procedure in the fluorescence method was omitted, crystallization of yeast extract on the slides occurs. The yeast extract appears to bind PL very well in a hapten specific way. Kamberger (26) also observed PL binding of yeast extract in an ouchterlony double diffusion assay. Therefore the use of a yeast extract depleted medium (B^-) was introduced. Also on a nitrate medium the PL mediated agglutination was restricted to the two pea nodulating strains, while FITC-PL binding on this medium and also on the glutamate medium of Dazzo (27) was found with all three strains. In addition, most tested strains of other cross-inoculation groups (Rh. meliloti, Rh. phaseoli, Rh. trifolii, Agrobacterium tumefaciens) revealed FITC-PL binding.
Washing the cells had no influence on PL binding. However when we removed the light microscope visible capsule from the cells by incubating them for 1hr in 0.15M NaCl (80°C) the number of PL binding cells decreased to almost zero.

Discussion

The here introduced agglutination assay, although useful to demonstrate lectin-bacteria interactions, has its limitations. Only strains which do not form flocs in a particular medium can be screened. Complex formation is dependent on the lectin/ receptor ratio. The influence of the lectin concentration on the agglutination has not been fully investigated and will be subject of further research.
The results of the agglutination assay are in agreement with the hypothesis that the host plantlectins interact specifically with their symbionts.
However the binding of FITC-PL to the non pea nodulating

strains counters this hypothesis. The same unspecific binding of FITC-PL to strains of different cross-inoculation groups, including the polarity of the binding, was found by G. Hombrecher with the use of a glutamate-mannitol medium (personal communication). These apparently contradictory results can be explained from the fact that with the agglutination assay one only detects lectin binding to cells which are capable of being cross-bridged by this binding to other cells. With FITC-PL the detection of more, deeper located, probably non-species specific, lectin receptors is also possible. The presence of glucans at the outer site of almost all Rhizobium species and even of Agrobacterium (28-30) can cause this host unspecific PL binding. Also the unspecific binding of the mannose/glucose specific lectin from Canavalia ensiformis seeds to several Rhizobium species (18,19) can be explained in that way.
The results presented here indicate that demonstration of binding of lectins to rhizobia is not enough to prove or disapprove the lectin hypothesis. In cases of positive lectin binding one should check if these cells, through this lectin binding, are capable of binding to other cells or roothairs. In cases of negative binding one should consider the age and culture medium of the bacteria used.

References

1. Dart, P.J.: In: The biology of nitrogen fixation. Ed.: A. Quispel, North-Holland/Elsevier, Amsterdam. 379-429 (1974).
2. Libbenga, K.R., Bogers, R.J.: In: The biology of nitrogen fixation, Ed.: A. Quispel, North-Holland/Elsevier, Amsterdam. 430-472 (1974).
3. Broughton, W.J.: J. Appl. Bact. 45, 165-194 (1978).
4. Hamblin, J.S., Kent, S.P.: Nature New Biol. 245, 28-30 (1973).
5. Bohlool, B.B., Schmidt, E.L.: Science 185, 269-271 (1974).
6. Dazzo, F.B., Hubbell, D.H.: Appl. Microb. 30, 1017-1033 (1975).

7. Bhuvaneswari, T.V., Pueppke, S.G., Bauer, W.D.: Plant Phys. 60, 486-491 (1977).

8. Bhuvaneswari, T.V., Bauer, W.D.: Plant Phys. 62, 71-74 (1978).

9. Kato, G., Maruyama, Y., Nakamura, M.: Agric. Biol. Chem. 43, 1085-1092 (1979).

10. Kato, G., Maruyama, Y., Nakamura, M.: Agric. Biol. Chem. 44, 2843-2855 (1980).

11. Kato, G., Maruyama, Y., Nakamura, M.: Plant Cell Phys. 22, 759-771 (1981).

12. Wolpert, J.S., Albersheim, P.: Biochem. Biophys. Res. Comm. 70, 729-737 (1976).

13. Van der Schaal, C.A.M., Kijne, J.W.: Proc. 7th North Am. Rhizobium Conf., College Station, Texas, 17 (1979).

14. Paau, A.S., Leps, W.T., Brill, W.J.: Science 213, 1513-1514 (1981).

15. Chen, A-P.T., Phillips, D.: Phys. Plant 38, 83-88 (1976).

16. Law, I.J., Strijdom, B.W.: Soil Biol. Biochem. 9, 79-84 (1977).

17. Pueppke, S.G., Freund, T.G., Schulz, B.C., Friedman, H.P.: Proc. 4th Int. Symp. Nitr. Fix., Canberra, 423 (1981).

18. Wong, P.P.: Plant Phys. 65, 1049-1052 (1980).

19. Dazzo, F.B., Hubbell, D.H.: Plant Soil 43, 713-715 (1975).

20. Mort, A.J., Bauer, W.D.: Plant Phys. 66, 158-163 (1980).

21. Dazzo, F.B., Urbano, M.R., Brill, W.J.: Curr. Microb. 2, 15-20 (1979).

22. Beynon, J.L., Beringer, J.E., Johnston, A.W.B.: J. Gen. Microb. 120, 421-429 (1980).

23. Kijne, J.W., Van der Schaal, C.A.M., De Vries, G.: Plant Science Lettr. 18, 65-74 (1980).

24. Johnson, G.D., De C. Nogueira Araujo, G.M.: J. Imm. Meth. 43 349-350 (1981).

25. Van der Schaal, C.A.M., Kijne, J.W.: Proc. 4th Int. Symp. Nitr. Fix, Canberra, 425 (1981).

26. Kamberger, W.: Arch. Microb. 121, 83-90 (1979).

27. Dazzo, F.B., Yanke, W.E., Brill, W.J.: Biol. Bioch. Acta 539, 276-286 (1978).

28. Zevenhuizen, L.P.T.M., Scholten-Koerselman, H.J.: Ant. v. Leeuwenhoek, J. Microbiol. 45, 165-175 (1979).

29. Zevenhuizen, L.P.T.M., Scholten-Koerselman, H.J., Posthumus, M.A.: Arch. Microbiol. 125, 1-8 (1980).

30. Planqué, K., Kijne, J.W.: FEBS Lettr. 73, 64-66 (1977).

POSSIBLE ROLE OF LECTINS IN BINDING RHIZOBIA TO HOST ROOTS.

Bjørn Solheim

Institute of Biology and Geology, Box 3085 Guleng, University of Tromsø, N-9001 Tromsø, Norway.

Introduction

Lectins are carbohydrate-binding proteins or glycoproteins that occur in many plants, especially in seeds of leguminous plants. Lectins have been suggested to play a role in determining specificity in the infection process of legume roots by the symbiotic bacteria from the genus *Rhizobium* (1); this possible lectin function has recently been reviewed by several authors (2,3,4,5) and models for the involvement of lectins in the infection process have been proposed (6,7,8). Common for these models is that lectins are mediating some sort of binding of the bacteria to the host surface. Studies of interactions of isolated lectins with isolated polysaccharides from fast-growing rhizobia indicate that an initial binding of the lectin to the capsular polysaccharide cannot account for the specificity of the infection process (9,10). Reactions between host lectin and symbiont lipopolysaccharide (LPS) correlate with infectivity in *R. leguminosarum* (10). A transient ability of the symbiont to bind host lectin correlates with the appearance of a specific LPS in the cell wall of *R. trifolii* (9). It is possible to specifically inhibit adsorption of *R. leguminosarum* to pea roots by LPS and by capsular poly-

Lectins, Vol. III

saccharide from infective bacteria (11), but this will not inhibit but rather induce infection in the host (to be published).

In this paper evidence is presented for host lectins being able to mediate a rapid unspecific binding of bacteria to the root surface. This binding is not involved in the infection process. Lectins might, however, be involved in other steps in the infection process.

Materials and Methods.

Lectin isolation. *Vicia hirsuta* seed was finely ground in a rotary mill. The resulting meal was extracted with 0.14M NaCl. Lectins in this seed extract were purified by ammonium sulfate precipitation, and protein precipitating between 0-80% saturation were collected by centrifugation, washed twice with saturated ammonium sulfate and dialysed against phosphate-buffered saline (PBS) (0.01M K_2HPO_4/KH_2PO_4, 0.14M NaCl, pH 7.2). *V. hirsuta* lectin was purified further from this solution by affinity chromatography on a Sephadex G-100 column in PBS. After washing the column with PBS, the bound proteins were eluted with PBS containing 0.1M glucose. Lectin activity in the fractions was assayed, after dialysis against PBS, by agglutination of washed horse erythrocytes.

Bacterial strains. The bacterial strains used in this experiment were as follows: *R. leguminosarum* 129-1 and 129-4, infective strains from the University of Tromsø; 1013 and 1016 non-infective strains from Rothamstead Experimental Station; *R. meliloti* 126-1, infective on *Medicago sativa*; *R. trifolii* 128-1, infective on *Trifolium repens*. All bacteria were grown on MGA-medium (contains, per liter, 1g K_2HPO_4, 1g KH_2PO_4, 0.36g $MgSO_4 \cdot 7H_2O$, 0.2g $CaCl_2 \cdot 2H_2O$, 5mg $FeCl_3 \cdot 6H_2O$, 1g yeast extract, 3g mannitol and 15g agar).

Host plants. Vicia hirsuta and Trifolium repens were used as host plants. In all infection experiments seeds were surface sterilized, germinated on water agar and transfered to modified Fåhraeus slides before inoculation (12). Infection was recorded indirectly by scoring number of nodules seen in a microscope at 100X magnification.

Plant exudates. Exudates from germinating clover plants were collected by germinating surface sterilized seeds in 2 ml sterile water in a 10 cm petridish for 24 h in the dark followed by 48 h in continous light. The exudates were filtered through filterpaper, concentrated to 1mg protein/ml (about 20 fold concentration) and sterile filtered.

Immunological methods. Rabbit antibodies against affinity-purified lectin were prepared (13), and made fluorescent by reaction with fluorescein isothiocyanate (FITC) (14).

Lectin binding. Binding of the affinity purified V. hirsuta lectin to different rhizobia was studied by suspending bacterial cells in a 0.1 mg/ml solution of lectin for 2 h, washing the bacteria three times with PBS, incubation with FITC-anti-lectin immunserum for 10 to 60 min, washing again with PBS, and examination by fluorescense microscopy the binding of antibodies to the bacterial cells.

Scanning electron microscopy. Plants to be examined with the scanning electron microscope were grown in modified Fåhreaus slides (12). Inoculum was prepared from 4 day old cultures. Bacteria were washed twice in Fåhreaus mineral solution (15) and suspended in the same solution by shaking for 2 h (10^8 cells/ml). Lectin treatment of bacteria was done by adding soluble affinity purified lectin (2 mg/ml) to the bacterial suspension to a final concentration of 0.1mg lectin/ml. After inoculation the final number of bacteria in the rhizosphere solution was about 10^7cells/ml. The rhizosphere solution was drained from the root at different times after inoculation, and the roots washed gently in Fåhreaus mineral solution before being fixed in 2.5% glutaraldehyde overnight and post-fixed for 3 h with 1% osmium tetroxide buffered with 0.2M

cacodylate, pH 7.4. The roots were dehydrated through a series of graded concentrations of ethanol and finally into amylacetate. Preparations were dried by the critical point method using liquid CO_2, coated in a vacuum evaporator with carbon (100Å) and with gold (100Å). Roots were then examined and photographed in the scanning electron microscope.
Polyacrylamide gel (PAG) electrophoresis. PAG electrophoresis in 7% disc gels with tris/glycine buffer pH 8.3 was routinely done at all purification steps. Proteins in the gel were stained with Coomassie Brilliant Blue after fixation with 12.5% trichloroacetic acid (TCA).

Results and Discussion

Two host-symbiont systems, *Vicia hirsuta* - *Rhizobium leguminosarum* and *Trifolium repens* - *R. trifolii* were studied. Lectin was isolated from seeds of *V. hirsuta* by buffer extraction, ammonium sulfate precipitation and affinity chromatography. In this way a mixture of 4 iso-lectins could be isolated (Fig. 1). The main iso-lectin has been purified to homogeniety and further characterized (to be published). Both infective and non-infective strains of *R. leguminosarum* and the non-homologous *R. meliloti* were found to bind fluorescent antibodies after reaction with *V. hirsuta* lectin. Bacteria not treated previously with lectin and *R. trifolii*, regardless of treatment, did not bind fluorescent antibodies. This lack of specificity in binding of lectin to bacteria has also been found for the closely related pea lectin (see van der Schaal et.al., this issue). It was not possible to see any difference in binding to host roots by infective and non-infective strains of *R. leguminosarum* when roots were examined in the scanning electron microscope. Binding of bacteria to the root

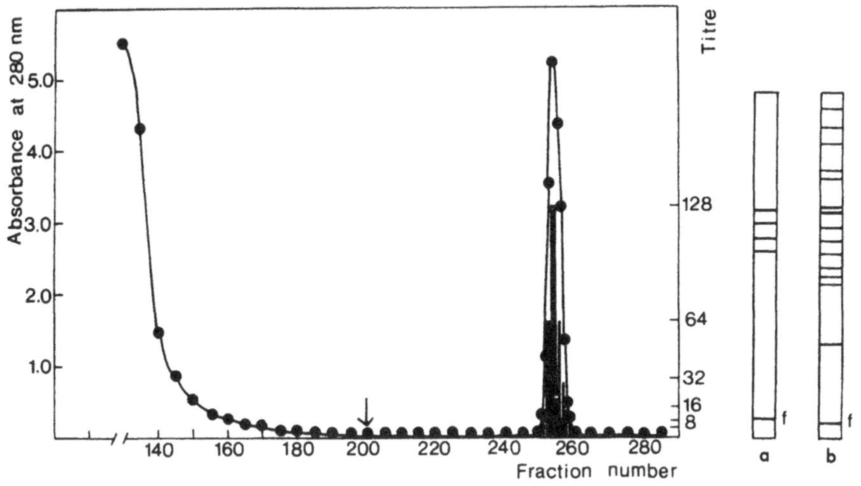

Fig. 1. Affinity chromatography of Vicia hirsuta lectins on a Sephadex G-100 column (65 cm x 2.6 cm). Proteins not adsorbed to the Sephadex eluted with PBS. The arrow indicates when 0.1M glucose is added to the elution buffer. Flow rate 15 ml/h, fraction volume 4 ml. To the right, PAG-electrophoresis of active lectin fractions (a) and ammonium sulfate precipitation (b). Migration front: f.

surface could be observed after 20 min incubation. When R. leguminosarum cells were treated with host lectin prior to inoculation, a massive binding of bacteria could be observed by 4 h after inoculation (Fig. 2A). Even after 24 h, without prior lectin treatment, comperable numbers of bacteria could not be seen (Fig. 2B). The rapid binding of lectin treated cells might be due to lectin receptors on the root surface. Indication of the presence of such receptors on roots of host plants of R. leguminosarum have been found (see Kijne et.al., this issue). These receptors might also react with R. leguminosarum agglutinin (see Kijne et.al., this issue). Lectin treated cells appeared to be shorter and have a more smooth surface than untreated cells, and treated cells did not agglutinate prior to inoculation. After 24 h incubation of host

Fig. 2. Rhizobium leguminosarum attached to Vicia hirsuta roots. A. Lectin treated bacteria 4 h after inoculation (X 4000). B. Not treated bacteria 24 h after inoculation (X 2000).

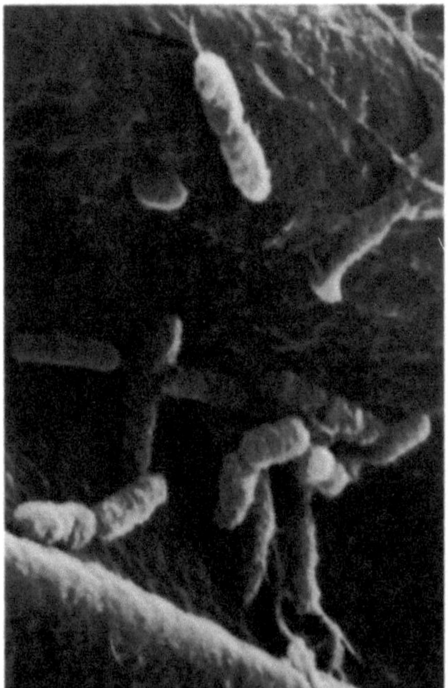

and symbiont, strands or fibrils of unknown nature and origin could be seen between the bacteria and host (Fig. 3). These observations support earlier proposed models (6,7) that a host component, probably a lectin, crossbridge the bacterial cell to the host cell. But in the V. hirsuta/R. leguminosarum interaction non-infective as well as infective strains bound to the host root. Lectin binding of bacterial cells to the host surface can not alone account for the specificity of the infection process. If the binding of rhizobia to the host surface with subsequent infection is dependent on a lectin-carbohydrate interaction, it should be possible to interfere with the reaction by adding surplus of the carbohydrate receptor or the lectin, analogous to antigen-antibody reactions. Germinating clover plants will excrete lectin into the rhizospehere medium (16). When clover seedlings in modified Fåhreaus slides are supplied with dialysed concentrated exudates from germinating clover plants or a dialysed, sterile culture filtrate of R. trifolii containing 0.1 mg/ml exopolysaccharide, initiation of infection was stimulated compared to control plants where water was added (Fig. 4). Especially

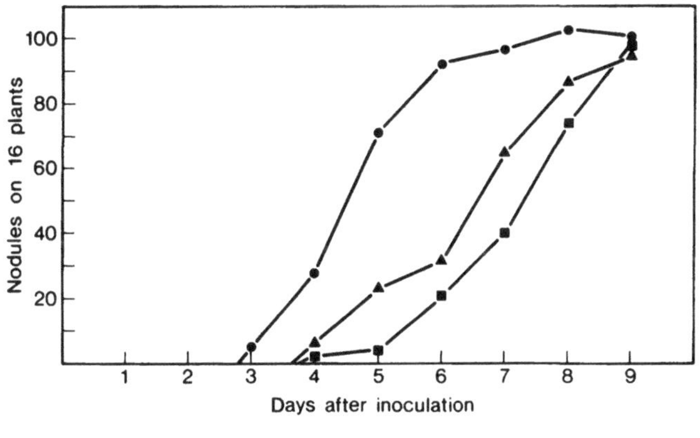

Fig. 4. Infection of White Clover by *Rhizobium trifolii* scored as number of nodules on 16 plants. One ml added to rhizosphere each day from 1 day before to 6 days after inoculation: dialysed exudates from germinating Clover seedlings (●), dialysed culture filtrate of *R. trifolii* (▲) and sterile water (■).

exudates from germinating plants induced the plants to become infected earlier then control plants. The mechanism of this stimulation is under investigation and appears not to be caused by a lectin in the exudates. Adding dialysed culture filtrate to the inoculated rhizosphere also induced the plants to be infected earlier. It has been shown by Kato et.al.(11) that symbiont polysaccharides such as LPS and capsular polysaccharides can specifically inhibit adsorption of bacteria to the host, but they did not examined the inhibition of the infection itself. We have found that symbiont polysaccharides stimulate infection (to be published), and oligosaccharides from the rhizosphere of inoculated clover plants induced new infection sites on root hairs (17). Bacterial carbohydrates that might interact with host lectins stimulate infection even if they inhibit lectin mediated binding of the bacteria to the host surface. This binding is not specific and probably not important for the infection process. Lectins might, however, be involved in other steps in the infection process. There is probably receptors on the root hair surface that can transmit signals over the cell membrane in response to symbiont produced products resulting in pre-penetration host reactions, such as root hair deformation. Lectins might function as receptors for such signal molecules rather then crossbinding the bacteria to the surface. A receptor for a signal molecule theoretically requires only one binding site. Host plants of Rhizobium lacking lectins (see Pueppke, this issue) might have such monovalent receptors.

References

1. Bohlool, B.B., Schmidt, E.L.: Science 185, 269-271 (1974).

2. Bauer, W.D.: Ann.Rev.Plant Physiol. 32, 407-449 (1981).

3. Bhuvaneswari, T.V.: Econ. Bot. 35, (1981).

4. Schmidt, E.L.: Ann.Rev.Microbiol. 33, 355-376 (1979).

5. Solheim, B., Paxton, J.: In Plant Disease Control, ed. Staples, R.C., pp. 71-83, J. Wiley & sons, New York, 1981.

6. Dazzo, F.B., Hubbell, D.H.: Appl.Microbiol. 30, 1017-1033 (1975).

7. Raa, J., Robertsen, B., Solheim, B., Tronsmo, A.: In Cell Wall Biochemistry Related to Specificity in Host-Plant Pathogen Interaction, eds. Solheim, B., Raa, J. pp. 11-30, Universitetsforlaget, Oslo 1977.

8. Hubbell, D.H.: Bio Science, 31, 832-837 (1981).

9. Hrabak, E.M., Urbano, M.R., Dazzo, F.B.: J.Bact. 148, 697-711 (1981).

10. Kamberger, W.: FEMS Microbiol.Lett. 6, 361-365 (1979).

11. Kato, G., Maruyama, Y., Nakamura, M.: Agric.Biol.Chem. 44, 2843-2855 (1980).

12. Solheim, B., Raa, J.: J.Gen.Microbiol. 27, 241-247 (1973).

13. Kabat, E.A., Mayer, M.M.: Experimental innunochemistry. Sec. ed. Charles C. Thomas, Springfield 1971.

14. King, C.A., Preston, T.M.: FEBS Lett. 73, 59-63 (1977).

15. Fåhraeus, G.: J. Gen. Microbiol. 16, 374-381 (1957).

16. Dazzo, F.B., Brill, W.J.: Appl.Environ.Microbiol. 33, 132-136 (1977).

17. Bhuvaneswari, T.V., Solheim, B.: Plant.Phys. (in press).

PURIFICATION AND CHARACTERIZATION OF AN ENDOGENOUS ROOT LECTIN FROM PISUM SATIVUM L.

Monique Hosselet[1], Edilbert Van Driessche[2], Marcel Van Poucke[1], Louis Kanarek[2].

1. Laboratorium voor Plantenfysiologie, Limburgs Universitair Centrum, Universitaire Campus, B-3610 Diepenbeek, Belgium.
2. Laboratorium voor Chemie der Proteinen, Vrije Universiteit Brussel, Paardenstraat 65, B-1640 Sint-Genesius-Rode, Belgium.

Introduction

The biological role of lectins is very intriguing. One of the suggested functions for Leguminosae lectins is their involvement in the specific recognition of Rhizobium (1). About the specificity of the interaction between legume lectins and Rhizobium, contradictory results are obtained (2,3,4). Part of these discrepancies can be explained by the dependence of lectin binding or agglutination of Rhizobium on the growth phase (5,6) or the composition of the nutrient medium (7). Moreover, little is known about the properties of root lectins (3,8-13) and their contamination with seed lectin that has leaked during imbibition (14,15) is not excluded. The present study deals with the occurrence of an endogenous pea root lectin that exhibits identical properties as the pea seed lectin.

Materials and Methods

1. Pea seed lectin is prepared as described previously (16).
2. Purification of root lectin. Surface-sterilized seeds of Pisum sativum L. (cvs. Wonder van Kelvedon) were germinated in vermiculite in the dark. Roots were removed after 4 days,

Lectins, Vol. III

washed with water, and the lectin was purified as described for seed lectin : precipitation with 60 % $(NH_4)_2SO_4$ followed by affinity chromatography on Sephadex G-100. The lectin was eluted as one peak with a solvent containing 0.3 M glucose.

3. Separation of endogenous root lectin and root surface lectin. 4 days old roots were harvested and shaken for 3 hours at room temperature either in 6 volumes of 1M NaCl + 20mM $MnCl_2$ + 20 mM $CaCl_2$ + 0.02 % NaN_3(I), or in the same solvent containing 0.3M glucose (II), or in bidistilled water (III). The roots were then rewashed twice with 2 volumes of the respective bathing solutions. Subsequently, the endogenous root lectin was purified as described in 2. The bathing solutions were concentrated by Amicon ultrafiltration (UM 10) and the proteins, including a root surface lectin, were precipitated with 60 % $(NH_4)_2SO_4$. The results were identical when the roots were shaken for 25 hours.

4. Protein was estimated either by absorbance measurement at 280nm, using $A^{1\,\%}_{280nm} = 15.9$, or by the Coomassie Brilliant Blue (Bio-Rad) method (17), using bovine serum albumin as a standard. For the same purified lectin samples the values by the latter method are slightly lower than those by the former.

5. Hemagglutinating activity was determined by mixing serial two-fold dilutions of lectin (50µl) and 50µl of untreated erythrocytes (1.5×10^8 cells/ml) in 0.15M NaCl + 0.02 % NaN_3. Agglutination was determined after 2 hours at 37° C. One hemagglutinating unit (HU) is defined as the reciprocal of the highest dilution of the lectin sample showing total hemagglutination. Sugar inhibition was measured by a serial dilution of the sugar at a fixed final lectin concentration (0.019mg/ml, i.e. 2 HU when tested against human erythrocytes).

6. SDS-polyacrylamide gel electrophoresis was performed in a linear 12-25 % gradient slab gel using the buffer system of Laemmli (18).

7. Antiserum against seed lectin was prepared as described elsewhere (19).

8. Immunodiffusion was carried out according to Ouchterlony.

The gels were made 0.2M in glucose to avoid interaction of the lectin with serum proteins.

Results

A lectin was isolated from a homogenate of pea seedling roots.

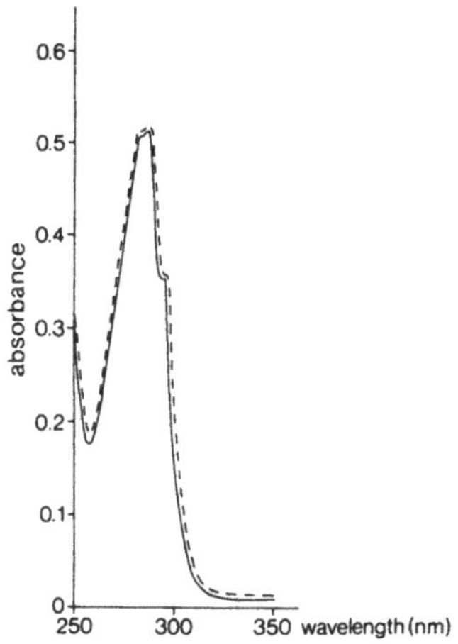

Fig. 1. UV absorption spectra of 0.3mg/ml of root lectin (--) and seed lectin (—).

erythrocyte origin	titer	
	seed lectin	root lectin
human A^+	32	32/16
B^+	32/16	16
AB^+	64/32	32/16
O^+	64/32	32
O^-	32	32/16
rabbit	2048/1024	1024
horse	2048	1024

Table 1. Blood group specificity was tested for lectin samples of 0.3mg/ml.

The yield of purified root lectin was about 5mg/100g dry weight of roots, varying slightly from experiment to experiment. Root lectin and seed lectin seem to be identical. Using the same concentrations of root lectin and seed lectin (assuming $A^{1\%}_{280nm}$ = 1.59 for both) the same UV absorption spectrum was observed (fig. 1). When tested in identical conditions, both lectins showed the same hemagglutinating activity against a panel of erythrocytes (table 1). All human blood groups were equally agglutinated, whereas rabbit erythrocytes were routinely used in further assays because of their high sensitivity. Root and seed lectin also were indistinguishable with respect to sugar specificity (table 2). Moreover, in radial immunodiffusion assays both lectins showed a reaction of total identity, when tested against anti-seed lectin antiserum (fig. 2) ; no reaction was observed with preimmune serum. The SDS-polyacrylamide gel electrophoresis patterns were essentially the same (fig. 3). The only difference is the absence in the root lectin of the 28kd band, which is considered to be a precursor chain from which the α- and β-chains are derived by proteolytic cleavage (20). Since total roots were used in our experiments, it seemed possible that our root lectin in fact was a root surface lectin. However, only small differences were observed in

sugar tested	seed lectin	root lectin
D-mannose	3.125	3.125
maltose	3.125	3.125
sucrose	6.25	6.25
D-glucose	12.5	6.25
D-fructose	12.5	12.5
L-sorbose	25	25
lactose	> 25	> 25

Table 2. Minimal sugar concentration (mM) necessary for complete inhibition of agglutination of human erythrocytes by 0.019mg lectin/ml (2 HU).

yield of purified lectin from roots washed either with water, or high molarity buffer, or high molarity buffer supplemented with glucose, whereas differences in HU/g fresh weight and in specific hemagglutinating activity fell within the limits of error of the assay (table 3). To check these results, the hemagglutinating activities of the bathing solutions were also

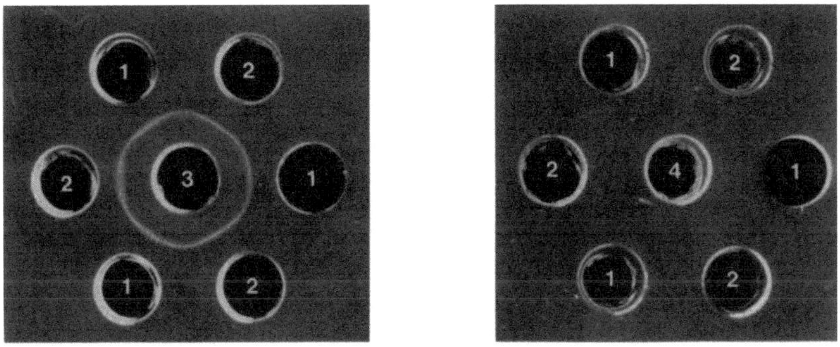

Fig. 2. Immunodiffusion of seed lectin (1) and root lectin (2) against rabbit anti-seed lectin antiserum (3) and preimmune serum (4).

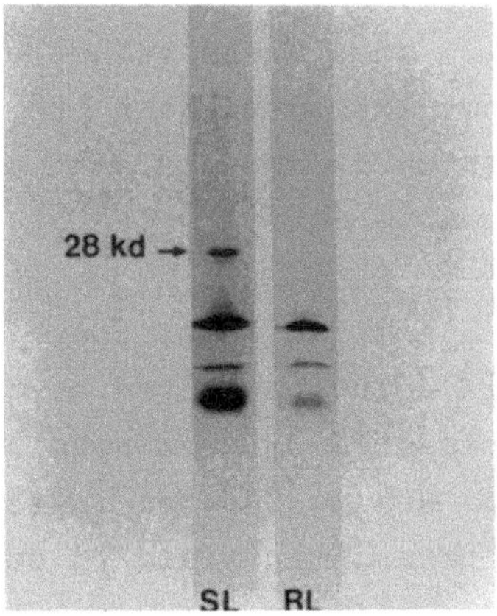

Fig. 3. SDS-polyacrylamide gradient gels (12-25 %) of root lectin (RL) and seed lectin (SL).

parameter	I	II	III
μg lectin/g fresh weight	2.2 ± 0.2	2.0 ± 0.1	2.3 ± 0.1
μg lectin/g dry weight	33.6 ± 3.2	29.1 ± 1.1	33.5 ± 1.1
HU/g fresh weight	320	200	330
HU/g dry weight	4800	3000	5000
HU/mg protein	$1.5x10^5$	$1.0x10^5$	$1.5x10^5$

Table 3. Comparison of affinity-purified lectin in roots washed with solutions I, II or III (see materials and methods). Protein was estimated by the Coomassie Brilliant Blue method. Values are given ± standard deviation.

parameter	RL			RSL		
	I	II	III	I	II	III
μg protein/g fresh weight	624	594	697	-	-	-
HU/g fresh weight	760	660	760	14	11	0
HU/mg protein	1300	1200	1100	-	-	-

Table 4. Comparison of root lectin (RL) and root surface lectin (RSL) of roots washed with solutions I, II or III (see materials and methods) after ammonium sulphate precipitation. Protein was estimated by the Coomassie Brilliant Blue method.

compared. With the high molarity buffer either with or without glucose a small hemagglutinating activity could be removed from the roots (table 4). When these values are compared to the values of the root extracts after the ammoniumsulphate precipitation step, it appears that about 1.5-2 % of the total hemagglutinating activity of the roots was eluted.

This root surface lectin showed a similar sugar specificity as seed lectin and root lectin in being inhibited by glucose, but not by lactose. The root surface lectin was not further purified because of its low yield and no further investigations were done. The fact that no difference in yield is seen with or without sugar in the bathing solution may indicate that the root surface lectin is not bound with its carbohydrate binding site.

Discussion

Lectins are well-known tools in glycoconjugate and mammalian cell research, but their biological role in plants remains unclear. Several functions have been suggested ; for example pea seed lectin may be a storage protein (19,21), it may play a role in cell wall structure and function (8,9) and/or it may be involved in the specific recognition of symbiotic Rhizobia (1,2,3,6,8,9).

This latter suggestion is the most studied one for legume lectins in general, because of the importance of nitrogen fixation in agriculture. However, most studies in which a specific recognition function was studied were performed with seed lectin, although the identity of root and seed lectin was not clearly established (1-4). Roots of some plants, such as soybean, appear not to contain the seed lectin (11), whereas Gietl and Ziegler (22) found a lectin with another sugar specificity than the seed lectin in the root exudate of Phaseolus vulgaris. In this paper a lectin was isolated from the roots of 4 days old pea seedlings by a method that is very successful for the seed lectin (16). The yield of purified root lectin was considerably lower than that of pea seed lectin, i.e. about 5mg/100g dry weight of roots as compared to 150mg/100g dry weight of seeds (16). A comparison of the properties of both lectins indicates that they are identical. The absence of the 28kd band in the root lectin may be due to a very low concentration of precursor. Even in preparations of seed lectin this band did not always show the same intensity relative to the other bands. On the contrary, Gatehouse and Boulter (10) purified a pea root lectin that had a similar SDS-polyacrylamide gel electrophoresis pattern and was antigenically identical to the seed lectin, but differed from it in sugar specificity and specific hemagglutinating activity. These discrepancies can be due to the isolation and storage conditions. Indeed, it has been shown previously in our group for seed lectin that such factors can effect its structural

features and sugar binding properties (16).
As our root lectin was isolated from total root extracts, it was possible that it was in fact a root surface lectin. This hypothesis is based on several arguments. Indeed, the isolation of a lectin from pea root surface was reported (3,8,9). This lectin had the same sugar specificity as the seed lectin (3,9), whereas non-SDS-polyacrylamide gel electrophoresis as well as analytical electrofocusing indicate that pea root surface lectin might be related to seed isolectin 2 (8,9). Furthermore it was shown that during imbibition lectin leaks from the seeds (14,15) and may adsorb onto the roots. However, it is found here that only 1.5-2 % of the total root lectin activity can be eluted from the root surface by high molarity solvent with or without inhibitory sugars. The rest of the hemagglutinating activity can be ascribed to endogenous root lectin, although this does not exclude its internal localization near the root surface. Part of the lectin may be localized in the root cell wall (9). Further localization of the endogenous root lectin and the possible occurrence of a related protein in other plant organs, as is reported for Dolichos biflorus (23,24,25), will give more information about its function in the plant.
The presence of endogenous root lectin raises the question whether the removable lectin is secreted by the root. In this case, root lectin may be actively participating in recognition phenomena, or, as was suggested by Kijne et al. (8), play a structural and/or enzymatical role in root slime and cell wall metabolism. Root origin of root exudate lectin was also suggested for Phaseolus vulgaris (22). On the other hand, the lectin may have leaked from the seeds. This has been reported for soybean (14,15), whereas the phenomenon of general protein leakage during imbibition is also known for pea (26). Further investigations of lectins and isolectins in seeds, root cells, root cell walls and on the root surface are necessary to reveal the relationships between these proteins. Whatever the origin of the removable lectin, it may play a role in defence

(27,28) or recognition mechanisms. This is supported by the finding of Causse and Rougé (15) that lectin, leaked from soybean seeds, is resistant to hydrolysis by soil extract proteinases, indicating that it survives long enough to play an ecological role.

References

1. Bauer, W. : Ann. Rev. Plant Physiol. 32, 407-449 (1981).
2. Kato, G., Maruyama, Y., Nakamura, M. : Agric. Biol. Chem. 43, 1085-1095 (1979).
3. Kato, G., Maruyama, Y., Nakamura, M. : Plant Cell Physiol. 22, 759-771 (1981).
4. Rougé, P., Labroue, L. : C.R. Acad. Sc. Paris 284 D, 2423-2426 (1977).
5. Dazzo, F.B., Urbano, M.R., Brill, W.J. : Curr. Microbiol. 2, 15-20 (1979).
6. van der Schaal, I., Kijne, J., Diaz, C., van Iren, F. : these proceedings.
7. Bhuvaneswari, T.V., Bauer, W.D. : Plant Physiol. 62, 71-74 (1978).
8. Kijne, J.W., van der Schaal, I.A.M., De Vries, G.E. : Plant Sci. Lett. 18, 65-74 (1980).
9. Kijne, J.W., van der Schaal, I.A.M., Diaz, C.L., van Iren, F. : these proceedings.
10. Gatehouse, J.A., Boulter, D. : Physiol. Plant. 49, 437-442 (1980).
11. Su, L.C., Pueppke, S.G., Friedman, H.P. : Biochim. Biophys. Acta 629, 292-304 (1980).
12. Stacey, G., Paau, A.S., Brill, W.J. : Plant Physiol. 66, 609-614 (1980).
13. Gade, W., Jack, M.A., Dahl, J.B., Schmidt, E.L., Wold, F. : J. Biol. Chem. 256, 12905-12910 (1981).
14. Fountain, D., Foard, D., Replogle, W., Yang, W. : Science 197, 1185-1187 (1977).
15. Causse, H., Rougé, P. : these proceedings.
16. Van Driessche, E., Smets, G., Dejaegere, R., Kanarek, L. : in "Lectins. Biology, Biochemistry, Clinical Biochemistry" (T.C. Bøg-Hansen, ed.) Vol. 2. Walter de Gruyter, Berlin, New York, p. 729-742 (1981).

17. Bradford, M. : Anal. Biochem. 72, 248 (1976).

18. Laemmli, U.K. : Nature (London) 227, 680-685 (1970).

19. Van Driessche, E., Smets, G., Dejaegere, R., Kanarek, L. : Planta 153, 287-296 (1981).

20. Lauwereys, M., Van Driessche, E., Dejaegere, R., Kanarek, L. : these proceedings.

21. Rougé, P., Plantefol, L. : C.R. Acad. Sc. Paris 282 D, 621-623 (1976).

22. Gietl, C., Ziegler, H. : Naturwissenschaften 66, 161-162 (1979).

23. Talbot, C.F., Etzler, M.E. : Plant Physiol. 61, 847-850 (1978).

24. Talbot, C.F., Etzler, M.E. : Biochemistry 17, 1474-1479 (1978).

25. Etzler, M.E., Borrebaeck, C. : Biochem. Biophys. Res. Commun. 96, 92-97 (1980).

26. Powell, A.A., Matthews, S. : J. Exp. Bot. 32, 1045-1050 (1981).

27. Mirelman, D., Galun, E., Sharon, N., Lotan, R. : Nature (London) 256, 414-416 (1978).

28. Père, D., Chavant, L., Rougé, P. : these proceedings.

LECTIN RELEASE FROM IMBIBED SOYBEAN SEED AND ITS POSSIBLE FUNCTION

H. Causse and P. Rougé

Laboratoire de Botanique et Biologie cellulaire,
Faculté des Sciences Pharmaceutiques, 35 chemin des Maraîchers,
31062 Toulouse Cedex, France.

The possible role of lectins either in the specific binding of Rhizobia to legume root hairs or in the defense of plants against pathogenic fungi or bacteria, presupposes the presence of these proteins outside seeds and roots. Recent observations are particularly interesting as regard this possible role. Fountain et al. (1) have shown that lectin is released from soybean seeds during water uptake and these findings were corroborated by Hwang et al. (2). Similarly, a lectin release from imbibed *Phaseolus vulgaris* L. seeds has been reported by Manen (3). In addition, the work of Gietl & Ziegler (4) provides evidence for the exudation of lectins by intact roots of seedlings from *Phaseolus vulgaris* L. These results suggest that lectins are not only intracellular or surface-attached components but that they are also released in the medium, that is, in soil in natural conditions. However, this exudation process is not restricted to lectins since release of other proteins (5), including proteinase inhibitors (2,6), by legume seeds during germination has been reported.
To assess the extent to which lectins can play an ecological role in soil by inhibiting pathogenic microorganisms or by stimulating beneficial symbionts, it is necessary to estimate the percentage of seed lectin released, to elucidate the mechanism of this lectin release and to follow the behaviour of lectin in soil. It is the purpose of the present work to investigate the above mentioned points.

MATERIAL AND METHODS

Material :
Soybean seeds (*Glycine max* (*L.*) *Merr.*, *var. Hodgson*) were used.

Seed imbibition :
Lots of 40 g of soybean seeds were surface-sterilized with 70% ethanol for 5 sec. and 5% sodium hypochlorite for 5 min. The sterilized seeds were subsequently rinsed with a large excess of sterile distilled water and incubated at 20C in 100 ml of sterile distilled water. After the desired length of time, the leachate was removed from the seeds which were subsequently washed with two 10 ml aliquots of water. The washes were added to the leachate and concentrated at 4C on Minicon (Amicon) to the final volume of 1 ml.

Protein and lectin isolation :
Proteins from seeds, cotyledons, embryo axis and seed coats were isolated from hexane-defatted meal extracts with 0.1M phosphate buffer pH 7.2 (7).
Seed globulin and albumin were separated by extensive dialysis against tap and distilled water of seed protein precipitated by acid treatment (1).
Globulin from protein bodies were isolated according to Tombs (8).
Soybean seed lectin (SBL) was isolated according to the procedure of Allen & Neuberger (9), by affinity chromatography on N-caproylgalactosamine-sepharose gel and its purity was checked by disc polyacrylamide gel electrophoresis.

Immunochemical techniques :
Antibodies against SBL were raised in rabbit using Freund's complete adjuvant, according to an immunization schedule already described (10). Antibodies against seed soluble proteins and seed coat proteins, both proteins being isalated from hexane-defatted meal extracts with 0.1M phosphate buffer pH 7.2 as stated above, were also prepared in the same conditions.
Immunoelectrophoretic analyses of SBL and proteins present in leachates from imbibed seeds or in seeds or protein bodies ex-

tracts, were performed on microscopic slides according to Scheidegger (11). Precipitin lines were fixed and stained with Amidoblack 10B according to Uriel (12).

The lectin content of seeds extracts and leachates was estimated by radial immunodiffusion in agarose gel (13) using specific antibodies against SBL.

Hemagglutination :

Hemagglutinating activities of seeds extracts or leachates was measured by hemagglutination using untreated rabbit red blood cells (2% suspension in phosphate buffered saline pH 7.0) by the serial dilution method on microtitration plates. The reciprocal of the greatest dilution, which caused full agglutination, was given as the titer.

Protein and lectin hydrolysis :

Soil proteinases were solubilized in 0.1M phosphate buffer pH 7.2 containing 0.15M NaCl and 5 mM mercaptoethanol. The resulting solution was treated with solid ammonium sulfate up to a 90% saturation and the resulting precipitate was taken up in phosphate buffer and extensively dialyzed against the same buffer. All these steps were carried out at 4C. The extract corresponding to 3.5 g of dried soil was diluted into 0.1 ml of phosphate buffer.

The hydrolysis of SBL, of seed proteins, and of casein (Merck) used as a standard, by proteinases from soil extract or trypsin (NBCo), was performed according to Ladd & Butler (14). Each protein (1 mg) was hydrolyzed with 0.2 ml of soil extract (corresponding to 7 g of dried soil) or 1.2 ug of trypsin : the soluble released products were estimated by the Folin reagent (15) while the non-hydrolyzed proteins were estimated by a micro-biuret technique (16). Results were expressed as nmoles of tyrosine or mg of non-hydrolyzed proteins.

Biochemical techniques :

Total soluble protein from seeds extracts was determined colorimetrically using the micro-biuret technique of Goa (16). A more sensitive modified Lowry method (17) was used to estimate the protein content of leachates from imbibed seeds. In all cases, a standard of bovine serum albumine (Sigma) was used. Polyacrylamide gel electrophoresis was performed according to Davis (18) in 7% acrylamide gel rods, at a constant current of 2 mA/tube. The gels were fixed and stained with Coomassie brilliant blue R (BioRad) according to Chrambach et al. (19). Thin-layer isoelectric focusing in polyacrylamide gels was performed with ready-made LKB 1809 Ampholine PAGplates, pH range 3.5-9.5. Isoelectric focusing, fixation, staining and destaining were carried out as indicated by the manufacturer instructions.

RESULTS

1. <u>Time course of lectin and proteins release from imbibed seeds</u> :

Soybean seeds are fully imbibed at 18 to 24 hours after they have been soaked in water (Fig. 1).

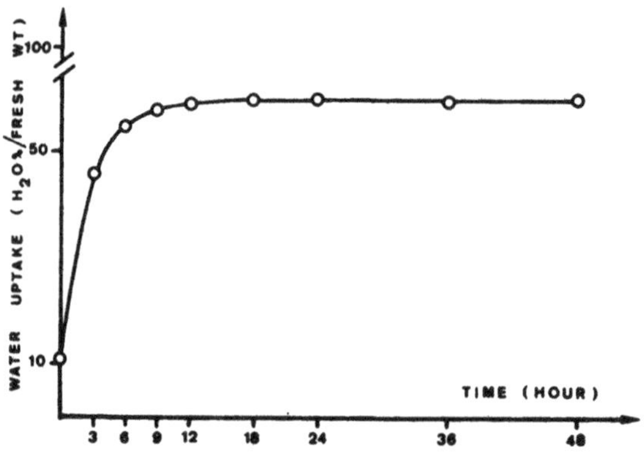

Fig. 1. Time course of water uptake by hydrating soybean seeds. Curve is mean of three determinations.

The time course of water uptake parallels the lectin release, the highest amounts of SBL being leached out 24 hours after the onset of imbibition (Fig. 2). The release of proteins reachs a maximum 36 hours only after the onset of water uptake. After this time, the release of lectin and proteins

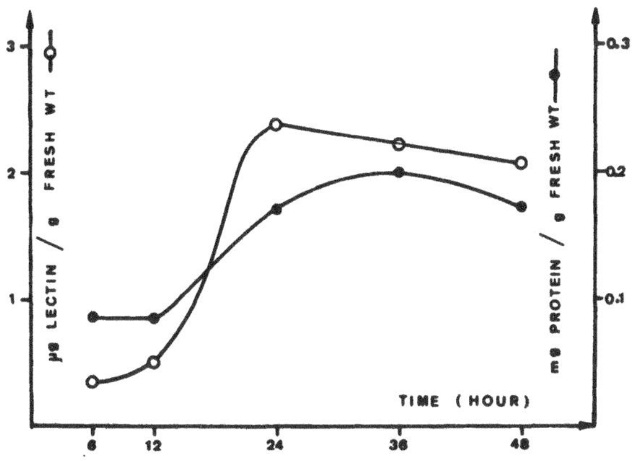

Fig. 2 : Time course of lectin release (o) and total protein release (•) from imbibed seeds. Curves are mean of three determinations.

slows down and such a decrease could be explained by an increasing activity of seed proteases. Similar results have been reported by Fountain et al. (1).

2. Origin of lectin and proteins leaking from imbibed seeds

A comparison between the amounts of lectin and proteins leaked from imbibed seeds (Fig. 3) and the lectin and protein levels of the different parts of the seeds (Fig. 4), suggests that all the lectin and most of the proteins exuded originated from the cotyledons and the embryo axis. However, a determination of the hemagglutinating activities of the different parts of the seeds (Fig. 5) clearly indicates that the seed coat is not completely devoided of lectin activity although this activity is very low as compared to those found in cotyledons and embryo axis. Therefore the lectin content of the seed coat is too small to account for the amounts of lectin exuded from soaked seeds, which indicates that the lectin is essentially issued from cotyledons and embryo axis.
Although the seed coat contains very low amounts of proteins, the concentration of these compounds is sufficient to explain that all the exuded proteins could be originated from the seeds coats. However, the following data indicate that exuded proteins are mainly issued from cotyledons and embryo axis. A comparison by isoelectric focusing of exuded proteins with

lectin	0.0037
proteins	0.180

Fig. 3 : Lectin and protein amounts (mg/g dry weight) released by seeds after a 18 hours imbibition period. Values are mean of three determinations.

	seeds	cotyledons	embryo axis	seed coats
lectin	3.40	3.38	0.26	undetected
proteins	163	161	400	3

Fig. 4 : Lectin and protein levels (mg/g defatted dry weight) of seeds, cotyledons, embryo axis and seed coats. Values are mean of three determinations.

seeds	131 072
cotyledons	131 072
embryo axis	4 096
seed coats	8

Fig. 5 : Hemagglutinating activities of seeds, cotyledons, embryo axis and seed coats measured with untreated rabbit red blood cells. In all cases, 2 g of defatted flour have been extracted with 15 ml of 0.1M phosphate buffer pH 7.2 and 0.1 ml of aliquots have been used for determinations.

those present in the different parts of the seed (Fig. 6) shows high similarities with proteins from the cotyledons and embryo axis. Similarly, a comparison by immunoelectrophoresis between exuded proteins and proteins from the different parts of the seeds, using antibodies against SBL, seed proteins and seed coat proteins, shows that almost of the proteins from the cotyledons and embryo axis are present in the exuded proteins (Fig. 7). In addition, the electrophoretic (Fig. 8) and immunoelectrophoretic (Fig. 9) analysis of the exuded proteins, of seed albumins and globulins, clearly indicates that leaked proteins are a very complex mixture of both seed albumins and globulins. These results suggest that exuded proteins are originated from both the protein bodies and the cytoplasm where globulins and albumins are respectively located. As shown in Fig. 9, exuded proteins

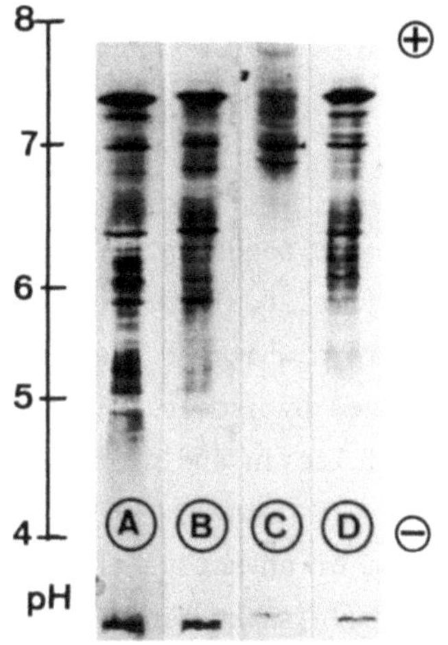

Fig. 6 : Comparison by isoelectric focusing on thin-layer polyacrylamide gel of proteins from cotyledons (A), embryo axis (B), seed coats (C) with those released by seeds (D) after a 18 h imbibition period.

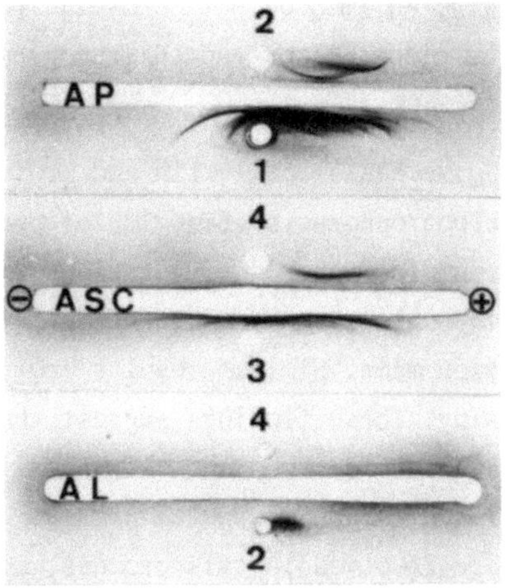

Fig. 7 : Immunoelectrophoretic analysis of proteins from seeds (1), seed coats (3), proteins released by imbibed seeds (2) and imbibed isolated seed coats (4) for a 18 h period,with antibodies against seed proteins (AP), seed coat proteins (ASC) and lectin(AL)

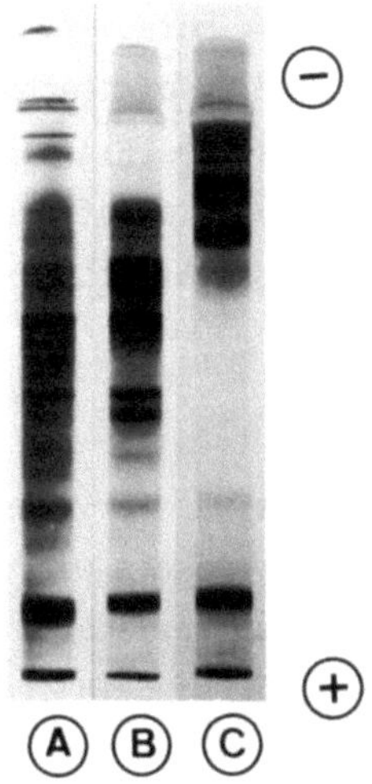

Fig. 8 : Comparison by disc electrophoresis on polyacrylamide gel of proteins released by imbibed seeds for a 18 h period (A) with seed albumin (B) and globulin (C).

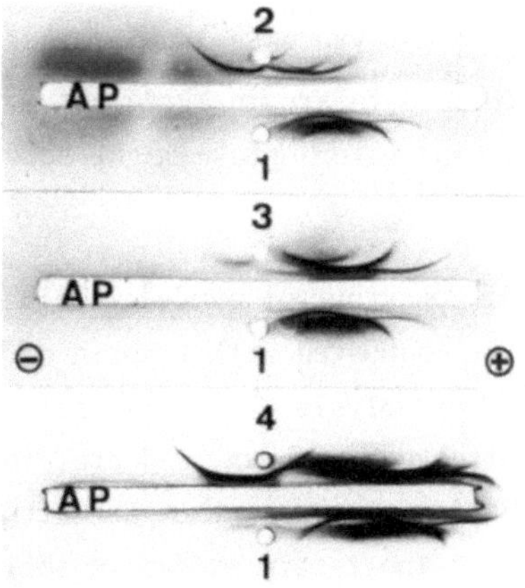

Fig. 9 : Immunoelectrophoretic analysis of proteins released from imbibed seeds for a 18 h period (1), seed albumin (2), seed globulin (3) and globulin isolated from protein bodies (4),with antibodies against seed proteins (AP).

give an immunoelectrophoretic pattern very similar to the one exhibited by globulins extracted from protein bodies.

3. Mechanism of lectin and proteins leakage :

Electrophoresis (Fig. 10) of proteins released by the seeds into the surrounding water during various imbibition periods (i.e. 6, 12, 24, 36 and 48 hours), shows an identical electrophoretic pattern whatever the imbibition time. Similar results (not shown) were obtained by isoelectric focusing. These findings suggest that the mechanism of lectin and protein leakage could be essentially passive.
To determine whether a passive diffusion mechanism or an active transport are involved in this process, soybean seeds have been soaked in water containing various inhibitors at given concentrations. The following inhibitors were used : sodium azide, which is a potent inhibitor of oxidative phosphorylation; cycloheximide (CHM), which acts as a protein synthesis inhibitor; N-ethylmaleimide (NEM) + mercaptoethanol (ME) and phenylmethylsulfonyl fluoride (PMSF), which respectively inhibit the thiol- and serine-containing proteases. In all cases, lectin and protein leakage from imbibed seeds was unaffected by these inhibitors (Fig. 11); moreover, isoelectric focusing analyses indicate that the protein patterns of leachates from treated seeds are similar to those of untreated seeds (Fig. 12). These data are in agreement with the fact that the exudation of proteins during water uptake might be due to an essentially passive mechanism.

4. Behaviour of exuded lectin and proteins in soil :

SBL incubated with trypsin or soil extract proteinases was very resistant to hydrolysis (Fig. 13) while, in the same conditions, casein was readily attacked. Soybean seed proteins (data not shown) appear resistant to the action of trypsin but are attacked by soil extracts. These data indicate that the lectin exuded in soil by germinating seeds could remain in an active form for a sufficient time to play an ecological role. In fact, previous results seem to indicate that various soybean seed proteins, including SBL, could interact *in vitro* with humic acids; this could lead to a better protection of lectin against inactivation by soil proteinases in natural conditions.

inhibitors	used conc.	proteins	lectin
standard		0.180	0.0039
sodium azide	3.08 mM	0.207	0.0041
CHM	1.0 mM	0.137	0.0039
NEM + ME	2.5 mM + 1.0 mM	0.167	0.0035
PMSF	2.0 mM	0.181	0.0039

Fig. 11 : Effect of sodium azide, cycloheximide (CHM), N-ethylmaleimide + mercaptoethanol (NEM + ME) and phenylmethylsulfonyl fluoride (PMSF) used at indicated concentrations according to (1)(20)(21), on lectin and protein and protein amounts released by seeds after a 18 h imbibition period. Each value is mean of three determinations.

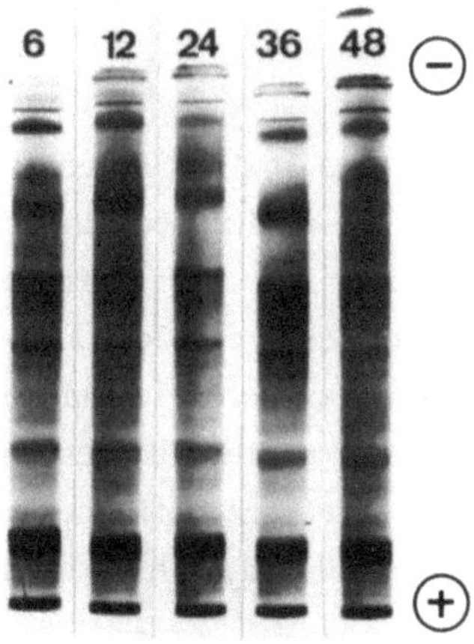

Fig. 10 : Comparison by disc electrophoresis on polyacrylamide gel of proteins exuded from imbibed seeds, respectively 6, 12, 24, 36 and 48 h after the seeds have been soaked in water.

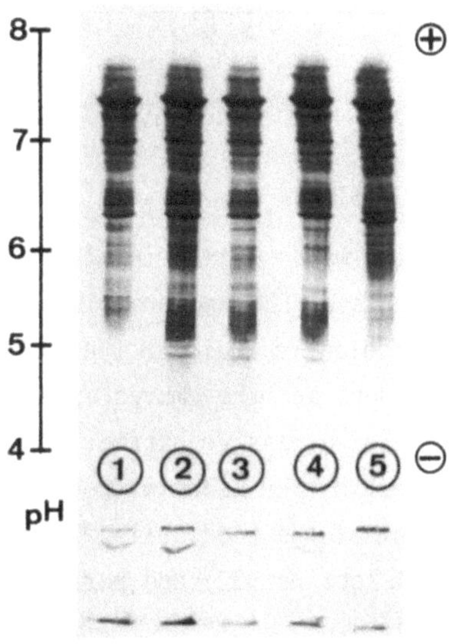

Fig. 12 : Comparison by isoelectric focusing on thin-layer polyacrylamide gel of proteins released by seeds imbibed for a 18 h period without (1) or with inhibitors : sodium azide (2), CHM (3), NEM + ME (4) and PMSF (5).

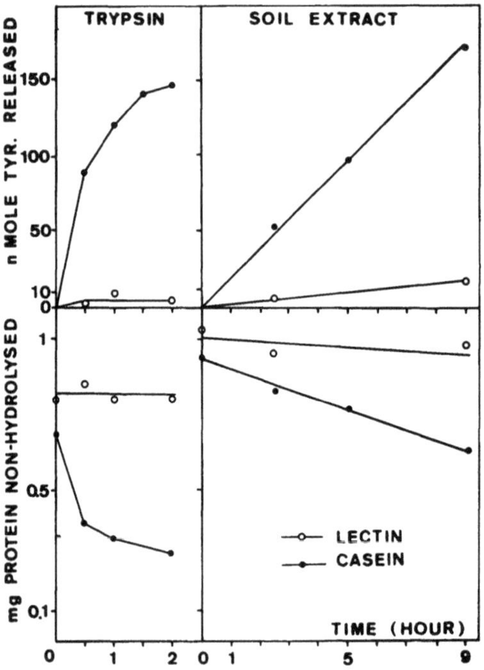

Fig. 13 : Time course hydrolysis of lectin (o) and casein (•) by trypsin (left part) or soil extract containing proteinases (right part). Hydrolysis was performed according to (14). Results are expressed as nm of tyrosine released by 1 mg of protein, measured with the Folin reagent (15), or as mg of remaining non-hydrolysed protein under the same condition, measured with the micro-biuret method (16).

DISCUSSION

During water uptake *in vitro*, soybean seeds release low but detectable amounts of SBL and proteins in the surrounding water. Immunochemical studies clearly indicate that most of these compounds are exuded from the cotyledons and the embryo axis while the seed coat seems to play only the role of a passive barrier. Exuded proteins appear as a complex mixture of globulins and albumins respectively located in the protein bodies and cytoplasm of the seed cells. These findings are in agreement with the fact that lectins (22) and particularly SBL (23) are located within the protein bodies.

Although the exact mechanism of this protein release is still poorly understood it implies membraneous changes of the plasmalemna and protein body membranes. An interesting hypothesis has been developed by Simon (24) who considers that the leakage of soluble compounds from imbibed seeds is a very rapid process in which the membrane organization plays a key role. In the dry state, membranes would exhibit a relatively porous hexagonal model identical to those exhibited by phospholipid-water mixtures containing less than 20% water, as described by Luzzati & Husson (25). In contrast, imbibed cells possess membranes composed of a phospholipid bilayer which impedes the free diffusion of solutes. During water uptake, the conversion of the hexagonal state into the bilayer state would require a sufficient time to allow the diffusion of solutes out of the cells. However, recent observations of McKersie & Stinson (26) and Seewaldt et al. (27) show no evidence for such a change from a bilayer to a hexagonal state associated with dehydration. Despite these conflicting opinions, modifications of protein bodies during seed water uptake, concerning both their shape (28)(29) and their fragmentation during the early stages of germination (8)(30), support the idea that the leakage of soluble compounds from imbibed seeds is associated with membraneous changes. The fact that various inhibitors are ineffective on protein exudation suggests that these changes correspond to a passive process. In soil, it is assumed that seed imbibition proceeds at a slower rate than *in vitro* and thereby, lectin and protein release in soil should be a very progressive process. Although the amounts of SBL leaked from imbibed seeds are very low as compared to the SBL levels of the cotyledons and the embryo axis, one can suppose that they are sufficient to account for an ecological role of lectin in soil. To play such a role SBL must remain unaltered and the fact that lectin is resistant to the action of both trypsin and soil proteinases is of particular interest. As already shown by Liener (31), SBL resists to the action of trypsin but is readily inactivated by pepsin. Such a resistance of lectins to proteolysis is not restricted to SBL since Font et al. (32) have shown that the horse gram *Dolichos biflorus* lectin is resistant to the action of both trypsin and chymotrypsin. A similar behaviour has been reported (33) for the *Robinia pseudoacacia* seed lectin. Proteinases with pepsin-like activity have been isolated (34) from soils but proteinases with trypsin- or chymotrypsin-like activities seem to be more widely distributed in soils (35). Thus, most of the lectins exuded could remain unaltered by soil proteinases. Moreover, preliminary results indicate that SBL could interact with humic acids, thus being protected against soil proteinases.

Although obtained *in vitro*, all the above data support the hypothesis that lectins leaked from germinating seeds into the surrounding soil could remain temporarily unaltered, perhaps being protected against soil proteinases by fixation on humic acids-containing colloids. It is suggested that this lectin exudation could allow the constant presence (and the constant function, if any) of lectins in the vicinity of seedlings and roots during the early developmental stages of the plant. As such, they could play a role either in the recognition of specific symbionts or in the defense against plant pathogenic fungi or bacteria.

Recent observations of Hosselet et al. (these proceedings) demonstrate the presence in garden pea (*Pisum sativum*) roots of an endogenous lectin which can be partly removed from the root surface by high molarity solvents or acid buffered solutions (36,37). The fact that the chemical properties and sugar specificity of the root lectin are the same as reported for pea seed lectins (Hosselet et al., Kijne et al., these proceedings), suggests that the root lectin could be one of the two isolectins present in the embryo axis of the seed (38) whose content decreases during seed germination (39). Van der Schaal et al. and Solheim (these proceedings) have shown that *Pisum sativum* and *Vicia hirsuta* seed lectins bind to rhizobia but it appears that such a binding is unspecific. As already observed with WGA by Mirelman et al. (40) and Barkai-Golan et al. (41), we have shown (Père et al., these proceedings) that PHA also binds to fungal wall and inhibits the hyphal growth and spore germination of various filamentous fungi including plant pathogenic fungi. However, the fact that several lines of *Glycine* (Pueppke, these proceedings), *Phaseolus arborigineus* (42), *Phaseolus vulgaris* (43), *Pisum sativum* (44) and *Vigna unguiculata* (45) are devoided of lectins suggests that these proteins have not any indispensable function in plant and perhaps, do not play a key role in the recognition of symbionts or in the defense against pathogenic fungi or bacteria.

REFERENCES

1. Fountain, D.W., Foard, D.E., Replogle, W.D. and Yang, W.K. : Science. 197, 1185-1187 (1977).

2. Hwang, D.L., Yang, W.K., Foard, D.E. and Lin, K.T.D. : Plant Physiol. 61, 30-34 (1978).

3. Manen, J.F. : Contribution à la caractérisation et à la biologie des lectines dans la graine de *Phaseolus vulgaris L.* var. Contender, Thèse Doct. ès-Sc., n° 1866, Genève (1978).

4. Gietl, Chr. and Ziegler, H. : Naturwis. 66, 161-162 (1979).

5. Amoros, M. : Imbibition des graines et spermosphère, Thèse Doct. spécialité, Toulouse (1967).

6. Wilson, K.A. : Phytochem. 19, 2517-2519 (1980).

7. Causse, H., Lemoine, A. and Rougé, P. (in preparation).

8. Tombs, M.P. : Plant Physiol. 42, 797-813 (1967).

9. Allen, A.K. and Neuberger, A. : FEBS Lett. 50, 362-364 (1975).

10. Rougé, P. : Biologie des hémagglutinines chez le Pois (*Pisum sativum L.*), Thèse Doct. ès-Sc., n° 746, Toulouse (1977).

11. Scheidegger, J.J. : Int. Arch. Allergy. 7, 103-110 (1955).

12. Uriel, J. : *in* Immuno-electrophoretic analysis, Grabar, P. and Burtin, P. ed., Elsevier Publ. Co., Amsterdam, 30-57 (1964).

13. Mancini, G., Vaerman, J.P., Carbonara, A.O. and Heremans, J.F. : *in* Protides Biological Fluids, Peeters, H. ed., Pergamon Press, Oxford, 370-373 (1964).

14. Ladd, J.N. and Butler, J.H.A. : Soil Biol. Biochem. 4, 19-30 (1972).

15. Lowry, D.H., Rosebrough, N.I., Farr, A.L. and Randall, R.J. : J. Biol. Chem. 193, 265-275 (1951).

16. Goa, J. : Scand. J. Clin. Lab. Invest. 5, 218-222 (1953).

17. Bensadoun, A. and Weinstein, D. : Analyt. Biochem. 70, 241-250 (1976).

18. Davis, B.J. : Ann. N. Y. Acad. Sc. 121, 404-427 (1964).

19. Chrambach, A., Reisfeld, R.A., Wyckoff, M. ans Zaccari, J. : Analyt. Biochem. 20, 1-15 (1967).

20. Nakano, M. and Asahi, T. : Plant Cell. Physiol. Jap. 14, 1205-1208, (1973).

21. Chrispeels, M.J. and Boulter, D. : Plant Physiol. 55, 1031-1037 (1975).

22. Manen, J.F. and Pusztai, A. (these proceedings).

23. Horisberger, M. and Vonlanthen, M. : Histochem. 65, 181-186 (1980).

24. Simon, E.W. : New Phytol. 73, 377-420 (1974).

25. Luzzati, V. and Husson, F. : J. Cell Biol. 12, 207-219 (1962).

26. McKersie, B.D. and Stinson, R.H. : Plant Physiol. 66, 316-320 (1980).

27. Seewaldt, V., Priestly, D.A., Leopold, A.C., Feigenson, G.W. and Goodsaid-Zalduondo, F. : Planta 151, 19-23 (1981).

28. Swift, J.G. and Buttrose, M.S. : J. Ultrastruct. Res. 40, 378-386 (1972).

29. Swift, J.G. and Buttrose, M.S. : Planta 109, 61-66 (1973).

30. Catsimpoolas, N., Campbell, T.G. and Meyer, E.W. : Plant Physiol. 43, 799-805 (1968).

31. Liener, I.E. : J. Biol. Chem. 233, 401-405 (1958).

32. Font, J., Leseney, A.M. and Bourrillon, R. : Biochim. Biophys. Acta 243, 434-446 (1971).

33. Bourillon, R. and Font, J. : Biochim. Biophys. Acta 154, 28-39 (1968).

34. Antoniani, C., Montanari, T. and Camoriano, A. : Ann. Fac. Agr. Univ. Milano 3, 99-101 (1954).

35. Ladd, J.N. : *in* Soil Enzymes, Burns, R.G. ed., Academic Press, London, 51-96 (1978).

36. Kato, G., Maruyama, Y. and Nakamura, M. : Plant Cell Physiol. 22, 759-771 (1981).

37. Kijne, J.W., Van der Schaal, C.A.M. and De Vries, G.E. : Plant Sci. Lett. 18, 65-74 (1980).

38. Entlicher, G., Kostir, J.V. and Kocourek, J. : Biochim. Biophys. Acta 221, 272-281 (1970).

39. Rougé, P. : C. R. Acad. Sc. Paris, 280, 2105-2108 (1975).

40. Mirelman, D., Galun, E., Sharon, N. and Lotan, R. : Nature 256, 414-416 (1975).

41. Barkai-Golan, R., Mirelman, D. and Sharon, N. : Arch. Microbiol. 116, 119-124 (1978).

42. Brücher, O., Wecksler, M., Levy, A., Palozzo, A. and Jaffé, W.G. : Phytochem. 8, 1739-1743 (1969).

43. Brücher, O. : Proc. Trop. Region Am. Soc. Hort. Sci. 12, 68-73 (1968).

44. Guldager, P. and Bøg-Hansen, T.C. : Protides of biological fluids, 27 colloquium, 401-404 (1979).

45. Uhlenbruck, G., Karduck, D. and Van Mil, A. : *in* Lectins, Biology, Biochemistry, Clinical Biochemistry, Bøg-Hansen, T.C. ed., De Gruyter Publ., Berlin, 1, 125-134 (1981).

STRUCTURAL RELATIONSHIPS AND PROPERTIES OF LEGUME LECTINS

Leland M. Shannon

Department of Biochemistry, University of California, Riverside, California 92521

Lectins are carbohydrate-binding proteins, of unknown physiological function, which have been isolated from many plant and animal sources (1-3). A large number of purified lectins have been characterized with respect to their physical and chemical properties (4). Most lectins possess a native molecular weight between 130,000-150,000, are tetramers composed of one or two different types of subunits, and possess 2 to 4 carbohydrate binding sites. The K_m for the preferred monosaccharide generally falls within the 2-5 mM range. Several lectins are dependent upon tightly associated manganese and/or calcium for sugar binding. Nearly all lectins are glycoproteins containing 1-5% carbohydrate.

Although many plant lectins have been characterized with respect to their physical and chemical properties, relatively little is known about their synthesis, localization, and function(s) in plants. A number of

Lectins, Vol. III

hypothetical roles for lectins in plants have been advanced, i.e., transport of carbohydrates (1), involvement in seed formation and/or seed germination (4), plant "immune" system (5), specific recognition processes (6), and control of cell growth (7). To date, however, there is no convincing evidence to support any of these as a general role for plant lectins. Research in my laboratory is directed toward achieving a better understanding of the physiology and biochemistry of plant lectins, with the ultimate goal of elucidating their biological function(s). Progress to date has been both exciting and substantial. The following is a brief summary of recent work from this laboratory.

Structural Similarities Among Legume Lectins

Immunochemistry provides a powerful method for determining relationships among proteins. Antibodies raised against a given lectin will react with a different lectin only when the two proteins share a common determinant(s). Different proteins can be functionally similar and yet antigentically totally unrelated. However, when proteins are found to be antigenically similar, there is a strong probability of both an evolutionary and a functional relationship. The following experiments were designed to assess the immunological relatedness among legume lectins.

Affinity purified lectins were screened by Ouchterlony double diffusion against normal rabbit sera and antisera prepared against pure lectins from six legume species. The tests were done on agar plates containing sugars to prevent any spurious lectin-glycoprotein interactions. Figure 1 shows the results of a typical set of tests using antisera raised against *Bandeiraea* lectin. Under these conditions no precipitin lines developed with preimmune sera. In all cases where a cross-reaction was observed, a single "spur" was present indicating that the cross-reacting lectin was immunologically related, but not identical to the homologous lectin. The results of this experiment are summarized in Table I. Note that 11 of the 13 galactose-specific lectins tested crossreacted with one or more of the antisera.

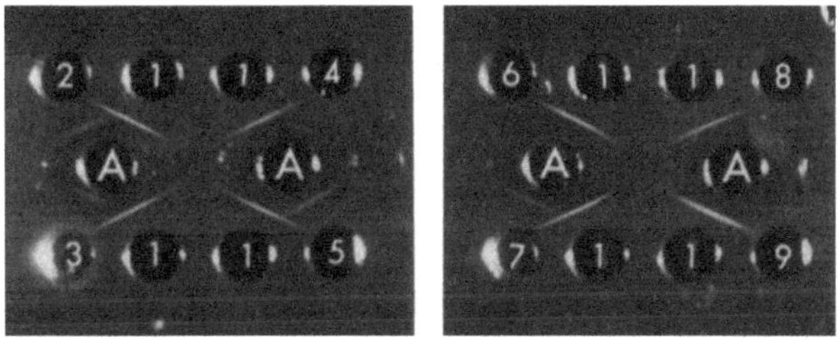

Figure 1. Typical immunological cross-reactions between Bandeiraea antiserum and assorted lectins. A: Bandeiraea antiserum, 10 µl; 1: Bandeiraea lectin, 10 µg; 2: Bauhinia lectin, 10 µg; 3: Lotus lectin, 100 µg; 4: Ulex lectin, 20 µg; 5: Sophora lectin, 10 µg; 6: P. vulgaris lectin, 50 µg; 7: Ricinus lectin, 90 µg; 8: Vigna lectin, 10 µg; 9: Soybean lectin, 10 µg. Photograph taken after 4 h at 37°C.

As a further test of both the strength and specificity of the cross-reactions, we examined the ability of the antisera to inhibit lectin-induced hemagglutination. The hemagglutinating activity of every lectin which cross-reacts with an antiserum, is also specifically inhibited by that antiserum. The inhibitory potency of each antiserum was for the most part closely related to the relative strength of the cross reactions seen by double diffusion.

All of the legume lectins that have been characterized to date can be placed into two main groups with respect to their carbohydrate-binding specificities: (a) those specific for galactose (and its derivatives such as fucose and GalNAc); and (b) those specific for glucose or mannose (and their derivatives). The overwhelming majority of the legume lectins that have been studied fall into the galactose-specific group (Table I).

Table I. Immunological cross-reactions between legume lectins

	Antisera					
Lectin	Bandeiraea simplicifolia	Bauhinia purpurea alba	Glycine max	Phaseolus lunatus	Sophora japonica	Vigna radiata
1. Arachis hypogaea (Gal)[a]	-	-	-	-	-	-
2. Bandeiraea simplicifolia (Gal)	⊕	+	-	-	+	+
3. Bauhinia purpurea alba (GalNAc)	+	⊕	-	+	+	+
4. Dolichos biflorus (GalNAc)	-	+	+	+	+	+
5. Glycine max (GalNAc)	-	-	⊕	-	-	-
6. Lotus tetragonolobus (fucose)	+	+	-	-	+	+
7. Phaseolus lunatus (GalNAc)	+	-	+	⊕	-	-
8. Phaseolus vulgaris (Gal)	+	+	-	-	-	-
9. Ricinus communis (Gal)	+	+	-	+	+	+
10. Sophora japonica (GalNAc)	+	+	-	-	⊕	+
11. Ulex europeus (fucose)	+	+	-	-	+	+
12. Vigna radiata (Gal)	+	+	-	-	+	⊕
13. Wisteria floribunda (GalNAc)	-	+	-	-	+	-
14. Concanavalia ensiformis (Glu/Man)	-	-	-	-	-	-
15. Lens culinaris (Glu/Man)	-	-	-	-	-	-
16. Pisum sativum (Glu/Man)	-	-	-	-	-	-
Total positive CRM	8	9	2	3	8	7

a. Sugar specificity
⊕ = homologous reaction; + = precipitin band observed;
- = no precipitin band observed.

The results presented here indicate that there is extensive immunological relatedness among the lectins in the galactose group. Of 13 galactose-specific lectins tested, only peanut and soybean agglutinins failed to show a cross-reaction with at least one of the sera used. Antisera against soybean lectin did, however, react with two other lectins. We believe that these results coupled with what is already known about these proteins, i.e. they are all lectins, they have similar chemical and physical properties, they are all glycoproteins, they are all from the same plant family, and they are all isolated from seeds, suggest that many of the galactose lectins from legumes are homologs. A similar conclusion has been reached (8-10) from comparative amino acid sequence analysis and, although the sequence studies involved a very limited number of lectins, the similarities in N-terminal amino acid sequences among several galactose lectins was striking. The evidence now available suggests that most of the well characterized legume lectins are members of a specific class of evolutionarily related legume proteins. Many of these lectins are likely to be homologs.

Legume Species Devoid of Lectins.

In view of the above findings and the fact that GalNAc specific lectins often account for 1% or more of the seed protein, and are present in species widely spread throughout the legume family (5), we wondered if lectins might be ubiquitious legume proteins. A review of the literature suggested that this possibility was unlikely, since many legume species, perhaps 50% or more of all species tested, appeared to be totally devoid of hemagglutinin activity of any kind.

The following experiment was designed to examine extracts of a number of GalNAc hemagglutinin "negative" species to determine if they contained proteins immunologically related to any of the GalNAc hemagglutinins. The results, given in Table II, reveal that every plant tested contained cross reacting material to antisera raised against one or more of the GalNAc hemagglutinins. Although many plants showed cross reacting material with antisera raised against several different lectins, it appeared likely that one major protein was responsible, since adsorption of the extracts by any one of the reactive antisera results in a loss of reactivity with all the other antisera.

Table II. Immunological cross reactions between crude seed extracts (which are devoid of hemagglutinin activity) and antisera raised against purified GalNAc specific legume lectins

Seed Extract	Antisera		
	Bauhinia	Dolichos	Sophora
1. Acacia	+	+	+
2. Amorpha	+	-	+
3. Cercis	±	-	+
4. Colutea	+	-	+
5. Delonix	+	±	+
6. Genista	+	-	+
7. Gleditsia	+	-	+
8. Lupinus	+	-	+
9. Mimosa	+	+	+
10. Parkinsonia	±	-	+
11. Poinciana	+	-	+
12. Pueraria	+	+	+
13. Spartium	+	-	+
14. Thermopsis	+	-	+
15. Lespedeza	+	-	+
Total positive CRM	13	3	15

+ = precipitin line observed; - = no precipitin line observed;
± = very weak reaction, questionable.

Sepharose, to which N-acetyl galactosamine has been covalently attached, provides an excellent adsorbant for the purification of most of the GalNAc specific hemagglutinins. It is also known that hemagglutinins may sometimes exist in forms, or can be converted to forms, which retain carbohydrate binding properties but are no longer hemagglutinins. This leads one to inquire whether or not the cross reacting materials, even though they do not display hemagglutinin activity, can be adsorbed by affinity matricies. To answer this question, a number of cross reacting material containing plant extracts were chromatographed on GalNAc sepharose columns. We found that very little, if any, of the cross reacting material was retained by the column. These findings suggest that the cross reacting materials, which do not display classic hemagglutinin activity, also do not possess the carbohydrate binding properties typical of GalNAc lectins.

α-Galactosidases with Hemagglutinin Activity.

The recent discovery that certain legume enzymes possess hemagglutinin properties (11-14) opened an exciting area of research. The evidence collected to date indicates that the α-galactosidase-hemagglutinins and the "classic" legume hemagglutinins are members of distinct classes of protein: they can be separated from each other (14); they can exist simultaneously in the same tissue (14); and they have different sugar specificity (12). Immunological studies reveal, however, that these two classes of proteins share at least some antigenic determinants (15). Antisera raised against several "classic" hemagglutinins cross react with mung bean α-galactosidase-hemagglutinin and antisera to the mung bean protein cross react with several "classic" hemagglutins. Thus, while the α-galactosidase-hemagglutinins and the "classic" legume hemagglutinins appear to be members of distinct classes of proteins, immunochemical data suggest they may be evolutionarily related.

The α-galactosidase-hemagglutinins (14,16) are reversibly converted by pH changes from a tetrameric form which displays both enzymic and hemagglutinin activities to a monomeric form which displays enzymic activity only. Although both the monomeric and tetrameric forms are enzymically active, they display different pH optima and carbohydrate specificities (14,16).

Summary

The above data provide immunochemical evidence that lectins isolated from different legume species may be structurally related proteins and are probably homologues. This generalization derives from the great overall similarities in physical, chemical and biological properties of these proteins. More specifically, these proteins appear to be immunologically closely related and possess a high degree of conservation in their primary structures. Acceptance of this generalization implies a common physiological function for these proteins. Hence, in the search for their function, serious consideration must be given to the group as a whole.

A number of legume species were shown to be totally devoid of proteins with hemagglutinin activity. Without exception, however, these plants contained cross reacting material which was immunologically closely related to one or more of the classic lectins. These results suggest that independent of whether a legume seed extract can agglutinate red blood cells, it appears to contain a protein physiologically homologous to the well studied legume lectins. Further, these observations suggest that "hemagglutinin activity" may *not* be a required manifestation of the "normal" *in vivo* activity of this group of proteins.

All legumes appear to contain a specific α-galactosidase, but only rarely does this enzyme possess hemagglutinin activity. We suspect the hemagglutinin activity associated with the α-galactosidase-hemagglutinin molecule may not be important with respect to the *general* functioning of this class of protein.

The possibility of *in vivo* changes in subunit equilibria, when combined with the accompanying alterations in activity, provides a new concept worthy of consideration with respect to the physiological role of phytohemagglutinins.

It is clear that a great many questions about these interesting proteins remain to be answered. We hope the observations reported here will stimulate additional work and help provide the foundation from which the physiological roles of lectins will ultimately emerge.

Acknowledgments: Major portions of this research were supported by National Science Foundation, grant PCM-7910450.

Literature Cited

1. Lis, H., Sharon, N.: Ann. Rev. Biochem. 42, 541-574 (1973).

2. Liener, I.E.: Ann. Rev. Plant Physiol. 27, 291-319 (1976).

3. Goldstein, I.J., Hayes, C.E.: Adv. in Carb. Chem. Biochem. 35, 127-340 (1978).

4. Toms, G.C.: Chemotaxonomy of Legumes, (Harbourne, J.B., Boulter, D., Turner, B.L. eds.), pp. 367-462, Academic Press, New York (1971).

5. Bird, G.W.G.: Brit. Med. J. 15, 165-168 (1959).

6. Bohlool, B.B., Schmidt, E.L.: Science 185, 269-281 (1974).

7. Nagl, W.: Exp. Cell Res. 74, 599-602 (1972).

8. Foriers, A., Wuilmart, C., Sharon, N., Strosberg, A.D.: Biochem. Biophys. Res. Commun. 75, 980-986 (1977).

9. Etzler, M.G., Talbot, C.F., Ziaya, P.R.: FEBS Lett. 82, 29-41 (1977).

10. Foriers, A., de Neve, R., Kanarek, L., Strosberg, A.D.: Proc. Natl. Acad. Sci. USA. 75, 1136-1139 (1978).

11. Albersheim, P., Wolpert, J.S.: Plant Physiol. 57, S-79 (1976).

12. Hankins, C.N., Shannon, L.M.: J. Biol. Chem. 253, 7791-7797 (1978).

13. Paus, E., Steen, H.B.: Nature 272, 452-454 (1978).

14. del Campillo, E., Shannon, L.M.: Plant Physiol. 69, 628-631 (1982).

15. Hankins, C.N., Kindinger, J., Shannon, L.M.: Plant Physiol. 64, 104-107 (1979).

16. del Campillo, E., Shannon, L.M., Hankins, C.N.: J. Biol. Chem. 256, 7177-7180 (1980).

SPECULATIONS ABOUT A PHYSIOLOGICAL ROLE OF SOME PLANT LECTINS

Willy Peumans, Hetty Stinissen and Albert Carlier
Laboratorium voor Plantenbiochemie, katholieke Universiteit Leuven, Vaartstraat 24, B-3000 Leuven, Belgium

Introduction

During the past decades, plant lectins have intensively been studied and steadily increasing numbers of them have been purified and characterized (1-4). Although their chemical, physico-chemical and biological properties are well documented, their physiological role in the plant is still a matter of controversy. Among their possible functions, a participation in the binding of nitrogen-fixing bacteria to roots of legumes received by far the most attention (5). Unfortunately, unambiguous and conclusive evidence that they play a major role in the recognition or binding of specific bacterial strains has not been presented yet (6). Another, completely different hypothesis has been proposed for wheat germ agglutinin. Because of its specific binding to N-acetylglucosamine and its oligomers, this lectin is able to bind to fungal cell walls and inhibit *in vitro* fungal growth under some circumstances (7). Here again, however, it appears that the proposed function is unlikely under natural conditions. As an alternative to a possible involvement of phytohemagglutinins in plant-symbiont or plant-pathogen interactions, we would propose an endogenous function. Evidence for a physiological role within the plant results from a study of cereal lectins and their resemblance to other lectin families. The aim of this contribution is not merely to give detailed experimental evidence about one or another aspect of cereal lectins, but to demonstrate that these lectins (and some others) fulfil a specific endogenous role in the plant. For that purpose all evidence available in favour of such a role is summarized and together with some recent findings and new ideas and concepts integrated into a working hypothesis of how cereal lectins act.

1. Evidence for a physiological role of cereal lectins

A general analysis of the biochemical, physico-chemical and biological aspects of cereal lectins as a group provides several lines of indirect evidence that these lectins have a role to fulfil in the plant.

I. We did a survey of the occurrence of lectins within the grass family. Thereby, it became clear that quite a large number of species contain WGA-like lectins in their embryos. Concomittant immunological studies further revealed that lectins from all species belonging to the *Triticeae* tribe are immunologically indistinguishable from eachother, and that even the lectins from rice and false brome grasses (*Brachypodium*) (which are taxonomically far distant from wheat) cross-react with antiserum raised against WGA. It is obvious, therefore, that WGA-like grass lectins represent a group of closely related proteins, which have highly been conserved during the evolution and divergency of the grass family. Since such a limited evolutionary change is characteristic for proteins, which fulfil a specific role, it is an argument in favour of a specific physiological role of WGA-like lectins.

II. In a screening experiment of a large number (more than 500) of rye and barley varieties we could not detect any lectin-deficient line. Analysis of isolectin patterns in different hexaploid wheat varieties, on the contrary, revealed that most of them lacked the isolectin coded for by the B-genome (8). It appears, therefore, that recessive mutants of lectin genes occur in cereal genomes, but that they can only be detected in allopolyploid species. This can be explained by the fact that varieties of a polyploid cereal species, which are deficient for one isolectin, still contain one or more other isolectins, whereas the inactivity of one set of lectin genes in a diploid species leads to a complete absence of (active) lectin and results apparently in lethality of the affected line. It is obvious, therefore, that cereal lectins are required for normal development at one or another stage of the life cycle of the plant, and that in order to ensure their proper function, the molecular structure of cereal lectins has been conserved rigorously. The exact nature of this function has not been demonstrated yet, but there are some indications in which direction it has to be looked for. Indeed, studies of some botanical and molecular-biological aspects of cereal lectins have given valuable information concerning their biology.

2. Biology of cereal lectins

Cereal lectins (at least those in the grains) are exclusively present in the primary axes, in which they appear to be rather loosely bound as they elute very readily from imbibing embryos. It is obvious, therefore, that they have a role to fulfil in the primary axes.

During seed formation cereal lectins accumulate exponentially in a period, which coincides with the development of the seed (9). Germinating embryos also synthesize some lectin. However, this synthesis depends exclusively on preformed messenger RNAs and as a consequence decreases exponentially as a function of germination time (9). It can be concluded, therefore, that cereal lectins are specifically synthesized during seed formation and thus, most probably, play a role in the developing primary axes.

Cereal lectins are synthesized as higher molecular weight precursors, which are subsequently "vectorially processed" and transported through the cytoplasmic membrane into an extra-cytoplasmic compartment (Stinissen et al., these proceedings). This evidence suggests that cereal lectins might be active outside the cell menbrane.

Summarizing it can be concluded that lectins in cereal embryos fulfil a physiological role in the developing primary axes and that they act most probably in an extra-cellular compartment.

3. Embryo and leaf lectins from *Agropyrum repens*

The search of the physiological function of cereal lectins has been hampered by the lack of any meaningful image of how these proteins could act at the molecular level. At present, we have indications that physiologically active lectins occur as lectin-receptor-complexes. Evidence for this supposition emerges merely from a study of lectins in *Agropyrum repens* embryos and leaves, and a comparison of one of them to Solanaceous lectins.

Agropyrum repens is a wild perennial grass species closely related to our common cereals. As could be expected, embryos of this grass species contain a WGA-like lectin immunologically indistinguishable from WGA and other cereal lectins (Table 1). Beside this typical embryo lectin, however, we could detect two different lectins in leaves and other vegetatively growing parts of *Agropyrum* plants. One of these lectins (which is further referred to as leaf lectin I) resembles the embryo lectin in all aspects, except that upon immunodiffusion against anti-WGA, its precipitine line does not fuse with that of the embryo lectin (Table 1). Leaf lectin II, on the contrary, differs strikingly from the embryo lectin and leaf lectin I with respect to its specificity, molecular weight, dissociation at low pH, agglutination properties and elution behavior on SP-Sephadex (Table 1). Nevertheless, leaf lectin II is immunologically related to the embryo lectin, and what is surprising, immunologically indistinguishable from leaf lectin I.

Table 1: Comparison of some biochemical, phisico-chemical and biological properties of *Agropyrum repens* lectins and potato lectin.

	Agropyrum			Potato
	Embryo lectin	Leaf lectin I	Leaf lectin II	Tuber lectin
Specificity	NAG, CH	NAG, CH	CH	CH
Molecular weight:				
-native lectins	36,000 d	36,000 d	100,000 d	110,000 d
-subunits	18,000 d	18,000 d	unknown	unknown
Carbohydrate content	0	0	n.d.	± 50 %
Sedimentation constant:				
-at pH 7.0	3.6	3.6	3.6	3.6
-at pH 2.8	2.1	2.1	3.6	3.6
Dilution endpoint (µg/ml)				
with RBC	0.2	n.d.	0.008	0.012
with TRBC	0.01	n.d.	0.008	0.012
Cross-reactivity against anti-WGA	l.i.	c.r.	c.r.	o
Elution from SP-Sephadex at pH 3.8 (M NaCl)	0.4	0.4	0.1	0.1

Abbreviations: NAG = N-acetylglucosamine; CH = chitin hydrolysate; RBC = red blood cells; TRBC = trypsin-treated red blood cells; l.i. = line of identity with WGA; c.r. = cross-reaction (no fusion of precipitine lines); n.d. = not determined; o = no cross-reaction

Although these results clearly demonstrate that *Agropyrum repens* contains 3 different, but related lectins, they give no indication about the exact nature of their mutual relationships. The key to the solution of this problem came from a totally different hook. During our investigations of the leaf lectins a striking resemblance between leaf lectin II and potato lectin became evident. Indeed, both lectins, from taxonomically unrelated species, share in common most of their physico-chemical and biological properties (Table 1). The amazing similarity between these two totally unrelated proteins suggested that both types of lectins represent a physiologically active form and that their mode of action might be identical.

4. Lectin-receptor-complexes

In order to explain the relationship between the 3 *Agropyrum* lectins and the resemblance between *Agropyrum* leaf lectin II and Solanaceous lectins the concept of lectin-receptor-complexes is introduced for these lectins. According to this concept the latter lectins are not lectins in a strict sense but represent lectins associated with receptor molecules. It should be mentioned here that lectin receptors, or at least lectin binding proteins from plants were described by several occasions (3, 11, 12, 13, 14). Several Solanaceous lectins have been isolated and characterized (15-19). They represent apparently a unique class of closely related lectins (19) differing from all other known lectins by their unusually high carbohydrate content (about 50 %; 15). Unfortunately, their exact molecular structure has not been determined yet. Moreover, apparently there is some additional confusion in the sense that there might be considerable differences between different preparations of the same lectin.

Because of some resemblances between cereal and Solanaceous lectins (e.g. specificity, high glycine and cysteine content)

(19, 20) it looked worth while to compare both types of lectins. For that purpose we used highly purified potato lectin (STA). However, soon it became evident that all preparations were consistently "contaminated" with a lower molecular weight component (about 20,000) and that the high molecular weight component also represented in fact a mixture of several molecular species (results not shown here). Therefore, we put forward the hypothesis that STA represents a lectin-receptor-complex. We made several attempts to dissociate the presumed complexes, but all were unsuccessful until recently we succeeded in dissociating STA in a number of lower molecular weight components under mild conditions (results not shown). Meanwhile, however, we became also aware of the amazing resemblance between Solanaceous lectins and *Agropyrum* leaf lectin II. In fact, the *Agropyrum* system offered the opportunity to compare in one species a WGA-like and a STA-like lectin, and a possible intermediate between both types. In addition, the immunological relationship between the 3 *Agropyrum* lectins assured that they were all related to eachother. The only problem inherent to the *Agropyrum* system is that one needs to explain both the similarities and differences between the 3 lectins. Our current view is that the embryo lectin represents free lectin whereas leaf lectin II represents a complex of free lectin and its receptor. Leaf lectin I is likely to be an intermediate between both extreme forms. The same should hold true for STA, with the exception that this lectin occurs (virtually) exclusively as lectin-receptor-complex. Fortunately, STA is biochemically very well characterized (14). After subtraction of a polypeptide equivalent to WGA from STA, there remains a hydroxyproline and arabinose-rich glycoprotein, which suggests that the receptor resembles strongly some peculiar types of plant cell wall glycoproteins. Therefore, the possibility has to be taken into consideration that lectin receptors are associated with cell walls.

Summarizing, the basic ideas of the lectin-receptor-complex

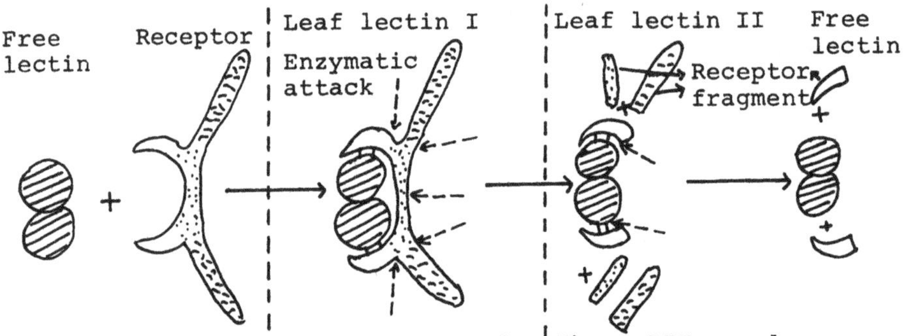

Fig.1: Schematic representation of lectin-receptor-complex formation in *Agropyrum*. Free lectin associates with a complex receptor giving rise to leaf lectin II. The receptors are attacked by enzymes and break up in leaf lectin I and receptor fragments, and finally in free lectin.

concept would give the following model (Fig. 1):

- Lectins are synthesized in the cell and subsequently transported into the periplasmic space (10) where they combine with specific receptor molecules (at least if receptors are available) (phase I).

- Thereby, the native lectins are firmly encapsulated by their receptors so that stable complexes are formed. This complex formation, however, masks the lectin so that (phase II):

    1) It is less accessible for carbohydrate ligands which may result in an altered specificity.
    2) Its antigenic determinants are partially covered resulting in a reduced antigenic activity.

- Lectin-receptor-complexes are biologically active. Their activity, however, can be abolished by a controlled degradation and or dissociation of the complexes. Thereby, different intermediates can be formed (phase III).

4. How do cereal and Solanaceous lectins act?

The lectin-receptor-complex concept sheds a new light on the search of the physiological function of at least some plant lectins. Since receptor molecules in the plant itself argue for a specific endogenous role our thoughts about functional aspects of lectins have to be renewed. On the analogy of the mode of action of other biologically active molecules, which require the presence of specific receptors (e.g. hormones), it can be expected that the lectin-receptor-complexes trigger in some way the membrane in order to pass a message to the nucleus.

Acknowledgements: W. Peumans is Bevoegdverklaard Navorser of the Belgian Nationaal Fonds voor Wetenschappelijk Onderzoek (N.F.W.O.). H. Stinissen receives a grant from the Belgian Instituut tot Aanmoediging van het Wetenschappelijk Onderzoek in Nijverheid en Landbouw (I.W.O.N.L.).

References

1. Liener, I.E.: Ann. Rev. Plant Physiol. 27, 291-319 (1976)
2. Goldstein, I.J., Hayes, C.E.: Adv. Carbohydr. Chem. Biochem. 35, 127-340 (1978)
3. Rudiger, H.: in: Lectins, Vol 1, Bøg-Hansen, T.C., Berlin, p. 3-10 (1981)
4. Kauss, H.: in: Tanner, W. and Loewus, F.A., Springer, Berlin, p. 627-657 (1981)
5. Barondes, S.H.: Ann. Rev. Biochem. 50, 207-231 (1981)
6. Pueppke, S.: these proceedings
7. Mirelman, D., Galun, E., Sharon, N., Lotan, R.: Nature (London), 256, 414-416 (1975)
8. Peumans, W.J., Stinissen, H.M., Carlier, A.R.: Planta: in the press (1982)
9. Peumans, W.J., Stinissen, H.M., Carlier, A.R.: Arch. Int. Physiol. Biochim. 90, 60 (1982)
10. Stinissen, H.M., Peumans, W.J., Carlier, A.R.: these proceedings

11. Rudiger, H., Gansera, R., Gebauer, G., Schurz, H.: in: Lectins, Vol 1, Bøg-Hansen, T.C., Berlin, P. 135-144 (1981)

12. Guldager, P.: in: Lectins, Vol 1, Bøg-Hansen, T.C., Berlin, p. 151-156 (1981)

13. Rougé, P., Chatelain, C.: in: Lectins, Vol 1, Bøg-Hansen, T.C., Berlin, p. 145-150 (1981)

14. Bowles, D.J., Marcus, S.: FEBS Lett. 129, 135-138 (1981)

15. Allen, A.K., Desai, N.N., Neuberger, A.: Biochem. J. 171, 665-674 (1978)

16. Kilpatrick, D.C., Yeoman, M.M.: Biochem. J. 175, 1151-1153 (1978)

17. Kilpatrick, D.C.: Biochem. J. 191, 273-275 (1980)

18. Kilpatrick, D.C.: Biochem. J.: 185, 269-272 (1980)

19. Kilpatrick, D.C., Jeffree, C.E., Lockhart, C.M., Yeoman, M.M.: FEBS Lett. 113, 129-133 (1980)

20. Peumans, W.J., Stinissen, H.M., Carlier, A.R.: Biochem. J. 203, 239-243 (1982)

IN VIVO SYNTHESIS AND PROCESSING OF CEREAL LECTINS

Hetty Stinissen, Willy Peumans and Albert Carlier
Laboratorium voor Plantenbiochemie, Vaartstraat 24,
B-3000 Belgium

Introduction

Since several decades, plant lectins have intensively been studied, and many of them been purified and characterized (1-3). At present, the research on lectins is no longer exclusively focused on their isolation and characterization, but tries to give an answer to the principal questions what lectins do and how they act. One of the main reasons for the poor understanding of the physiological role of plant lectins is the fact that some fundamental processes as their synthesis and possible processing and transport are too scarcely documented. Nevertheless, an insight in these processes is a prerequisite for the unraveling of the function of plant lectins. Until now, two leguminous lectins (one of peas and another of kidney beans) have been translated *in vitro* (4, 5). Unfortunately, however, the results of these *in vitro* experiments do not allow any conclusion concerning possible functions of leguminous lectins. In a recent report on the biosynthesis of *Ricinus communis* lectin a combination of *in vivo* and *in vitro* experiments revealed that both lectin polypeptides are synthesized on membrane-bound polysomes as higher molecular weight precursors and cotranslationally processed (6). Thereby, they are discharged in the lumen of the endoplasmic reticulum and eventually further modified. In this case, all evidence indicates that the castor bean lectin behaves like typical storage proteins (7) and suggests that its function is related, in one way or another, to storage proteins or protein bodies.

This report deals with the *in vivo* synthesis and processing of lectins in germinating cereal embryos, which proved to represent a favorable system for this sort of work.

Material and methods

Germination and labelling of embryos. Embryos (100) were germinated in 0.4 ml water containing 25 µCi ^{35}S cysteine (1160 Ci/mmol) for 150 min, washed free of labelled precursor and further incubated in 1 mM unlabelled cysteine for different (chase) times. Then, they were exhaustively washed with ice-cold water, blotted dry and frozen in liquid nitrogen until use.

Lectin isolation and analysis. Frozen embryos were homogenized in 1 ml Na-acetate buffer (50 mM, pH 3.8) containing 0.1 M NaCl. The homogenates were heated at 50 °C for 2 min and cleared by centrifugation (2,000 *g*, 5 min). The resulting supernatants were passed through small affinity columns of immobilized N-acetylglucosamine (0.2 ml bed volume of Selectin 1 from Pierce Chemical Company). After washing off unbound proteins, the bound lectin was desorbed with 0.5 ml 0.1 M N-acetylglucosamine (in water), concentrated by evaporation for subsequent analysis on 12.5-25 % polyacrylamide gradient gels (in SDS) using a discontinuous system. Labelled polypeptides were detected by fluorography.

Ion-exchange chromatography. Affinity purified lectin fractions were loaded onto sulphopropyl-Sephadex columns (1 x 10 cm), preequilibrated with Na-acetate buffer (pH 3.8) containing 0.1 M NaCl. Columns were eluted with a 100 ml gradient from 0.1 to 0.7 M NaCl buffer. Fractions (1 ml each) were collected and counted. The position of the authentic lectin was determined by agglutination assays of the fractions.

Results

1. Wheat embryo lectin is initially synthesized *in vivo* as a higher molecular weight precursor.

An analysis of the lectin fraction isolated from wheat embryos germinated in the presence of ^{35}S cysteine by highly specific affinity-chromatography revealed that these fractions were always contaminated with a (labelled) polypeptide of higher molecular weight than the lectin. So the possibility became evident that cereal lectins are synthesized as higher molecular weight precursors. In order to test this possibility wheat embryos were incubated for 2.5 h in a medium containing ^{35}S cysteine, and after washing off the labelled precursor further incubated in the presence of 1 mM unlabelled cysteine.

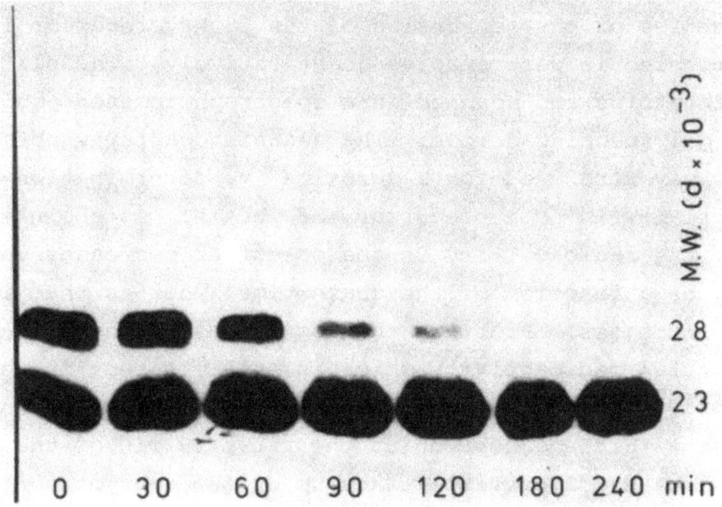

Fig. 1: SDS-PAGE and fluorography of lectin recovered from embryos incubated with ^{35}S cys for 150 min, followed by a chase for 0, 30, 60, 90, 180 and 240 min respectively.

Samples were taken with regular intervals during the chase period, and the lectin isolated by affinity chromatography for subsequent analysis by SDS-polyacrylamide gel electrophoresis and fluorography. Thereby, it turned out that the radioactivity in the polypeptide with apparent MW of 28,000 d (28 Kd polypeptide) decreased progressively as a function of chase time, whereas that in the lectin (apparent MW 23,000 d) increased correspondingly (Fig. 1). It is evident from these results that wheat lectin is synthesized as a precursor polypeptide of considerably higher MW. Similar results were obtained with rye lectin (not shown).

It should be noted here, that the precursor is biologically active since it binds to the immobilized N-acetylglucosamine on the affinity columns. Thus the sequence, which is cleaved off, does not interfere with the carbohydrate binding properties of the lectin precursor.

2. Kinetics of the conversion of the 28 Kd precursor into authentic lectin.

The kinetics of the conversion of the 28 Kd precursor into authentic lectin were studied quantitatively. The plate was cut into strips and scanned in a spectrophotometer equipped with a gel scanning device. The scanning patterns obtained in this way were used for a quantitative determination of the radioactivity in both precursor and lectin. As shown in Fig. 2, the radioactivity in the precursor decreases very rapidly as a function of the chase time, whereas that in the lectin increases. Plotting the log values of the percentage of the total radioactivity which is still present in the 28 Kd polypeptide against chase time yields a linear relationship, at least with the exception of the first 30 min of the chase (during which ^{35}S cysteine taken up by the embryos continues to be incorporated into lectin). It can be concluded, therefore, that the conversion of the 28 Kd polypeptide into lectin follows exponential kinetics. The half-life of the precursor

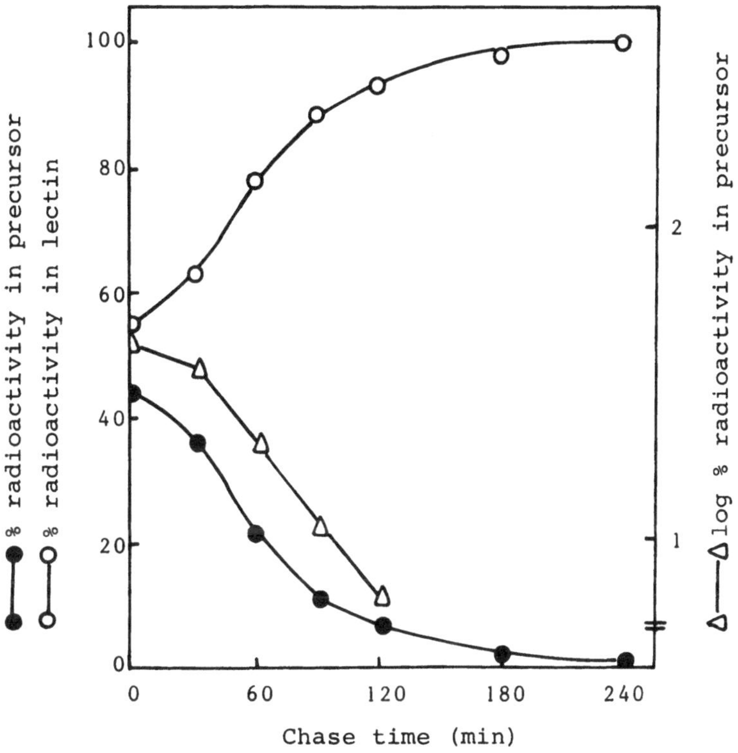

Fig. 2: Kinetics of the conversion of wheat lectin precursor into authentic lectin. The results shown in the fluorogram of Fig. 1 were quantitated. Results are expressed as % of the total radioactivity (in precursor + lectin) in both precursor (O——O) and lectin (O——O). The log values of the % in the precursor are indicated by (Δ——Δ).

was calculated to be about 36 min.

Discussion

A major conclusion which can be drawn from the results presented in this paper is that wheat and rye lectins are synthesized as higher molecular weight precursor polypeptides, which are subsequently post-translationally processed.

Germinating wheat embryos were chosen as a model system since they offer several advantages. The material is easy to prepare and handle; the lectins can be purified in one single and simple step; the lectin is particularly suitable for labelling with ^{35}S cys because of its unusually high cys content; the lectin mRNAs can easily be translated in homologous cell-free extracts; and finally, wheat lectins are very well characterized.

The presence of extra-peptides on initial translation products is usually associated with the so-called signal mechanism, a type of topogenic process whereby polypeptides are synthesized with an extra signal sequence of some 15-30 amino acids, which contains the information for the translation of these polypeptides into the lumen of the endoplasmic reticulum (10-12). Such a signal mechanism has been found to be operative for the synthesis of castor bean lectin (6) and several other types of plant proteins which are synthesized on membrane-bound polysomes (13-17). As a general rule it can be stated that all proteins which are destined to be transported into the lumen of the endoplasmic reticulum are synthesized according to the signal mechanism (10-12). Their final destination, however, depends on the peculiar kind of modification the endoplasmic vesicles undergo. In the case of legume storage proteins, the vesicles are involved in protein body formation and the storage proteins become compartmented in these organelles (7). Other types of proteins become sequestered lysosome-like organelles (e.g. lytic enzymes) or in secretory vesicles (e.g. digestive enzymes) (10-12). Since in the signal mechanism, translation and processing are essentially simultaneous events, this type of cotranslational processing has been termed "vectorial translation" (22). Because of the extremely short half-life of the precursor polypeptides, their synthesis cannot be followed *in vivo* but only by *in vitro* translation of their corresponding mRNAs. In the case of castor bean lectin, it was demonstrated that the A chain occurs *in vivo* exclusively as a 31,000 d polypeptide. However, when polysomal mRNA is translated in a reticulocyte lysate the A chain immunoreactive product has a MW of 33,500 d. Upon addition of canine pancreas microsomes (which are known to remove signal peptides *in vitro*) this 33,500 d polypeptide was converted into authentic A chain (6). It is evident therefore, that castor bean lectin A chain is synthesized with an additional signal peptide, which is cleaved off concomitantly with its transport into the endoplasmic reticulum.

Post-translation processing of proteins represents another type of topogenic mechanism of some plant proteins. Some chloroplast and mitochondrial proteins are synthesized on free polysomes as higher MW precursors (18-22). These primary translation products are post-translationally transported into chloroplasts or mitochondria (in an energy-requiring process) (23,24) in which the extra sequence (usually some 40-50 amino acid residues) is removed by some soluble enzyme (25). Because in this type of protein modification transport and processing are closely linked, it has been termed "vectorial processing" (22).

According to the results shown in Fig. 1 and 2, wheat lectin is synthesized unlike signal peptide-containing precursors (the 28 Kd precursor is clearly not cotranslationally processed as it has a half-life of 36 min). A comparison of the *in vivo* synthesis and processing of wheat lectin to the vectorial processing of organellar proteins (18-22) indicates that there is a striking similarity between both mechanisms. In both cases the processing is an essentially post-translational event, the size of the extra sequence is nearly the same, and the half-life of the precursor is of the same order of magnitude. These striking resemblances indicate that the processing of cereal lectins could have a function analogous to that of the vectorial processing. However, earlier we showed that cereal lectins are not localized in subcellular organelles (26). Therefore, it is tempting to speculate that their vectorial-like processing allows them to reach a compartment outside the cytoplasmic membrane. Most probably, this compartment corresponds to the periplasmic space. In plants, the periplasmic space cannot be considered as extracellular space, which is in good argument with the differences in synthesis / processing of lectins and extracellular proteins.

Summarizing, it can be concluded that a study of the synthesis of cereal lectins at the molecular level yielded some interesting information about their physiological function in the sense that it becomes evident that they may be destined for a function outside the cytoplasm. In addition, the lectin system represents a nice example of *in vivo* protein synthesis and processing in higher plants.

Acknowledgements: H. Stinissen receives a grant from the Belgian Instituut to Aanmoediging van het Wetenschappelijk Onderzoek in Nijverheid en Landbouw (I.W.O.N.L.). W. Peumans is Bevoegdverklaard Navorser of the Belgian Nationaal Fonds voor Wetenschappelijk Onderzoek (N.F.W.O.).

References

1. Goldstein, I.J., Hayes, C.E.: Adv. Carbohydr. Chem. Biochem. 35, 127-340 (1978).
2. Barondes, S.H.: Ann. Rev. Biochem. 50, 207-231 (1981).
3. Kauss, H.: in: Tanner, W. and Loewus, F., A. eds., Springer, Berlin, p. 627-657 (1981).
4. Peumans, W.J., Delaey, B.M., Manickam, A., Carlier, A.R.: Planta 150, 286-290 (1980).
5. Bollini, R., Chrispeels, M.J.: Planta 146, 487-501 (1979).
6. Roberts, L.M., Lord, J.M.: Eur. J. Biochem. 119, 31-41 (1981).
7. Boulter, D.: Adv. in Bot. Research 9, 1-31 (1981).
8. Peumans, W.J., Stinissen, H.M., Carlier, A.R.: Planta: in the press (1982).

9. Peumans, W.J., Stinissen, H.H., Carlier, A.R.: Planta: in the press (1982).

10. Blobel, G., Dobberstein, B.: J. Cell Biol. 67, 835-851 (1975).

11. Blobel, G.: Proc. Natl. Acad. Sci. USA 77, 1496-1500 (1980).

12. Kreil, G.: Ann. Rev. Biochem. 50, 317-348 (1981).

13. Boston. R.S., Miller, T.J., Mertz, J.E., Burgess, R.R.: Pl. Physiol. 69, 150-154 (1982).

14. Burr, F.A., Burr, B.: J. Cell Biol. 90, 427-434 (1981).

15. Higgins, T.J.V., Spencer, D.: in: Genome organization and expression in plants, Leaver C, ed., Plenum, New York 245-258 (1980).

16. Nelson, C.E., Ryan, C.A.: Proc. Natl. Acad. Sci. USA 77, 1975-1979 (1980).

17. Weber, E., Ingversen, J., Manteuffel, R., Püchel. M.: Carlsberg Res. Comm. 46, 383-393 (1981).

18. Chua, N-H., Schmidt, G.W.: J. Cell Biol. 81,461-483 (1979).

19. Ellis, R.J.: Ann. Rev. Plant Physiol. 32, 111-137 (1981).

20. Highfield, P.E., Ellis, R.J.: Nature (London) 271, 420-424 (1978).

21. Maccecchini, M.L., Rudin, Y., Blobel, G., Schatz, G.: Proc. Natl. Acad. Sci. USA 76, 343-347 (1979).

22. Schatz, G.: FEBS lett. 103, 203-211 (1979)..

23. Grossman, A., Bartlett, S., Chua. N-H.: Nature (London) 285, 625-628 (1980).

24. Nelson, N., Schatz, F.: Proc. Natl. Acad. Sci. USA 76, 4365-4369 (1979).

25. Smith, S.M., Ellis, R.J.: Nature (London) 278, 662-664 (1979).

26. Stinissen, H.M., Peumans. W.J., Carlier, A.R.: Arch. Int. Physiol. Biochim. 89, 132 (1981).

THE α- AND β-SUBUNITS OF PEA LECTIN ARE THE RESULT OF A POST-TRANSLATIONAL CLEAVAGE OF A PRECURSOR CHAIN

Mark Lauwereys[‡], Edilbert Van Driessche[‡¶], Arthur Strosberg[‡], Robert Dejaegere[¶] and Louis Kanarek[‡]

[‡]Laboratorium voor Chemie der Proteïnen, and
[¶]Laboratorium voor Plantenfysiologie;
Vrije Universiteit Brussel, Instituut Moleculaire Biologie, Paardenstraat 65, B-1640 Sint-Genesius-Rode, Belgium

Introduction

Based on the molecular weight of their subunits, Leguminous lectins can be divided into two groups : the one-chain type lectins are tetramers or dimers of one type of subunits of similar molecular weight; the two-chain type lectins contain two different subunits (usually called α- and β-chains) held together by non-covalent forces and form tetramers $\alpha_2\beta_2$. Sequence data suggest (1, 2) that the two-chain type lectins are originally synthesized as a larger precursor polypeptide which is subsequentially cleaved into two or more fragments to yield both chains. We noticed in pea lectin preparations, purified from dry, mature seeds, that besides the α-chain and the β-chain a minor component with a molecular weight of 28 kD was consistently present (3) and which was supposed to correspond to the native chain. However, the quantities that could be prepared from this chain were too small to allow further investigations. On the contrary, when the lectin is isolated from seeds before drying on the plants, the 28 kD polypeptide is present in appreciable amounts on SDS-gels, indicating that its presence may be developmentally regulated. In this paper it is shown that this chain corresponds to the precursor polypeptide from which both the α- and β-chains are formed during the maturation of the seed.

Materials and Methods

1. Lectin isolation and separation of the subunits :
Plants (var. Wonder van Kelvedon) were grown in the field and seeds were collected before maturation. The seeds were immediately homogenized and the lectin was prepared as described previously (3). The subunits were separated by gel filtration on Sepharose-4B in 5 M guanidine-HCl. Fractions containing protein were desalted on a Biogel P-2 in 1 M acetic acid and lyophilized. SDS-electrophoresis was performed on linear 12.5%-25% polyacrylamide gradient gels using the buffer system as described by Laemmli (4). Molecular weights were determined from the migration relative to standard proteins with known molecular weight.

2. Characterization of fractions I and II :
Automated sequence determinations were performed on a Beckman 890C sequencer using a 0.25 M Quadrol program. The phenylthiohydantoin-amino acids were identified by high-pressure liquid chromatography on an Ultrasphere ODS RP18 (Altex) column, using a phosphate-methanol gradient. Detection was carried out at 254 and 320 nm.
Lyophilized chains were digested with trypsin in 0.05 M potassium phosphate buffer at 37°C for 24 hours. The tryptic peptides were purified by high-pressure liquid chromatography on a Hibar RP-8 column using 0.05 M ammonium-acetate as starting buffer and applying a linear gradient (0%-75%) of methanol in 1 hour. The fractionated peptides were sequenced by the 4-NN-dimethylaminoazobenzene 4'-isothiocyanate/phenylisothiocyanate (DABITC/PITC) double coupling method as described by Chang et al. (5). The 4-NN-dimethylaminoazobenzene 4'-thiohydantoin-amino acid derivatives were identified by two-dimensional thin-layer chromatography on micro-polyamide sheets (3 cm x 3 cm) (5).

Results

When affinity purified pea lectin, isolated from immature seeds, is chromatographed on Sepharose-4B in 5 M guanidine-HCl, two fractions are obtained (Fig. 1.).

When lectin from immature seeds is submitted to electrophoresis in the presence of sodium dodecylsulphate (Fig. 2, lane D) essentially three bands are resolved, i.e. the α-chain (molecular weight 5.8 kD), the β-chain (molecular weight 18 kD) and a γ-chain (molecular weight 28 kD). When compared to the band pattern of lectin isolated from dry seeds (Fig. 2, lane A) it is clear that the γ-band in the latter preparation is only present as a minor component. SDS-electrophoresis of the separated chains (fractions I and II from Fig. 1.) revealed that fraction I contains the β-chain and the γ-chain and is completely devoid of α-chain (Fig. 2, lane E). Fraction II consists

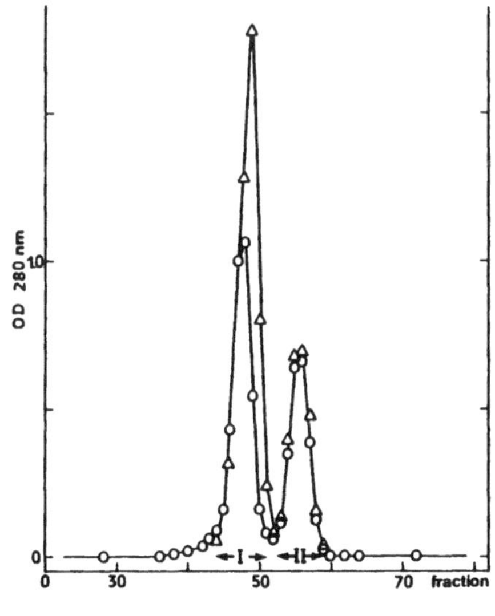

Fig. 1. Gel filtration of affinity purified pea lectin isolated from mature (o) and immature (Δ) seeds on Sepharose-4B in 5 M guanidine-HCl.
Fraction I : mixture of β- and γ-chains from immature seeds; β-chain and trace amounts of γ-chain from mature seeds.
Fraction II : α-chain.

of the α-chain (Fig. 2, lane F) and two minor polypeptides which were previously shown to be natural proteolytic fragments of the β-chain (3). Although fraction I is a mixture of β- and γ-chains, sequence analysis revealed only one sequence which is identical to that of the purified β-chain (3, 6). The sequence data showed no trace of α-chain contamination.

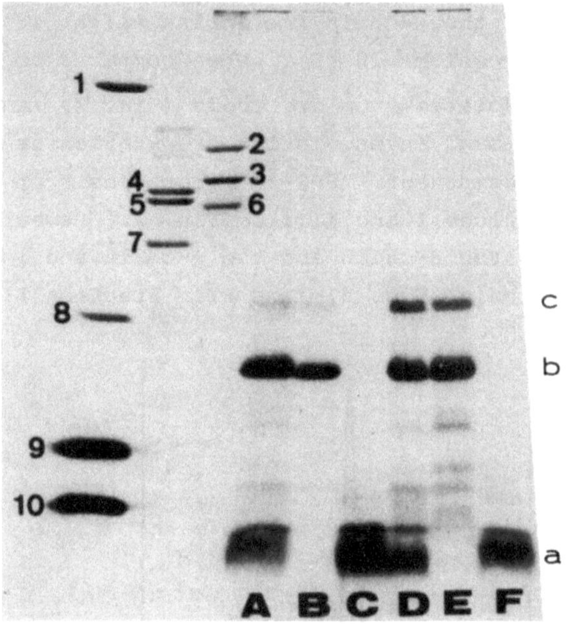

Fig. 2. SDS-electrophoresis on a linear 12.5%-25% polyacrylamide gel.

Lane A: affinity purified pea lectin from mature seeds
B: fraction I (see Fig. 1.) from mature seeds
C: fraction II (see Fig. 1.) from mature seeds
D: affinity purified pea lectin from immature seeds
E: fraction I (see Fig. 1.) from immature seeds
F: fraction II (see Fig. 1.) from immature seeds

Numbered bands : reference proteins of known molecular weight, i.e. phosphorylase b, 92 kD (1); catalase, 60 kD (2); fumarase, 48.5 kD (3); citrate synthase, 46 kD (4); aspartate aminotransferase, 43 kD (5); aldolase, 40 kD (6); malate dehydrogenase, 35 kD (7); trypsin, 23.3 kD (8); cytochrome c, 12.5 kD (9); aprotinin, 6.5 kD (10).

Lettered bands : a = α-chain; b = β-chain; c = γ-chain

Fraction I was digested with trypsin (1% w/w) for 24 hours in 0.05 M potassium phosphate buffer and, after removal of non-soluble material by centrifugation, the digest could be separated by high-pressure liquid chromatography into 6 fractions (Fig. 3B). Five of these fractions were characterized by manual Edman degradation (Table 1). From these sequence data and from its retention time on the high-pressure liquid chromatography column (Fig. 3A and 3B), peptide B2 could be unequivocally identified as an internal α-chain tryptic peptide, i.e. A2, corresponding to residues 14-21.

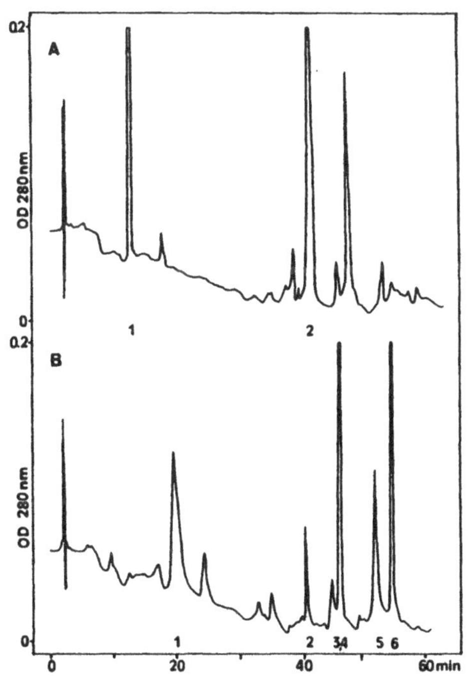

Fig. 3. Fractionation of tryptic peptides by high-pressure liquid chromatography of : (A) pea lectin α-chain and (B) the mixture of β- and γ-chains (fraction I from Fig. 1., immature seeds). See text for experimental details. The numbered peptides were subjected to sequence analysis (Table 1).

Table 1. N-terminal sequence determination of the separated tryptic peptides of fractions I and II (see Fig. 1.) from immature pea seeds.

Peptide	Sequence	Homologous region in favin (8) or pea lectin (6)
B 1	No sequence	-
B 2	D-V-V-P-	α 14-21
B 3	T-E-P-	-
B 4	F-S-P-D-	β 11- ?
B 5	F-S-P-	β 11- ?
B 6	A-L-Y-S-	β 44-55
A 1	V-T-S-	α 1-11
A 2	D-V-V-P-	α 14-21

Tryptic peptides were separated by high-pressure liquid chromatography as shown in Fig. 3. The peptides are named according to the terminology of Fig. 3.

Discussion

In this paper, evidence is given that the α- and β-chains of pea lectin are derived from a precursor polypeptide with a molecular weight of 28 kD. Post-translational cleavage, without severe damage of the native conformation or loss of biological activity, has also been demonstrated for ConA (7) which consists of a mixture of intact subunits and natural occuring fragments. Likewise the favin β-chain (molecular weight 18.7 kD) is most probably lacking some part at the -COOH terminal region as a result of proteolysis and may be derived from the favin β'-subunit (8). A similar modification was suggested by Foriers et al. (2) to explain the absence of 24 amino acid residues at the -COOH terminal part of the lentil β-chain when compared to favin. Howard et al. (9) found that the native lentil lectin is built up of two identical subunits, whereas in later literature reports this lectin was clearly shown to have a $\alpha_2\beta_2$ structure. Highly purified preparations of soybean agglutinin contain low molecular weight polypeptide chains, resulting probably from proteolytic cleavage of the intact subunit (10). For pea lectin, we were able to isolate and characterize two natural fragments of the β-chain (3). These fragments are most probably associated with native molecules. From the leaves and stems of Dolichos biflorus, Talbot et al. (11) isolated a protein that cross-reacts with antibodies to the seed lectin ("Cross-reacting material"). These authors suggested that all the subunits of the cross-reacting material and the seed lectin might represent different stages of completion or proteolytic modifications at the -COOH terminus of a common polypeptide chain. For Vicia sativa lectin, the conflicting literature data on the molecular weight of the subunits (12-14) as well as on the quaternary structure can be explained by post-translational proteolytic cleavage of precursor chains. The few examples given above clearly demonstrate that post-synthetic proteolysis may be common place in seed lectins.

Our data on pea lectin can be explained in the following way. A one-chain precursor is first synthesized. This chain has a molecular weight of 28 kD and is subsequentially proteolytically cleaved to yield the α-subunit and a second fragment which we call the pre-β-chain. Developmentally regulated loss of fragments at the -COOH terminus of the pre-β-chain finally results in the β-chain with a molecular weight of 18 kD. This model can also explain the ontogeny of other two-chain Leguminous lectins such as favin, lentil lectin and Vicia sativa lectin.

The N-terminal sequences of some one-chain lectins (15) have been found to be homologous to the β-chains of the two-chain type lectins. The lack of processing of these lectins may originate from conformational restraints rather than from the lack of processing enzyme(s) itself. This hypothesis could be confirmed by the results of Baumannet al. (16) who showed that Vicia cracca contains the two types of lectin, which, apart from their subunit structure, also differ by their primary structure and agglutinating properties.

References

1. Cunningham, B.A., Hemperly, J.J., Hopp, T.P. and Edelman, G.M.: Proc. Nat. Acad. Sci. USA 76, 3218-3222 (1979)
2. Foriers, A., Lebrun, E., Van Rapenbusch, R., De Neve, R. and Strosberg, A.D.: J. Biol. Chem. 256, 5550-5560 (1981)
3. Van Driessche, E., Smets, G., Dejaegere, R. and Kanarek, L.: Lectins, Biology, Biochemistry, Clinical Biochemistry Vol. 2 (T.C. Bøg-Hansen, ed.) (W. De Gruyter, Berlin) pp. 729-741 (1982)
4. Laemmli, U.K.: Nature (Lond.) 227, 680-685 (1970)
5. Chang, J.Y., Brauer, D. and Wittmann-Liebold, B.: FEBS Lett. 93, 205-214 (1978)
6. Van Driessche, E., Foriers, A., Strosberg, A.D. and Kanarek, L.: FEBS Lett. 71, 220-222 (1976)
7. Cunningham, B.A., Wang, J.L., Pflumm, M.N. and Edelman, G.M.: Biochemistry 11, 3233-3239 (1972)
8. Hemperly, J.J., Hopp, T.P., Becker, J.W. and Cunningham, B.A.: J. Biol. Chem. 254, 6803-6810 (1979)
9. Howard, I.K., Sage, H.J. and Stein, M.D.: J. Biol. Chem. 246, 1590-1595 (1971)
10. Lotan, R., Lis, H. and Sharon, N.: Biochem. Biophys. Res. Comm. 62, 144-150 (1975)
11. Talbot, C.F. and Etzler, M.E.: Biochemistry 17, 1474-1479 (1978)
12. Falasca, A., Franceschi, C., Rossi, C.A. and Stirpe, F.: Biochim. Biophys. Acta 577, 71-81 (1979)
13. Van Driessche, E., Vandenbranden, S., Dejaegere, R. and Kanarek, L.: Arch. Intern. Physiol. Bioch. 88, B50-B51 (1980)
14. Gebauer, G., Schiltz, E., Schimpl, A. and Rüdiger, H.: Hoppe-Seyler's Z. Physiol. Chem. 360, 1727-1735 (1979)
15. Foriers, A., De Neve, R. and Strosberg, A.D.: Physiol. Veg. 17, 597-606 (1979)
16. Baumann, C., Rüdiger, H. and Strosberg, A.D.: FEBS Lett. 102, 216-218 (1979)

LIGHT AND ELECTRON MICROSCOPIC LOCALIZATION OF LECTINS IN PLANT CELLS AND TISSUES

Jean-Francois Manen

Université de Genève, Institut de Botanique. 1, ch. de l'Impératrice, CH-1292 Chambésy, Suisse

Arpad Pusztai

The Rowett Research Institute, Bucksburn, Aberdeen AB2 9SB, Scotland U.K.

Introduction

Despite the great surge of interest and intensive study in the last decade, the biological function(s) of plant lectins are still not known. However, as any physiological role postulated for the lectins should be consistent with their cellular location, it is hoped that studies on the localization of these highly reactive cellular components will lead to a better understanding of their functions in the plant.

Most of the methods used for lectin localization are the variants of two different approaches and based on either subcellular fractionation or on cytochemical techniques. In the first, different subcellular compartments are separated by density, while in the second, thin sections of plant tissues are reacted with specific antibodies which are then visualized by either a marker attached to them or by a second reaction. In a powerful and selective variant of the latter method the lectins in the tissues are detected with a specific glycosylated marker (17).

Thus it has now been firmly established that seed lectins from legumes and other plant families studies so far are associated with protein bodies, the storage organelles of seed cells.

Lectins, Vol. III

Lectins from Ricinus communis and Phaseolus vulgaris were shown by subcellular fractionation to be located in the protein body matrix (3,4,5,23,27,35,37). There were also indications for the existence of a weak association between protein body membranes and some lectin components in the seeds of Phaseolus vulgaris (23,28). Lectins in the seeds of Phaseolus vulgaris (6,25), Canavalia ensiformis (6) and Datura stramo nium (16) were also shown by Light microscopy to be located in the protein bodies. The same was shown to be true by electron microscopy for Glycine max (14) and Pisum sativum (36). However with Datura stramonium cytoplasmic membranes were also specifically stained for lectins (16).
Seed lectins from a number of Phaseolus vulgaris cultivars are known to be composed of two different subunits: E and L, the erythroagglutinating and the leucoagglutinating subunits. From the possible combinations of these two subunits five tetrameric lectins exist in these seeds (19,22). As previous studies have indicated that the E subunits is synthesized early during seed development while the L subunit is towards the end of seed maturation (22,29,33), it was thought possible that lectins containing the two different subunits should be found in different cellular compartments. Indeed the results of previous subcellular fractionation studies (23,28) suggested that there were differences in lectin content and composition of the various protein body fractions, including the matrix and the membranes of these organelles. As with the increased resolution of the E and L subunits monospecific antibodies could now be raised to the two lectin subunits, an attempt was made by immunocytochemical methods to refine the results of earlier lectin localization studies. In the followings these results are described and their biological implications discussed as part of a short review of other relevant lectin localization studies to date.

Materials and methods

Isolectin purification:

Lectins from Phaseolus vulgaris seeds (Contender or Processor cultivars) were purified by affinity chromatography (21). The five tetrameric lectins were separated by ion-exchange chromatography on SP-Sephadex in 0,05M acetate buffer, pH 3.6 (22). The NaCl concentration during elution was increased from 0.3 to 0.8M (Fig. 1a). Purity of the fractions was tested by disc gel electrophoresis at pH 4.5 (30) (Fig. 1b) and by SDS gel electrophoresis (20) (Fig. 1c).

Immunochemistry:

Anti-L_4 IgG and anti-E_4 IgG were obtained by rabbit immunisation against pure L_4 and E_4. Monospecific anti-L_4 IgG were obtained by immunoabsorption of antibodies reacting with E_4 on a E_4- Sepharose-4B column. Similarily, monospecific anti-E_4 IgG were obtained with a L_4-Sepharose-4B column. Antibodies solutions (E^{1cm}_{280} = 10) were tested by the Ouchterlony technique against the five purified isolectins solutions (0.5% W/V).

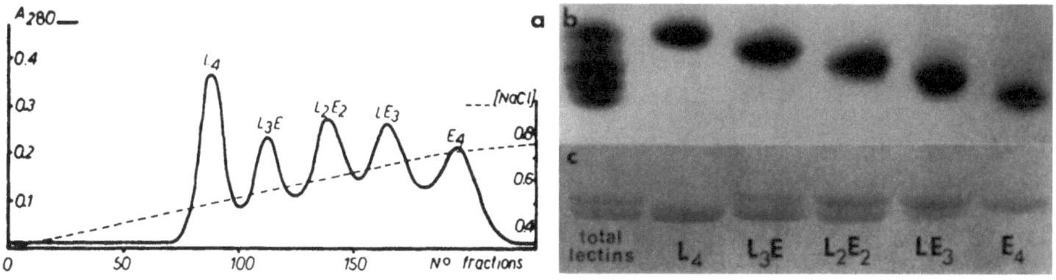

Fig. 1- Separation of the five tetraemeric lectins from a "contender" lectin preparation. (a) ion-exchange chromatography on SP- Sephadex. (b) disc gel electrophoresis at pH 4.5. (c) SDS gel electrophoresis.

Indirect immunofluorescence (7) :
Cotyledonary parenchyma cells imbibed for 3h at 30°C were fixed in 4% paraformaldehyde in 0.025M phosphate buffer, pH 7.0 at room temperature, dehydrated and embedded in glycolmethacrylate (17). Sections of 1 µm thick were cut by a dry glass knife, lead on a water drop on a glass slide and dried. Three serial sections were incubated with monospecific anti-L_4, anti-E_4 antibodies ($E^{1cm}_{280} = 5$) and non-immune IgG as control for 30 mn and afterwards with tetramethylrhodamin isothiocyanate labelled goat anti-rabbit IgG as secondary fluorescent antibodies (1/40 diluted in PBS) for 10 mn.

Electron microscopy with the gold method (15):
Seeds were imbided for 3h at 30°C. Cotyledonary parenchyma and vascular cells and axis cells were fixed with 4% paraformaldehyde, then by 1% OsO_4 in 0.025M phosphate buffer, pH 7.0 and embedded in Epon. Monospecific anti-L_4, anti-E_4 and non-immune antibodies were bound to gold particles of the average size of 20 nm (15). Grids with ultrathin sections were floated on drop of gold suspension for 30 mn, washed free of unbound reagents and finally contrasted with lead citrate.

Results

Rabbit anti-E_4 and anti-L_4 IgG obtained after injection of pure E_4 and L_4 lectins cross-react (Fig. 2a, b). When monospecific anti-L_4 and anti-E_4 antibodies were produced (see materials and methods) no such cross-reaction was visible (Fig. 2c, d).
When sections of cotyledonary parenchyma were stained with monospecific anti-E_4 or anti-L_4 immunoglobulins and visualized by indirect immunofluorescence the results (Fig. 3) demonstrated that both lectins were located in the protein bodies and that the same protein bodies contained both lectins. Similar-

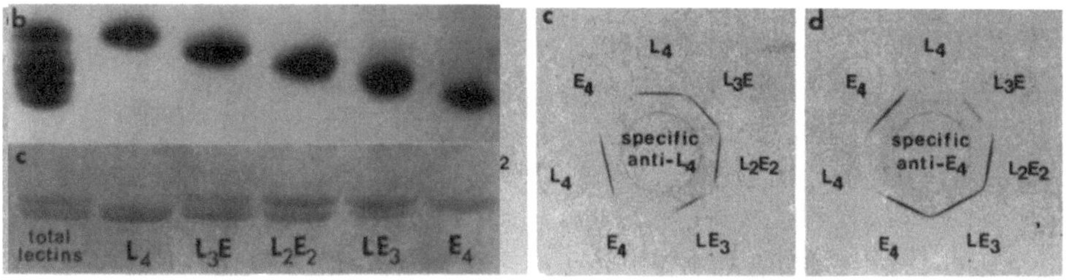

Fig. 2 - Double gel diffusion patterns of (a) anti-L_4, (b) anti-E4 IgG, (c) monospecific anti-L4 IgG, and (d) monospecific anti-E4 IgG, when reacted with the five lectins components (see materials and methods). Centre well : 40 µl of antibodies. Outer well : 20 µl of the respective lectin solutions.

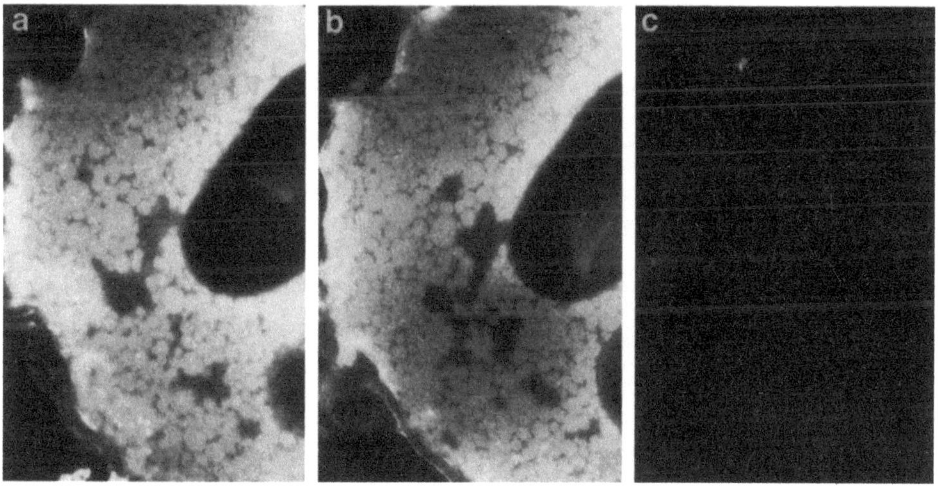

Fig.3 - Localization of E-type and L-type lectins by indirect immunofluorescence. Three serial sections were stained with specific anti-L_4 (a), anti-E_4 (b) and non-immune antibodies as control (c).

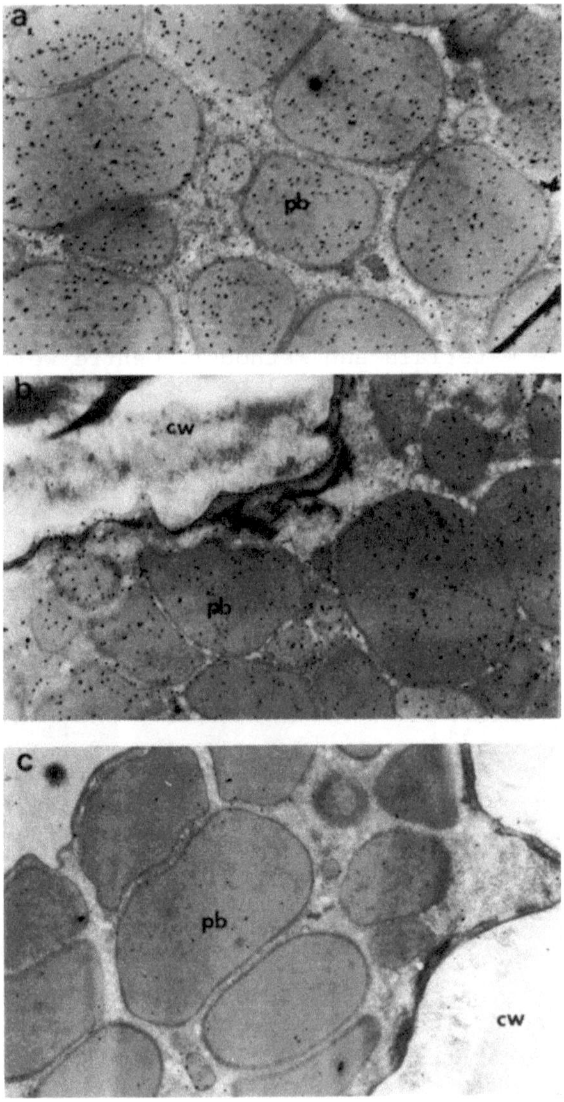

Fig. 4 - Immunoelectron microscopic localization of lectins by the gold method of cotyledonary parenchyma cells. Gold particles coated with specific anti-L4 (a), specific anti-E4 (b) and non-immune JgG (c). Key : pb, protein bodies ; cw, cell walls.

ly, the deposition of anti-E_4 or anti-L_4 coated gold particles on ultrathin sections of cotyledonary tissue also indicated that the location of the two type of lectin subunits inside the protein body matrix was the same (Fig. 4). Gold particles were apparently absent in the starch granules or the cell walls and there was only a low level of staining observed in cytoplasmic spaces between the protein bodies.

In contrast, both E- and L-type subunits were found to be mainly outside the protein bodies in the cytoplasm of cotyledonary vascular cells or cells from axis tissues (Fig. 5). In addition, these cells were rather small, locked secondary cell wall, contained very few starch granules and contained electron-transparent and nearly empty protein bodies.

Discussion

The two L and E subunits of Phaseolus vulgaris lectins have a

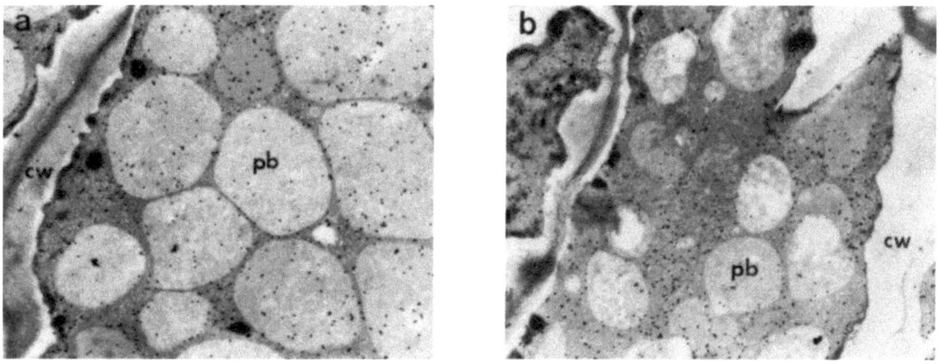

Fig. 5 - Immunoelectron microscopic localization of lectins in axis (a) and in cotyledonary vascular cells (b) with specific anti-L4 IgG by the gold method.
The staining with specific anti-E4 IgG gave the same results.

close similarity in the amino-acid sequence (26). Thus, cross-reaction between anti-E_4 and anti-L_4 antibodies was expected. However the fact that monospecific anti-L_4 and monospecific anti-E_4 antibodies can be produced by immunoabsorption respectively on E_4-and L_4- columns means that the two L and E subunits show different antigenic determinants.

The results obtained in the present study confirmed previous findings which indicated that lectins in the seeds of Phaseolus vulgaris are located in the protein bodies of the cotyledonary parenchyma. Furthermore, by the use of monospecific anti-E_4 and anti-L_4 immunoglobulins at both the light and at the electron microscope level, the results showed clearly that both the E-type and the L-type subunits were contained in the same protein bodies. Thus the existence of a weak association of the pH 5.8 isoelectric point lectin component (mainly E_4) with isolated protein body membranes (28) and other similar interactions was not supported by the results of the present immunocytochemical experiments. Preferential interaction of lectins such as the E-type lectins, with components of protein body or other internal membranes after the disruption of cellular integrity appears to be the most plausible explanation for contradictory results obtained by previous subcellular fractionation (23, 28) and the present immunocytochemical experiments. However an explanation is still to be found as to how the two different lectin subunits, which are synthesized at different times during seed development, could be located in the same cellular compartment.

On the contrary, lectins from vascular and axis cells were found located in the cytoplasm, mainly outside the protein bodies. This was previously also established for Glycine max (14), although the opposite was found for Pisum sativum (36). It is possible that this finding of the bulk of the lectins outside the protein bodies could be the result of an osmotic shock produced during imbibition before the tissues were fixed.

If this is true, the almost complete escape of lectins from the protein bodies suggests a high degree of fragility for their limiting membrane. Protein bodies from the parenchyme cells of the cotyledons of Phaseolus vulgaris are also known to have limited stability in aqueous environment (8) and release small amounts of lectins from soaked seeds (10). However, in contrast to the near complete emptying of the protein bodies from vascular tissue and axis cells, gold-coated anti-E and anti-L immunoglobulin-stained sections of cotyledonary parenchyma revealed a small lectin leakage only (Fig. 4).
The plant physiological significance of this difference in the stability of the protein bodies in different types of seed cells remains to be established.
Although there appear to be differences in the subcellular location of lectins in different types of cells from the same plant, the bulk of the legume seed lectins were found in the protein bodies of the main storage tissue of the seed in the cotyledonary parenchyme. As this cellular compartment is the site of storage for various reserve materials of the seed (proteins, phytin, certain salts, etc.) it is difficult to disregard the role for the lectin as a storage protein. Indeed, lectins in certain seeds account for a large proportion of total seed protein. Thus about 30% of the proteins from Canavalia ensiformis seeds consist of ConA lectin (9, 12). Similarly, lectins in some Phaseolus vulgaris cultivars can reach nearly 20% of the total seed proteins (29). In some seeds lectins appear during seed maturation and disappear during germination exactly like other storage proteins (30), although in some other seeds, such as Ricinus communis (37), Phaseolus vulgaris (unpublished) and Canavalia ensiformis (9) the disappearance of the lectins during germination is slightly retarded in comparison to that of other storage proteins.
Thus all these properties suggest a mainly storage role for a number of the lectins. However this does not exclude the

possibility that, in addition to this storage protein role, the lectins or certain components of the lectins, may have some other functions during seed maturation or food mobilization on germination. One such role is suggested by the discovery of lectin interactions with certain endogenous seed proteins (2, 11, 32), the so-called lectin-binders. It is also possible that lectins derive from originally functionally reactive proteins, for example from enzymes, which are then amplified during evolution and the increased amounts are used for nigrogen storage. The discovery of extensive similarities between some lectins and α-galactosidases (13, 24) suggests such a possibility. Similar materials, related to lectins are also found in other plant tissues. Thus, the immunologically cross-reactive leaf protein in Dolichos biflorus (34) and the small amounts of root lectins possibly involved in legume-Rhizobium interactions may all be functionally related to some components of seed lectins.

Finally, lectins may well belong to the class of the numerous "disposable" plant proteins (1), with biological activity but without any biological function. However the very fact that lectins are "disposed off" and in the process generate large amounts of re-usable amino acids for supporting growth, as other storage proteins do, militate against the concept of the lack of biological functions.

References

1. Adams, C.A.A., Rinne, R.W.: New Phytol. 89: 1-14 (1981).

2. Basha, S.M.M., Roberts, R.M.: Plant Physiol. 67: 936-939 (1981).

3. Barker, R.D.J., Derbyshire, E., Yarwood, A., Boulter, D.: Phytochemistry 15: 751-757 (1976).

4. Begbie, R.: Planta 147: 103-110 (1979).

5. Bollini, R., Chrispeels, N.J.: Planta 142: 291-298 (1978).

6. Clarke, A.E., Knox, R.B., Jermyn, M.A.: J. Cell. Sci. 19: 157-167 (1975).

7. Craig, S., Goodchild, D.J., Millard, A.: J. Histochem. Cytochem. 27: 1312-1316 (1979).

8. Croy, R.R.D.: Ph.D. Thesis University of Aberdeen (1977).

9. Etienne, A., Morgan, M.: Travail de diplôme, Université de Genève (1981).

10. Fountain, D.W., Foard, D.E., Replogle, W.D., Yang, W.K.: Science 197: 1185-1187 (1977).

11. Gansera, R., Schurz, H., Rüdiger, H.: Hoppe-Seyler's Z. Physiol. Chem. 360: 1579-1585 (1979).

12. Hague, D.R.: Plant Physiol. 55: 636-642 (1975).

13. Hankins, C.N., Kindinger, J.I. and Shannon, L.M.: Plant Physiol. 66: 375-378 (1980).

14. Horrisberger, M., Vonlanthen, M.: Histochemistry 65: 181-186 (1980).

15. Horrisberger, M.: Biol. Cell. 36: 253-258 (1979).

16. Jeffree, C.E., Yeoman, M.M.: New Phytol. 87: 463-471 (1981).

17. Kieda, C.M., Roche, A.C., Delmotte, F., Monsigny, M.: FEBS Letters 99: 329-332 (1979).

18. Köhle, H., Kauss, H.: Biochem. J. 184: 721-723 (1979).

19. Leavitt, R.D., Feldsted, R.I., Bachur, N.R.: J. Biol. Chem. 252: 2961-2966 (1977).

20. Laemmli, U.K.: Nature 227: 680-685 (1970).

21. Manen, J.F., Miège, M.N.: Physiol. Veg. 15: 163-173 (1977).

22. Manen, J.F.: Saussurea 9: 23-44 (1978).

23. Manen, J.F., Miège, M.N.: Physiol. Veg. 19: 45-58 (1981).

24. McPherson, A., Smith, S.C.: Phytochemistry 19: 957-959 (1980).

25. Mialonier, G., Privat, J.P., Monsigny, M., Kahlem, G., Durand, R.: Physiol. Veg. 11: 519-537 (1973).

26. Miller, J.B., Hsu, R. Heinrikson, R., Yachnin, S.: Proc. Nat. Acad. Sci. USA 72: 1388-1391 (1975).

27. Pusztai, A., Stewart, J.C., Watt, W.B.: Plant Sci. Lett.: 12: 9-15 (1978).

28. Pusztai, A., Croy, R.R.D., Steward, J.C., Watt, W.B.: New Phytol. 83: 371-378 (1979).

29. Pusztai, A., Croy, R.R.D., Grant, G., Stewart, J.C.: Seed lectin distribution, location and biological role. In "seed protein" Ed. by Phytochem, Soc. Eur. Symp. Versailles (1981).

30. Reisfied, R.A., Lewis, U.J., Williams, D.E.: Nature 195: 281-283 (1962).

31. Rougé, P.: C.R. Acad. Sci. Paris 282: 621-623 (1976).

32. Rougé, P., Père, M.: Bull. Soc. Bot. 126: 387-398 (1979).

33. Sun, S.M., Mutschler, M.A., Bliss, F.A., Hall, T.C.: Plant Physiol. 61: 918-923 (1978).

34. Talbot, C.F., Etzler, M.E.: Biochemistry 17: 1474-1479 (1978).

35. Tully, R.E., Beevers, H.: Plant Physiol. 58: 710-716 (1976).

36. Van Dreissche, E., Smets, G., Dejaegere, R., Kanarek, L.: Planta 153: 287-296 (1981).

37. Youle, R.J., Huang, A.H.C.: Plant Physiol. 58: 703-709 (1976).

A SPECTROPHOTOMETRIC STUDY OF THE CARBOHYDRATE BINDING SITE OF PEANUT LECTIN

Jacqueline Ohanessian, Michel Caron
Laboratories of Biomolecular Structural Chemistry and Immunochemistry
UER Biomédicale, 74, rue Marcel Cachin, 93012 Bobigny Cedex, France.

Introduction

There is a need for rapid and sensitive assays of the interactions of lectins with carbohydrates (1,2). The difference spectrophotometric method is one of the useful tools for the study of specific binding of sugars to a lectin. As pointed out earlier (3-6) changes in the spectrum of the lectin from Arachis hypogaea (peanut) may be used to calculate association constants. Difference spectra intensities dependent on the amount of sugars bond are obtained with specific sugars, but not with non specific sugars. The binding of lactose to the lectin is accompanied by pronounced changes dependent on the temperature, the affinity decreasing with increasing temperature (5). The interactions between the lectin and different mono- and oligo-saccharides are related to the structure of the sugars (6). The present study is undertaken to ascertain whether these sugars interacts with the binding site of the lectin.

Materials and Methods

Peanut lectin is purified by affinity chromatography on desialylated red cell ghosts as previously described (7) and dialyzed extensively to remove the eluting sugar. Sugars are commercial products of analytical grade except for methyl-7-deoxy-L and D-glycero-β-D-galacto-heptopyranosides synthetized as described by Lemieux and al. (8). Measurements are performed with a concentration of lectin in the range 0.5-1 mg/ml, concentrations of sugars of 1-10.10^{-3} M, in NaCl 0.15 M at 10°C.

TABLE I Configuration and association constants of mono and oligosaccharides

SUGARS	K (M^{-1})	SUGARS	K (M^{-1})
D-Talopyranose	410	Methyl-D-Galactopyranoside	α 1800 β 1400
Lactose	1500	Lactulose	1200
Lactobionic acid	no measurable	Mélibiose	210
Stachyose			310

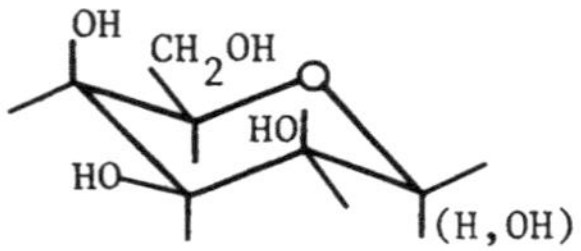

D-Talopyranose — 410

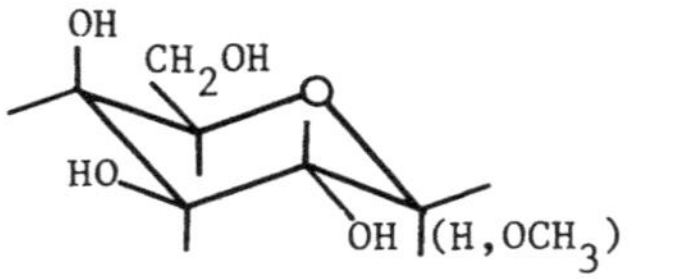

Methyl-D-Galactopyranoside — α 1800, β 1400

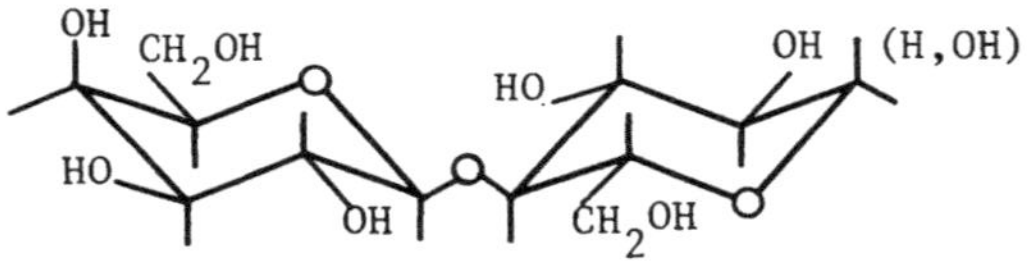

Lactose — 1500

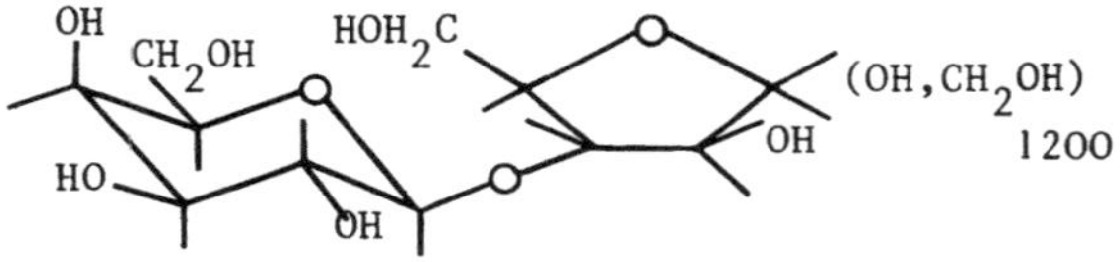

Lactulose — 1200

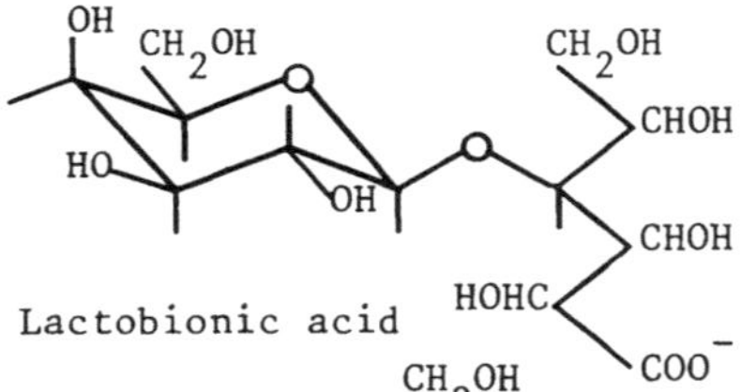

Lactobionic acid — no measurable

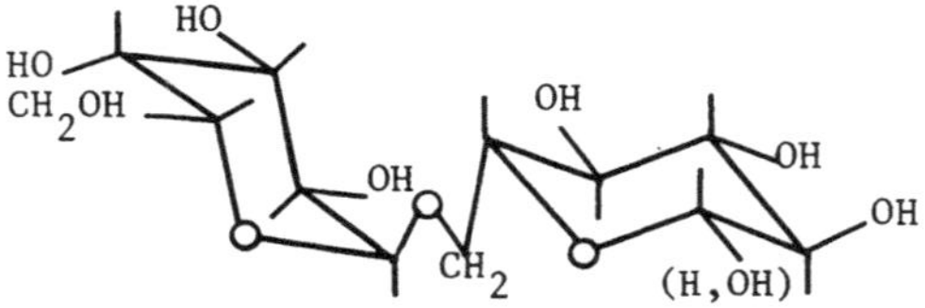

Mélibiose — 210

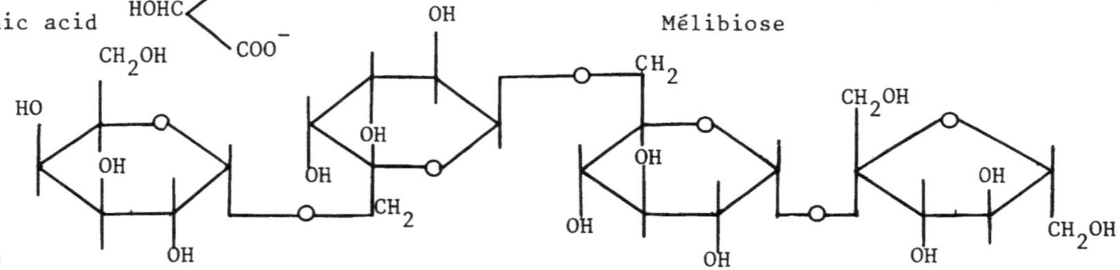

Stachyose — 310

SUGARS	K (M^{-1})
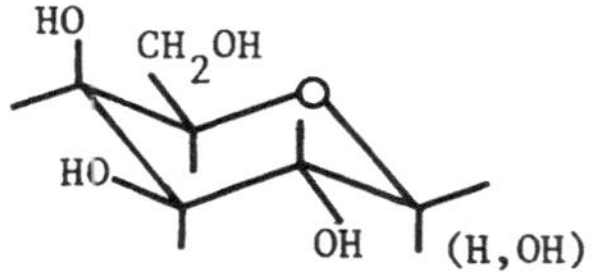 D-Galactopyranose	1200
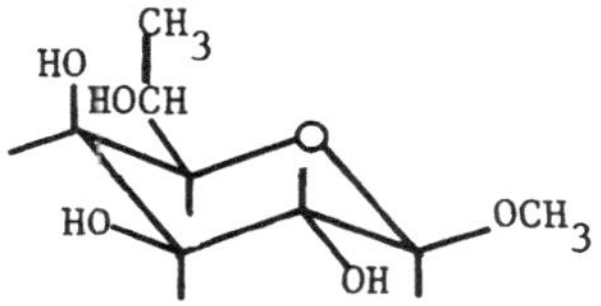 Methyl-7-deoxy-L-glycero-β-D-galacto-heptopyranoside	280
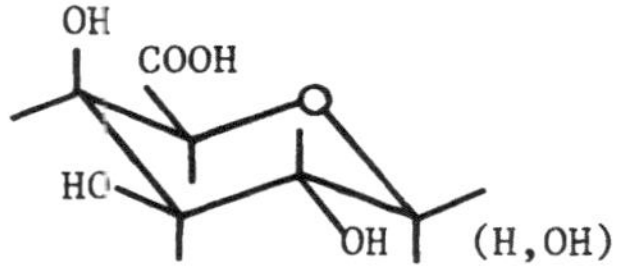 Galacturonic acid	no measurable
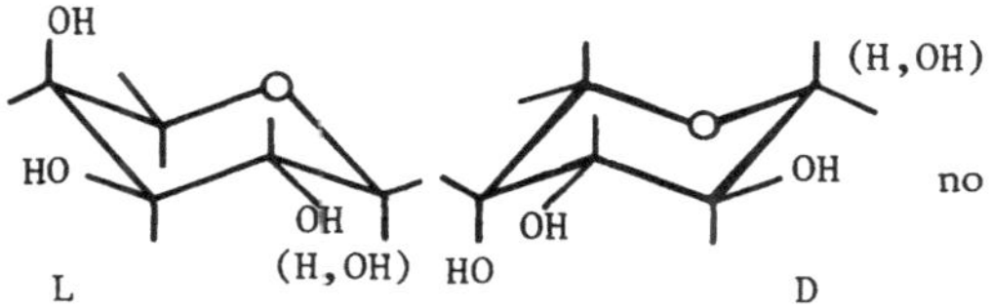 D,L-Arabinopyranose	no measurable

SUGARS	K (M^{-1})
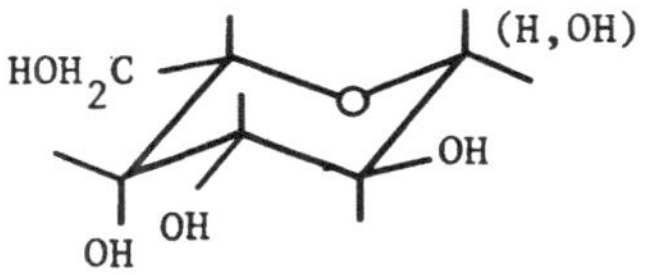 L-Galactopyranose	170
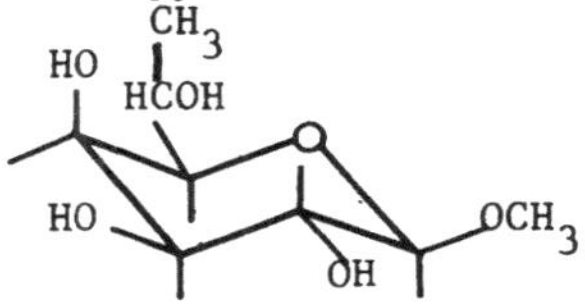 Methyl-7-deoxy-D-glycero-β-D-galacto-heptopyranoside	no measurable
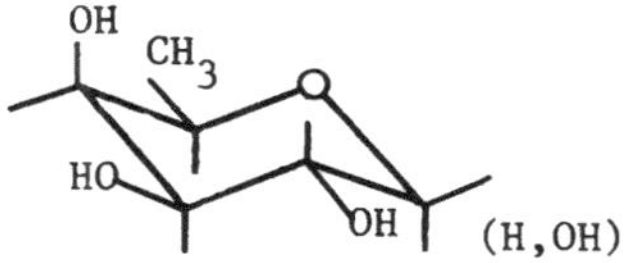 D-Fucopyranose	no measurable
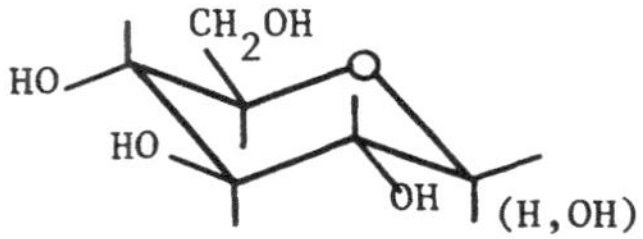 D-Glucopyranose	no measurable

Absorbance and difference spectra are recorded between 260 and 300 nm using a temperature-controlled Cary 219 spectrophotometer. The intensities of difference spectra are determined between the typical positive and negative peaks (3,5). By plotting the change in difference spectrum ΔA versus the sugar concentration, a typical saturation curve is obtained. Linearisation according to Matsumoto (3) results in a straight line. The slope and the intersections obtained with this representation, calculated by the least-squares method, permit the determination of the binding constant (6).

Results and Discussion

By plotting [sugar]/ΔA as a function of [sugar] we investigated the association constants for sixteen mono- and oligosaccharides at a constant temperature of 10°C. The resulting data are summarized in Table 1.

With D-galactopyranose (conformation $^4C_1$) $K = 1200\ M^{-1}$ while the constant measured for L-galactopyranose is $170\ M^{-1}$. The other saccharides studied possess structural configuration differences and are substituted on C(6), C(4), C(2) and C(1) (see Table 1). The following conclusions were drawn with respect to the importance of the configuration of these atoms for the interaction with the carbohydrate binding site of peanut lectin.

Atom C(6)

An extracyclic chain is necessary for the carbohydrate lectin interaction and the orientation of the hydroxyl group on C(6) is essential. A substitution of the hydroxymethyl group decreases the strength of the association.

Atom C(4)

A free axial hydroxyl on C(4) is needed for binding.

Atom C(2)

A C(2) equatorial hydroxyl group is not essential for the interaction but the axial position of this group decreases the binding.

Atom C(1)

O(1) methylated : Methylation favorises association. Among methyl-D-

galactosides differing only in the configuration about the anomeric C(1), there seems to be a slight preference for the α anomer.
O(1) involved in glycosidic bond : for the oligosaccharides studied the second residue appears to be involved in the association. The linear chain of the gluconic acid prevents the binding. Furthermore, the β(1-4) linkage seems to give better binding than the α(1-6) linkage. These results suggest that the configuration on C(1) is important. Comparing their effectiveness to interact with the carbohydrate binding site of peanut lectin, the only sugars to have a more marked affinity than D-galactose appeared to be galactosides substituted on C(1).

Conclusion

Our results show the following details : the sequence C(4)-O(4), C(5), C(6)-O(6) of the galactopyranosyl ring is involved in the interactions in the carbohydrate binding site of peanut lectin as postulated by Lotan et al.(9). Furthermore, for di- and oligosaccharides, it appears that the configuration on C(1), the nature of the osidic bond and of the second residue may influence the strength of the association.

References

1. Takeo, K., Fujimoto, M. and Kuwahara, A. (1982). These proceedings.
2. Wu, A.M. (1982) These proceedings.
3. Matsumoto, I., Jimbo, A., Kitagaki, H., Golovtechenko-Matsumoto, M. and Seno, N. (1980) J. Biochem. 88, 1093-1095.
4. Neurohr, K.J., Young, N.M. and Mantsche, H.M. (1980) J. Biol. Chem. 255, 9205-9209.
5. Caron, M., Ohanessian, J. and Becquart, J. (1981) 1st European symposium on Carbohydrates and Glycoconjugates (Vienna) Abstract G.S. 3.
6. Caron, M., Ohanessian, J., Becquart, J. and Gillier-Pandraud, H. Biochim. Biophys. Acta (In press).
7. Caron, M., Ohanessian, J., Faure, A. and Felon, M. (1981) C.R. Acad. Sci. Paris 292, 183-186.

8. Lemieux, R.U., Boullanger, P.H., Bundle, D.R., Baker, D.A., Nagpurkar, A. and Venot, A. (1978) Nouv. J. Chim. 2, 321-329.

9. Lotan, R., Skutelsky, E., Danon, D. and Sharon, N. (1975) J. Biol. Chem. 250, 8518-8523.

A CRITICAL STUDY ON THE PURIFICATION OF POTATO LECTIN (SOLANUM TUBEROSUM L.)

Edilbert Van Driessche[‡¶], Sonia Beeckmans[‡], Robert Dejaegere[¶] and Louis Kanarek[‡]

[‡]Laboratorium voor Chemie der Proteïnen, and
[¶]Laboratorium voor Plantenfysiologie;
Vrije Universiteit Brussel, Instituut Moleculaire Biologie, Paardenstraat 65, B-1640 Sint-Genesius-Rode, Belgium

Introduction

The purification of the lectin from potato tubers has been described by different groups (1-5). Most of these methods are based on affinity chromatography, taking advantage of the sugar specificity of the lectin (NAG_x; x 2-5). During the isolation of proteins from phenol-rich tissues such as potatoes, special attention has to be paid to the inhibition of phenol oxidases. Indeed, during the extraction, endogeneous phenols are quickly enzymatically oxidized to quinones which can irreversibly modify proteins by covalent binding to reactive groups such as sulfhydryl-, amino- and imino groups. These modifications result in most instances in reduced solubility and stability of the proteins or in complete loss of activity due to precipitation and/or inhibition of the proteins (6-8). The problems mentioned above can best be avoided either by eliminating the phenols or by inhibiting the phenolase activity with reagents such as sodium metabisulphite ($Na_2S_2O_5$), diethyldithiocarbamate (DIECA) or thiourea (6, 7).
Until now, exclusively $Na_2S_2O_5$ was used as antioxidant (1, 2, 4, 5). Previously we noticed (2) that when $Na_2S_2O_5$ (2 mM) had been added during the extraction of the tubers, the purified lectin displayed some unexpected spectral properties,

i.e. the UV-absorption spectrum had a maximum at 271 nm, and more curiously the ratio of the absorbances at 280 to 250 nm was 1.1 - 1.3. Later we found that all the hemagglutinating activity was lost when $Na_2S_2O_5$ (2 mM) was added during the dialysis of the ammonium sulphate precipitate of the extract. In this paper, we describe a new preparation method for potato lectin and we compare the effect of the use of different phenolase inhibitors on the activity and the spectral properties of the lectin, model peptides, enzymes and amino acids.

Materials and Methods

1. Purification of potato lectin : potatoes were washed, peeled, cut in fine pieces and homogenized in 0.1 M sodium-acetate buffer containing 9 g/l NaCl and 0.2 g/l NaN_3. The pH of the buffer was adjusted to 3.8 or 5.4 and either 2 mM $Na_2S_2O_5$, or 1 mM DIECA, or 5 mM thiourea were added. Extractions were allowed to proceed at 4°C for 6 hours, the slurry was passed through cheesecloth and centrifuged for 30 minutes at 7000 r.p.m. Solid ammonium sulphate (243 g/l) was added to the clear supernatant in order to get 40% saturation at 4°C. The ammonium sulphate pellet was collected by centrifugation and dialyzed against the respective extraction buffers at room temperature. Precipitated proteins were removed by centrifugation and the clear supernatant was slowly applied to a chitin column which was equilibrated with the same buffer. The column was then washed with buffer until the absorbance of the effluent had dropped to zero. The column was then washed with 1 M $CaCl_2$ in buffer and finally with saline. The lectin was eluted with 0.1 M acetic acid.

2. SDS-electrophoresis was performed on 10% polyacrylamide slab gels according to Laemmli (9), or on linear 12.5%-25% polyacrylamide gradient gels. Molecular weights were determined from the migration relative to standard proteins with known molecular weight.

3. Hemagglutinating activity was tested using rabbit red blood cells.

4. Chitin was prepared as follows : commercial (Sigma Chem. Comp.) chitin was exhaustively washed with 0.1 N HCl, water, 0.1 N NaOH, water, and was deproteinized with pepsin and trypsin.

Results

1. The effect of $Na_2S_2O_5$, DIECA and thiourea on N-acetyl-tryptophanamide, N-acetyl-tyrosinamide, model peptides and proteins :

The deleterious effect of $Na_2S_2O_5$ on the UV-absorption spectrum of N-acetyl-tryptophanamide is shown in Fig. 1. Similar

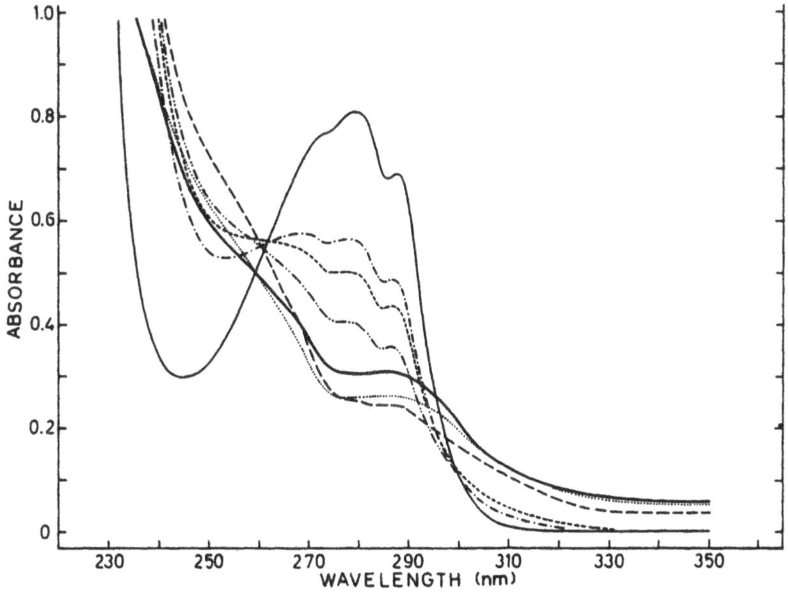

Fig. 1. The effect of 20 mM metabisulphite on the UV-absorption spectrum of N-acetyl-tryptophanamide in 0.1 M sodium-acetate buffer, pH 5.4.
Incubation times : Δt = 0 h (initial spectrum) (—); 1 h 30 (– · –); 3 h (---); 4 h 30 (– ·· –); 10 h (– –); 25 h (....) and 46 h (final spectrum) (▬).

changes were observed in a broad pH-range with tryptophane containing model proteins such as melittin (1 tryptophane, no tyrosine) (10), lysozyme, pig heart fumarase and purified potato lectin. No spectral changes were observed with N-acetyl-tyrosinamide or bovine pancreas ribonuclease A (6 tyrosines, no tryptophane) (11). Concomitant with the induction of changes in the UV-absorption spectrum, loss of activity of lysozyme, fumarase and potato lectin was observed. On the other hand, neither DIECA, nor thiourea had any effect on the spectral properties of the above mentioned probes.
The action of the three phenolase inhibitors was also tested on a model S-S-bridge (5,5'-dithio-bis(2-nitrobenzoate)) and on a cysteine containing peptide (glutathion). The extent of modification of cysteinyl groups (-SH) was estimated (according to Ellman (12)) from the measurement of the residual -SH groups of glutathion after incubation with phenolase inhibitor (Scheme 1 and Table 1). The extent of modification of the

Scheme 1.

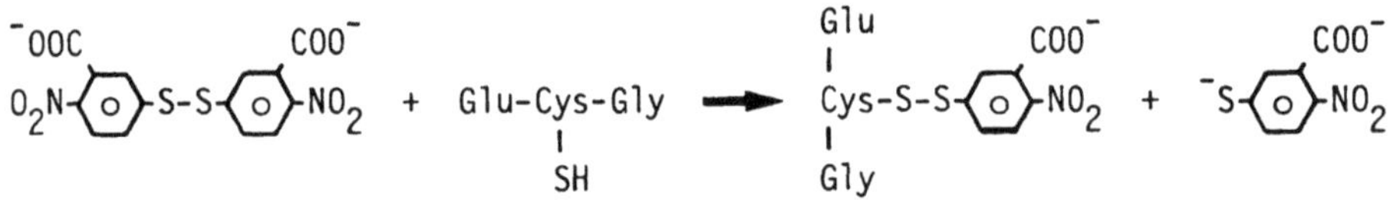

5,5'-dithio-bis (2-nitrobenzoate) — glutathion — 2-nitro-5-thio benzoate anion

model S-S-bridge with phenolase inhibitor (R) was estimated directly from the generation (Scheme 2) of 2-nitro-5-thiobenzoate anion (see Table 2).

Scheme 2.

5,5'-dithio-bis (2-nitrobenzoate) — 2-nitro-5-thio benzoate anion

From these results it is clear that only thiourea does not interfere with S-S bridges or sulfhydryl groups. These results are further confirmed by the fact that thiourea is the only phenolase inhibitor tested which does not inactivate lysozyme, an S-S containing enzyme (13), or fumarase, an enzyme with essential SH-groups (14).

Table 1. The effect of phenolase inhibitors on cysteine.

	[2-nitro-5-thiobenzoate anion] (mM)			
pH	no addition	thiourea	DIECA	$Na_2S_2O_5$
3.8	6.56	6.51	3.68	6.25
5.4	6.56	6.49	3.53	6.38
7.0	6.47	6.28	4.41	6.35
8.0	6.16	6.02	5.44	6.09

Glutathion (6.5 mM) in buffer (0.1 M sodium-acetate pH 3.8 and 5.4, or 0.1 M Tris-acetate pH 7.0 and 8.0) was incubated with 5 mM of respectively thiourea, DIECA or metabisulphite during 3 hours. The concentration of the remaining free -SH groups was determined spectrophotometrically at 412 nm according to Ellman (12) by adding 30 μl of the glutathion solutions to 3 ml of 5 mM DTNB in 0.1 M potassium phosphate buffer pH 7.0; a solution containing 5 mM DTNB in 0.1 M potassium phosphate buffer pH 7.0 and 50 μM of the corresponding phenolase inhibitor was used as blanco. The concentration of 2-nitro-5-thiobenzoate anion was estimated from its specific absorbance : $\epsilon_{412\ nm} = 13600\ M^{-1}cm^{-1}$ (12).

Table 2. The effect of 50 μM phenolase inhibitors on 5 mM DTNB (5,5'-dithio-bis(2-nitrobenzoate)) in 0.1 M potassium phosphate buffer pH 7.0.

	[2-nitro-5-thiobenzoate anion] (μM)
Thiourea	0
Diethyldithiocarbamate	30
Sodium metabisulphite	90

The concentration of 2-nitro-5-thiobenzoate anion was estimated from its specific absorbance : $\epsilon_{412\ nm} = 13600\ M^{-1}cm^{-1}$ (12).

2. Comparison of the effect of $Na_2S_2O_5$, DIECA and thiourea on the activity of potato lectin in extracts :

In order to compare the effect of adding $Na_2S_2O_5$, DIECA or thiourea during the extraction on the activity of potato lectin, tubers were homogenized in saline and extracted for 30 minutes. The coloured extract was passed through cheese-cloth and centrifuged at 15000 r.p.m. for 20 minutes. The supernatant was devided into different fractions which were adjusted to pH 3, 4, 6 (0.1 M sodium-acetate, final concentr-ation), 8 and 9 (0.1 M Tris-acetate, final concentration) and incubated with 5 mM $Na_2S_2O_5$, 5 mM DIECA or 5 mM thiourea. After respectively 2 1/2 and 24 hours of incubation, the phenolase inhibitors were removed by dialysis against the corresponding buffer and finally against saline containing 0.02% NaN_3.

From the results in Table 3 it is clear that $Na_2S_2O_5$ should be avoided as anti-oxidant for the preparation of potato lectin and should be replaced by DIECA or by thiourea.

Table 3. The effect of different phenolase inhibitors on the agglutinating activity of potato lectin in extracts.

	Titer			
	Thiourea	Diethyldithio-carbamate	Metabisulphite	
pH	2.5 h and 24 h incubation	2.5 h and 24 h incubation	2.5 h	24 h incubation
3.0	128	128	16	1
4.0	128	128	8 - 16	0
6.0	128	128	0	0
8.0	128	128	0	0
9.0	128	128	0	0

3. Purification of potato lectin :

The lectin was prepared as described in Materials and Methods. The progress in the purification is shown in Fig. 2.

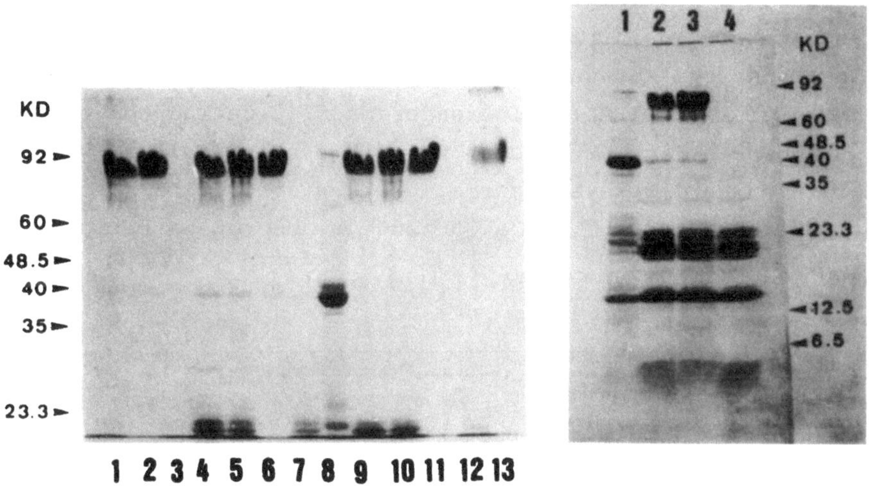

A B

Fig. 2. The progress in the purification of potato lectin as demonstrated by SDS-PAGE on (A) 10% polyacrylamide and (B) a linear 12.5%-25% gradient slab gel.

Lane in gel A	B	Sample
		Preparation with DIECA (pH 5.4)
1		elution from chitin (0.1 M acetic acid)
2		fraction I from Fig. 3A (pure lectin)
3		fraction II from Fig. 3A
		Preparation with thiourea (pH 5.4)
4	2	elution 1 from chitin (0.1 M acetic acid)
5	3	elution 2 from chitin (0.1 M acetic acid)
6		fraction I from Fig. 3B (pure lectin)
7	4	fraction II from Fig. 3B
8	1	1 M $CaCl_2$ wash
		Preparation with thiourea (pH 3.8)
9		elution 1 from chitin (0.1 M acetic acid)
10		elution 2 from chitin (0.1 M acetic acid)
11		fraction I from Fig. 3C (pure lectin)
12		fraction II from Fig. 3C
		Preparation with metabisulphite (pH 3.8)
13		purified lectin

The lectin is strongly bound to the chitin matrix and is eluted as two fractions. A first lectin containing peak is eluted immediately with 0.1 M acetic acid (elution 1), while a second peak can only be released after an overnight incubation period in 0.1 M acetic acid (elution 2).
When the eluted fractions are further chromatographed on Sephadex G-150 (Fig. 3), highly pure potato lectin is obtained as can be seen in Fig. 2A, lanes 2, 6, 11.
The UV-absorption spectrum of potato lectin which has been prepared with DIECA, thiourea or $Na_2S_2O_5$ during the extraction is shown in Fig. 4. Lectin which has been isolated with DIECA or thiourea displays a normal protein absorption spectrum with a maximum at 280 nm and a 280/250 nm absorption ratio of 2.

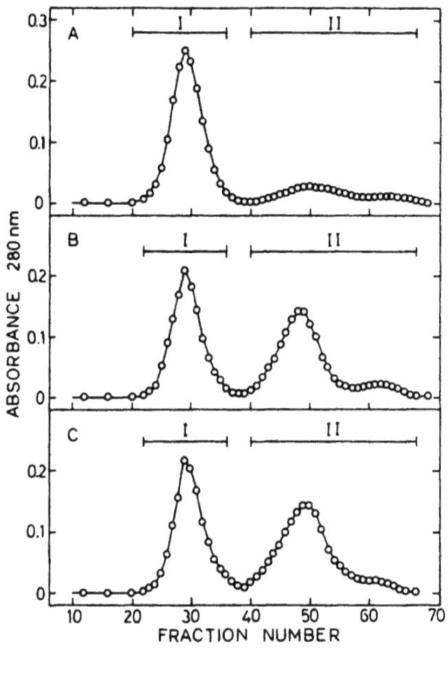

Fig. 3. Gel filtration of affinity purified potato lectin on Sephadex G-150 in 0.1 M sodium acetate containing 9 g/l NaCl and 0.2 g/l azide, pH 5.4 (A: 1 mM DIECA and B: 5 mM thiourea used in the extract at pH 5.4) or pH 3.8 (C: 5 mM thiourea used in the extract at pH 3.8).

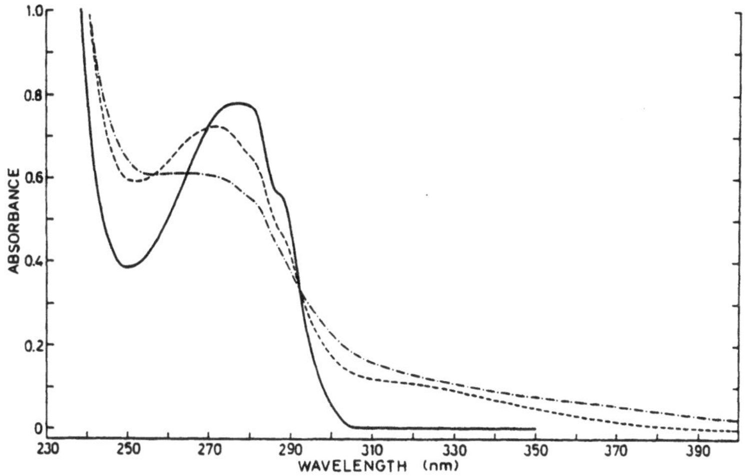

Fig. 4. The UV-absorption spectrum of potato lectin in phosphate buffered saline.
Ultra-pure lectin prepared with DIECA or thiourea (—); lectin prepared with metabisulphite (---); pure lectin prepared with DIECA and completely inactivated with metabisulphite (-·-).

Discussion

In this paper we clearly demonstrate that $Na_2S_2O_5$ should be avoided as anti-oxidant for the preparation of proteins from phenol-rich tissues and should be replaced by thiourea which was shown to have no effect on either tryptophane, tyrosine, S-S bridges or cysteine. Although DIECA can react with both cystine and cysteine, it can be successfully used for the preparation of potato lectin which is free of SH-groups, but rich in S-S bridges (1). Since DIECA does not affect the activity of purified potato lectin, it may be assumed that the cystines are well buried in the interior of the molecule. Although potato lectin can be isolated to purity according to available literature procedures (using $Na_2S_2O_5$), the specific activity of such preparations is only about 25% when compared to lectin prepared as described above. The reduced specific activity could be ascribed to the modification of an essential tryptophane residue whose presence has been demonstrated by Ashford et al. (15) in the carbohydrate binding site. Although two active peaks are consecutively eluted from the chitin column with 0.1 M acetic acid, electrophoretic analysis demonstrated that these two fractions are indistinguishable, suggesting that the difference in elution has to be attributed to heterogeneity of the chitin matrix rather than to the presence of isolectins which differ in their affinity for carbohydrate.

References

1. Allen, A.K. and Neuberger, A.: Biochem. J. 135, 307-314 (1973)
2. Van Driessche, E. and Kanarek, L.: Arch. Intern. Physiol. Bioch. 83, 414-415 (1975)
3. Delmotte, F., Kieda, C. and Monsigny, M.: FEBS Lett. 53, 324-330 (1975)
4. Desai, N.N. and Allen, A.K.: Anal. Bioch. 93, 88-90 (1979)
5. Owens, R.J. and Northcote, D.H.: Phytochem. 19, 1861-1862 (1980)
6. Anderson, J.W.: Phytochem. 7, 1973-1988 (1968)
7. Mayer, A.M. and Harel, E.: Phytochem. 18, 193-215 (1979)
8. Patil, S.S. and Zucker, M.: J. Biol. Chem. 240, 3938-3943 (1965)
9. Laemmli, U.K.: Nature (Lond.) 227, 680-685 (1970)
10. Habermann, E. and Jentsch, J.: Hoppe-Seyler's Z. Physiol. Chem. 348, 37-50 (1967)
11. Richards, F.M. and Wyckoff, H.W.: The Enzymes 4 (3rd. ed.) (P.D. Boyer, ed.) pp. 647-806 (1971)
12. Ellman, G.L.: Arch. Biochem. Biophys. 82, 70-77 (1959)
13. Imoto, T., Johnson, L.N., North, A.C.T., Philips, D.C. and Rupley, J.A.: The Enzymes 7 (3rd. ed.) (P.D. Boyer, ed.) pp. 665-868 (1972)
14. Hill, R.L. and Teipel, J.W.: The Enzymes 5 (3rd. ed.) (P.D. Boyer, ed.) pp. 539-571 (1971)
15. Ashford, D., Menon, R., Allen, A.K. and Neuberger, A.: Biochem. J. 199, 399-408 (1981)

A GALACTOMANNAN PRECIPITATING LECTIN IN SPHENOSTYLIS STENOCARPA SEEDS

Theresa Animashaun

Biochemistry Department, University of Ife,
Ile-Ife, Nigeria

Introduction

Haemagglutination activity was reported in extracts of *Sphenostylis stenocarpa* seeds by Schertz *et al* (1). In the present paper the agglutination of red blood cells by saline extracts from *Sphenostylis stenocarpa* (Hochst. ex A. Rich) Harms (*Papilionaceae*) seeds was examined further. The haemagglutination activity of this extract was shown to be inhibited by D-galactose and its derivatives. Some D-galactose binding lectins, e.g from *Ricinus communis* (2) and *Bandeiraea simplicifolia* (3), precipitate various galactomannans. *S. stenocarpa* extract was found to precipitate a galactomannan from *Afzelia africana* Sm (*Caesalpiniaceae*) seeds. The potential use of this polysaccharide in purifying the lectin from *S. stenocarpa* seeds is discussed.

Materials and Methods

Preparation of S. stenocarpa extract. *S. stenocarpa* seeds were picked locally and identified by Mr. I.B. Faremi. The seeds were ground, defatted with petroleum ether and stirred for 18 hours at 4°C with 5 volumes (wt/vol) of 0.85% saline. The supernatant, obtained after spinning the suspension at 10,000g for 20 min, is referred to as the extract.

Preparation of A. africana galactomannan. *A. africana* seeds

Lectins, Vol. III

were kindly donated by the Seed Section, Forestry Research Institute of Nigeria, Ibadan. After removal of the aril and husk, *A. africana* seeds were ground and defatted. Two grams of the meal were stirred in 100 ml water with heating. The polysaccharide was precipitated from the hot-water extract by 40% ethanol and resuspended in water. The carbohydrate concentration was determined by the anthrone procedure (4).

Haemagglutination studies. Haemagglutination studies were performed in V-shaped micro-titre plates with 1.5% formalinised red blood cells (5). The titre value was determined with doubling dilutions of *S. stenocarpa* extract. The concentration of sugar which reduced the titre value by one was said to have inhibited haemagglutination by 50% (6).

Agar-gel double diffusion studies. Agar-gel plates contained 1.5% agar (Oxoid Ltd.) and 0.01% sodium azide (Difco Laboratories) in PBS (0.1 M sodium chloride, 0.025 M phosphate, pH 7.4). The plates were incubated in a humid atmosphere at room temperature for 48 hours.

Examination of affinity chromatography media. Sepharose 6B was purchased from Pharmacia while immobilised-lactose was from Pierce Chemical Co. Acid-treated Sepharose was prepared according to Ersson *et al*(7). 0.5 ml of the affinity chromatography media being tested was incubated with 0.5 ml of *S. stenocarpa* extract for 1 hour. The titre value of the supernatant was determined and compared with a control containing the same dilution of extract.

Table I:- Titre Values of *S.stenocarpa* Extract

	Blood Type				
	A_1	A_2	B	O	Bat
Titre Value	2^5	2^5	2^5	2^6	0

Table II:- Concentration of Sugar Producing 50% Inhibition of Agglutination of O Red Blood Cells by S. stenocarpa extract

	mM
D-galactose	3.0
N-acetyl-D-galactosamine	0.2
Lactose	0.8
Raffinose	3.0

Results

Saline extracts of S. stenocarpa seeds agglutinated O red blood cells with a titre value of 2^6 (Table I). The extracts agglutinated A_1, A_2 and B cells to the same or to a lesser extent, but did not affect the sedimentation pattern of chicken and fruit-bat (Eidolon helvum) erythrocytes. There was no clear-cut difference between the agglutination of $A_1$and A_2 red blood cells by the extract. The agglutination of O erythrocytes by a 2^2 dilution of S. stenocarpa extract was totally abolished by 50 mM N-acetyl-D-galactosamine, D-galactose, lactose or raffinose. 50 % inhibition of O red blood cells was achieved in the presence of 0.2 mM N-acetyl-D-galactosamine, 0.8 mM lactose, 3 mM raffinose or 3 mM D-galactose (Table II).

The polysaccharide isolated from A. africana seeds has been shown to be a galactomannan (8). S. stenocarpa extract formed precipitin bands in agar gel double diffusion plates with 13 to 0.4 mg/ml A. africana galactomannan (Figure I). The galactomannan did not precipitate dilutions of the S. stenocarpa extract.

Various reagents were examined for their potential to purify the lectin from S. stenocarpa extract by affinity chromatography. The lectin did not bind to Seharose or acid-treated Sepharose. The amount of lectin bound to immobilised-lactose was very low (approximately 20 μg protein per ml of gel).

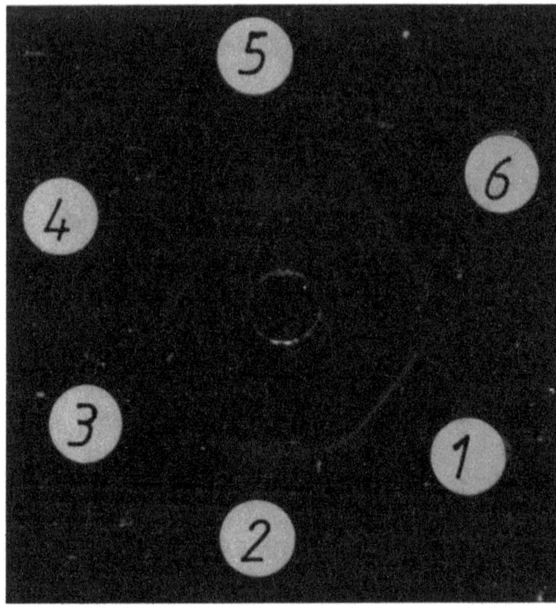

Figure I:- Diffusion of S. stenocarpa extract and A. africana galactomannan in agar gel. Centre well - S. stenocarpa extract, outside wells (1 - 6) - doubling dilutions of A. africana galactomannan (13 - 0.4 mg/ml).

Discussion

It has been confirmed that S. stenocarpa extracts agglutinate human erythrocytes (Table I). This haemagglutination is relatively non-specific and unlike the findings of Schertz et al (1) the extract did not preferentially agglutinate A_2 rather A_1 cells. Haemagglutination was inhibited not only by N-acetyl-D-galactosamine and D-galactose but also by the β- and the α-D-galactoside containing disaccharide lactose and trisaccharide raffinose. Some galactose-binding lectins have been purified by affinity chromatography on Sepharose e.g. ricin (9), abrin (10) and lectin from Pseudomonas aeruginosa (11). Tomita et al (12) found that only three of the four galactose-binding

lectins tested bound to Sepharose. Acid-treated Sepharose has been used to purify lectin from Erythrina corallodenron (13). However, the lectin from S. stenocarpa did not bind to Sepharose or acid-treated Sepharose and the yield from immobilised-lactose was poor.
Lectins from R. communis (2) and B. simplicifolia (3) precipitate galactomannans from several sources such as guar gum, locust bean and Cassia alata. Moreover, cross-linked guaran has been used to purify both these lectins and that from Echinocystis labata (14). Guaran immobilised in polyacrylamide gel was used to prepare a high-capacity adsorbent for B. simplicifolia lectin (15). Since S. stenocarpa extracts interact with A. africana galactomannan (Figure I), the possibility of isolating S. stenocarpa lectin employing cross-linked or immobilised A. africana galactomannan will be explored.

Acknowledgements

I am grateful to Messrs A. Abimbola, O.O.O. Binutuand O.M. Udeagu for technical assistance and to Mr. B.M. Salami for the photography

References

1. Schertz, K.F., Boyd, W.C., Jurgelsky, W., Cabanillas, E.: Econom. Botany 14, 232-239 (1960)
2. Van Wauwe, J.P., Loontiens, F.G., De Byuyne, C.K.: Biochim. Biophys. Acta 313, 99-105 (1973)
3. Hayes, C.E., Goldstein, I.J.: J. Biol. Chem. 249, 1904-1914 (1974)
4. Viles, F.J.Jr., Silverman, L.: Anal. Chem. 21, 950-953 (1949)
5. Butler, W.T.: J. Immunol. 90, 663-671 (1963)
6. Schimizu, S., Ito,M., Niwa, M.: Biochim. Biophys. Acta 500, 71-79 (1971)

7. Errson, B., Aspberg, K., Porath, J.: Biochim. Biophys. Acta 310, 446-452 (1973)

8. Ikhile, P., Animashaun, T.: Submitted to XIth International Carbohydrate Symposium (1982)

9. Olsnes, S., Pihl, A.: Biochemistry 12, 3121-3126 (1973)

10. Olsnes, S., Pihl, A.: Eur. J. Biochem. 35, 179-185 (1973)

11. Gilboa-Garber, N., Mizrahi, L., Garber, N.: FEBS Letts. 28, 93-95 (1972)

12. Tomita, M., Kurokawa, T., Onazaki, K., Ichiki, N., Osawa, T., Ukita, T.: Experientia 28, 24-26 (1972)

13. Gilboa-Garber, N., Mizrahi, L.: Can. J. Biochem. 59, 315-319 (1981)

14. Lonngren, J., Goldstein, I.J.: FEBS Letts. 68, 31-34 (1976)

15. Horisberger, M.: Carbohydrate Res. 53, 231-237 (1977)

FURTHER INVESTIGATIONS ON MISTLETOE LECTIN I (ML I): EFFECT OF A-CHAIN ON CELL-FREE PROTEIN SYNTHESIS

[1] H. Franz, Annemarie Kindt and P. Ziska

[2] H. Bielka, R. Renndorf and Liselotte Venker

[1] Staatliches Institut fur Immunpraparate und Nahrmedien, 1120 Berlin-Weißensee, DDR

[2] Zentralinstitut fur Molekularbiologie, Abteilung Zellphysiologie, Akademie der Wissenschaften der DDR, 1115 Berlin-Buch, DDR

On the 4 th lectin meeting we presented a method for simultaneous reduction of mistletoe lectin I (ML I) and affinity chromatography of its A- and B-chains (1). In analogy to ricin and abrin we expected toxophoric properties of the A-chain of ML I. It is the aim of this paper to describe the inhibition of protein synthesis in cell-free system by the A-chain of ML I.

Material and methods

1. Preparation of the A- and B-chains of ML I

ML I was prepared as described earlier (2) the procedure for the isolation of both chains was slightly modified in comparison to our first communication. The lectin was dissolved in phosphate buffered saline (PBS), pH 7.2, (1 mg ML I/ml) and applied to a column of partially hydrolized Sepharose 4B. Non--bound proteins were washed out with PBS. About 10 mg of ML I were bound to 50 ml of partially hydrolized Sepharose 4B. Then β-mercaptoethanol (5%) was applied to the column. After appearence of the β-mercaptoethanol in, the effluent the flow was stopped and the reduction of ML I in the column was allowed to proceed overnight at room temperature. Thereafter the column was washed with PBS and the effluent was applied to a second column of partially hydrolized Sepharose 4B in order to eliminate residual amounts of the B-chain. The A-chain passed the columns together with β-mercaptoethanol.

The B-chain remained bound and was eluted with 0.1 M D-galactose. The fractions were dialysed against PBS or 0.05 M potassium acetate to remove β-mercaptoethanol and D-galactose. The chains were identified by SDS-PAGE according to the method of WEBER and OSBORN (3).
If the B-chain fractions contain small amounts of A-chain the whole procedure has to be repeated. The protein content of the preparations was measured by the method of LOWRY et al. with bovine serum albumin as standard (4). Cell-free protein synthesis systems.

Two different translation systems have been used in this study:

a. A polysomal system from rat liver as described in (5) (but 100 μl samples only) containing 10 μg of polysomes (prepared with some modifications according to (6)), 20 μl of rat liver cytosolic fraction (corresponding to 20 μg protein; 105 000 x g supernatant in 50 mM Tris/HCl, pH 7.6, 1 mM EDTA, 6 mM β-mercaptoethanol by Sephadex G-10 filtration) and, besides the other ingredients as described in (8), 1 μC (^{3}H) leucin (68 Ci/mmol). The samples were incubated at 37°C for 30 min without (controls) or in the presence of varying amounts of the A-chain samples (see Table 1) and then furthermore treated and counted for radioactivity as described in (5).

b. A rabbit reticulocyte lysate system.
The lysate was prepared with some modifications following the procedure of CRYSAL et al. (7) (the buffer for lysis of the reticulocytes contained additionally 5 mM Hepes/KOH, pH 7.0). Immediately before use the lysate was filtered through Sephadex G-25, equilibrated and eluted with a buffer containing 25 mM KCl, 10 mM NaCl, 3 mM $Mg(OAc)_2$, 1 mM ATP, 1 mM GTP, 0.25 mM

dithioerythritol, 1 mM glycose and 20 mM Hepes/KOH, pH 7.2, following JAGUS and SAFER (8). To 1.0 ml of the filtered lysate were added 25 µl of a hemin stock solution and 20 µl of a creatine phosphokinase stock solution as described by SAFER et al. (9) to give final concentrations of 25 µM and 0.2 mg/ml, respectively. 2.5 µl of 2 M KCl plus 10 mM Mg $(OAc)_2$, 1 µl of 400 mM creatine phosphate, 1 µl of 0.5 M putrescine acetate, pH 7.0, 1 µl of 0.2 mM spermine plus 2 mM spermidine, 2 µl of a mixture containing all amino acids (1 mM each), and 3 µl of U (^{14}C) leucine (0.6 µC; 213 mCi/mmol) were added to 40 µl of the gel filtered lysate. A-chain preparations were added as specified in Table 2. The volume was completed to 57.5 µl by addition of aqua bidest. Incubation was at 37°C for 30 min; thereafter, two 20 µl samples were removed and pipetted into filter paper discs. The discs were processed for determination of hot TCA-insoluble radioactive material by liquid scintillation counting.

Results and discussion

1. Reduction of ML I under mild conditions:

Attempts to split ML I using β-mercaptoethanol only in absence of denaturating reagents were not successfull. Therefore, we tried to separate A- and B-chains by reducing Sepharose-bound ML I. This procedure is based in the hypothesis that only the B-chain is responsible for lectin activity.
The result was the liberation of a single protein, which passed the column and another one, which was eluted by D-galactose solution. This leads to the following conclusions: ML I consists of two different chains. The A-chain does not bind D-galactose. The B-chain is fixed to the carrier and therefore represents the virtual lectin component in ML I. Binding to a D-galactose bearing carrier effects alterations in the conformation of ML I which permit to reduce the S-S-bound. Denaturating reagents like 6 M urea can be avoided. This method

Table 1

Effect of A-chain preparations of mistletoe lectin I in protein synthesis in a cell-free rat liver polysimal system.

Preparation I			Preparation II			Preparation III		
ng/assay	^{3}H leu incorp. (dpm)	% of control	ng/assay	^{3}H leu incorp. (dpm)	% of control	ng/assay	^{3}H leu incorp. (dpm)	% of control
			0	31320	-	0	31320	-
			20	31890	101.8	16	29550	94.3
			40	27700	88.4			
			0	70950		0	56300	-
			200	21780	30.6	160	19250	34.2
0	55530	-	0	56300	-	0	93850	-
300	13150	23.6	400	17550	31.2	300	30540	32.5
0	45000	-	0	93850	-			
600	10440	23.2	800	17105	18.1			

Table 2

Effect of A-chain preparations of mistletoe lectin I on protein synthesis in a cell-free rabbit reticulocyte lysate system

	Preparation II			Preparation III	
ng/assay	incorporated (dpm)	% of control	ng/assay	incorporated (dpm)	% of control
-	14850	-	-	14850	-
100	6190	41.6	40	5670	38.2
-	13700	-	-	13700	-
300	4535	33.1	125	4945	36.0

may be convenient for preparation of A- and B-chains from ricin and abrin, too. It must be mentioned that both the A- and B-chain of ML I tend to aggregate in concentration higher than 0.1 mg/ml.

2. The effect on protein synthesis in cell-free systems:

A-chain samples of all three preparations (I, II, III) isolated and tested as described exhibit strong inhibitory effects on the biosynthesis of proteins in the cell-free systems as demonstrated in Tables 1 and 2. The effects are significantly dose-dependent as can be seen from the studies with the polysomal system from liver (Table 1). 160 ng to 600 ng per assay inhibit Leu-incorperation into protein to about 70 to 80%, while doses of 30 ng or less were found to be nearly or totally uneffective (16 ng). In some experiments we have studied also the effect of B-chain preparations on cell-free protein synthesis. While the B-chain of preparation I and II, which were A-chain contaminated, inhibited protein synthesis by about 50 to 40% at concentrations of 1 μg to 0.5 μg per assay, no effect could be measured for the B-chain sample of preparation III, which was electrophoretically free of A-chain impurities. From these findings it can be concluded that the A-chain moiety of the mistletoe lectin I is toxic for protein synthesis at the ribosomal level.

References

1. Franz, H., Kindt, A. and Ziska, P.: in Lectins-Biology, Biochemistry Clinical Biochemistry Vol II, 1982 Edit.: T. C. Bøg-Hansen W. de Gruyter, Berlin, New York

2. Ziska, P., Franz, H. and Kindt, A.: Experientia 34, 123-124 (1978)

3. Weber, K. and Osborn, M.: J. biol. Chem. 244. 4406-4412 (1969)

4. Lowry, O.H., Rosenbrough, N.J., Farr, A.L. and Randall, R.J.: J. biol. Chem. 193, 265-275 (1951)

5. Junghahn, I. and Bielka, B.: Acta Biol. Med. Germ. 32, 267-269 (1974)

6. Junghahn, I. and Scholz, I.: Acta Biol. Med. Germ. 31, 691-699 (1973)

7. Crystal, R.G., Elson, N.A. and Anderson, F.W.: Methods Enzymol. 30, 101-127 (1974)

8. Jagus, R. and Safer, B.: J. Biol. Chem. 256, 1317-1323 (1981)

9. Safer, B., Jagus, R. and Kemper, W.M.: Methods Enzymol. 60, 61-87 (1979)

Purification and Partial Characterization of a Streptococcus mutans Cell-agglutinating Substance from Carrot, Daucus carota

Bo Mattiasson and Matts Ramstorp
Department of Pure & Applied Biochemistry, Chemical Center, University of Lund, Lund, Sweden

Peter Carlsson and Douglas Bratthall
Department of Cariology, School of Dentistry, University of Lund, Malmö, Sweden

Introduction

The interaction between bacteria, especially Streptococcus mutans, and the surface of teeth is of utmost importance for the initiation of caries.
Studies of this system have for a long period of time, been focused on the external factors that directly or indirectly cause alterations in the tendencies for bacterial colonization to occur on tooth surfaces. In these studies sucrose have been of major interest since it is transformed into macromolecules which are directly involved in the interaction.
The molecular mechanisms behind the binding of cells to teeth revealed that sucrose was converted into dextran by the bacterial cells, which concomitantly became adhesive (1).
It was earlier reported that extracts of carrot, Daurus carota, contained a macromolecular substance capable of agglutinating Streptococcus mutans cells (2). It was suggested that the substance in the carrot homogenate was of lectin nature. The present paper presents some further studies on this system in order to elucidate the mechanisms behind cell adhesion to tooth surfaces.

Materials

Streptococcus mutans, strain OMZ 65, grown on glucose medium was used throughout this study.
Swedish carrot, Daurus carota, was homogenized in phosphate buffered saline 0,1 M with 0,15 M NaCl, pH 7.0 (PBS).
All chemicals used were of analytical grade.

Methods

Immobilization of Streptococcus mutans cells.
The cells were entrapped in polyacrylamide using a two-step bead polymerization procedure described in more detail elsewhere (3).
Bacterial cells, 0.34 g wet weight, were suspended in a 14.6 ml monomer solution, consisting of acrylamide, 77.4 g/L, and N,N'-methylene-bisacrylamide, 31.2 g/L. The catalyst system composed of 50 µl N, N, N',N' -tetramethylethylenediamine and 75 mg ammonium peroxodisulphate, dissolved in 1 ml PBS, was added and the suspension was mixed before being quickly poured into a vigorously stirred hydrophobic phase. The hydrophobic phase consisted of toluene/chloroform (2.64/1.00, v/v) containing 0.7 % (v/v) of the surfactant, Arlacel 83.
Polymerization was initiated in the small droplets, but before it was complete, the stirring speed was reduced, thereby allowing the still sticky particles to aggregate into clusters.
After completion of the polymerization process the particles were collected by decantation and thoroughly washed with PBS.

Results and Discussion

The entrapped bacterial cells were subsequently packed into a column (1.8 x 10.0 cm) and after equilibration with PBS buffer carrot homogenate was applied. Washing with PBS continued until the effluent showed no absorbance at 280 nm. Elution of bound material was performed by changing to a buffer consisting of 0.2 M glycine-HCl pH 2.2.

Fractions were collected and dialyzed against PBS. Their ability to agglutinate Streptococcus mutans cells was studied.

Figure 1 shows both the elution profile and a corresponding histogram illustrating each fractions ability to agglutinate bacterial cells. Agglutination measurements were performed in plastic tubes using a serial dilution technique.

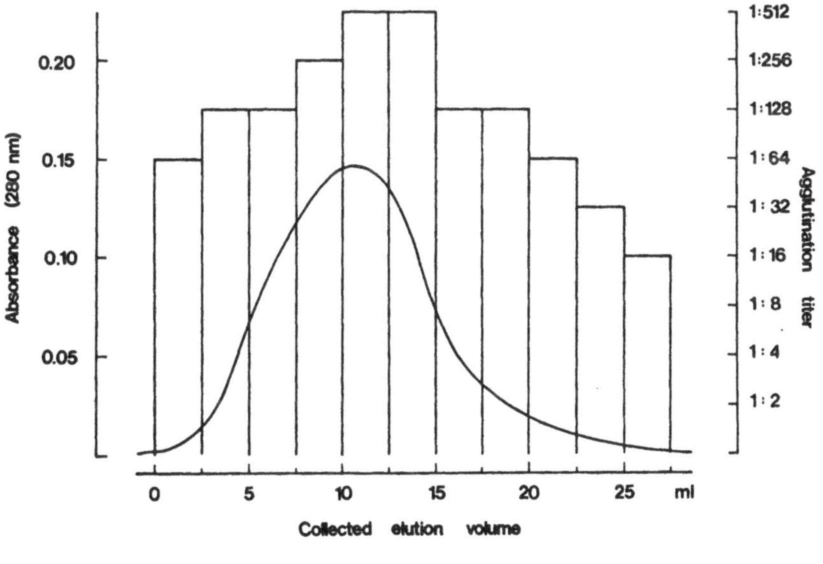

Figure 1. Elution profile of carrot agglutinin from affinity chromatography on immobilized Streptococcus mutans (redrawn from ref. 4.)

The purified agglutinating substance was further characterized by treatment with trypsin, heat or dextranase, respectively. The effect of the treatment was evaluated by comparing the agglutinating ability with that of an untreated sample, Table I. It was found that trypsin and heat had no effect indicating that the agglutinin was not of protein nature. Treatment with dextranase ruined the agglutinating ability totally.

Table I. Effects of various treatments of purified carrot agglutinin on the agglutination of Streptococcus mutans OMZ 65.

Agglutination titer	Agglutinin treated with			Untreated
	heat 90 C	trypsin	dextranase	
1:2	+	+	-	+
1:128	+	+	-	+

A similar series of treatments was performed on Streptococcus mutans cells. The results were evaluated as agglutinating ability towards untreated agglutinating substance from carrot, Table II.

Trypsin- and heat treatment destroyed the agglutination totally, whereas dextranase treatment had no effect.

Table II. Effects of various treatments of *Streptococcus mutans* OMZ 65 cells on the agglutination by carrot agglutinin.

Agglutination titer	Cells treated with			Untreated
	heat 90 C	trypsin	dextranase	
1:2	-	-	+	+
1:128	-	-	+	+

It was further demonstrated that treatment of the cells with low molecular weight dextran prior to exposure to the purified agglutintin reduced agglutination in a competitive manner.

Conclusion

The conclusions drawn from these studies are then:

* The agglutinating molecules from carrot is of non-protein origin -- presumably a dextrane.

* The binding positions of the cell surface is probably of protein nature.

References

1. Hamada, S., Gill, K., Slade, H. D.: Infect. Immunity 18, 708-716 (1977).

2. Bratthall, D.: Secretory Immunity and Infection, pp. 327-333 (Plenum Press, New York) (1978).

3. Mattiasson, B., Ramstorp, M., Wideback, K., Kronvall, G.: J. Appl. Biochem. 2, 321-335 (1980).

4. Ramstorp, M., Mattiasson, B., Carlsson, P., Bratthall, D.: Caries Research, in press.

Appendix

RED CELL GHOST: A TOOL FOR THE PURIFICATION OF LECTINS

Alain FAURE, Josiane DELRIEU

Centre National de Transfusion Sanguine BP 100
F. 91943 - LES ULIS Cédex - France

Michel CARON

Université Paris Nord - U.E.R. Biomédicale F. 93012 Bobigny
Cédex - France

Introduction

Affinity methods are widely used at the laboratory scale and for analytic methods and tests (1,2). They begin to appear in processes at an industrial scale (3). We describe here a procedure for the preparation of red cells ghosts and their use as an affinity matrix for the purification of lectins.

Material and Methods

The process used is based on a first step of slow osmotic haemolysis linked with a second step of separation between ghosts and cytosol.

Three points have been optimized :

Duration of the process
Level of residual hemoglobin
Volume of red cells prepared.

Lectins, Vol. III

1/ Dilution of the red cell suspension:

Outdated human red cell concentrates (hematocrit 90) are used as starting material. They are diluted in the following buffer (4):
Sodium Chloride 60 mM, disodium phosphate 25 mM - potassium phosphate 8 mM pH 7,2 (200 m Osm) containing sodium azide (1 °/₀₀ w/v) and aprotinin (3.10^{-3} cm^3 °/₀₀ v/v).
Temperature of the buffer is + 4°C. The final dilution of red cell concentrate is 1.18 (v/v).

2/ Haemolysis and separation between ghosts and cytosol:

For the haemolysis the same buffer is used at a tenfold dilution in distilled water. A closed system including the diluted red cell suspension, a tangential microfilter Plasmaflo* (Asahi Medical), two peristaltic pumps, and a vessel containing the diluted buffer is used. Macromolecules salts and water are separated in the microfilter (pore diameter 0,2 microns) and found in the ultrafiltrate. Red cells ghosts are kept in the retentate. The diluted buffer is added to the red cell suspension at a rate equivalent to the ultrafiltration rate. This allows to obtain a gradient of osmolarity and to maintain the same dilution of the red cells (fig 1).

3/ Concentration:

When the ultrafiltrate is colourless, the addition of 20 m Osm phosphate buffer is stopped. The decrease of the total volume of the ghosts suspension is stopped when their volume is 10 % lower than the initial red cell concentrate. Tubings and Plasmaflo* are then washed with diluted buffer. The washings are added to the retentate containing the ghosts.

Results:

1/ Preparation of red cells ghosts:

This method enables the preparation of large quantities of ghosts (1 liter of red cell concentrate give in 4 hours between 7,8 and 9,4 g of freeze dried ghosts).
The yield and quality of the ghosts are given in the table 1.

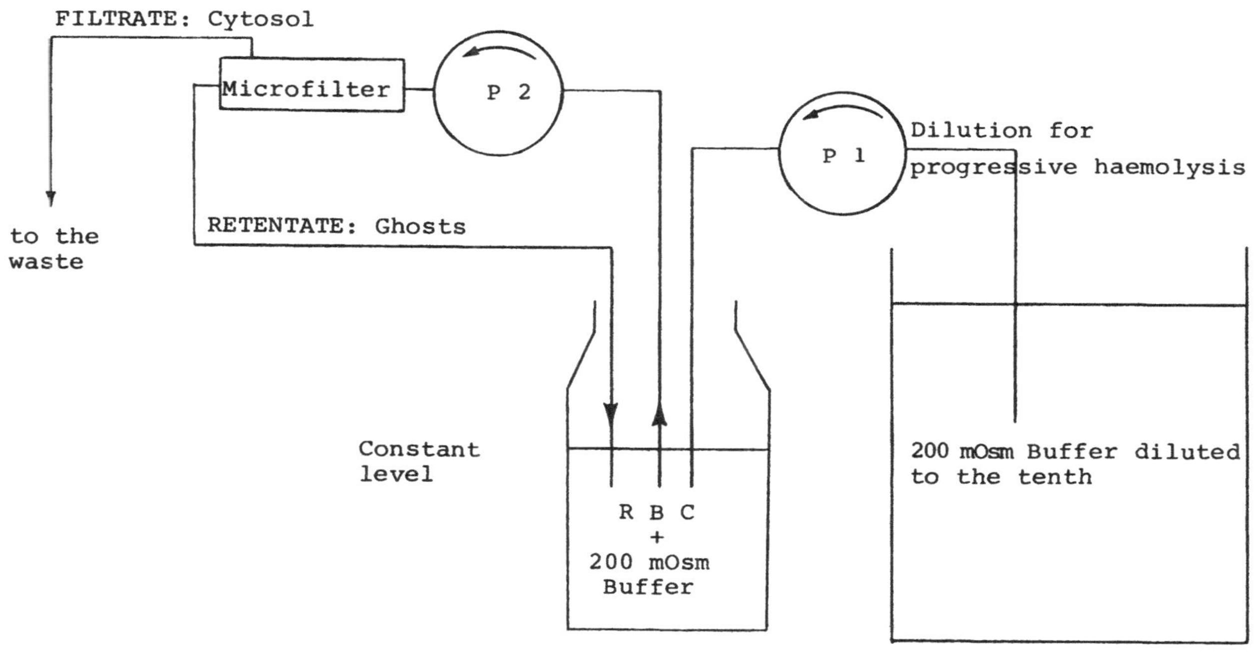

Figure 1: Haemolysis and separation of ghosts and cytosol

Erythrocytes (ml)	Dilution	Hemoglobin Final/Initial %	Ghosts (ml)	Weight of the ghosts (g per liter of erythrocytes after freeze drying)
650	10	0,20	650	9,4
1.160	18	0,15	1.500	7,8
1.200	10	0,30	1.200	7,8
2.050	20	0,10	2.000	-
550	19	0,004	960	-
750	14	0,06	1.120	8,3

Table 1: Yield and quantity of ghosts obtained with the proposed method.

2/ Use for the purification of lectins

The purification of peanut lectin has been previously obtained with polymerized ghosts (5,6,7), but these ghosts can be used without polymerization. For that purpose the contact of the desialylated ghosts and the crude extract containing the activity is maintained for 10 minutes under mild agitation. Ultrafiltration with Plasmaflo* or centrifugation are used to separate cells and unbound molecules. After several washings, the elution is performed by competition with galactose 0,1 M and the lectin is found in the supernatant, the pellet of ghosts being re-usable.

Discussion

1/ Preparation of ghosts. This method permits in sterile conditions the preparation of large quantities of ghosts, with low concentration of hemoglobin, very good conservation of membrane antigens in a relatively rapid manner.

2/ Purification of lectins. Red cells ghosts after the appropriate treatment enable to purify peanut lectin rapidly and with a good yield.

References

1. Lowe, C.R., Dean, P.D.G.: Affinity Chromatography J. Wiley and Sons, Ed. London (1974)

2. Sundaram, P.V., Ekcstein, F.: Theory and Practice in Affinity Techniques. Academic Press London (1978)

3. Gribnau, R.C.J., Visser J., Nivard R.J.F.: Affinity Chromatography and related techniques. Theorical Aspects. Industrial and Biochemical Applications. Analytical Chemistry Symposia Series. Vol. 9, Amsterdam (1982)

4. Weed R.I., Weed C.F., Berg G.: JL. Clin. Invest. 42 581 - 588 (1963)

5. Lawny, F., Bot, M.H., Lentwojt, E., Segard, E.: Affinity Chromatography. O. Hoffmann. Ostenhof, M. Breintenbach, F. Koller, D. Kraft, O. Scheiner, ed. Pergamon Press, 299 - 303 (1978)

6. Faure, A., Caron, M., Leroy, M.H., Duval, D., Segard, E.: Separation of Cells and subcellular Elements. J. Peeters Ed. Pergamon Press, Oxford 99 - 102 (1979)

7. Caron, M., Ohanessian, J., Faure, A., Felon, M.: C.R. Acad. Sc. Paris, 292 - 183 (1981)

THE MOLECULAR BASIS OF BACTERIAL ADHERENCE TO EPITHELIAL CELLS

Yuval Eshdat*+ and Nathan Sharon+
*The Agricultural Research Organization, Bet Dagan and +Department of Biophysics, The Weizmann Institute of Science, Rehovoth, Israel

Introduction

The attachment of bacteria to mucosal surfaces is presumed to be an essential early step in the development of most infectious diseases. The knowledge that can emerge from investigation of the detailed molecular mechanism of this process can be useful not only in attempts to design drugs which will inhibit this attachment and thus prevent diseases, but also in understanding what determines intercellular recognition in general.

Already in 1955, Collier and Miranda (1) demonstrated that agglutination of human red blood cells by Escherichia coli can be inhibited specifically by D-mannose. In another study, started in the same year, Duguid and his collaborators (2) have shown that the presence of pili (fimbriae) on the surface of enteric bacteria could be correlated with their ability to adhere to a variety of cells of human origin by a mannose-specific mechanism.

In 1977, Ofek et al. suggested that lectin-like molecules, present on the surface of the bacteria, enable it to bind to mannosyl residues located on epithelial cells (3). In the same year, Salit and Gotschlich (4) showed that isolated type 1 pili from E. coli were the bacterial constituents which mediate agglutination of cells by binding specifically to them. These bacterial cell surface lectins are a distinctive group which differ from the mannose-specific lectins present in the periplasmic space of the bacteria (5). The nature of the glycosidic receptors of surface bacterial lectins has been only partially characterized (6).

In this paper we review our recent studies which give a better understanding of the structure-function relationship of type 1 pili, as well as providing evidence for the involvement of other bacterial surface constituents in their mannose-specific adherence.

Results and Discussion

Type 1 pili of E. coli are hair-like appendages 7 nm in diameter protruding from the outer surface of the bacteria, and are comprised of very stable arrays of protein subunits (pilin) of mol wt approx 17000 (7). A detailed study of the structure, organization and biological properties of these pili is crucial for the understanding of the interaction between the bacteria and the host cell surfaces, and of the role of pili in microbial pathogenicity. Studies of type 1 pili have been greatly hampered by difficulties encountered in their solubilization and dissociation into biologically functional subunits. For example, type 1 pili are not readily denatured by sodium dodecylsulfate (SDS) even with boiling, and their dissociation into subunits has been achieved under conditions such as boiling at low pH or chemical modification (4,7,8), which usually lead to irreversible protein denaturation. We have found (9) that type 1 pili, isolated from E. coli 346, can be dissociated by saturated solutions of guanidine hydrochloride into pilin monomers, as revealed by their apparent molecular weight determined by gel filtration in the presence of saturated guanidine hydrochloride. After dialysis against 5 mM Tris buffer (pH 8.0), the monomers appeared to undergo aggregation into dimers as evidenced by their behaviour on gel filtration and ultracentrifugation.

The exposure of pili to 8.6 M guanidine hydrochloride was sufficiently mild so that up to 60% of the resultant subunits could reassemble into structures resembling short fragments of intact pili; this reassembly required the presence of $MgCl_2$.

It is possible that complete reassembly was not obtained because some subunits were irreversibly denatured by the treatment with guanidine hydrochloride or because the conditions used (such as buffer type, ionic strength, temperature and concentrations of subunits) were not optimal.

The subunits formed by dissociation with saturated guanidine hydrochloride retained not only their ability to reassemble into intact pili but also their ability to bind to D-mannose. About 25% of the subunit dimers obtained after removal of the guanidine hydrochloride by dialysis were bound to immobilized mannan and could be eluted by methyl α-D-mannoside. We cannot adequately explain why only part of the subunits bound specifically to mannan. One possibility is that the majority of the subunits, after removal of guanidine hydrochloride, were not renatured to molecules capable of binding mannosyl residues. On the other hand, it is possible that only a portion of the pilin monomers in each intact pilus has mannose-binding activity, implying biological heterogeneity of the individual pili subunits.

Whatever the significance of the low proportion of binding activity, the fact that there was binding at all is additional, and perhaps definitive, proof that type 1 pili mediate mannose-binding activity of E. coli. Earlier attempts at establishing that intact pili bind to D-mannose could be criticized on the grounds of insufficient purity of the preparations used. In such preparations, the mannose-binding activity might have resided in contaminant proteins. In our study, the isolation of soluble pili subunits, which behave chromatographically and electrophoretically as a single protein species and possess the ability to bind specifically and reversibly to immobilized mannan, provides strong evidence for the involvement of pili in the mannose-binding activity of E. coli.

In a qualitative screening study of 393 clinical bacterial isolates, tested for the presence of mannose-specific lectin

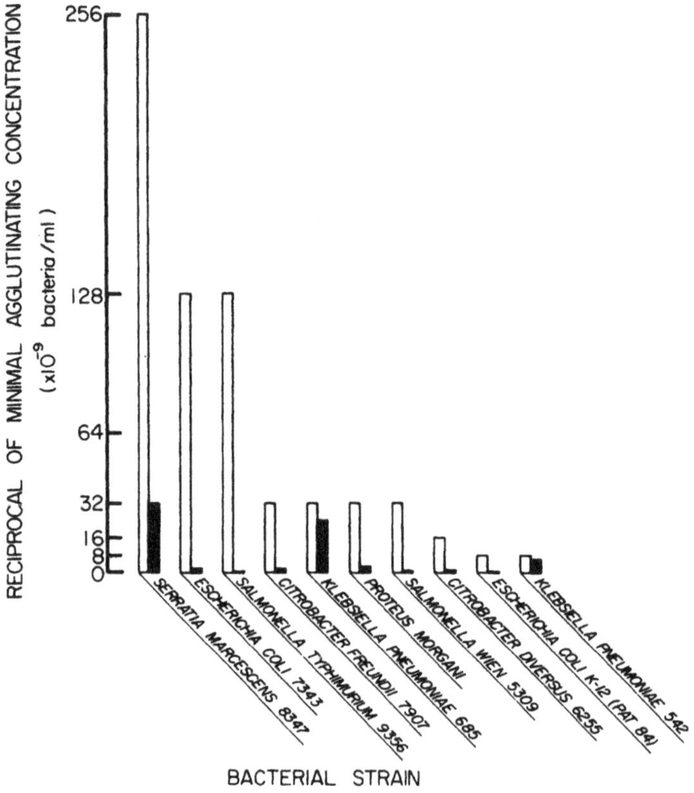

Fig.1 Agglutination of yeast cells by various bacterial isolates in the absence (left column) or presence (right column) of 0.125 M methyl α-D-mannoside. [Based on data taken from Mirelman, Altman and Eshdat (10).]

on their surface, 38% of them were positive in their ability to agglutinate Saccharomyces cerevisiae, which could be inhibited by methyl α-D-mannoside (10). Quantitative assay of a few selected strains revealed that Serratia marcescens 8347 and E. coli 7343 possess the highest mannose-specific agglutination of the yeast cells (Fig.1). Surprisingly, observation of these two strains in an electron microscope revealed that they had practically no pili on their surfaces and were heav-

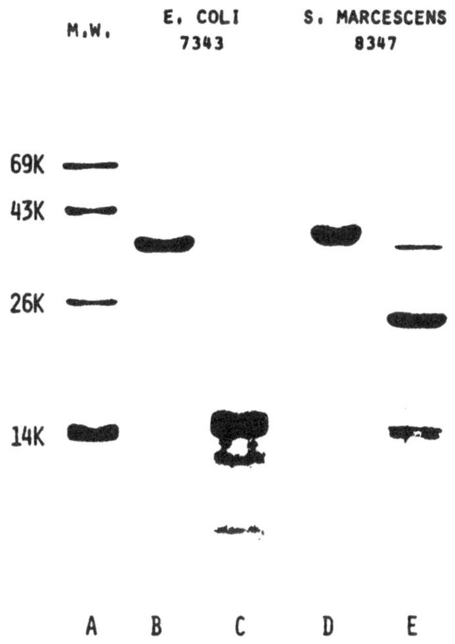

Fig.2 Polyacrylamide gel electrophoresis in the presence of dodecylsulfate of the mannose-sensitive flagellar lectins of E. coli 7343 and S. marcescens 8347, before and after their cleavage by CNBr. (A) Molecular weight markers; (B) E. coli lectin; (C) CNBr cleavage pattern of B; (D) S. marcescens lectin; (E) CNBr cleavage pattern of D.

ily flagellated. Agitation of the bacteria with an omnimixer resulted in removal of their flagella and loss of their yeast-agglutinating activity (11). The bacteria were removed by centrifugation at 7000 x g and the flagella were isolated by further centrifugation at 50000 x g for 2 h and gel filtration on a Sepharose 6B column. The mannose-specific lectin activity could be demonstrated only in the flagella preparations, which were observed as pure by electron microscopy. SDS-polyacrylamide gel electrophoresis showed that the mannose-specific flagella of S. marcescens 8347 and E. coli 3743

Table 1
AMINO ACID COMPOSITION OF BACTERIAL MANNOSE-SPECIFIC FLAGELLAR LECTINS
(mole percentage)

Amino acid	E. coli 7343 lectin	S. marcescens 8347 lectin	Flagellin[a]
Lysine	5.1	5.8	5.0
Histidine	0.5	0.4	0.5
Arginine	4.1	3.7	3.6
Aspartic acid	14.7	17.2	16.0
Threonine	8.5	8.4	9.4
Serine	9.5	10.2	7.4
Glutamic acid	10.1	11.1	11.5
Proline	2.2	1.2	1.0
Glycine	6.9	7.9	7.6
Alanine	13.2	11.3	13.2
Cysteine	0.0	0.0	0.0
Valine	6.5	5.0	5.9
Methionine	1.2	1.0	1.3
Isoleucine	5.2	4.7	6.1
Leucine	8.2	8.7	8.2
Tyrosine	1.9	1.4	1.5
Phenylalanine	2.2	2.2	1.8
Tryptophan	0.0	0.0	0.0

[a] Values for gram-negative bacterial flagellin are from reference 12.

are comprised of identical protein subunits of molecular weight 38500 and 36500 respectively (Fig.2). Although the amino acid composition of the two flagellar lectins were similar to each other and to that of common bacterial flagellin (Table 1) while different from that of other bacterial appendages participating in bacterial adherence, the CNBr cleavage pattern of the two lectins differed from each other (Fig.2), suggesting differences in their primary structures. The two lectins also differ in their resistance toward proteolytic digestion by trypsin: after 24 h at 37°C, the E. coli lectin was not inactivated by trypsin while the S. marcescens lectin lost more than 50% of its mannose-specific agglutinating activity.

In an additional study (13), the participation of the bacterial outer membranes in the mannose-specific adherence was investigated. *E. coli* 2699, which had been isolated from a patient with urinary tract infection, agglutinated *S. cerevisiae* cells as well as guinea pig and human erythrocytes, and this agglutination was inhibited by methyl α-D-mannoside (14). Examination of the bacteria in an electron microscope revealed that the bacteria, which did not possess pili after 4 h of growth in Luria broth at 37°C, were densely piliated after prolonged growth in liquid culture or on agar at 37°C. Growth at 20°C largely prevented the formation of pili. Since the bacteria were also devoid of flagella, we have investigated the possibility that constituents of the outer membrane are directly involved in the mannose-specific attachment of the bacteria to the yeast cells. Indeed it was found that outer membranes isolated from depiliated *E. coli* 2699 agglutinated specifically *S. cerevisiae* cells. Electron microscopic and gel electrophoretic examinations did not reveal pili in the outer membrane preparation. The fact that pili isolated from these bacteria (when grown on agar at 37°C) exhibited mannose-specific agglutinating activity and, on the other hand, the lack of incomplete stubs of pili in unpiliated *E. coli* 2699, raised the possibility that the cell wall agglutinins are pilus monomers inserted into the outer membrane without polymerization. Alternatively, the cell wall agglutinin could be a protein different from pilin which contains a similar mannose-binding site as pilin. These two possibilites were also in accord with the high sensitivity of the *E. coli* 2699 mannose-specific lectin towards tryptic digestion, which is opposite to the stability of both *E. coli* 346 pili and *E. coli* 7343 flagellar lectins.

The agglutinating activity of *E. coli* 2699 was lost after short (<1 h) incubation of the bacteria with trypsin; trypsinization also abolished the agglutinating activity of purified pili and outer membranes. The SDS-polyacrylamide gel electrophoretic pattern of trypsin-treated pili was the same as that of untrypsinized pili. When the pili were dissociated into their pilin monomers prior to their treatment with trypsin, complete enzymatic digestion occurred. Analysis of the isolated outer membranes by SDS-polyacrylamide gel electrophoresis after incubation with trypsin, revealed a major outer membrane protein which had disappeared and a new smaller protein band which was observed. However, it is not yet known whether or not this protein is the mannose-specific lectin.

Fig.3 Cell surface components involved in mannose-specific bacterial adherence: pili isolated from E. coli 346 (9); flagella isolated from S. marcescens 8347 (Eshdat, Izhar, Sharon and Mirelman, to be published); and outer membranes isolated from E. coli 2699 (13). Bar = 0.1 mμ.

Our results show that E. coli 2699 possesses mannose-specific lectins both in the pili and in the outer surface of the cell wall, and that both exhibit the same specificity for attachment. This may indicate the operation of a two-stage adhesion of E. coli 2699 to target cells, a long-distance interaction via pili and a short-distance interaction via the cell wall causing final adherence.

In conclusion, mannose-specific surface lectins of gram-negative bacteria were found to be associated with pili, flagella and the outer membranes (Fig.3). These three types of lectins may differ in their subunits' molecular weight, their structural organization on the bacterial surface, and their stability towards detergents and proteolytic enzymes.

References

1. Collier, W.A., de Miranda, J.C.: Antonie van Leeuwenhoek J. Microbiol. Serol. 21, 133-140 (1955).
2. Duguid, J.P., Smith, W.I., Dempster, G., Edmunds, P.N.: J. Pathol. Bacteriol. 79, 3335-3348 (1955).
3. Ofek, I., Mirelman, D., Sharon, N.: Nature (London) 265, 623-625 (1977).
4. Salit, I.E., Gotschlich, E.C.: J. Exp. Med. 146, 1169-1181 (1977).
5. Gilboa-Garber, N.: this symposium.
6. Wadstrom, T., Faris, A., Lindahl, M.: this symposium.
7. Brinton, C.C. Jr.: Trans. N.Y. Acad. Sci. 27, 1003-1054 (1965).
8. McMichael, J.C., J.T. Ou.: J. Bacteriol. 138, 969-975 (1979).
9. Eshdat, Y., Silverblatt, F.J., Sharon, N.: J. Bacteriol. 148, 308-314 (1981).
10. Mirelman, D., Altmann, G., Eshdat, Y.: J. Clin. Microbiol. 11, 328-331 (1980).
11. Eshdat, Y., Ofek, I., Yashouv-Gan, Y., Sharon, N., Mirelman, D.: Biochem. Biophys. Res. Commun. 85, 1551-1559 (1978).
12. Smith, R.W., Koffler, H.: Adv. Microb. 6, 219-339 (1971).
13. Eshdat, Y., Speth, V., Jann, K.: Infect. Immun. 34, 980-986 (1981).
14. Blumenstock, E., Jann, K.: Infect. Immun. 35, 264-269 (1982).

CHANGES IN LECTIN RECEPTORS DURING LYMPHOCYTE DIFFERENTIATION: APPLICATION TO BONE MARROW TRANSPLANTATION

Yair Reisner
Department of Biophysics, The Weizmann Institute of Science, Rehovoth, Israel

I Introduction

I.A Transplantation of histoincompatible marrow depleted of immunocompetent lymphocytes

In the immunologically deficient or immature recipient, the transplantation or transfusion of cells from another individual can result in the persistance or proliferation of the foreign cells without risk of host-mediated rejection. If the transplanted cells are immunologically competent, they may react to the histocompatibility antigens of the host tissues. This graft-versus-host (GVH) reaction is controlled by genetic differences between donor and host and is similar in its basic mechanism to the host-versus-graft reaction that led to rejection of transplanted tissues. Transplantation of immunologically competent cells to syngeneic recipients produces no GVH reactions, whereas the transplantation of immunologically competent cells between allogeneic individuals can produce GVH activity.

In the last 12 years, marrow transplantation has emerged as the treatment of choice for lethal congenital immunodeficiencies, aplastic anemia and possibly for certain forms of leukemia (1,2). Application of this approach is largely limited to the 30-40% of patients who have an HLA-identical sibling, since transplants of marrow from HLA-haploidentical or fully allogeneic donors have regularly induced lethal GVH disease (2). Moreover, despite the use of matched sibling donors for marrow transplantation, GVH disease develops in 50-70% of the transplant recipients and is lethal in 15% of cases (3-5).

Experience with transplantation of marrow from non-sibling related donors who are HLA genotypically or phenotypically identical, or who are HLA-D compatible but mismatched with respect to the HLA-A and HLA-B determinants, is still very limited. It seems that GVH disease has been more severe in these cases.

Studies in murine models of marrow transplantation have demonstrated that GVH disease is initiated by thymus-derived T lymphocytes of donor origin. Transplantation of allogeneic hematopoietic tissues deficient in T lymphocytes, such as early fetal liver (6), adult spleen or marrow cells from mice thymectomized in the neonatal period (7), or marrow or spleen cells pretreated with a cytotoxic antiserum specific for T lymphocytes (8), have resulted in stable chimeras without GVH disease. Such transplants have regularly produced immunologically vigorous, long-lived chimeras when donor and recipient are haploidentical (e.g. parent → F_1). Production of long-lived chimeras has been less consistent when fully allogeneic transplants have been performed, due in part to a reduced incidence of engraftment and also, possibly, to genetic restrictions on cooperation between donor lymphoid cells and host elements in the generation of effective immune responses (9).

Several techniques have now been developed which are effective in depleting alloreactive T lymphocytes from mouse marrow or spleen cells. These include: (a) physical separation of the stem cells either by discontinuous gradient of bovine serum albumin or by velocity sedimentation (10); (b) in vitro "suicide" techniques (11); (c) pretreatment with antibodies specific for the θ antigen present on thymic and circulating T lymphocytes (8); (d) selective culture of hematopoietic precursors (12); and (e) our approach using selective fractionation of hematopoietic precursors with plant lectins (13). The adequacy of the T cell depletion achieved has been assessed by in vitro analyses demonstrating absence of allospecific cell mediated lympholysis and mixed lymphocyte culture responses, and confirmed in vivo by the capacity of these T lymphocyte-depleted cell fractions to reconstitute lymphoid and hematopoietic function in lethaly irradiated H-2 incompatible recipients without inducing GVH disease.

While certain of the techniques used in mice have also been shown to be effective in depleting human alloreactive T lymphocytes detectable by in vitro mixed lymphocyte culture or cell mediated lympholysis assays, attempts to utilize such techniques to transplant human patients with marrow from histoincompatible donors have been limited, and uniformly unsuccessful. Most of the failures can probably be ascribed either to differences between the responses of inbred mouse and outbred human cells to the fractionation procedures used, or to technical problems in separating the large volumes of marrow necessary to achieve a cell dose adequate for transplantation in man.

I.B Fractionation of mouse bone marrow stem cells by lectins

During the past few years we have studied in detail the interaction of lymphocytes with two galactose binding lectins: peanut agglutinin (PNA) and soybean agglutinin (SBA). PNA is highly specific for the disaccharide D-galactose $\beta(1\rightarrow3)$-N-acetyl-D-galactosamine, a structure commonly found in the glycoproteins and glycolipids located on cell surfaces. SBA is specific for N-acetyl-D-galactosamine and, to a lesser extent, for D-galactose (Fig.1). Of special relevance is the fact that the native forms of PNA and SBA are not mitogenic and therefore even their prolonged contact with lymphocytes does not cause cell stimulation (14).

We have found that receptors for these lectins change in an orderly manner during lymphocyte differentiation and maturation and that they can serve as markers for lymphocyte subpopulations in both mouse and man (15). We have also demonstrated that the differences in receptors for PNA and SBA permit the effective separation of lymphocytes into biologically distinct subpopulations by selective agglutination with these lectins (Fig.2). Thus with this technique, immature and

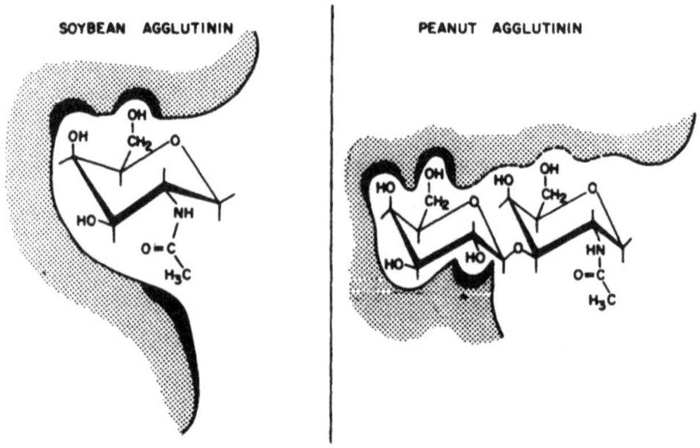

Figure 1. Proposed sugar binding sites on SBA and PNA (14).

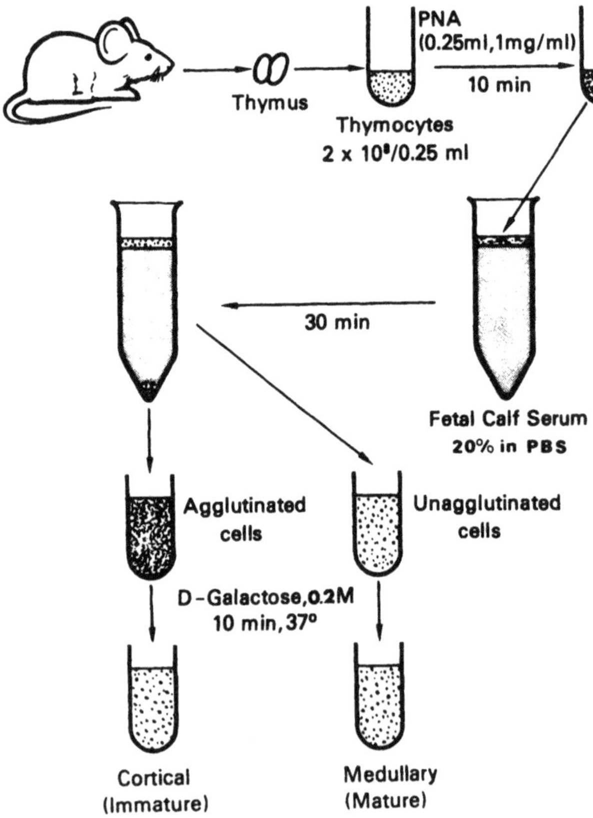

Figure 2. Procedure for fractionation of mouse thymocytes by differential agglutination with PNA.

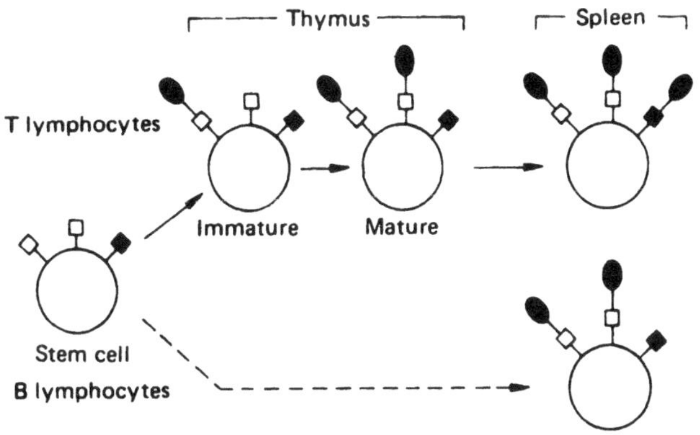

Figure 3. Model for changes in lectin receptors on murine lymphocyte surfaces during differentiation and maturation. □, Receptor for SBA and PNA; ■, receptor for SBA only; ●, sialic acid.

mature thymocytes were separated by PNA (16) and mouse splenocytes were fractionated into B and T cells using SBA (17). In both cases, the agglutinated cells were devoid of GVH activity.

Studies of the binding of SBA and PNA to these subpopulations, both before and after neuraminidase treatment, have led to the conclusion that the receptors for both PNA and SBA are present on the surface of immature thymocytes, that in mature thymocytes only the SBA receptor is present (the PNA receptor being masked by sialic acid), and that peripheral blood T lymphocytes, as well as most spleen T cells, have both the PNA and SBA receptors masked by sialic acid. On the basis of these findings we have suggested that hematopoietic stem cells may also possess PNA and SBA receptors unmasked by sialic acid on their surfaces (Fig.3). To test this assumption, we isolated a cell fraction from mouse bone marrow and spleen by sequential agglutination with SBA and PNA and demonstrated, by spleen colony assay and by reconstitution of irradiated allogeneic mice, that this fraction is depleted of GVH activity and enriched with hematopoietic stem cells (Figs.4,5).

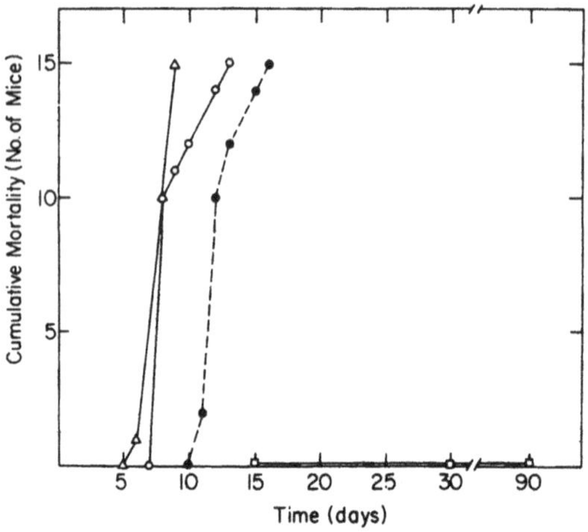

Figure 4. Cumulative mortality of irradiated (BALB/c x C57BL/6)F_1 mice after transplantation with splenocytes (10^7 cells per animal) from BALB/c mice starting with 15 mice in each group. Grafts: △—△, unfractionated splenocytes; □—□, splenocytes agglutinated by SBA; o—o, splenocytes unagglutinated by SBA; ●---●, control without graft.

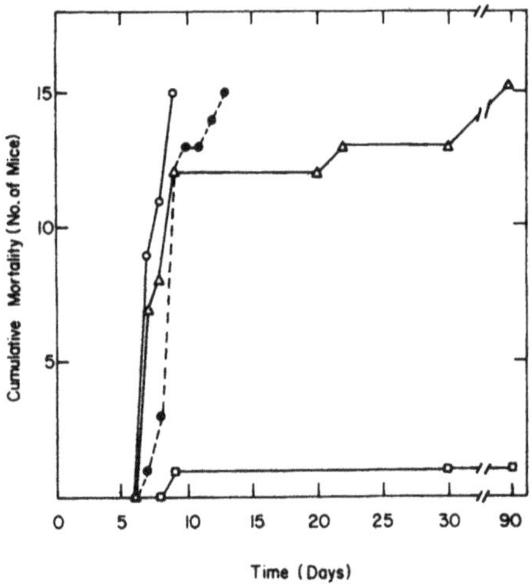

Figure 5. Cumulative mortality of irradiated (BALB/c x 57BL/6)F_1 mice after transplantation with splenocytes (10^7 cells per animal) from SWR mice, starting with 15 mice in each group. Grafts: △—△, unfractionated splenocytes; □—□, splenocytes sequentially agglutinated by SBA and PNA; o—o, splenocytes unagglutinated by SBA; ●---●, control without graft.

The reconstitution of lethally irradiated mice was tested in a parent-to-F_1 system [BALB/c → (BALB/c x C57BL/6)F_1] and in a complete allogeneic system [SWR → (BALB/c x C57BL/6)F_1] (13).

II In Vitro Experiments with Human Bone Marrow Cells

II.A Purification of hematopoietic stem cells from human bone marrow, using differential agglutination with SBA and E-rosetting with sheep red blood cells

The majority of human bone marrow cells (60-80%) are agglutinated by SBA similarly to mouse bone marrow cells. However, in contrast to the results with mouse bone marrow cells, the agglutinated human bone marrow subpopulation is remarkably depleted of myeloid progenitors (CFU-c) whereas the minor unagglutinated subpopulation contains 10 times more myeloid progenitors than the unseparated bone marrow cells (18-20).

We have also found that SBA agglutinates a major subpopulation of human peripheral blood T cells which helps activity for antibody production by B cells ("helper") while none of the mouse T cells interact with SBA (21).

These results suggest that: (i) most of the human myeloid precursors, unlike mouse stem cells, are not agglutinated by SBA; (ii) human hematopoietic stem cells cannot be isolated from T cells by one-step procedure with SBA.

Considering these limitations on the direct application of SBA for human marrow fractionation, we tested additional cell-separation procedures for their efficiency in removal of T cells from human bone marrow. The T cell contamination of the separated bone marrow cell fractions was estimated by various assays of immunological functions in vitro [cell mediated lympholysis, mixed lymphocyte culture, mitogenic response to PHA (the phytohemagglutinin from the red kidney bean Phaseolus vulgaris) and concanavalin A], as well as by E-rosette formation with sheep red blood cells.

This study enabled us to design a new procedure for the purification of human bone marrow myeloid progenitors (Fig.6), suitable for allogeneic bone marrow transplantation in man.

There are four main steps in the procedure (19): (i) Selective removal of red blood cells by gravity sedimentation in "Hetastarch", yielding a leukocyte-rich fraction which contains 68-76% of the original bone marrow nucleated cells. (ii) Agglutination of bone marrow cells with SBA and subsequent differential sedimentation of agglutinated cells (SBA^+). This step removes 70-90% of the bone marrow nucleated cells, which include, as previously shown, B cells, monocytes and helper T cells, as well as most of the polymorphonuclear leukocytes. The erythrocytes remaining after the "Hetastarch" separation are also agglutinated by SBA and removed from the SBA^- cell fraction. The average cell recovery in the unagglutinated SBA^- cell fraction was 17.1% (range 8.6-29.1) of the nucleated cells in the original leukocyte-rich fraction. (iii) Removal of E-rosette forming T cells from the SBA^- fraction by centrifugation over Ficoll-hypaque. In this step residual bands and polymorphonuclear leukocytes are also removed, together with the rosetted T cells during centrifugation over ficoll. The average cell recovery of non-rosetting cells (SBA^-E^-) from the SBA^- cell fraction was 30.5% (range 28.6-40.5%). (iv) Removal of the trace of residual T cells (<0.5%) in the SBA^-E^- fraction by separation of cells forming E-rosettes with neuraminidase-treated sheep red blood cells. No more than 5% of the SBA^-E^- cells were lost during this final step.

Whereas the final cell recovery in the $SBA^-E^-E_N^-$ cell fraction averages approximately 5% of the starting fraction, more than 80% of the colony forming cells are usually retrieved (19).

As can be seen in Table 1, the average concentration of myeloid progenitors in the final $SBA^-E^-E_N^-$ cell fraction ($352/10^5$ cells, range 246-547) is 10-fold greater than that found in the starting leukocyte-rich fraction ($30/10^5$ cells, range

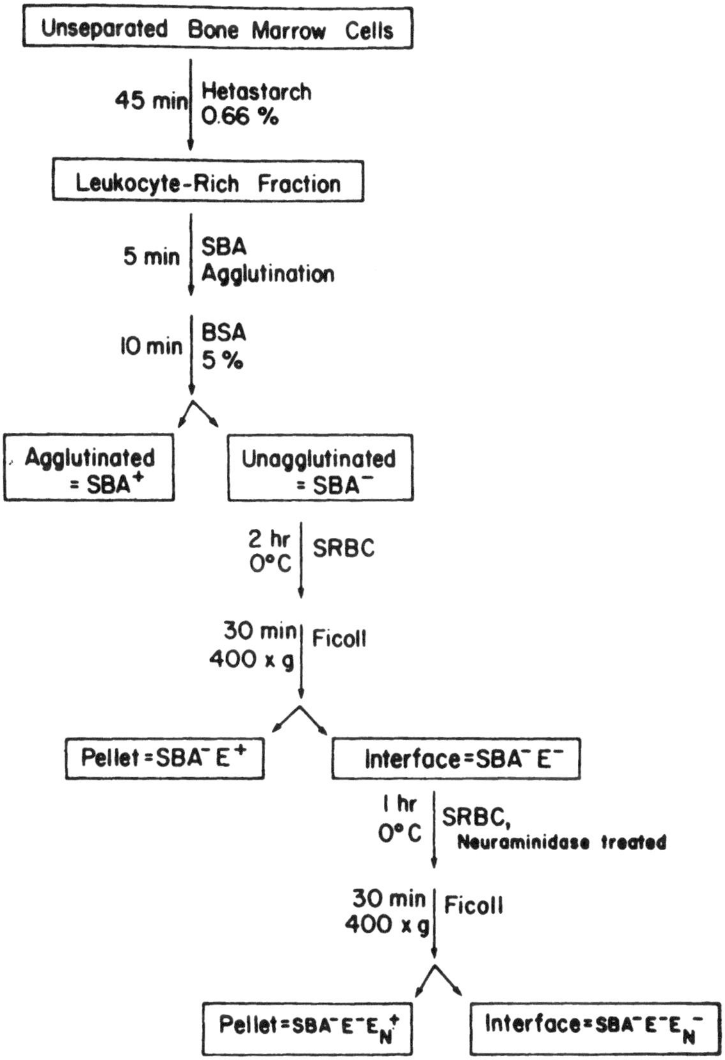

Figure 6. Procedure for the fractionation of human bone marrow cells by SBA and sheep red blood cells. BSA = bovine serum albumin.

Table 1
Recovery of Myeloid Progenitors (CFU-c) in Human Bone Marrow Cell Fractions Separated by SBA and Sheep Red Blood Cells

Bone marrow cell fraction	Number of cells ($x10^6$)	Average number of CFU-c/10^5 cells	Total number of CFU-c (10^6)	Percentage recovery of CFU-c
Donor 1				
Unseparated	19,000	N.D.	-	-
Post-Hetastarch	13,400	77	10.32	100
SBA^-	3,900	159	6.20	60.1
SBA^-E^-	920	162	1.49	14.4
$SBA^-E^-E_N^-$	860	270	2.32	22.5
Donor 2				
Unseparated	15,240	N.D.	-	-
Post-Hetastarch	11,675	20	2.33	100
SBA^-	1,600	225	3.60	150
SBA^-E^-	440	421	1.85	79.3
$SBA^-E^-E_N^-$	420	541	2.27	97.3
Donor 3				
Unseparated	17,000	N.D.	-	-
Post-Hetastarch	11,700	9	1.05	100
SBA^-	1,000	68	0.68	64.6
SBA^-E^-	405	310	1.25	119.2
$SBA^-E^-E_N^-$	380	240	0.91	86.6

9-77) and the final $SBA^-E^-E_N^-$ cell fraction of bone marrow donors 2 and 3 contained more than 80% of the original number of myeloid progenitors present in the leukocyte-rich fraction.

Although for donor 1 only 22.5% of the original myeloid progenitors were retained in the $SBA^-E^-E_N^-$ fraction, the total number of myeloid progenitors in this fraction (2.32 x 10^6) was no less than that isolated from the other two donors (2.27 and 0.91 x 10^6) and was sufficient to fully engraft a lethally irradiated patient with acute lymphocytic leukemia in second remission.

II.B Determination of T cells in the separated bone marrow cell fractions

As shown by the mixed lymphocyte culture assay (Table 2),

Table 2
Alloreactivity[a] and E-Rosette-forming Cells Before and After Purification of Human Bone Marrow by SBA and Sheep Red Blood Cells

Bone marrow	Autologous stimulators	Allogeneic stimulators	Stimulation index	E-Rosettes
Donor 1				
Whole	4,537±544	8,548±1,157	1.90	22%
$SBA^-E^-E_N^-$ fraction	73,648±419	55,349±10,482	0.75	0%
Donor 2				
Whole	1,236±292	2,234±677	1.81	10%
$SBA^-E^-E_N^-$ fraction	72,852±7,529	26,408±7,206	0.36	0%
Donor 3				
Whole	4,770±1,665	12,602±2,919	2.64	16%
$SBA^-E^-E_N^-$ fraction	24,129±1,986	21,956±524	0.91	0%

(a) As determined from mixed lymphocyte cultures with autologous and allogeneic stimulators.

alloreactive T cells were present in the leukocyte-rich fraction of all donors tested (stimulation index: 1.90, 1.81 and 2.64), whereas such reactivity was not detectable in the final $SBA^-E^-E_N^-$ cell fraction (stimulation index: 0.75, 0.36 and 0.91, respectively). Alloreactive T cells were detectable in all fractions of donor 2, except the $SBA^-E^-E_N^-$ fraction. For donors 1 and 3, the SBA^-E^- cell fraction (obtained after one cycle of E-rosetting with sheep red blood cells untreated with neuraminidase) appeared to be already depleted of T cell reactivity as detectable in mixed lymphocyte culture (Table 3). In the case of donor 3, however, the subsequently separated $SBA^-E^-E_N^+$ cells exhibited a marginal response in mixed lymphocyte culture, suggesting that perhaps a minute number of alloreactive T cells was still present in the SBA^-E^- fraction.

The E-rosette test, using neuraminidase-treated sheep red blood cells and overnight incubation at 4°C, could detect a minute number of T cells (1.5%) in the $SBA^-E^-E_N^+$ cell fraction

Table 3
Alloreactivity and E-Rosette-forming Cells Before and After the Second Cycle of E-Rosetting with Neuraminidase-Treated Sheep Red Blood Cells

	SBA^-E^-		$SBA^-E^-E_N^+$		$SBA^-E^-E_N^-$	
Donor	S.I.[a]	Rosettes	S.I.[a]	Rosettes	S.I.[a]	Rosettes
1	0.87	0	1.01	1.5	0.75	0
2	1.52	0	2.08	0	0.36	0
3	0.97	0	1.13	0	0.91	0

(a) S.I. = Stimulation index of alloreactivity as determined by the mixed lymphocyte culture assay.

of only one of the donors (donor #1), whereas none could be found in the SBA^-E^- cell fraction (19).

The total number of T cells in the separated bone marrow cell fractions was also determined by staining with a pan-T monoclonal antibody, termed anti-Leu-4, or by simultaneous staining with anti-Leu-2 (specific for suppressor/killer T cells) and anti-Leu-3 (specific for helper T cells) (22). Our results (total of 5 experiments, a typical experiment is presented in Fig.7) demonstrated conclusively, in agreement with the mixed lymphocyte culture assay and the assay of E-rosette forming cells, that (i) no T cellls are detectable in the final $SBA^-E^-E_N^-$ cell fraction used for transplantation; and (ii) in most instances a single step of E-rosetting efficiently removes detectable T cells, although a marginal contamination with T cells may still be present in the SBA^-E^- fraction as reflected by the presence of T cells in the $SBA^-E^-E_N^+$ cell fraction which results from a secondary rosette forming step using neuraminidase-treated sheep red blood cells.

The significance of the second E-rosetting step is indicated by the finding that T cell contamination of murine bone marrow preparations should not exceed 0.3% in order to avoid GVH disease (23). The sensitivity threshold of indirect immunofluorescence analysis on the Fluorescence Activated Cell Sorter

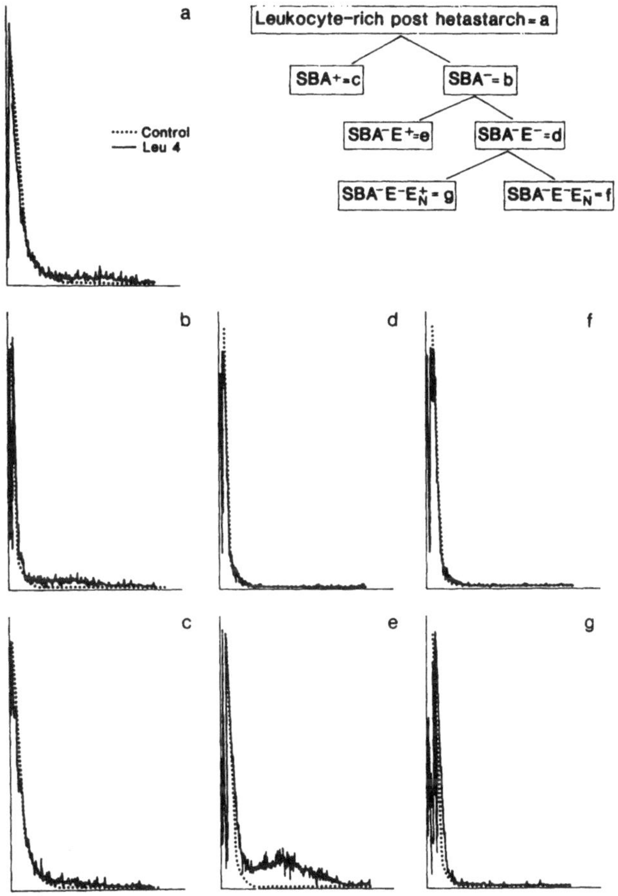

Figure 7. Anti-Leu-4 staining profile of human bone marrow cells separated by SBA and sheep red blood cells. Analyses of all cell populations were performed by indirect immunofluorescence with affinity-purified, fluorescein-conjugated, goat anti-mouse IgG on a Fluorescence Activated Cell Sorter. Abscissas: fluorescence intensity; ordinates, relative cell numbers.

is in the range of 1-2% due to background non-specific staining, and it cannot accurately detect minute T cell concentrations which may still be capable of producing GVH disease. In this respect, the assay for E-rosette forming cells may be superior in that it can detect the presence of smaller T cell concentrations. Using this assay, no detectable T cells could be found per count of 1000 cells of the $SBA^{-}E^{-}E_{N}^{-}$ bone marrow fraction.

III Pre-Clinical Studies in Primates

Our experience with in vitro analysis of mouse spleen and bone marrow cell fractionation by lectins, which was used successfully for allogeneic transplantation, indicated that the new procedure for the purification of human bone marrow might provide a cell fraction sufficiently purified to allow transplantation across a major histocompatibility barrier in man. However, since all attempts to apply new methods of allogeneic bone marrow transplantation directly from mouse to man have failed in the past, we chose to test the new procedure, prior to clinical use, in an outbred animal model having stem cells resembling human stem cells in their expression of SBA receptors and whose T cells form E-rosettes with sheep red blood cells.

We found that such similarity exists between Cynomolgus monkey and human bone marrow cells (18). Monkey peripheral blood lymphocytes as well as bone marrow cells can be depleted of T cells by E-rosetting with sheep red blood cells by the same technique which is used for the separation of human T cells. Also, as with human cells, the majority of monkey bone marrow cells (80-90%) are agglutinated by SBA and the unagglutinated fraction is enriched with myeloid progenitors relative to the unseparated cells. The unagglutinated fraction of both human and monkey marrow cells is also highly enriched with spontaneously dividing cells, as shown by uptake of tritiated thymidine after 68 hours of culture in RPMI-1640 medium supplemented only with fetal calf serum (5%) (18).

This remarkable similarity between monkey and human bone marrow cells suggested that the outbred Cynomolgus monkey may provide a particularly accurate model for pre-clinical investigations of the use of SBA fractionated marrow for transplantation between histoincompatible individuals. Together with Drs. Kapoor, Csurny, O'Reilly and Good, such in vivo studies were started at the Memorial Sloan Kettering Cancer Center (manuscript in preparation). The experimental design was as follows: bone marrow cells of male donors were transplanted into female recipients (by such pairing the origin of the regenerating bone marrow and peripheral blood lymphocyte cells can be defined from karyotype analysis of the Y chromosome). The mismatching of donor and recipient was established prior to transplantation by evidence of strong transformation responses in reciprocal mixed lymphocyte culture.

Our results are summarized in Table 4. Four animals received SBA^-E^- bone marrow cells after 1000 rad total body irradiation. Engraftment was not achieved in these unsensitized animals, suggesting that in monkey, as in mouse and man, there is evidence of a non-immune mechanism of resistance to hematopoietic engraftment which is radio-resistant. This resistance can be partially overcome in Cynomolgus monkey, as in mouse and man, by pretreating the host with cyclophosphamide. Thus, of 14 monkeys pretreated with 1000 rad total body irradiation, or 11x120 rad total body irradiation (hyperfractionated radiation), in combination with cyclophosphamide, 11 animals achieved engraftment. Karyotype analysis, which was technically successful in 8 of the engrafted animals, revealed that only 4 animals achieved engraftment with complete conversion to donor karyotype (Fig.8). It seems that radioresistant host stem cells, as well as host T cells, may persist in some animals even after heavy cytoreductive regimens. Although these animals ultimately succumbed to malaria and/or infections with gram negative pathogens, the period of follow up post-transplant (13-51 days) was adequate to assess for GVH disease in the four animals which had been fully engrafted with donor cells (Fig.8). As documented by Woodruff et al. (24) and confirmed in our GVH positive controls (a group of three animals), GVH disease develops in monkeys of the Rhesus family (of which the Cynomolgus is a member) within 8-12 days of

Table 4
Survival of Lethally Irradiated[a] Cynomolgus Monkeys Transplanted with SBA^-E^- Bone Marrow Cells from Allogeneic Donors

Engraftment (days post-transplant[b])	GVH disease	Survival (days)	Karyotype[c]
1000 rad TBI			
No	No	34	
No	No	25	
No	No	18	
No	No	12	
1000 rad TBI + CTX[d]			
Yes (11)	No	25	100% donor
Yes (11)	No	13	100% donor
Yes (10)	No	12	N.D.
Yes (29)	No	39	100% host
No	No	22	N.D.
Yes (14)	No	15	N.D.
Yes (32)	No	38	11/14 host, 3/14 donor
1320 (11x120) rad TBI hyperfractionated radiation + CTX[d]			
No	No	16	
Yes (18)	No	202	10/22 donor
Yes (50)	No	70	100% host
± (18)	No	22	N.D.
Yes (14)	No	51	100% donor
Yes (16)	No	16	100% donor
Yes (14)	No	15	N.D.
Control: unseparated bone marrow cells, 1000 rad TBI + CTX[d]			
No	Yes	18	
±	Yes	8	
±	Yes	11	

(a) Animals were exposed either to a single dose of 1000 rad total body irradiation (TBI) or to 1320 rad TBI in 11 doses of 120 rad each over a period of four days. (b) Day in which leukocyte count in peripheral blood was $1000/mm^3$ or more. (c) Spontaneously dividing metaphases and PHA stimulated blasts in bone marrow aspirates were evaluated. (d) Cyclophosphamide (CTX) was administered 2 x 60 mg/kg.

transplantation, and can be diagnosed by pathological changes in the skin, tongue, liver and intestinal tract. None of these animals developed clinical or pathological evidence of GVH disease, despite engraftment of marrow from a fully allogeneic donor.

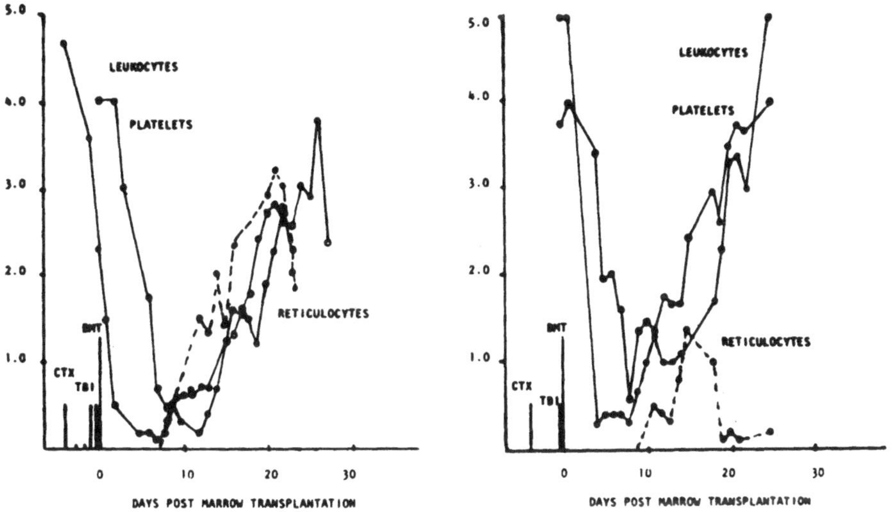

Figure 8. Hematopoietic reconstitution following $SBA^{-}E^{-}$ bone marrow transplantation (BMT) in Cynomolgus monkeys. Symbols: o——o, platelets x 10^{-5}; •——•, leukocytes x 10^{-3}; •---•, reticulocytes %.

IV Clinical Aspects

The technique developed has recently been applied (in collaboration with Drs. Kapoor, Good and O'Reilly, at the Memorial Sloan Kettering Cancer Center) as an experimental transplantation approach for a restricted group of patients afflicted with lethal congenital immune deficiences and leukemia, refractory to other forms of therapy, who lacked HLA identical sibling donors. To date, we have applied the lectin separation technique for transplants in nine patients, including three patients with severe combined immune deficiency and six patients with late stage leukemia resistant to chemotherapy.

As can be seen in Table 5, our preliminary experience strongly suggests that histoincompatible marrow, depleted of T cells by

Table 5
Transplantation of Mismatched Parental Marrow Cells Fractionated by SBA and Sheep Red Blood Cells for Severe Combined Immune Deficiency (SCID) and Acute Leukemia[a]

Patient No.	Disease	Engraftment (days post-transplant)	GVH disease	Survival	Outcome, cause of death
1	SCID	10	No	82 days	Adeno virus lung infection
2	SCID	18	No	>12 months	Alive and well
3	SCID	82	No	>11 months	Alive and well
4	ALL, early secondary relapse	15	No	132 days	Leukemic relapse
5	AML, secondary remission	10	No	24 days	Intestitial pneumonia
6	ALL, third relapse	10	No	29 days	Pre-transplant cardiomyopathy (following TBI and CTX)
7	AML, early secondary relapse	No	No	46 days	No engraftment, herpes zoster
8	ALL, third remission[b]	Partial[c]	No	41 days	Infection
9	ALL, second remission[d]	18	transient[e]	>5 months	Alive and well

(a) Acute lymphocytic leukemia (ALL) and acute myelocytic leukemia (AML) patients were prepared for transplantation with 1320 rad total body irradiation and 120 mg/kg cyclophosphamide. (b) A second graft was given following additional immunosuppression with procarbazine, antithymocyte globulin and low dose (300 rad) total body irradiation. (c) Only the second graft was engrafted. (d) Pretreatment with anti-thymocyte globulin (40 mg/kg x 3) and procarbazine (12.5 mg/kg x 3) was added to the regular cytoreductive regimen. (e) Mild transient GVH disease, limited to the skin, was observed.

SBA agglutination and E-rosetting, can be transplanted into patients with lethal blood diseases without risk of severe or lethal GVH disease, and that the cells transplanted are capable of reconstituting durable hematopoietic and lymphoid function in the transplanted host. Clearly however we saw better results with the severe combined immune deficient patients who do not require heavy cytoreduction regimens prior to transplantation; two out of the three patients with severe combined immune deficiency are alive and well at home one year post-transplant. The problems of engraftment in patients 7 and 8 and our success with patient 9, strongly suggest that tolerable cytoreductive regimens in excess of that provided by cyclophosphamide and lethal (>1000 rad) total body irradiation may well be needed to ensure engraftment of HLA-disparate cells in multitransfused immunocompetent leukemic hosts.

It is hoped that continued progress of this line of work should permit us in the near future to apply marrow transplants to all individuals and not just to the 40% of individuals possessing matched sibling donors. Use of such transplants will also enable us to explore the allogeneic advantage for promoting leukemia resistance in diseased human hosts. Furthermore, abrogation or amelioration of GVH disease as a complication of marrow transplantation, should reduce the acute morbidity and mortality associated with this procedure to between 5 and 10% and, therefore, may ultimately permit us to consider transplants for a number of other blood diseases and congenital enzymopathies which are not acutely lethal but rather lethal over the course of years (25).

References

1. Camitta B.M., Rappeport J.M., Parkman R., Nathan D.G.: Blood 45, 355 (1975).
2. Thomas E.D., Buckner C.D., Clift R.A., Fefer A., Johnson F.L., Neuman P.E., Sale G.E., Sanders J.E., Singer J.W., Shulman H., Storb R., Weiden P.L.: New Eng. J. Med. 301, 597 (1979).
3. Gale R.P., Feig S., Opelz G., Territo M., Young L., Sarna G., Fahey J., Cline M.J.: Transplant. Proc. 8, 611 (1976).
4. Neiman P.E., Thomas E.D., Reeves W.C., Ray C.G., Sale G., Lerner K.G., Buckner C.D., Clift R.A., Storb R., Weiden P.L., Fefer A.: Transplant. Proc. 8, 663 (1976).
5. Thomas E.D., Storb R., Clift R.A., Fefer A., Johnson F.L., Neiman P.A., Lerner K.G., Glucksberg H., Buckner C.D.: New Eng. J. Med. 292, 832 (1975).
6. Lowenberg B., Dicke K.A., van Bekkum D.W., Dooren L.J.: Transplant. Proc. 8, 527 (1976).
7. Yunis E.J., Good R.A., Smith J., Stutman O.: Proc. Natl. Acad. Sci. USA 71, 2544 (1974).
8. Von Boehmer H., Sprent J., Nabholz M.: J. Exp. Med. 141, 322 (1975).
9. Zinkernagel R.M., Althage A., Callahan, G., Welsh R.M.Jr.: J. Immunol. 124, 2356 (1980).
10. Dicke K.A., van Bekkum D.W.: Transplant. Proc. 3, 666 (1971).
11. Rich R.R., Kirkpatrick, C.H., Smith T.K.: Cell. Immunol. 5, 190 (1972).
12. Dexter T.M., Spooner E.: Nature 275, 135 (1978).
13. Reisner Y., Itzicovitch L., Meshorer A., Sharon N.: Proc. Natl. Acad. Sci. USA 75, 2933 (1978).
14. Lis H., Sharon N.: in Marcus A. (ed.): The Biochemistry of Plants Vol. 6 (New York, Academic Press, 1981), p. 371.
15. Reisner Y., Sharon N.: Trends Biochem. Sci. 5, 29 (1980).
16. Reisner Y., Linker-Israeli M., Sharon N.: Cell. Immunol. 25, 129 (1976).
17. Reisner Y., Ravid A., Sharon N.: Biochem. Biophys. Res. Commun. 72, 1585 (1976).
18. Reisner Y., Kapoor N., O'Reilly R.J., Good R.A.: Lancet (2), 1320 (1980).

19. Reisner Y., Kapoor N., Kirkpatrick D., Pollack M.S., Dupont B., Good R.A., O'Reilly R.J.: Lancet (2), 327 (1981).

20. Reisner Y., Kapoor N., Hodes M.Z., O'Reilly R.J., Good R.A.: Blood 59, 360 (1982).

21. Reisner Y., Pahwa S., Chiao J.W., Sharon N., Evans R.L., Good R.A.: Proc. Natl. Acad. Sci. USA 77, 6778 (1980).

22. Friedrich W., O'Reilly R.J., Koryiner B., Gebhard D.F., Good R.A., Evans R.L.: Blood 59, 696 (1982).

23. Korngold R., Sprent J.: J. Exp. Med. 148, 1687 (1978).

24. Woodruff J.M., Eltringham J.R., Casey H.W.: Lab. Invest. 20, 499 (1969).

25. Good R.A.: New Engl. J. Med. 306, 421 (1982).

REPORT ON THE THIRD SESSION OF THE INTERNATIONAL WORKING PARTY ON STANDARDIZATION OF LECTINS FOR DIAGNOSIS.

T.C. BØg-Hansen (secretary)
The Protein Laboratory, University of Copenhagen, Denmark

The International Working Party on Standardization of Lectins for Diagnosis at the Fourth Lectin Meeting in Copenhagen, June 1981, was outlined in the first report (1). The ultimate goal would be to draw up guidelines for lectin production, the first phase of the program being a well-controlled test scheme for a few widely used lectins. It was voted to test various commercial and non-commercial con A preparations in as many chemical and biological assays as possible, and most major lectin producers responded to my request to provide con A. The preliminary test program is listed in (1).

The Third Session was held during Interlec 5 and coded con A samples were distributed for testing. There was a motion for a more thorough test program and the following additional groups volunteered to take part in the test program to make it very comprehensive:
M.J.Waxdal and B.J.Mathieson (Bethesda), J.Y.Lin (Taiwan), J.A.Kint (Gent), A.M.Wu (College Station), I.Lorenc-Kubis (Wroclaw), B.Ersson (Uppsala), J.Kohn (London) and M.Caron (Bobigny), and G. A. Spengler (Bern).

A number of participants wished to provide non-commercial con A for inclusion into the test program. This was wellcomed, especially since this would mean a controlled comparison of a lectin provided as a solution and as a freeze dried powder. It is anticipated that all con A samples should be distributed during September 1982.

It was decided that commercial companies could take part in the program on the same terms as other partners, i.e. promarily that the highest scientific standards should be applied and that all test results of the coded samples should be made available for the Working Party.

After the Third Session of the Working Party a letter was received from R.E. Ritts, chairman of International Union of Immunological Societies Standardization Committee Communicating Minutes of the IUIS/WHO Standards Executive Committee, WHO, Geneva, 27-28 May, 1982. The Minutes contain the following paragraph under the heading New Subcommittees: "Several comments or requests have been received concerning standardization of lectins. Correspondence has also been transmitted by WHO from the International Working Party in Standardization of Lectins. After reading Secretary Bøg-Hansen's report of this group, the committee members voted to ask Profs. J. Breborowicz and Bøg-Hansen if they would be willing to have their group act as an official subcommittee of the WHO/IUIS Standardization Program with the assurance that we would assist them with review of protocols to assure biostatistical acceptibility for eventual WHO submission for the candidate preparations, annual budget support of $ 2,000 - 3,000 if available and assistance in raising funds for this effort."

On behalf of the International Working Party for Standardization of Lectins for Diagnosis, I have accepted to serve as an IUIS/WHO subcommittee.

1. T.C.Bøg-Hansen, J.Breborowicz, H.Franz: A report from the International Working Party on Standardization of Lectins for Diagnosis. On the necessity of standardized products. Lectins - Biology, Biochemistry, Clinical Biochemistry, Vol. 2, pp. 789-791.

AUTHOR INDEX

SUBJECT INDEX

Bei Fragen zur Produktsicherheit wenden Sie sich bitte an:
If you have any questions regarding product safety,
please contact:

Walter de Gruyter GmbH
Genthiner Straße 13
10785 Berlin
productsafety@degruyterbrill.com